SKEENA RIVER
FISH AND THEIR HABITAT

Allen S. Gottesfeld and Ken A. Rabnett

Foreword by
Jim Lichatowich

ecotrust

ecotrust

Skeena Fisheries Commission

Ecotrust
721 NW Ninth Ave, Suite 200
Portland, OR 97209
tel 503.227.6225
fax 503.227.1517
info@ecotrust.org
www.ecotrust.org

Skeena Fisheries Commission
Box 166
Hazelton, B.C. V0J 1Y0
tel 250.842.6780
fax 250.842.6709
info@skeenafisheries.ca
www.skeenafisheries.ca

Ecotrust's mission is to inspire fresh thinking that creates social equity, economic opportunity and environmental well-being. For nearly two decades, Ecotrust has created, capitalized and catalyzed innovative ways to restore environmental conditions while fostering economic opportunities in the temperate rain forest–Pacific salmon region that stretches from Alaska to California.

The Skeena Fisheries Commission is an aboriginal organization that focuses on fisheries management, science, and conservation. The SFC, directed by signatory First Nations with traditional territory in the Skeena drainage and the adjacent North Coast of British Columbia, includes the Tsimshian, Gitxsan, Gitanyow, Wet'suwet'en, and Lake Babine Nations.

Skeena River Fish and Their Habitat
by Allen S. Gottesfeld and Ken A. Rabnett
© 2008 Skeena Fisheries Commission
Published by Ecotrust in Portland, Oregon, USA.
All rights reserved. Except in brief quotations in critical articles or reviews, no part of this book may be reproduced in any manner without prior written permission from the publisher.

ISBN 978-0-9779332-5-9

Cover and book design by Andrew Fuller
Cover photo by Doug Demarest
Back cover maps by Analisa Noel McKay
Copyedit by Malka Geffen

Printed by Centerpoint Graphics in Portland, Oregon on 80# White Exact Opaque Recycled Cover and 60# White Exact Opaque Recycled Book, acid-free, 30% post-consumer recycled content papers.

To order copies of this publication:
for individual orders
www.ecotrust.org/publications

for bulk orders
www.oregonstate.edu/dept/press

CONTENTS

Preface

Ecotrust is honored to help the authors and the Skeena Fisheries Commission publish this book. It represents the first extensive description of the anadromous and freshwater fish populations of the magnificent five million hectare Skeena River drainage basin.

Together with the Wild Salmon Center we have looked at the characteristics, distribution, and status of all species of Pacific salmon and steelhead across their entire global range – from Korea, Japan, eastern China, and Russia across the Bering Sea to Alaska, British Columbia, Washington, Oregon, and California. And, with so many others, we have catalogued the northern march of declining wild salmon populations on both sides of the Pacific. Measured by abundance, diversity of species, and life histories, as well as integrity of habitat, we believe that the Skeena is one of the most important remaining wild Pacific salmon and steelhead systems. It has the highest population of wild steelhead of any river drainage in the world.

We gratefully acknowledge William Adams, Richard C. Barker, Christopher C. Brand, Matthew B. Brand, Robert Colman, Charles R. Conn III, Christine and Derek Denniston, The Charles Engelhard Foundation, Mark T. Gates, Jr., Kerr Pacific Corporation, Deborah J. and Peter A. Magowan Family Foundation, Jennifer L. Miller, Paul D. Miller, David E. Mundell, William H. and Sally Neukom, Foster Reed, The Jim and Patty Rouse Family Foundation, Swift Foundation, Thendara Foundation, The John and Frances von Schlegell Family Fund, Albert W. Solheim, Sam Vierson Family Foundation, Lawrence J. White, and William P. Wolcott for their generous support.

Spencer B. Beebe
Founder & President, Ecotrust

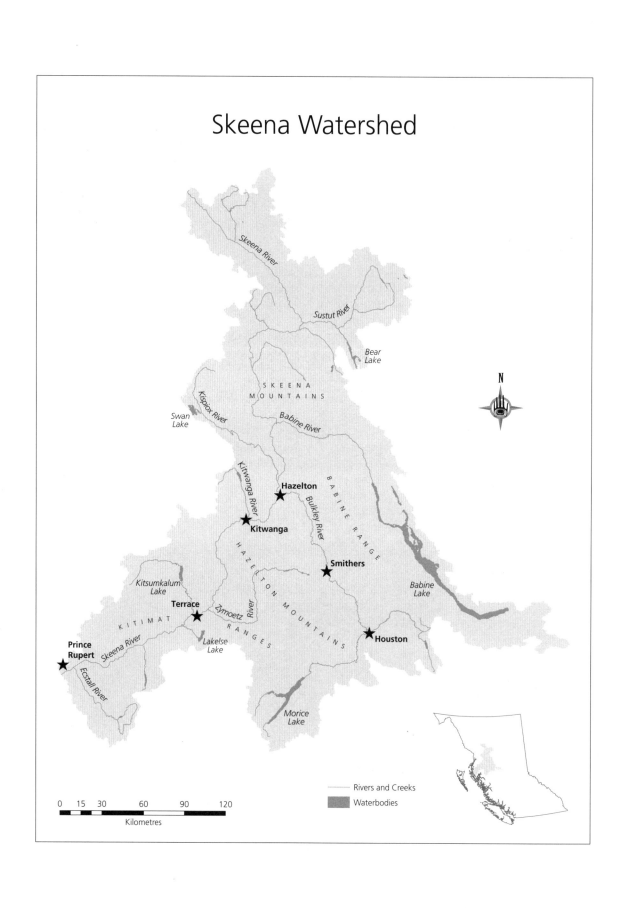

Skeena Watershed

Foreword
Salmon, People, and Place

Although they undertake long ocean migrations, adult salmon return to the river, tributary and even the specific stream reach where they started life. For thousands of years, the salmon coevolved with the physical and biological landscape of their home stream and, as a result, their life histories are tuned to the natural environmental rhythms of their natal river or specific stream reach. The salmon's life history binds them to their home stream. Because the First Nations have a strong relationship with the salmon, their culture and economies also coevolved with specific rivers and their salmon — a relationship between salmon, people, and place that persists today.

European settlers in North America cared little for the ancient relationship between salmon, people, and place, and that attitude is reflected in their fisheries. Euro-American fishermen moved from the rivers to the ocean where they harvested salmon in mixed aggregates of stocks from many different rivers and regions. First Nation's fisheries were dependent on the health of the local salmon population. Mixed stock fisheries of Euro-Americans were dependent on and responded to distant markets and, as a result, the relationship between salmon, people, and place was strained to the breaking point. When their approach to salmon management led to the inevitable overharvest and decline of the salmon, Euro-Americans either moved to new, unexploited regions or turned to fish factories to replenish the supply of salmon. The fish factories or hatcheries were managed in ways that contributed to the separation of salmon from place. The salmon fishery and its management became placeless. Throughout this change and its tragic consequences, the First Nations retained their historical, cultural, and spiritual ties to the specific places, the rivers or tributaries that both the people and the salmon called

home. They understood that the survival of salmon and salmon people depended as much on healthy relationships as on the number of fish.

There are signs, however tentative and fragile, of a reawakening to the wisdom of the past, a wisdom rooted in place. In the Skeena, for example, there is talk of shifting the fisheries back up the river and away from the mixed-stock fisheries. Concern for specific places or habitats has increased and is being expressed in policies directed at the conservation of wild salmon and the places they inhabit. However, we really can't protect and conserve wild salmon and their habitats unless we develop a caring attitude toward them, and we cannot care for that which we do not know. To develop a caring approach to wild salmon we need knowledge of the salmon and their habitats; we need to really know the river and its fish. This simple truth is the reason I found the book *Skeena River Fish and Their Habitat* such an encouraging document. Allen Gottesfeld and Ken Rabnett obviously care about the Skeena and its salmon, and that caring is based on extensive knowledge of the river and its fish — knowledge that they share in this book. To me this book is an important first step in the process of healing and strengthening the relationship between salmon, people, and place.

Jim Lichatowich
Columbia City, Oregon

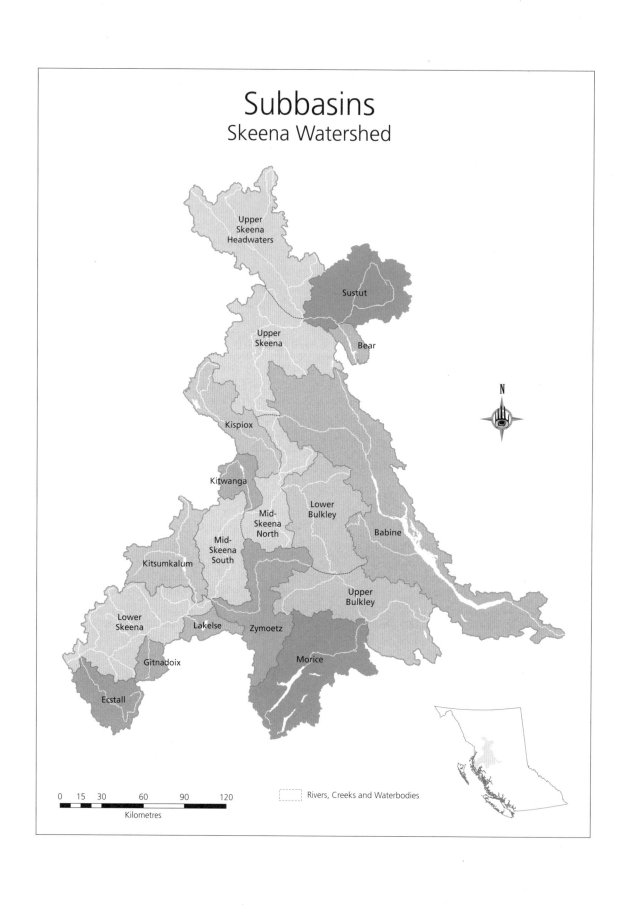

Subbasins
Skeena Watershed

Upper
Skeena
Headwaters

Sustut

Upper
Skeena

Bear

N

Kispiox

Kitwanga

Lower
Bulkley

Mid-
Skeena
North

Babine

Mid-
Skeena
South

Kitsumkalum

Upper
Bulkley

Lower
Skeena

Lakelse

Zymoetz

Gitnadoix

Morice

Ecstall

0   15   30   60   90   120
Kilometres

Rivers, Creeks and Waterbodies

Introduction

The Skeena River is the second largest river system in British Columbia. Abundant rainfall and snowmelt create a discharge that ranks highly among the world's salmon rivers, while landscape and flow conditions make for high-quality fish habitat. The Skeena is currently one of the most productive salmon rivers in North America, and yet Skeena fish and fish habitats are relatively poorly understood. This book summarizes both the published and unpublished literature pertaining to this watershed, as well as decades of personal experience and the advice of colleagues.

Our focus is on the freshwater part of the salmon life cycle. We look at the physical and biological qualities of the Skeena watershed as a whole and then sequentially examine the subwatersheds of the Skeena, from the coast to the headwaters. We summarize freshwater ecology, fish biology, abundance where known, habitat information, enhancement efforts, freshwater aboriginal and sport fisheries, and the history of salmon escapement and human occupation.

This is a time of change in global salmon populations. Wild Atlantic salmon populations and abundance have declined throughout the northern Atlantic Ocean habitats. North American Pacific salmon populations have experienced serious declines in the United States outside of Alaska. In Canada, salmon populations and abundance have been suffering in the Fraser River watershed and the British Columbia Central Coast. Efforts to restore salmon to their native habitats are underway, and there is some slow but steady progress. The restoration and protection of productive habitat, alongside the scaling back of excessive harvest rates, are key to this process. Sustainable fishery management systems need to be prioritized and implemented. In the end, the level and fluctuation of Pacific salmon abundance will largely be determined by human society.

An abundance of salmon was once taken for granted in the Skeena River system. First Nations across the watershed relied heavily on the salmon they used and managed for thousands of years. Over the course of the last century, in the Skeena and its tributaries, fish populations have suffered consequences from overfishing and habitat alteration. It appears that the most serious effects of overfishing were experienced during the rise of industrial salmon fishing in the late 19th and early 20th century. Severe habitat changes resulted from railroad construction in the first decades of the 20th century and from log drives and logging expansion along the valley bottoms in the middle of the 20th century.

Skeena salmon populations continue to face environmental challenges from high rates of logging, and to lesser extents from highway, mining, farming, and urban development. Future salmon stocks will likely have to cope with the anticipated growth of energy developments such as run-of-river hydro, oil, and gas pipelines, and tanker traffic off the coast. The Skeena will also likely witness changes in water quality and stream habitats due to human-caused climatic change – possibly limiting salmonid productivity in some portions of the watershed. Even more serious may be the effects of climatic change on the salmon productivity of the North Pacific Ocean.

For now, industrial development of the Skeena region is still in its early stages. Skeena salmon stocks remain relatively healthy and productive. We need to make a concerted effort to rebuild weak fish populations, maintain the stronger stocks, and ensure they have the conditions necessary to thrive – for the fish, for ourselves, and for future generations.

Many people contributed their knowledge and experience toward this summary volume, and a partial list is included here. This work was funded in part by the Skeena Fisheries Commission and grants from the Department of Fisheries and Oceans; however, the opinions expressed are our own and do not necessarily reflect the funding agencies.

We offer our appreciation and thanks to the following individuals and organizations for their interest, support, and contributions. From Fisheries and Oceans Canada we would like to thank Angela Addison, Rob Dams, Mitch Drewes, Barry Finnegan, Dale Gueret, Lana Miller, Tom Pendray, Bruce Shepherd, Brian Spilsted, Barb Spencer, Ivan Winther, and Peter Woloshyn. Special thanks to Dave Peacock and Steve Cox-Rogers for enduring endless questions, contributing figures and their concise reviews. Particular thanks to Lidia Jaremovic for her critical ideas and patience. From the Ministry of Environment we would like to acknowledge and thank for their time, expertise, and constructive reviews: Dana Atagi, Mark Beere, Darren Fillier, Paul Giroux, and Jeff Lough.

We would also like to thank First Nations' leaders, biologists, and elders who contributed their knowledge and expertise to this project. Some of these people include: Mark Cleveland, Peter Hall, Rod Harris, Walter Joseph, Derek Kingston, Donna McIntyre, Charlie Muldon, Bill Spence, Stephan Schug, Cristina Soto, and Tim Wilson. We extend our thanks to Gitanyow Fisheries Authority, Wet'suwet'en Fisheries, Ned'u'ten Fisheries, Kitselas Fisheries, Kitsumkalum Fisheries, and Gitksan Watershed Authorities.

We offer big thanks to the consultants, retired government officers, and people with specialized local knowledge and photos that contributed to this project: Jim Allen, Melinda Bahr, Dave Bustard, Dave Gordon, George Kofoed, Mike McCarthy, Mike Morrell, Rod Palmer, Brad Pollard, Irene Weiland, and Mike Whelpley. This book would not be possible without a century of fishery biologists, officers, and workers.

Special thanks and appreciation is due to Lance Williams and Gordon Wilson for help with the GIS mapping and production of the report from which this book is adapted. Lastly, thanks to Mariusz Wroblewski, Howard Silverman, Andrew Fuller, and Malka Geffen of Ecotrust for their assistance in publishing this book.

Allen S. Gottesfeld
Ken A. Rabnett
July 2008

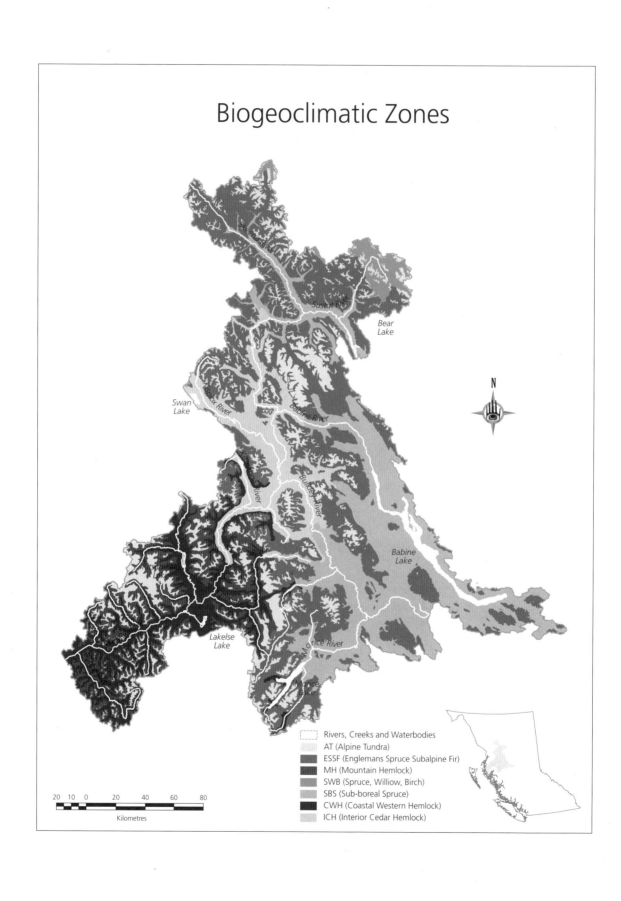

Biogeoclimatic Zones

Rivers, Creeks and Waterbodies
AT (Alpine Tundra)
ESSF (Englemans Spruce Subalpine Fir)
MH (Mountain Hemlock)
SWB (Spruce, Williow, Birch)
SBS (Sub-boreal Spruce)
CWH (Coastal Western Hemlock)
ICH (Interior Cedar Hemlock)

Sustut River
Bear Lake
Swan Lake
Kispiox River
Babine River
Bulkley River
Babine Lake
Lakelse Lake
Morice River
Morice Lake

N

20 10 0   20   40   60   80
Kilometres

1
Skeena Watershed Biophysical Profile

The Skeena watershed is the second largest watershed in British Columbia (54,432 km²). It is located in the northwestern portion of B.C. with its mouth at 54° N, just south of the Alaska panhandle. In its eastern headwaters, the Skeena River extends through the Coast Ranges to drain part of the Nechako Plateau. Consequently, different portions of the watershed experience different climatic and hydrological conditions.

In this book we use the term "watershed" to specify the area above the mouth of a stream that produces the flow observed at the mouth and extending to the drainage divide of adjacent watersheds. In this hydrological sense, a watershed is narrowest at its mouth, where only the width of the channel is included. Thus, in the Skeena watershed, a few streams adjacent to the Skeena River but not tributary to it, such as Kloiya Creek near Prince Rupert, are excluded from our analysis. Note that these streams are included in the Department of Fisheries and Oceans (DFO) Statistical Area 4. This strict definition of watersheds has the advantage of biogeographical rigor. However, when one divides a large watershed into major subwatershed portions, areas along the mainstem and small tributaries that flow directly into the mainstem are omitted by necessity. We therefore compromised in dividing up the Skeena and discuss several subbasins along the Skeena mainstem as "watersheds." These are, in an upstream progression, the lower Skeena, mid-Skeena south, mid-Skeena north, and upper Skeena. Mostly these areas cover the remainder of smaller streams entering the Skeena and the main channel, which functions largely as a transit route for salmon, although the lower Skeena watershed unit has important spawning and rearing capacity along the multichannel mainstem.

The Skeena watershed is composed of a series of northwest trending mountain ranges separated by broad valleys. The coastal Kitimat Ranges are mostly composed of granite and granitoid rocks. Low-grade metamorphic rocks make up the bulk of the Hazelton Mountains in the central part of the watershed. To the east, the Skeena Mountains and Babine Range are composed primarily of Mesozoic sedimentary and volcanic rocks. Because of the difference in bedrock composition, coastal drainages produce mostly sand as a rock breakdown product, while interior drainages produce abundant clay and silt as well as sand.

The entire Skeena watershed was intensely glaciated during the Ice Age, 28,000 to 11,000 years ago (Clague 1984). Valleys crossing the Coast Mountains were intensely scoured. Interior valleys accumulated up to tens of metres of glacial till. During the deglaciation phase, 11,000 to 10,000 years ago, large volumes of gravel accumulated in the Skeena valley and its major tributaries. Down-cutting by the Skeena River in the few thousand years after glaciation left remnants of these gravels as large terraces. Glacial lakes in several valleys accumulated fine sediments. During deglaciation, depressed coastal areas were flooded and marine conditions extended as far inland as the mouth of the Zymoetz River.

Ice margin lakes formed in several areas including the upper Babine watershed (Hastings et al. 1999), the Sustut valley, portions of the Bulkley valley, and the Skeena valley near Kitwanga, with fine sediments being deposited in these basins. These varied Ice Age deposits are the source of most sediment movement in the modern rivers of the Skeena watershed. Salmonids apparently invaded the Skeena basin during deglaciation, taking advantage of short-lived connections between watersheds, and have been an important part of the freshwater fauna ever since (McPhail and Carveth 1993b).

Climate and Hydrology

The climate of the Skeena watershed varies greatly. On the coast there is abundant precipitation, cool summers, and mild winters with average temperatures near 0° C. Usually, precipitation reaches a maximum in the fall and early winter, generally in October and November, with intense cyclonic storms from the North Pacific moving across the coast every day or two (Environment Canada 1993). Rainfall amounts of 50 to 100 mm per day occur annually. The interior has a more boreal climate with relatively low precipitation, warm summers, and prolonged cold winters with average temperatures less than -10° C for two or three months. Precipitation amounts are rather uniform throughout the year. Summer convective storms are common but rarely deliver more than 20 mm of rainfall in a day.

Precipitation decreases regularly from the coast to the interior. The coastal drainages of the Skeena watershed receive at least 2,500 mm of precipitation per year, with higher amounts in the mountains. The Kitimat-Kitsumkalum trough (Terrace and vicinity) receives half or less of this amount. East of the coastal mountains, the Smithers area receives about 600 mm, while further east on the Nechako Plateau annual precipitation is less than 500 mm. With these variations in precipitation and winter climate, the interior hydrological pattern differs greatly from that of the coast.

Most years, coastal drainages have one or more brief fall or winter floods. Figure 1 shows how snowmelt contributes to higher flows on the Exchamsiks River from April to August, while August to January has short peak flows from intense rainstorms. Nearly all large coastal floods are caused by intense rainfall, which are often rain-on-snow events. Interior drainages, such as the Bulkley River above Houston, or the Babine River, usually have a single dominant flood occurring annually at the peak of snowmelt in May or June and lasting for several weeks. Figure 2 shows how flood flows on Buck Creek are restricted to the April – June snowmelt period and late summer low flows increase with fall rains. In coastal-interior transitional areas such as the Kispiox, Kitwanga, and Zymoetz Rivers, a mix of fall rain floods and spring snowmelt floods is found. Throughout the watershed, prolonged freezing conditions yield low flows in the late winter. In the interior, higher summer temperatures and abundant evapotranspiration by forests contribute to low flows during the late summer and early fall.

Coho, chinook, pink, sockeye, and chum are fall spawners. In coastal watersheds, intense floods may destroy much of the egg production. To some extent coho avoid this problem by spawning late, in November and December, in coastal waters. Coho can also take advantage of the fall flood pattern by using the high flows to access small headwater areas, which may often be inaccessible due to shallow water depths and beaver dams.

In the interior, salmon spawning redds (nests) are generally undisturbed by flood events. Late summer and early fall low-flow conditions may restrict salmon access to potential spawning beds, especially in dry years. This is a particularly serious problem in the upper Bulkley drainage (Figure 3). Winter low flows may leave spawning gravels partially dewatered. Extremely low temperatures, especially early in the winter promote the formation of anchor ice on the streambed, which may restrict intergravel flow or even result in freezing of gravel patches

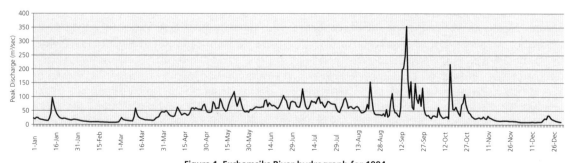

Figure 1. Exchamsiks River hydrograph for 1994

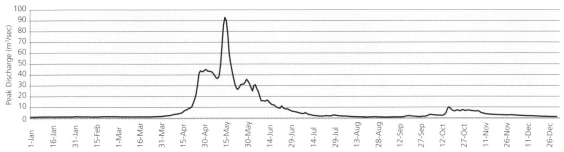

Figure 2. Buck Creek near Houston hydrograph for 1997

containing developing eggs and alevins. In the spring, high water flows from snowmelt aid the migration of salmon fry and smolts to the sea by providing higher flow velocities. These floods also provide some degree of protection from predators by hiding the presence of fry and smolts in the generally muddy waters.

Average annual flows, and especially late summer flows, have declined in the interior (eastern) part of the Skeena since the 1930s. In Figure 4, the fitted LOESS smoothed curve (f = 0.8) shows the general decline in the Bulkley River's late summer water discharge, with flows recorded in m^3/sec. The coastal portions of the Skeena have had near normal water discharge.

The decrease in late summer and early fall flows is a particularly strong trend in the most interior portions: the Babine River and the Bulkley River east of Houston. In the upper Bulkley, seven of the past 13 years have had average September flows of less than 1 m^3/sec, and nine out of the 13 years have had average September flows equal to or less than 1.1 m^3/sec. These September low flows decrease the ability of coho and sockeye adults to reach spawning areas and utilize spawning gravels. The migration of coho adults into spawning streams is also severely impaired, and some streams are dewatered. Intensified summer drought in the interior of B.C. may be a result of global climate change. Canadian Institute for Climate Studies research predicts a continuing decline in summer precipitation in the interior portion of the Skeena for the next 70 years (Price et al. 2001). Finer scale modeling shows a decrease also in September precipitation (Boer et al. 2000).

Stream Channels

Descriptions of stream channel information in the following chapters are mostly based on the aquatic biophysical 1:50,000 mapping that was compiled in the mid-1970s throughout a large part of the Skeena basin (Resource Analysis Branch 1975 – 79). One of the mapping components produced information on stream morphology including stream reaches, longitudinal and cross-section profiles of the channel, and the stream reach gradient.

Longitudinal profile is characterized as being either stepped or regular. Stepped profile is described as a repetitious sequence of forms or slopes; regular is defined as smooth, homogenous, or continuous profile. Cross-section profiles are described as confined, bounded, or unconfined. Confined means the channel is entrenched or lateral movement is controlled by the banks. Bounded means channel movement is limited by valley walls near the edge of the floodplain. Unconfined means the channel is not bounded by valley walls and lateral movement, and extensive overbank flooding is possible at high flows.

Water Quality

Water quality is defined as the natural physical, chemical, and biological characteristics of water. Water appears as rivers, streams, lakes, ponds, and wetlands, as well as underground storage. Forests, water, fisheries, wildlife, and humans are linked together by the hydrologic cycle. Water quality criteria are policy guidelines concerning the range of conditions, usually defining acceptable levels for particular kinds or classes of water use. Setting water quality objectives involves taking the set of criteria and adapting them to a specific body of water.

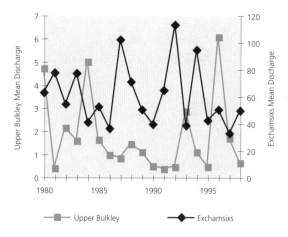

Figure 3. September mean flow for the upper Bulkley and Exchamsiks Rivers

For example, water quality objectives or guidelines are or may be applied to drinking water, fish and aquatic life, agricultural and mining activities, or forest development. Province-wide ambient water quality criteria include pH; substances that degrade water quality such as nutrients, algae, particulate matter; low-level toxic substances; high-level toxic substances such as cyanide, PCBs, and metals; as well as microbiological indicators of risks to humans (fecal coliforms, *Giardia*). Other common criteria include dissolved oxygen, total suspended solids, water stage, and biochemical oxygen demand (BOD).

Within the Skeena watershed, the major water use sectors are domestic, agricultural, municipal, and industrial. Industrial forestry is active in almost all subbasins of the watershed, while agriculture is the major land use in the Bulkley valley. There are three decommissioned mines that have various levels of acid rock drainage requiring maintenance and monitoring into the distant future. Communities throughout the drainage discharge waste waters into the Skeena and Bulkley Rivers. Remington (1996) noted that although point-source water quality data are available from various agencies, there has been little assessment from nonpoint sources or assessment of cumulative loading on water quality within the Skeena basin.

Turbidity is generally low in late summer and through winter in streams of the Skeena basin. Fine sediment content rises episodically during spring and fall floods in the smaller

streams. The pattern in the Skeena River is simpler as it integrates the contributions of many upriver source areas. Turbidity is moderate to high during the spring snowmelt floods and following fall rainstorms. In addition, glacial melt and redistribution of silts from alpine areas keeps the turbidity levels moderate for the summer season. The Zymoetz River is a major fine sediment source. The reddish clays derived from weathering of the Telkwa and Nilkitkwa Formations give a brown color to the Skeena following floods. The Telkwa River upstream of Smithers in the Bulkley valley drains bedrock of a similar nature and has a similar effect on fine sediment loading. Following floods, the difference in suspended sediment load is conspicuous at the confluences of these rivers with the Skeena.

Water quality of streams and streamflows responds to human activity in the terrestrial portion of the ecosystem. A key to understanding impacts is remembering that forest streams are not separate and distinct from the forestlands they drain, but are an integral part of the forest ecosystem. In the broader context, the small forest streams within the tributary subbasins contribute to the much larger Skeena watershed. Streams flowing from undisturbed watersheds generally have excellent water quality. It is this characteristic that makes them so valuable, not only for fish production but also for human consumption.

Small streams provide benefits in the form of fish production and domestic water production, and also hold considerable social, economic, and political significance for a large portion of society. Turbid water or loss of water quality, as well as reduced fish spawning and rearing production in small forest streams, are problems as socially significant as pollution problems on large rivers such as the Fraser. Maintenance of water quality and quantity is a major criterion for judging the skill and knowledge of foresters applying land-use management strategies. However, cumulative effects from logging operations in this area are little studied. The foremost concerns include the integrity and importance of small streams, hydrologic change, temperature change,

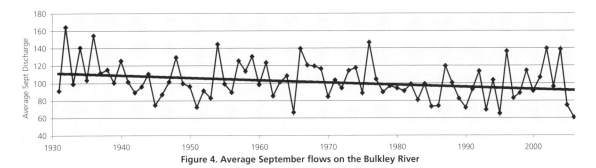

Figure 4. Average September flows on the Bulkley River

sedimentation, and the effects of changes to the streams' physical structures.

In summary, water quality inputs on the mainstem of the Skeena River are in large part controlled by inflowing tributaries fed by turbid glacial meltwater, shallow flow through forest soils, wetlands, and groundwater aquifers replenished by snowmelt. Based on these hydrologic flow-path regimes, each basin carries its own chemical signature, and the geology of the basin determines which constituents are available for mobilization.

As the Skeena River flows down through the Intermontane Belt, it may pick up the highly turbid flows of glacial meltwater from one source, which could then be diluted by clean-flowing snowmelt from the next large tributary. The typical timing of the tributary contributions creates strong yearly patterns in water chemistry. These yearly chemical patterns shape and interact with the biological community, determining size and quality of algal, bacterial, and fish populations, and providing seasonal cues for migrating and spawning fish. Biotic populations add the final layer of complexity to the biogeochemical cycle of nutrients and other solutes, as organisms grow, die, and degrade. The water quality of the Skeena River is intricate and highly complex, but it provides a crucial foundation for understanding how the health of the ecosystem is maintained (Henson 2004).

Vegetation

The Skeena watershed is densely forested throughout. In the coastal portions, rainforests of large Sitka spruce, hemlock, and red cedar cover all slopes, even cliffs. In the central Skeena watershed, hemlock forests dominate although spruce and red cedar are found on moist sites as far east as the lower Bulkley valley. In the interior, hybrid Engelmann and white spruce stands dominate, but large, low-elevation areas are composed of mixed forests and trembling aspen (poplar) forests. Some south-facing slopes are parkland with grass understory. Early successional stage lodgepole pine stand areas are common, especially on gravelly soils and dry sites. The subalpine forest that generally lies above 800 m elevation is primarily subalpine fir and Engelmann spruce, although it is replaced by mountain hemlock on the coast.

While boundaries between ecosystems can be abrupt, ecosystems more often tend to grade slowly from one another (Banner et al. 1993). Climate is the most important factor influencing the development of terrestrial ecosystems; however, ecosystems also vary because of differences in topography and soil. Vegetation is an important indicator of ecosystem type because it is readily visible. Changes in vegetation over time, a process called succession, can take multiple pathways. The coastal/interior transition forests are reflected in the major ecological zones that consist of two predominant climatic gradients: a west-to-east gradient of decreasing moisture and an elevation gradient of decreasing temperatures, increasing precipitation, and increasing snowfall.

Riparian vegetation is important for fish habitat. In smaller streams, overhanging vegetation shades streams and keeps them cooler in the summer and warmer in the winter (Beschta et al. 1987). In these streams, leaf and twig litter and insects provide the bulk of the food energy input (Swanson et al. 1982). In all parts of the watershed, trees are large enough so that when they fall into streams they produce stable structural features. In small streams, these are critical for providing fish habitat and controlling the storage and movement of sediment. In intermediate-sized streams, log accumulations are the principal structural features. Log debris seems less important for fish habitat in the largest streams (Sedell and Swanson 1984), such as the main channel of the Skeena and its principal tributaries, because transport tends to leave logs on gravel bars and obstructions above normal flow levels (Gottesfeld 1996).

2
Skeena Watershed Fish Populations

The Skeena watershed supports five salmon species: sockeye *(Oncorhynchus nerka)*, pink *(O. gorbuscha)*, chum *(O. keta)*, chinook *(O. tshawytscha)*, and coho *(O. kitsch)*. Trout and char presence includes steelhead/rainbow trout and cutthroat trout, Dolly Varden char, bull trout, and lake trout. In general, salmon populations are healthy, with relatively few threatened populations and no known recently extinct populations (Morrell 2000). The threatened populations are of high concern, and strong efforts need to be undertaken to prevent their loss. They are discussed more fully in the species sections that follow.

There are about 26 nonsalmonid/trout fish species in the Skeena. No species are classified in the B.C. Conservation Data Center's "moderate" to "high risk" range, but few nonsalmonid species have been adequately sampled, particularly at the local level, for conservation status to be known.

The richest source of data on the status of salmon stocks is the Area 4 Salmon Escapement Database System (SEDS) maintained by DFO. This data set consists of annual spawning ground observations of about 196 census areas collected since 1950. The census areas vary in size from whole river systems, such as the Kitwanga River, to short reaches of productive systems such as on the upper Babine River or Club Creek in the upper Kispiox watershed. The SEDS database is most reliable for the larger and more consistent spawning stocks. The bulk of salmon spawning areas appear to be represented; however, very small and infrequently used spawning areas may be overlooked. For example, in years of large pink salmon runs, spawning may occur at numerous sites that are not utilized under most conditions and may even appear to be unsuitable. These sites are mostly not in the SEDS database.

Spawning area counts are made with different techniques, including aerial counts, ground counts, counts from boats, swimming counts, counting weirs, and mark and recapture experiments. Most counts are simple estimates made from one or more ground visits or aerial surveys. Because of this variety in technique and varying natural conditions (visibility, etc.), the data vary in quality in often unknown ways. However, this dataset is the best available and far exceeds the quality of data available for steelhead, trout, and other nonanadromous species. In general, the number of stocks counted increased from 1950 to 1990 and declined after 1992.

The vast majority of spawning escapement data for the Skeena watershed that was used in this text was obtained from DFO's SEDS database. The data quality varies from observer to observer and place to place. While appreciating the great value of the data records, they can only be utilized as indicators of general trends and, at best, reflect relative abundance rather than actual values. Coho are probably the most poorly estimated fish.

Data for steelhead are more dispersed. Since steelhead spawn in the spring at high water conditions, direct counts are usually not possible. Catches in the Tyee Test Fishery give aggregate abundance indices for the whole Skeena River. More detailed data sources are discussed in the steelhead section. In general, enough is known to infer the order of importance of spawning streams. Data on the population size and status of nonanadromous fish species of the Skeena are scarce and of little use for determining population trends through time.

Salmon Species Status
The following six sections review the habitat and status of the five Pacific salmon species

and steelhead in the Skeena. For each species, the nature of their habitat and life history is described, endangered stocks are reviewed, the genetic structure of the population is described where available, and major stocks are identified.

Chinook Salmon

Skeena chinook are the largest salmon species and have the most complex and diverse life histories. Chinook are commonly characterized by their stream or ocean life-history variations. Stream-type chinook typically spend the first year or two in fresh water before migrating to the sea, while ocean-type, post-emergent juvenile chinook typically have a limited freshwater phase of 60 to 150 days before migrating to sea. In the Skeena, chinook fry spend one year in freshwater. A small number of mostly coastal fish go to sea shortly after hatching. Overall in the Skeena, only a few percent of chinook go to sea in their first summer (Peacock et al. 1997). Other prominent variations include seasonality of adult return migrations and size or age-at-maturity of adults.

In general, chinook are fish of larger streams and spawn in faster moving water with coarser gravel than other salmon. Typical chinook stocks are relatively small (Healey 1991). The largest Skeena chinook stocks in the Bear, Kitsumkalum, and Morice Rivers contribute 65% to 75% of the overall Skeena escapement and have average escapements under 20,000 in their most productive decade. The Kitwanga, Babine, and Kispiox Rivers typically contribute 10% to 15% of the overall escapement.

Chinook have been reported spawning in 85 localities since 1950 and infrequently in about eight others. Winther (2006) reported that 20 Skeena tributaries contain less than 1,000 chinook and 30-odd tributaries contain less than 100 chinook. Most escapement estimates are based on visual counts except in the Kitsumkalum where mark-recapture sampling is conducted. Skeena mainstem spawners are not counted, but it is suspected to be a relatively large population.

Chinook are the first salmon species to return to freshwater, resulting in their popular name of "springs." Early stocks arrive in May and June, late stocks in June and July. The early stocks are usually upriver (headwaters) stocks such as upper Bulkley and upper Kitsumkalum. Late stocks tend to be more coastal and/or tend to spawn downstream of lakes.

Most chinook spawning occurs in August and September in larger tributaries, on tributary alluvial fans in the mainstem, or in strong association with large lake outlets. When spawning occurs at high densities, as in the Morice, Bear, and Babine Rivers, spawning dunes are created by the merger of adjacent redds. Fry emerge from the gravel in early March to late May, depending on the ambient water temperature during incubation and the time of spawning.

After hatching, many fry move or are displaced downstream (Healey 1991). Chinook fry are often territorial, and as they grow, individual territories expand with the excluded fish displaced downstream. In the Kitsumkalum River, generally less than 10% leave before the first winter. The rest of the smolts migrate to sea during spring high flows in the following May and June, when turbid water and faster flows serve to reduce predation. Presently it is not clear what the proportion of fry, particularly fry originating in the headwaters, move to downstream habitat such as the multi-channeled reach below Terrace to rear. Chinook return after one to five years at sea, though most return after three seasons. Chinook with longer ocean residence times are typically larger as adults.

Chinook originating from Oregon through Alaska are widely mixed along the Pacific coast. Coastal fisheries therefore intercept fish originating in many rivers. This classic mixed-stock fishery has been difficult to manage without serious impacts on less-productive stocks. Since the 1950s, chinook stocks generally declined until the mid-1980s. Restriction of North Coast chinook fisheries and in-river sports fishing began in the mid-1970s (Ginetz 1976). The 1985 Canada-U.S. Pacific Salmon Treaty and its subsequent amendment in 1998 further restricts the coastal commercial fishery in B.C.

Since 2000, the chinook North Coast seine catch has been zero, while the gillnet catch in Area 4 has fluctuated from a few hundred to

10,000 chinook. The North Coast troll catch for the Skeena, as determined by DNA, averaged 4,200 chinook for 2004 and 2005. The Alaskan catch of Skeena chinook, as determined by Kitsumkalum coded wire tags, was 11,200 in 2004 and 14,335 in 2005. The tidal sport chinook catch was 6,000 in 2004 and 3,300 in 2005. The sport fishing catch primarily located in the lower Skeena is a moderate constituent of the overall exploitation and appears to fluctuate annually between 5,000 to 10,000 chinook.

Chinook salmon are also an important part of aboriginal fisheries, being next in importance to sockeye. Significant harvests are taken in the Skeena River above and below Hazelton and at Moricetown, with the catch since 1950 relatively steady at 5,000 – 10,000 chinook annually. The effort and catch variation often reflects in-river sockeye abundance. In the past there was a large chinook fishery at Bear River that continues today to a lesser degree. The total Skeena chinook catch in 2004 was 46,000, and the aggregate Skeena escapement was 47,000. The total Skeena chinook catch in 2005 was 38,500, and the aggregate Skeena escapement was 31,000 (Winther 2006). Current information indicates that Skeena chinook stocks are fished at significant levels.

Information on chinook stocks prior to 1950 is available only from catch data. Catches from 1899 to 1930 in the Skeena River fishery averaged over 100,000 chinook with peak catches exceeding 200,000 (Ginetz 1976, Riddell and Snyder 1989). Chinook catches declined steadily from 1930 to the 1970s; this was mostly due to fishing effort moving out of the Skeena

River. Escapement data based on spawning ground counts have been collected since 1950. Total Skeena River escapement was about 50,000 in the 1950s, declining to about 25,000 from 1965 to 1985. The overall aggregate Skeena chinook escapement has been recovering in the past twenty years and is now increased 150% above 1950s levels. Figure 5 shows chinook escapement in the Skeena River tributaries from 1977 to 2006. It is likely that the increase in escapement from 1985 to the present is due to the restrictions on chinook harvest in Alaska that took effect with the Pacific Salmon Treaty.

Overall, exploitation rates in all fisheries from the mid-1980s to mid-1990s, were 10% to 25% for the early upper Bulkley stock and 35% to 65% for the mid-season, lower Kitsumkalum stock (Peacock et al. 1997). Hankin and Healy (1986) suggested a maximum sustained yield (MSY) exploitation rate for chinook of 40%. The Kitsumkalum exploitation rates for the mid-1980s to mid-1990s were close to the limit proposed by Hankin and Healy, and the upper Bulkley stocks were well below it. The Kitsumkalum chinook stock generally exceeded the MSY escapement target. Nevertheless, it is likely that the long-term depression in chinook production and the recent increase in stock productivity are due not only to changes in exploitation rate, but also to changes in ocean survival. Figure 6 shows Alaskan and Canadian exploitation of the Kitsumkalum summer chinook stock that is monitored under the chinook "key stream" program in response to objectives set out in the Pacific Salmon Treaty.

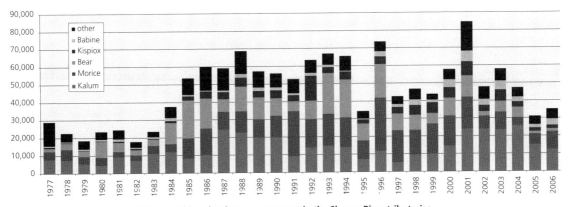

Figure 5. Chinook salmon escapement in the Skeena River tributaries

The genetic structure of chinook stocks as discussed by Beacham et al. (1996) focuses on the separation of regional stocks and defines Vancouver Island, Fraser, and North Coast aggregates, though individual stocks are also separable. Chinook from Skeena tributary rivers such as the Kitwanga River and the Bulkley River are clearly separable. Even within a watershed such as the Kitsumkalum, early-run chinook from above Kitsumkalum Lake are quite distinct from the late-run, lower Kitsumkalum stock. Early-run chinook which spawn in the upper Bulkley River are distinct from the more numerous late run, which spawn mostly in the Morice River. More recent unpublished work confirms these separations. Conservation units need to be narrowly drawn in order to conserve the smaller stocks as well as the more productive larger stocks. Therefore, genetically distinct spawning stocks will require appropriate-scale approaches to conservation.

Chinook enhancement via hatchery production began in 1978 at Fulton River. It developed rapidly in the 1980s and 1990s with the spread of small hatcheries to Terrace, Kispiox, Toboggan Creek, Emerson Creek, and Fort Babine (Peacock et al. 1997). Between 1978 and 1995, about nine million fry and smolts were released. The coded-wire tagged fry and smolts from the Deep Creek Hatchery (Kitsumkalum River) have contributed to the understanding of ocean survival rates and interception in various coastal fisheries. Estimated survivals to adult stage are variable: 0.0% to 0.9% for fry, 0.4% to 0.6% for smolts, and 0.0% to 5.5% for yearlings (Peacock et al. 1997).

The problem with mixed-stock fishing of chinook stocks is that at times of poor overall returns to the Skeena system, the smaller stocks

may be eliminated. When stock numbers fall below several hundred, there is a significant chance of loss of genetic diversity, which adversely affects the chances of long-term survival (Franklin 1980, Waples 1990). Despite the overall recovery of chinook stocks in the Skeena basin, some small stocks have not recovered and are a cause for concern.

Deep Creek is located on the northern edge of Terrace, and for about 40 years, a portion of its flow has been diverted to the Terrace water supply system; although, in recent years Terrace has used well water supplies, with only occasional Deep Creek use. Deep Creek chinook escapement was in the hundreds in the 1970s, in the tens in the 1980s, and near zero in the past decade. Chinook spawners have been noted in the past few years, but they might be escapees from the Deep Creek Hatchery. The run timing is apparently similar to the adjacent lower Kitsumkalum River and this stock may be part of the larger, lower Kitsumkalum run.

The upper Bulkley River had low chinook escapements from the 1960s to the 1980s, with substantial recovery since 1988. There were record high runs in 2000 and 2001, years of high summer flows. The upper Bulkley stock is an early-run stock that is genetically distinct from the larger and later Morice run (Beacham et al. 1996). The trend of lower late-summer streamflows probably has contributed to the decline of this stock. As of 2007, commercial fishing exploitation rates have been modest, which may have contributed to the recovery of this population. The exploitation in the food fishery by the Wet'suwet'en First Nation at Moricetown is probably modest because of the early migration timing. The upper Bulkley stock was enhanced with hatchery releases from 1985 through 1993.

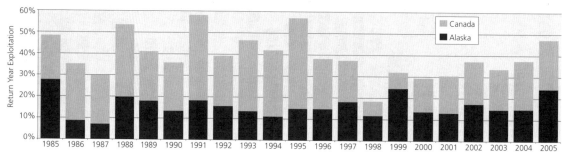

Figure 6. Kitsumkalum chinook exploitation by Canada and Alaska

The Zymagotitz (Zymacord) River has a very small chinook stock with stream characteristics posing enumeration difficulties. As far as the visual field counts can be trusted, there were about 100 spawners at the peak in the 1970s. Escapement estimates for the 1990s have declined to about 25. The Zymagotitz tributary Erlandsen Creek has had escapements of about 20 fish, but had a record high escapement of 140 in 2001. Since 2000, the annual average has been 85, indicating a trend in increasing abundance. The low escapement numbers for the Zymagotitz stocks are reason for concern about long-term survival.

The Gitnadoix River chinook stock has declined from about 400 in the 1960s to an average of less than 40 in the 1990s. The overall escapement has been poor for the past 20 years. The Gitnadoix watershed is a provincial park and is undeveloped with no known anthropogenic environmental changes. In the past two decades, Magar Creek, a Gitnadoix tributary, has been supporting increasing numbers of chinook. Escapements since 2000 have annually averaged 144 chinook, indicating a gently increasing trend.

The Shegunia River mouth, which is located across from the Skeena-Kispiox confluence, supported an important traditional Gitxsan chinook fishery. Escapement data are poor and confound clarity, but the stock trend appears to have been depressed since the 1950s. Hatchery-reared fry and smolts from this stock were released from 1987 to 1995. Increases from this outplanting may have contributed to a short-term increase in escapement, but long-term recovery is not apparent. Contributing factors affecting the chinook population include the watershed's high sediment producing regime, the logging that occurred from the 1970s to 1990s, and the road access that provided anglers easy entry to a critical holding area since the late 1950s. The lack of escapement observations confounds the situation.

Babine River chinook escapement has been poor since the 1960s. The major subpopulations that spawn upstream of the counting fence and on the Tsezakwa Creek fan (Rainbow Alley) have not fully recovered in the past two decades as many other stocks have, despite supplementation

by hatchery production. The peak of this run in late July and early August coincides with the sockeye and pink salmon fisheries.

Lakelse River and Coldwater Creek chinook are a small stock with peak escapements of about 400 in the 1970s and 1980s. The adult spawning population has declined strongly since 1990. The average count for the 1990s is about 100. Coldwater Creek is a tributary of the Lakelse River adjacent to the chinook spawning area below Lakelse Lake. In the 1980s and early 1990s, the escapement of chinook to Coldwater Creek exceeded that of the Lakelse River. Counts are not available for either locality for the last decade, which saw strong stock recovery in many other Skeena watershed streams. This is an early chinook run, with spawning mostly occurring in mid-July and August. Chinook releases from Deep Creek Hatchery were made from 1986 to 1991, although they do not appear to have helped increased escapement. The present population size of this stock, based on available data, appears alarmingly small. It is possible that the Lakelse and Coldwater chinook enumerations are counting the same stock that changed its spawning location in response to habitat conditions. If so, the total watershed chinook count would appear relatively stable. This situation possibly exists with the Gitnadoix River and Magar Creek chinook visual counts as well.

The Ecstall River system includes two census areas for chinook: Johnston Creek and the Ecstall River. This river was one of the top five producers in the 1950s through the 1970s, but has since declined severely. Sports fishing impacts were noted in the 1970s and may have contributed to the decline. A high commercial fisheries exploitation rate for this stock would restrict the ability of the Ecstall River chinook to recover. In the 1990s, escapement averaged 650, which is about one-seventh of the 1950s and 1960s escapement and one-third of the 1970s escapement. Although population levels are severely depressed, this stock has adequate escapement for maintenance. A lack of recent observations restricts status comments.

The Suskwa River has a small chinook stock. Peak escapements during the last 50 years were several hundred. Recorded escapement values in the period from 1988 to 1992 ranged from 0 to 60; observations from 2001 to the present indicate an annual average of 62 chinook. This watershed had heavy logging impacts in the 1970s and 1980s, and continues to have moderate-scale landslide activity. The present level of chinook escapement does not suggest a stable population level or long-term survival.

Pink Salmon

Pink salmon are the most abundant of the salmon species in the Skeena watershed. The majority of pink salmon production occurs in a few large populations within the Lakelse, Kitwanga, Kispiox, Babine, and Morice Rivers as well as the lower Skeena mainstem area downstream of Terrace including the Ecstall, Khyex, and Kasiks Rivers. Their wide distribution, which includes about 120-odd spawning streams, along with the large amounts of marine nutrients they deliver, make them an important component of the Skeena salmon ecosystem.

Pink salmon are exclusively two years old at spawning time. This means that odd- and even-year stocks are genetically separate. In many watersheds, either the odd- or even-year runs are dominant. Most discussions of pink salmon stock assessment treat the odd and even years separately. Skeena pink salmon do not have a well-developed dominance. Therefore, for ease of discussion in this text, the two cohorts are dealt with together.

Pink salmon arrive in the Skeena River from late July to early September; there is a strong overlap with sockeye in late July through early August. The largest spawning populations are in the coastal portion of the Skeena watershed, although significant numbers reach headwater areas such as the Kitwanga, Kispiox, Babine, and Morice Rivers. Pink salmon spawn in gravel areas in late August and September, soon after ascending the river. Pink salmon fry emerge in April and May and go to sea immediately upon hatching. They may begin to feed during their river migration, especially upstream stocks. Pink salmon return at a smaller size than other salmon due to their short life cycle. In the ocean, they grow faster than other salmon species

(Heard 1991). Their marine migrations are extensive with seasonal northern and southern ranges in the North Pacific.

Pink salmon tend to stray at higher rates than other salmon (Horrall 1981). Heard (1991) summarized mark and recapture experiments showing approximately 10% straying in pink salmon. Most straying is to nearby streams. In years of large escapement, many pinks wander into previously unused spawning areas and even spawn in places that appear to be unsuitable. The genetic structure of pink salmon populations reflects this pattern of straying. Only regional patterns of stock separation have been described. Beacham et al. (1985) reported on allozyme studies that resulted in identification of three stock groups: Fraser River, Puget Sound, and B.C. non-Fraser. In general, the odd- and even-year lineages of pink salmon are more different genetically than stream populations over large areas (Heard 1991).

There is probably only one biologically definable pink salmon population that occupies the Skeena River. For this aggregate there is an exceptional variability in stock recruitment from year to year, perhaps due to variation in survival of fry early in their ocean residence. These characteristics, and the general robust size of spawning stocks, mean that there are no identifiable stocks at risk.

Pink salmon are the one species of Pacific salmon that has been spatially expanding in the Skeena River. Escapements in the 1980s are the highest in the escapement record (Figure 7). Total escapement doubled from the 1950s to the 1990s. The total pink salmon run size before harvest averaged over five million in the 1980s and 1990s. Escapement in the last decade is an annual average estimate of 877,000 pinks.

In the 1980s, pink salmon increased their range in the Morice River. The population expanded rapidly and exceeded 600,000 in 1991. The lower Skeena River escapement also grew rapidly during this interval. The Lakelse River continues to be the leading pink salmon producer, with spawning occurring mostly in the few kilometres below the lake outlet. These gravels are consequently the most productive natural spawning areas in B.C. Because of the generalized population structure in pink salmon,

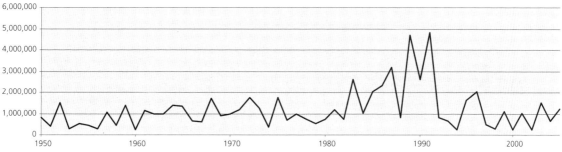

Figure 7. Pink salmon escapements in the Skeena region

and the relatively high rate of straying, there is little concern about the continued survival of small pink salmon populations. We assume that these areas are repopulated in years of high returns when straying is maximal.

Chum Salmon

Chum salmon are the least abundant of the five Pacific salmon species in the Skeena watershed. They are much more abundant in southern B.C. and in southeast Alaska where hatcheries production enhances some of the stocks. In the Skeena watershed, chum salmon live three to five years. Returning four-year-old fish are most abundant, but some three- and five-year-old fish are generally present at spawning time (Spilsted 2005). Chum salmon arrive in the Skeena system from late July to early September. Their migration coincides with the much larger runs of pink salmon, and they often spawn adjacent to pink salmon spawners.

Unlike coho and sockeye salmon, which may hold for a month or two before spawning, many chum salmon spawn soon after traveling up the Skeena River. Fry emerge early in the spring and migrate to the Skeena estuary immediately upon hatching. Chum salmon smolts typically remain in the estuary for one to several months, growing rapidly before dispersing in the ocean (Healey 1980). There is apparently a high degree of variability in the survival rate of chum early in their marine life.

Chum salmon are most common in the coastal portion of the Skeena watershed. The most important spawning areas are the Ecstall River and the multi-channeled reach of the Skeena River below Terrace. Other spawning areas are near the mouths of large tributary

streams and in back channels along the Skeena River from Terrace to Kispiox. Significant spawning occurs in the Kitwanga and Kispiox Rivers. Smaller stocks are present in the lower portions of several other Skeena River tributaries. Chum salmon are rare in the Bulkley River or in the Skeena River above the Kispiox River confluence.

There is an extraordinary variability from year to year in chum salmon returns. Annual escapement estimates (SEDS database) have varied one hundred fold over the past 50 years, as shown by Figure 8. Chum salmon stocks were apparently much larger early in the 20th century; the commercial catch of chum salmon in the Skeena area between 1916 and 1928 was over 200,000 per year (Argue et al. 1986). This suggests an escapement about 10 times larger than that of the recent past. Chum salmon escapements have been low for the past 50 years. With the exception of the spectacular high escapement in 1988, the average escapements have been declining over this period. Even with the 1988 data, a smoothed (LOWESS) trend line shows a decreasing escapement, especially in the last decade.

Field observations suggest that chum are highly specialized in their selection of spawning sites. Several of the Skeena River spawning sites used every year are less than a few hundred metres long. Chum continue to use these patches of gravel even when channel reorganization separates them from their former source of flow. At Andesite Creek and Coyote Creek, former tributary mouth spawners are still using the same patch of gravel after avulsions moved the stream mouths some distance away.

There are no genetic studies aimed at separating stocks of chum at the river tributary

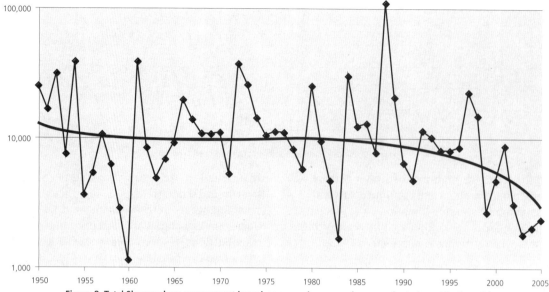

Figure 8. Total Skeena chum escapement based on spawning ground surveys (note logarithmic scale)

level. Beacham et al. (1987) used electrophoretic analysis to distinguish five large-scale population assemblages in chum salmon: the Queen Charlotte Islands, the North and Central Coast, the west coast of Vancouver Island, the South Coast, and the Fraser River system. Kondzela et al. (1994) used similar techniques to divide southeast Alaska and northern British Columbia stocks into six groups.

The decline in chum salmon stocks is basin wide, suggesting that much of the problem is in the marine realm. This suggests that one component of the decline in chum salmon is decreased ocean survival. Overall, the lack of escapement records since 1990, as a result of sampling, do not allow any clear trends to be stated.

Skeena River chum salmon are taken as incidental catches in the sockeye and pink salmon fisheries of Area 3, 4, and 5 and, to a lesser extent, in the Noyes Island and Cape Fox fisheries in Alaska. They are also taken in small numbers in First Nations food fisheries. Charles and Henderson (1985) calculated an overall exploitation rate of between 50% and 83% for the years between 1970 and 1982. This relatively high exploitation rate has probably contributed to the decline of the Skeena stocks.

The average overall escapement to Area 4 in the 1990s was only 10,000 to 14,000 chum. Since 2000, Area 4 annual escapements show

3,744 chum, but this number represents a count on only a small proportion of chum-spawning streams. The Ecstall and the lower Skeena areas contain the only strong stocks, accounting for over 9,000 of this total. In the 1990s, 29 censused stocks had escapements below 200, of which 26 had average escapements below 100. In general, one can conclude that chum are probably the Skeena watershed salmon species in greatest danger of significant loss of spawning stocks and genetic diversity.

On the Skeena River, several areas regularly used in the past, such as the back channel near Thornhill above Terrace, no longer support spawning fish (Kofoed 2001). On the other hand, several field workers suggest that the mainstem Skeena populations below Terrace are under-reported because of turbid water conditions at spawning time (Kofoed 2001, Bustard 2002, and C. Culp 2002).

Of note are the extremely low levels or loss of the once large Lakelse River chum stock. The formerly large Kispiox River stock has also declined dramatically, but there are still chum present. Other stocks with very poor recorded returns are Kleanza Creek, Fiddler Creek, Deep Creek, the Zymoetz River, and the Zymacord River. Of special note is the chum spawner fence enumerations on the Kitwanga River; annual averages since 2003 indicate 1,850 chum.

This facility provides the only high-quality escapement data in the Skeena.

Sockeye Salmon

Skeena sockeye salmon are the most valuable commercial fish, and within Canada, are second only to the Fraser River in terms of overall abundance. They have consequently received relatively significant research and management attention. Important sources of information are found in Brett 1952, Cox-Rogers et al. 2004, Larkin and McDonald 1968, Smith et al. 1987, Rutherford et al. 1999, Shortreed et al. 1998, Shortreed et al. 2001, and Wood et al. 1997.

A key theme of sockeye salmon (*Oncorhynchus nerka*) is that they are anadromous and semelparous, meaning they spend a portion of their life in the ocean and return to freshwater to spawn, after which they die, similar to other Pacific salmon. Their habitat includes the freshwater subbasin of origin and a large portion of the northeastern Pacific Ocean. Sockeye salmon are widely recognized for their distinctive migratory behavior, their unique reproductive and juvenile rearing life histories, and their intimate connection with indigenous and contemporary human cultures.

Each fall, drawn by natural forces, sockeye salmon return to the rivers which gave them birth. Sockeye typically spawn in tributaries to or adjacent to large lakes, though river spawners and lake spawners are not uncommon. These streams can vary in size and type, ranging from small streams to large mainstem rivers and side channels. Once the salmon reach their spawning grounds and hold for appropriate conditions, they deposit thousands of fertilized eggs in the gravel.

Adult salmon die following their long journey and spawning. Their carcasses provide nourishment and winter food for birds and wildlife, and provide nutrients to the river or lake for the next generation of salmon and other fish. As the salmon eggs lie in the gravel they develop an eye, and, over months, the embryo develops and hatches as an alevin. The alevin carries a yolk sac that provides food for two to three months. Once the nutrients in the sac are absorbed, the free-swimming fry move up and emerge into the water.

Sockeye fry usually migrate downstream to the rearing lake at a size of approximately 25–30 mm. At this small size, sockeye fry are vulnerable to predation by other fishes and birds. Once free-swimming, fry usually live in their rearing lakes for one to three years. Primary and secondary production that shapes the zooplankton community in sockeye nursery lakes is positively correlated to fry food availability, growth rates, and survival. Most productive sockeye stocks, such as those of Babine Lake, Lakelse Lake, and Alastair Lake, spend one year as lake residents. These lakes have relatively higher biological production with abundant plankton populations, the main food source for sockeye fry. Some sockeye populations based in colder subalpine lakes, such as Morice Lake and Bear Lake, have significant proportions of two- and three-year-old fry.

Skeena sockeye rearing lakes vary in size and productivity and are comprised of one very large rearing lake, Babine-Nilkitkwa, and 28 smaller ones. Babine Lake possesses approximately 67% of the total Skeena sockeye rearing area. Ten other Skeena nursery lakes comprising about 29% of the total rearing area are important sockeye producers and include Alastair, Bear, Johanson, Kitsumkalum, Kitwanga, Lakelse, Morice, Morrison, Sustut, and Swan Lakes (Shortreed et al. 1998). There are 18 other smaller Skeena lakes, which comprise about 4% of the total nursery area; they are Aldrich, Asitka, Atna, Azuklotz, Club, Damshilgwit, Dennis, Johnston, Kluatantan, Kluayaz, McDonell, Motase, Sicintine, Stephens, Slamgeesh, Spawning, Maxan, and Bulkley Lakes.

A number of the smaller lakes, which are part of larger lake systems, include Aldrich-Dennis-McDonnell in the upper Zymoetz River, Azuklotz-Bear in the Bear drainage, Atna-Morice in the Morice basin, Club-Swan-Stephens in the upper Kispiox, Damshilgwit-Slamgeesh, and the Morrison-Babine-Nilkitkwa in the Babine subbasin. Studies by Shortreed et al. (1998, 2001) and Cox-Rogers et al. (2004) note that the major factors limiting juvenile sockeye production in non-Babine nursery lakes is paucity of returning adult spawners and low in-lake growth and survival. In addition, Kitwanga and Lakelse Lakes have degraded habitat resulting from forestry and transportation development activities.

Fry ready to enter saltwater are called smolts. Smolts migrate seaward from April through June after having spent at least one full summer in their rearing lakes. Smolts feed and physiologically adapt to the marine environment in the Skeena estuary before typically traveling northward along the coast and offshore, where they spend one to three years in the North Pacific. The vast majority of Skeena sockeye return as four- and five-year-old fish, although three-year-old males (jacks) are common in some years.

Total annual sockeye run size before fishery harvests averages several million fish. The total Skeena in-season sockeye escapement is estimated by an annual test fishery at Tyee in the Skeena estuary, while total sockeye escapement at the spawning grounds is represented with the SEDS estimates. Escapement data for Skeena tributaries are of variable quality. The most reliable tributary counts are the fish weir counts on the Babine, Kitwanga, and Slamgeesh Rivers. By comparison, less intensive spawning ground counts on other streams, and especially lakes, may under-represent the true escapement. This is suggested by the apparently larger proportion of non-Babine sockeye in the Tyee Test Fishery when compared to stream counts on smaller spawning populations (McKinnell and Rutherford 1994).

Sockeye return to the Skeena River in late June through late August and are mostly four and five years old. Skeena sockeye stock run timing information is supported by marine tagging data (Aro and McDonald 1968, Smith and Jordan 1973), daily counts at the Kitwanga, Babine, Sustut, and Slamgeesh counting fences, the mark-recapture sampling program at Moricetown Canyon, and DNA data procured from the Tyee Test Fishery. Migration timing of various non-Babine sockeye stocks overlaps with the enhanced, productive mid- and late-timed Babine sockeye runs. Fisheries primarily directed to the enhanced Babine sockeye stocks have caused substantial impacts to these non-Babine

stock escapements. Skeena sockeye stocks are placed into three main timing groups:

- Early July: Pinkut, Alastair, early Babine wild, Morice-Nanika, Lakelse, and other early non-Babine

- Mid-July: Pinkut, Fulton, Morrison, Zymoetz, Swan-Stephens, and other mid non-Babine

- Late July: Kitwanga, Fulton, late Babine wild, Kitsumkalum, and other non-Babine

Total aggregate Skeena sockeye numbers have increased in the past 40 years due to the success of large spawning channels constructed on Babine Lake in the 1960s and the high capacity and productive rearing conditions in the Babine-Nilkitkwa lake system. As enhanced Babine stocks have increased, non-Babine wild sockeye populations have correspondingly decreased. Since the 1970s, wild sockeye stocks have declined in response to sustained high-exploitation rates supported by the success of the Babine Lake enhancement (Figure 9).

The commercial value of sockeye salmon led to a high exploitation rate in a series of southeast Alaskan, B.C. coastal, and in-river fisheries; since 1970, exploitation has ranged from 5% to 14%, 22% to 76%, and 7% to 13% respectively, depending on abundance. Early-20th-century harvests were high but declined from 1920 to the 1950s, paralleling sockeye abundance. Harvest rates in the 1940s were estimated at 50% (Anonymous 1964). Annual harvest rates rose to exceed 60% after 1970 and have since exceeded 70% four times (Rutherford et al. 1999.) Overall, sustained high-aggregate stock exploitation rates have led to the decline and impact on less-productive wild sockeye stocks. The timing of the declines has not been synchronous and the extent of the declines equal.

Research advances over the past 15 years have identified genetic markers in sockeye that can separate sockeye from different spawning areas and provide a tool to help understand their population structure (Wood et al. 1994, Varnavskaya et al. 1994, Beacham and Wood 1999, Beacham et al. 2004, Beacham et al. 2005). This information has important conservation

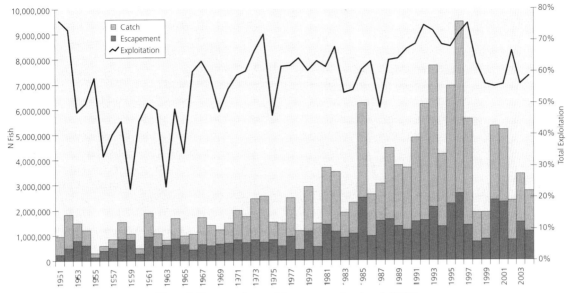

Figure 9. Skeena sockeye catch, escapement and exploitation

implications. Sockeye salmon are highly specific to individual lakes, and each lake system is genetically distinct. Different spawning areas in a single lake system have sockeye subpopulations that are similar genetically to one another and have modest amounts of genetic interchange between different spawning streams (Varnavskaya et al. 1994, Wood and Foote 1996, Withler et al. 2000).

Waples (1995) postulated that each sockeye rearing lake complex is an evolutionary significant unit and hence is an important fisheries management unit. In contrast, river dwelling sockeye are relatively similar genetically (Wood 1995, Beacham and Wood 1999, Beacham et al. 2004). Consequently, the preservation of even small lake sockeye populations is important to the preservation of species diversity. In recent decades, over 90% of the aggregate Skeena sockeye escapement came from Babine Lake stocks (West and Mason 1987, McKinnell and Rutherford 1994); this proportion has increased from pre-1960 levels of about 80% (Brett 1952).

Most wild Skeena sockeye populations are fluctuating at low levels of abundance, but are sufficiently stable that there are no short-term concerns about survival of the stocks. After the partial blockage of the Babine River by a 1951 landslide (cleared in 1953) and the partial recovery of the Zymoetz River sockeye stocks

after blockage in 1891 and in the 1960s, the recovery of the Babine Lake sockeye stocks is encouraging.

However, the decline of several other sockeye stocks is a serious conservation concern. The following sockeye populations are somewhat or seriously depressed:

- Kitwanga Lake: Escapement from 2004 to 2006 has improved with a range from 1000 to 5000 adults after decades of very low escapements.

- Bulkley and Maxan Lakes: Escapement has been less than 20 adults since 1990.

- Lakelse Lake: Escapement surveys for Lakelse Lake from 1992 to 2003 indicate that the sockeye stock has experienced a 92% decline over the last three cycles.

- Bear Lake: A declining escapement trend since the 1950s requires monitoring.

- Sustut-Johanson Lake system: A relatively low-escapement trend needs to be monitored.

- Morice-Nanika Lake system: Lack of adult spawners and the ultra-oligotrophic lake conditions has led to poor escapement in the last decade.

Kitwanga Lake sockeye abundance has declined from historic escapements in the 1940s that averaged in the 20,000 range to as low as several hundred fish in recent years. Constraints to sockeye production stem from a combination of stock and habitat problems that include high exploitation rates in the Skeena River mixed-stock fishery and cumulative impacts to critical spawning and rearing habitat. The Gitanyow Fisheries Authority (GFA) has been conducting research on this stock and has made substantial progress on their sockeye recovery plan efforts.

Bulkley Lake and Maxan Lake sockeye are two small stocks in the headwaters of the Bulkley River east of Houston. Historically, spawning occurred downstream of Maxan Lake and most likely in Maxan and Bulkley Lakes. Escapement ranged from 50 to 600 until 1978. The stocks then appear to have collapsed, and recent records show few or no fish returning. The upper Bulkley River and Maxan Creek often do not have sufficient flow to allow sockeye passage. High water temperatures could also cause a lack of movement and access problems. The upper Bulkley River has a variety of environmental impacts including transportation corridor problems that reduced floodplain habitat as well as agriculture-related nutrient and erosion problems (Remington 1996). This information strongly suggests that the Bulkley Lake sockeye are at high risk of extirpation.

Lakelse Lake sockeye populations that primarily include Williams and Scully Creeks, spawners were up to the 1970s levels of some of the most productive and resilient stocks in the Skeena basin. Since the mid-1990s, sockeye escapements (spawners) into Lakelse Lake, as well as the number of juveniles rearing in the lake itself, have fallen precipitously. Current information suggests the major problem is poor spawning success in the tributary streams of severely degraded spawning habitat. In the fall of 2003, concerned local stakeholders, First Nations, and provincial and federal governmental agencies began to jointly assemble the Lakelse Lake Sockeye Recovery Plan; significant progress has been accomplished to date.

Sockeye populations of Sustut, Johanson, and Bear Lakes – located in the upper and mid-Sustut drainage – show a recent declining escapement trend which is being monitored in order to alter management decisions, if required.

The Morice-Nanika sockeye stock is the largest sockeye run in the Bulkley basin. Since the mid-1950s, Morice-Nanika sockeye has mostly fluctuated at levels of abundance below historical escapements, and fry densities are low in relation to Morice Lake juvenile sockeye rearing capacity. Constraints to sockeye production stem from sustained high-exploitation rates in the combined Alaskan and Canadian coastal fisheries as well as First Nation in-river fisheries, and from the low production of the ultra-oligotrophic Morice Lake. The Morice-Nanika Sockeye Recovery process is moving forward with establishing a plan.

Evidence in early fisheries reports (DFO 1905 – 1949) point to two sockeye stocks that are probably extinct. In fishery officer reports dating from 1929 to 1934, Seeley Lake sockeye were described as an early migrating stock arriving in the beginning of July, then spawning at the outlet of Seeley Lake near Hazelton. Since the spawning locality is at the current crossing of Highway 16, it is unlikely that recent occurrences would be overlooked. Reports by fisheries inspector A. R. MacDonnell in 1932 and 1934 described sockeye returns to Canyon Lake, about 130 km north of Hazelton, and referred to runs in 1928 and 1929. There are no known recent records for this stock.

Coho Salmon

Coho are the most widely dispersed salmon throughout the Skeena River and its tributaries and show the least amount of concentration in forming few large productive stocks. There are approximately 25 major populations in the watershed. Coho typically spawn in small headwater streams and usually spend one to two winters in freshwater before migrating to the ocean. They typically return as two- or three-year-olds after spending one winter or about 16 months in the surface waters off the coast (Holtby et al. 1994). Approximately half of the Skeena coho remain in the coastal waters off B.C., while the other half migrate north off southeast Alaska. Some males (jacks) return to spawn after only a few months at sea. Marine diets range from euphausiids and various plankton to squid and small fish such as herring. Coho jacks are only present in the coastal portion of the Skeena from Kitwanga downstream.

Coho migrate into the Skeena River between late July and the end of September, as recorded by the Tyee Test Fishery. The annual peak of the migration is in late August. In general, the fish destined for upstream tributaries arrive first because they spawn earlier in coldwater tributaries and have longer river travel times. The early arrivals pass through the various coastal fisheries along with the large enhanced sockeye run destined for Fulton River, a tributary of Babine Lake. The timing of the coho run is nearly coincident with that of the pink salmon runs. Coho are usually the last salmon to spawn in the fall – from the end of September through December, with late spawning being especially common in coastal areas.

Coho rearing typically takes place in streams, ponds, and lakes. In ponds and lakes juveniles inhabit the near-shore littoral zone (Irvine and Johnston 1992). In streams they prefer habitat with structural complexity including stones, logs, and overhanging vegetation. River side channels and small streams often provide these conditions. Coho are dependent on low-gradient streams (<2%) for rearing habitat (Nass et al. 1995), and they frequently occupy small upstream habitats. To get there, the adults have to migrate into small streams to spawn. Coho often move into these small spawning streams when heavy fall rains increase water flows, allowing them to get over obstacles such as beaver dams.

The vast majority of coho return to their natal stream. However, when compared to other species like sockeye and chinook, coho typically have a higher amount of straying. Several recent accounts suggest that typical straying rates are less than 1% (Sandercook 1991). Coho that do not return to their natal stream most likely stray to similar nearby streams. In years of low flows they may stray to other nearby streams or spawn further downstream. The regional genetic structure of coho reflects this straying pattern of adult fish (Beacham et al. 2001). Coho straying rates of less than 1% are sufficient to ensure

gene flow between nearby streams (Wood and Holtby 1999).

Coho of the Skeena and Nass watersheds constitute a genetically distinct regional group of populations (Small et al. 1998). Within the Skeena watershed, variation appears to be roughly proportional to the distance apart of spawning streams. Wood and Holtby (1999) suggested that the geographic scale of subpopulations is approximately 100–400 km. Important evolutionary units are then at the major tributary level of separation. This suggests that there are several functional subpopulations in the Skeena. Typical units would be the coastal tributaries, the lower Skeena around Terrace, the Kispiox River and its tributaries, and the Bulkley River. The implication of this model is that decline of coho in a single stream is not an evolutionary concern if nearby streams retain healthy populations.

Current stock assessment primarily utilizes catch for the four major fisheries that exploit Skeena coho and indices of abundance to determine stock status. The first index is the Babine counting fence, the second is generated from the Tyee Test Fishery coho catch, the third index is derived from visual spawner counts, and the fourth is juvenile coho densities. Indicator stocks that provide adult escapement, freshwater survival and production, and marine survival and exploitation rates include Babine, Slamgeesh, and Toboggan. Overall, coho populations in the Skeena have been in a long decline. Coho escapement counts begin about 1950 and show a trend of continuing decline to 1997; however, there has been a strong rebound in recent years with escapements being generally higher than their brood years.

The declines were especially severe in upriver stocks such as the Bulkley River (Figure 10), the Babine River, and the Bear-Sustut Rivers. Likely causes of the decline in coho stocks are high rates of capture in Alaska and B.C. fisheries, a decline in ocean survival rates, and habitat damage. Skeena coho stocks are taken in a series of fisheries that include commercial fisheries in southeast Alaska and on the B.C. coast, aboriginal food fisheries on the coast and in-river, and sports fisheries on the coast and in-river.

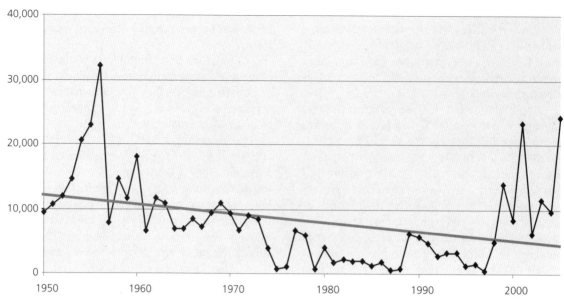

Figure 10. Spawning escapements of coho in the Bulkley River and its tributaries, including the Morice River

Overall exploitation rates for four Skeena stocks are shown in Figure 11. The Lachmach River coho stock is a wild coastal stock from north of the Skeena estuary; the Toboggan Creek and Fort Babine stocks are interior hatchery enhanced stocks, and the Slamgeeth is a small northern upper Skeena stock. Total exploitation rates before 1998 ranged for the most part from 60% to 80%. Few if any of the Skeena coho stocks can be expected to thrive at these exploitation rates. One third to one half of the total exploitation during this period was in Alaska, where Skeena coho are harvested as an incidental catch along with the much larger Alaskan hatchery-produced component. Much of the Canadian harvest was incidental catch in large sockeye and pink salmon coastal fisheries. In recent years, selective fishery changes have eliminated most incidental catch.

The low escapements of Skeena coho in the 1980s and 1990s raised concerns about coho survival, especially survival of stocks spawning upstream of Terrace. DFO responded to this management crisis by instituting substantial changes to the commercial and sports fisheries in 1998 and 1999, directed at reducing the catch to zero. Severe restrictions on commercial fishing continued through 2001. These actions have met with some success as escapements increased in 1999, 2000, and 2001, with the added benefit of better-than-average ocean survivals.

Smolt-to-adult survival rates are a measure of ocean survival. The general pattern in the past decade in Oregon, Washington, and British Columbia is a decline in ocean survival. Mortality is probably highest in the first months at sea. Ocean survival rates for Skeena coho are extremely variable (from 0.2% to 20%) and seem to have decreased in the 1980s through 1996. Survival rates for the few coho index stocks in the Skeena are shown in Figure 12. The increase in survival of 1998 smolts appears to have continued with 1999 and 2000 smolts, which returned in 2000 and 2001 (data not shown).

Coho are fish of small streams and are often dependent on off-channel habitat such as beaver ponds, back channels, and seasonally flooded areas. These small stream and flood plain habitats are highly susceptible to damage from logging. Prior to the Forest Practices Code (1994), protection of small streams was often inadequate. Floodplain logging has severely impacted several portions of the Skeena watershed including the Kitsumkalum, Lakelse, Zymoetz, and lower Skeena watersheds.

In contrast, agricultural impacts are minor and localized. In the upper Bulkley area, riparian damage, fine sediment production, increased nutrient loading, and water withdrawal are associated with cattle ranching (Remington 1996). Similar impacts may occur on a few tributary streams in the Bulkley valley.

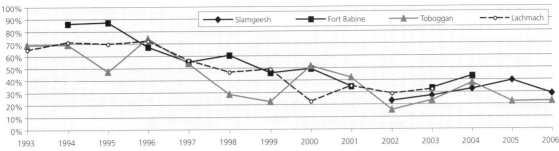

Figure 11. Total exploitation rates calculated for four Skeena index stocks

Agriculture elsewhere in the Skeena watershed is extremely limited, and therefore impacts will be small and local.

Coho enhancement was popular in the 1980s and 1990s. Small hatcheries were established in the vicinity of Terrace, Hazelton, Kispiox, Smithers, Houston, and Fort Babine. Some of these are still active. The effectiveness of these efforts is variable. The hatchery at Toboggan Creek north of Smithers has apparently been effective, such that most coho in the Bulkley watershed now originate from Toboggan Creek (SEDS database). However, the benefits of outplants to the upper Bulkley and Morice River are not apparent. For the most part, the effectiveness of other projects, such as outplanting from the Chicago Creek, Kispiox, Eby Street, and Fort Babine hatcheries, is unknown.

Following the overall decline of coho in the Skeena, which bottomed out in 1997, most individual stocks show a healthy recovery. The selection of stocks of particular concern that follows is based primarily on evaluation of the SEDS database. These records are in large part derived from stream escapement estimates. Counts of coho spawners in streams are notoriously difficult and often underestimate true escapement numbers. The concern for individual coho stocks should be assayed against the pattern of generalized genetic differentiation and the ability of coho to rapidly reoccupy available habitat.

Steelhead

Typically, Skeena watershed streams support both sea-going and resident populations of *Oncorhynchus mykiss*. The anadromous populations are known as steelhead and freshwater residents as rainbow trout. There are both summer-run and winter-run populations of steelhead in the Skeena. The summer-run steelhead comprise approximately 80% of the total run returning in July and August; while the winter-run steelhead return from October to May (Beere 2003). The summer-run steelhead spend the fall and winter in rivers or lakes, usually not far from their spawning areas. The gonads mature during this residence. Winter-run steelhead are limited to the coastal portions of the Skeena that are downstream of Terrace. Spawning in both groups takes place in the spring, generally in April and May, but may extend until late June in the Zymoetz and Kitsumkalum systems. Winter-run steelhead move onto the spawning beds soon after migration, and therefore migrate in a sexually mature state.

Steelhead fry emerge in July and August and spend between one and four winters in fresh water. Most steelhead in the Skeena watershed spend three years in fresh water; however, steelhead from the Morice River and Sustut River tend to reside in fresh water for four years (Whately and Chudyk 1979, Cox-Rogers 1985). Most steelhead smolts migrate to sea in the spring during the annual snowmelt flood. Steelhead spend one to three years in the ocean before returning to freshwater. The most common value is two years, although most Morice River steelhead return after one year in the ocean and significant proportions of Babine and Kispiox fish return after three years (Cox-Rogers 1985).

Unlike other Pacific salmon species, steelhead are iteroparous—they can spawn more than once. Shortly after spawning, kelts migrate back to the ocean and may return the following year. Repeat spawners are less common, usually

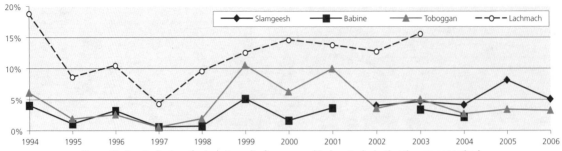

Figure 12. Survival rates of migrant smolts from several index stocks in the Skeena watershed

less than 10%. The Babine River has a repeat spawning rate of 2% (Whately and Chudyk 1979, Cox-Rodgers 1985). Among the steelhead returning to the Kitwanga River in 2001, 7% were repeat spawners (Cleveland 2002b). However, Cox-Rodgers (1985) reported repeat spawning rates of 22% and 14% for the Zymoetz and Kispiox River, respectively.

Overall, the Tyee Test Fishery is the best index of steelhead escapement in the Skeena watershed. The province of B.C. tends to rely more on results of angler surveys. A comparison of the results from these two sampling strategies is included in Smith et al. 2000, which establishes the utility of the mail-in angler survey for the Skeena by comparing the results with the Tyee Test Fishery. They concluded that the trends in wild steelhead abundance are mostly a response to environmental changes in climate and ocean circulation.

The Tyee Test Fishery provides a useful estimate for the summer-run portion of the steelhead. The summer-run steelhead arrive relatively late in the Skeena, along with coho salmon, and continue through the fall. In the lower Skeena, steelhead continue to arrive throughout the winter. In the Tyee Test Fishery, the earliest part of the steelhead run overlaps the much larger sockeye run. Most of the steelhead arrivals take place while pink salmon are entering the Skeena. The timing of the summer-run is shown in Figure 13. The index values are approximately equal to the daily percentage of the total summer-run. Few data are available to estimate the winter-run steelhead component.

Estimates of the total summer-run escapement based on the Tyee index data are available starting in 1956 (Figure 14). Data are shown for the period of July 1 to August 27 to provide an equivalent year-to-year comparison.

Steelhead declined from about 1985 to 1992. The low escapements in the early 1990s led to changes in the Area 4 commercial fisheries to decrease the impact on steelhead, as well as to the beginning of mandatory catch and release in the sports fishery. The total closure of the Area 4 fishery in 1998, as well as improving ocean survival, contributed to the high escapement of that year. Spense and Hooton (1991) suggested a minimum escapement target of 26,500 for summer-run steelhead, assuming no upriver harvest. Allowing for aboriginal food fisheries, the minimum escapement should be at least 28,000. Figure 14 shows that only nine of the last 45 years have met this criterion.

There are few good data on steelhead escapements at individual streams. This is in large part because steelhead spawn in spring at high water conditions when counts are usually not possible and because they are typically well dispersed within a stream. It is clear that the Bulkley-Morice is the most important spawning system. Based on population estimates for the Bulkley River (Mitchell 2001a, Saimoto 2002) from the Tyee Test Fishery data using the Ministry of Water, Land and Air Protection (WLAP) index conversion values, the Bulkley-Morice likely accounts for at least half of the total escapement in recent years. If the WLAP values underestimate the total steelhead escapement, then the proportion of steelhead spawning in the Morice River would be somewhat lower. The Babine, Bear, Kitwanga, Kispiox, and Zymoetz Rivers also contribute sizable numbers (Whately and Chudyk 1979, Heath et al. 2002). It is likely that the six identified watersheds produce at least 80% of the Skeena watershed steelhead.

Steelhead characteristically have low levels of straying (Quinn 1993, Heath et al. 2001), and, where straying occurs, it is likely to streams close to the natal stream. This pattern results in a moderate degree of genetic separation of steelhead. Heath et al. (2001) analyzed the genetics of steelhead stocks in the Skeena and Nass Rivers and found significant differences between stocks from the various tributary watersheds of these two rivers. Steelhead from the Morice, Babine, Kispiox, Sustut, and Zymoetz Rivers are genetically distinct, and differences increase proportionally with the geographic separation of the watersheds. A low level of within-population variance (heterozygosity) in the Zymoetz River stock was noted. The decline of within-stock diversity has occurred since 1960 (Heath et al. 2002). This may reflect decreased population size in the recent past due to the effects of logging disturbance, landslide blockage in the Zymoetz watershed, or over-fishing of the steelhead stock.

Changes to steelhead populations in Skeena tributary watersheds are hard to identify due to a shortage of relevant information. The most useful source of data is the Steelhead Harvest Analysis (Ministry of Water, Lands and Air Protection 2001), which reports the result of questionnaires mailed to steelhead anglers each year since 1968. While the reliability of catch per unit effort (CPUE) data and mail surveys in sports fisheries is debatable, the pattern of catch reported for the major steelhead streams – the Morice, Kispiox, and Babine Rivers – generally shows an increase in fishing effort and catch over the past 35 years. On these rivers, the pattern of total catch, including released fish, resembles the Tyee estimates for summer-run steelhead. In the more coastal systems dominated by winter steelhead, the pattern of sports fishing catches is different. The Lakelse River steelhead sports fishing catch has been declining since 1988; in 2000, catches were estimated at about one-quarter of the 1988–1990 catch. The Gitnadoix River, which has only winter-run steelhead, shows a similar pattern, but with a marked improvement in 1999 and 2000.

Resident Freshwater Fish

In comparison to salmon, information is sparse on resident (nonanadromous) freshwater fishes in both fluvial (rivers) and lacustrine (lake) habitats of the Skeena watershed. Indeed, much of the watershed is poorly known and may contain populations of special interest or status that are now unknown. Ecological and life-history information that permits good conservation planning is simply not available. There are 28 known species of freshwater fish in the Skeena system (McPhail and Carveth 1993a); of these, 18 regularly enter the sea and are widespread along the North Coast.

Freshwater species and populations inhabiting the Skeena system include rainbow trout *(Oncorhynchus mykiss)*, cutthroat trout *(Oncorhynchus clarki clarki)*, kokanee *(Oncorhynchus nerka)*, bull trout *(Salvelinus confluentus)*, Dolly Varden char *(Salvelinus malma)*, lake trout *(Salvelinus namaycush)*, river lamprey *(Lampetra ayresi)*, Pacific lamprey *(Lampetra tridentata)*, western brook lamprey *(Lampetra richardsoni)*, green sturgeon *(Acipencer medirostris)*, white sturgeon *(Acipencer transmontanus)*, American shad *(Alosa sapidissima)*, lake whitefish *(Coregonus clupeaformis)*, pygmy whitefish *(Prosopium coulteri)*, mountain whitefish *(Prosopium williamsoni)*, longfin smelt *(Spirinchus thaleichthys)*, eulachon *(Thaleichthys pacificus)*, peamouth chub *(Mylocheilus caurinus)*, northern

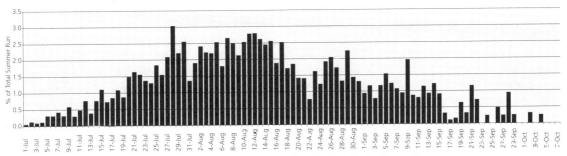

Figure 13. Timing of summer-run steelhead entry into the Skeena River

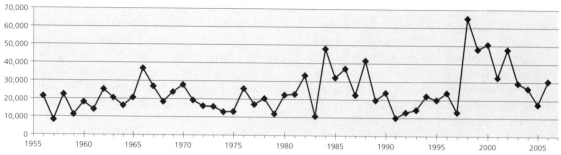

Figure 14. Skeena steelhead escapement estimated at the Tyee Test Fishery

pikeminnow *(Ptychocheilus oregonesis),* longnose dace *(Rhinichthys cataractae),* redside shiner *(Richardsonius balteatus),* longnose sucker *(Catostomus catostomus),* white sucker *(Catostomus commmersoni),* largescale sucker *(Catostomus macrocheilus),* burbot *(Lota lota),* threespine stickleback *(Gasterosteus aculeatus),* coastrange sculpin *(Cottus aleuticus),* and prickly sculpin *(Cottus asper).*

The lower reaches of the Skeena River are dominated by species that are part of the salt-tolerant fauna typical of most coastal rivers, and that presumably colonized the Skeena River from the coast. Some of the fish populations in the upper reaches of the watershed, such as lake trout, white sucker, and lake whitefish most likely entered the Skeena from the east by way of water connections between the Skeena, Fraser, and Peace Rivers, which functioned briefly at the end of the Ice Age (McPhail and Carveth 1993b).

Landlocked fish populations, many with interesting and special characteristics, are known to exist within the Skeena watershed. The only known introduced fish species to the Skeena system is the American shad *(Alosa sapidissima),* introduced from the Atlantic to the Sacramento River in the 1870s and likely reaching the Skeena by the turn of 20th the century (Hart 1973). Within the Skeena watershed, shad are rare but of regular occurrence in coastal waters, and, in 2006, shad spread upriver as far as Kitwanga.

There are three resident freshwater fish species of conservation concern: cutthroat trout, Dolly Varden, and bull trout. Cutthroat trout are blue listed by the B.C. Conservation Data Centre (CDC) as a species of concern. Dolly Varden char are blue listed by the B.C. CDC as a species of concern, but are not listed as identified wildlife by the Forest Practices Code or by the Committee

on the Status of Endangered Wildlife in Canada (COSEWIC). Bull trout, actually a char, are blue listed as a species of concern by the B.C. CDC, as well as by COSEWIC, primarily due to limited global distribution and threatened status in the southern part of their range in the U.S. They are also listed as an identified wildlife species under the B.C. Forest Practices Code (FPC).

Responsibility and jurisdiction for nonanadromous fish species and seagoing populations of predominantly fresh water fish such as steelhead (rainbow trout), cutthroat trout, and Dolly Varden trout rests with the provincial government. Recently, the focus of provincial government management efforts has shifted from a narrow range of game fish to interest in the characteristics and conservation of all species. Defining conservation levels requires understanding fish values and the biology of the species including their critical habitats and population status. What and where are critical areas for fish? What are the capability and constraints for fish production? How does the fish habitat function and how can it be maintained? The following section briefly reviews the habitat and status of selected trout and char species in the Skeena watershed.

Lake Trout

Lake trout *(Salvelinus namaycush),* actually a char, are known to exist in 22 lakes in the Skeena watershed (FISS 2005). Lake trout is a cold-water fish, usually frequenting deep lakes with distribution limited to the upper Skeena tributaries. They have been occasionally observed in streams tributary to lakes.

Within the Babine watershed, lake trout have been recorded in Babine Lake, Morrison Lake and Creek, Chapman Lake, Doris Lake, Fulton Lake, Augier Lake, Nilkitkwa Lake, Pinkut

Lake, Taltapin Lake, Tanglechain Lake, and several unnamed lakes. Within the Sustut River drainage, lake trout have been observed at Bear Lake, Asitka Lake, Johanson Lake, and at least one unnamed lake. Lake trout locations recorded within the Bulkley system include Bulkley River, Atna Lake, Maxan Lake, McBride Lake, Morice Lake, Nanika Lake, Owen Lake, and one unnamed lake. Within the Kispiox drainage, lake trout have been documented in Kispiox River, Swan Lake, and Stephens Creek (FISS 2005).

The lake trout life cycle takes place entirely within lakes, with spawning occurring in late summer or early fall in relatively shallow areas. Eggs are usually deposited on large rubble substrate and incubate for four to five months over the winter and early spring. Fry emerge and usually remain at inshore nursery areas adjacent to the spawning beds to feed on insects and crustaceans for a period ranging from several weeks to several months. Diets change as the juveniles grow and the fish move into deeper waters offshore. Lake trout are the top aquatic predator in most lakes where they are found (Martin and Oliver 1980). On average, maturity first occurs at age 11, with mature adults leaving offshore waters, and return inshore to spawn.

The lake trout may prey on kokanee and whitefish while in deep water, and aquatic insects and shore-dwelling minnows while in shallow water, usually in the spring and fall (Griffiths 1968). Lake trout are capable of reaching ages in excess of 50 years and achieving weights over 20 kg. Due primarily to their large size and palatable flesh, they are a prized sport fish by many anglers and are vulnerable to overexploitation. They are also taken in First Nations fisheries at Babine Lake and Bear Lake. DeLeeuw et al. (1991) reported reduced abundances of lake trout in road accessible lakes because of increased angler effort. Fourteen of the 22 lake-trout-bearing lakes in the Skeena watershed are known to be road accessible and are therefore likely suffering from reduced abundances.

Rainbow Trout

Rainbow trout have recently undergone a name change from *Salmo gairdneri* to *Oncorhynchus mykiss* (Smith and Stearly 1989), which recognizes their closer relationship to the Pacific salmon than to the Atlantic salmon and trout. Anadromous forms of rainbow trout are commonly referred to as steelhead, or steelhead trout, and were discussed earlier. Skeena watershed populations of the resident form, rainbow trout, exhibit three different life-history strategies with considerable variation depending on geographic location and habitat: populations that live their entire lives in small streams, those that spawn in small streams and migrate to rivers to rear and mature, and those that spawn in small streams and move into lakes to rear and mature.

In the Skeena watershed, rainbow trout are most commonly lake residents, but they enter streams in the spring to spawn. Females construct redds in fine gravel into which the eggs are deposited. Young emerge from the gravel in the summer and usually migrate into rearing areas of streams or lakes in the first year. Normally, the fish remain in the rearing lake, or in some populations in a river, until they reach maturity in two to four years, before moving back to natal streams for spawning. Scott and Crossman (1973) reported that survival after spawning is usually low and the number of repeat spawners is often less than 10% of the total spawning population.

Rainbow trout exhibit a wide range of growth rates dependent on habitat, food type and availability, and life-history strategy. Generally the growth of rainbow trout is slower in streams than in lakes and greatest in marine environments (Carlander 1969). The fish show seasonal movement to access suitable habitat for feeding and overwintering. Generally, the type of food eaten reflects the size of rainbows and the season, with principal prey being zooplankton, benthic invertebrates, terrestrial insects, and fish. Small rainbows may eat zooplankton, crustaceans, and small insects, while larger trout may take leeches, larger insects, molluscs, and a variety of juvenile fish (Griffiths 1968). Griffiths documented growth and feeding habits, primarily by stomach analysis of rainbow trout in Babine Lake.

Within the Skeena watershed, Babine and Nilkitkwa Lakes host rainbow sport fisheries of considerable size (Bustard 1987). Bustard (1990) described the importance of the Sutherland River as the natal stream for Babine Lake rainbow trout. The Morice River is also renowned by anglers for rainbow trout.

Hatchery-raised rainbow trout are the predominant species used for stocking lakes in the Skeena watershed. The majority of hatchery-produced fish are released into lakes that either cannot support rainbow trout or have insufficient natural production to satisfy sport fishing demands.

Cutthroat Trout

Cutthroat trout in the Skeena watershed are coastal cutthroat trout *(O. clarki clarki)*, which are blue listed by the B.C. CDC as a species of concern. It is not, however, an identified wildlife species under the Forest Practices Code, nor is it listed by COSEWIC. Lacustrine populations of cutthroat trout exist throughout the Skeena watershed, but are rare in Skeena tributaries upstream of the Babine-Skeena confluence and are not documented in the uppermost tributaries that include the Slamgeesh, Kluatantan, and Sustut Rivers.

Cutthroat trout are very adaptable to their environment, resulting in considerable variation in life history and appearance. Anadromous life forms of cutthroat exist in the Skeena and are both poorly studied and understood. For the purposes of regulations, populations upstream of Cedarvale (20 km below Kitwanga) are not considered to be anadromous. There are three life-history variations: adfluvial populations that spawn in tributary streams and migrate to lakes to grow to maturity; fluvial populations that move between mainstem and headwater streams; and resident populations that remain in headwater tributaries for their entire lives.

Cutthroat trout exhibit considerable variation in spawning time, which normally occurs from mid-May to mid-June (Bilton and Shepard 1955, Hart 1973, Imbleau 1978, Hatlevik et al. 1981). The fish usually spawn in small streams with gravel substrate that are tributary to rivers and lakes. Redds are constructed by the females. Emergent fry spend variable lengths of time in their natal streams, while migratory populations may spend as little as a few months to as long as four years in their original streams (Liknes and Graham 1988). Once in rearing areas, the river and tributary dwelling populations may make minor migrations to access preferred food and appropriate winter habitats. Cutthroat trout in lakes generally grow faster than those in streams, and, as Carlander (1969) suggested, the smaller the stream, the slower the growth.

Cutthroat trout are opportunistic feeders that consume a variety of freshwater invertebrates and may feed heavily on other fishes, crustaceans, and freshwater insects. In their study of cutthroat habitat preferences in tributary streams, Moore and Gregory (1988) found that fry abundance is proportional to the area of lateral habitat: stream margins, backwaters, and isolated pools. Cutthroat living in association with other trout species generally alter their feeding behavior to minimize competition with other species.

Cutthroat trout are targeted by sport fisheries in limited instances in the Skeena River drainage. The Lakelse River, the Kasiks River, and tributaries to the lower Skeena River are notable fisheries. Cutthroat trout are targeted in a winter ice fishery in Kitwanga Lake. Bilton and Shepard (1955), Imbleau (1978), Hatlevik et al. (1981), and DeLeeuw (1991) described the Lakelse River cutthroat trout sport fishery.

Dolly Varden Char

Dolly Varden char *(Salvelinus malma)* are blue listed by the B.C. CDC as a species of concern, but not listed by COSEWIC or defined as identified wildlife by the Forest Practices Code (FPC). Dolly Varden are common in the Skeena watershed. Small resident Dolly Varden are widely distributed in the upper reaches of small streams throughout the watershed, whereas anadromous populations, which are comprised of larger fish, exist primarily in close proximity to the coast. Dolly Varden char also exist in lacustrine-adfluvial populations. Beyond knowledge of distribution and general life history, Dolly Varden char have not received extensive management attention or biological study in the Skeena watershed.

Spawning takes place in streams in the autumn with maturity usually reached in the fifth year. Regular seaward migrations may take place in spring with return migrations in the fall. Hart (1973) suggested that in the Skeena watershed, coastal populations may spend three years in fresh water and two to three years in the ocean, with males tending to stay longer at sea. Generally, food consists of salmon eggs, molluscs, insects, crustaceans, and fishes that include herring, sticklebacks, and juvenile salmon (Hart 1973).

Cedarvale constitutes the regulatory upper limit of Dolly Varden anadromy; the remaining populations are fluvial or adfluvial residents. Dolly Varden char are only targeted as sport fish in the lower Skeena and its coastal tributaries, due primarily to their small size in upper watershed drainages. A rare monoculture Dolly Varden population exists in Netalzul Meadow Lake, a small lake within the Suskwa River watershed (FISS 2002).

Bull Trout

Bull trout *(Salvelinus confluentus)*, are actually a char that are blue listed as a species of concern by the B.C. CDC, as well as by COSEWIC, due primarily to their limited global distribution and their threatened status in its southern (U.S.) range. They are also listed as an identified wildlife species under the FPC. Bull trout are common within the Skeena River and its tributaries, and are suspected to be found throughout the drainage. The Annual Environmental Trends, published by B.C. WLAP, lists the Bulkley watershed as the only Skeena drainage identified as a conservation concern.

Studies on bull trout in the Skeena watershed are limited to the Morice watershed (Bahr 2002) and the Shelagyote River (Giroux 2002). Even with differences in life-history traits and morphometry, confusion between bull trout and Dolly Varden is common, and much of the available information on distribution is suspect (Haas 1998). Seaward populations are poorly studied, though bull trout are thought to have a low tolerance for salt water. Records of their occurrence exist for the Kitsumkalum River, the Zymoetz River, and lower Skeena tributaries including the Ecstall River (FISS 2002). The Gitnadoix River has also been reported through anecdotal sources to contain bull trout. Bull trout occurrence is considered common in

Skeena tributaries upstream of Cedarvale. Both bull trout and Dolly Varden char occur in many Skeena tributaries. In these situations they appear to have divergent life-history patterns (Bahr 2002), but hybridization occurs in some streams.

Bull trout spawn in small tributary streams and adults over-winter in larger rivers (fluvial populations) or lakes (adfluviual populations). Maturity is generally reached at five years of age, though precocious males may mature by age three (Shepard et al. 1984). Recent observations by Giroux (2002) and Bahr (2002) show that upper Skeena bull trout typically spawn in gravel and cobble pockets in streams during late summer and early fall. Eggs in the gravel usually hatch before the end of January, with fry emerging in late spring. After hatching, bull trout fry rear in low-velocity backwaters and side channels and avoid riffles and runs (McPhail and Murray 1979). Juveniles tend to utilize a variety of stream and lake habitats and are most abundant where water temperatures are 12° C or less. Their intrawatershed distribution patterns indicate they are sensitive to water temperatures, preferring cold natal streams.

Bull trout are a long-lived repeat spawning fish that can exceed 20 years of age and 10 kg in weight; however, in general terms, most bull trout captured by anglers range between 45 and 60 cm in length, and are 8 – 17 years old. Bull trout are a popular sport fish and are frequently harvested by sport anglers as by-catch during recreational fisheries targeted on summer-run steelhead, chinook, sockeye, and coho. As adults, they are an aggressive piscivorous fish and vulnerable to overharvest by anglers. Limiting angler access and critical habitat identification and protection remain the most significant issues for the protection of bull trout in the Skeena River drainage.

Kokanee

Kokanee salmon *(Oncorhynchus nerka)* are a landlocked form of sockeye salmon living their entire life cycle in freshwater. Kokanee are generally believed to have evolved independently from anadromous populations of sockeye salmon within many lake systems. In the Skeena system, kokanee have been observed in the Babine, Sustut, Lakelse, Zymoetz, Kispiox, Kitwanga, Morice, and Bulkley watersheds (Table 1). There is little recent information concerning kokanee distribution (Foote et al. 1989), and kokanee may be found in other Skeena lakes. Kokanee usually mature at smaller sizes than sockeye, and where the two forms occur together, they exhibit other morphological differences such as in gill raker number, male secondary sexual characteristics, and coloration (Nelson 1968). Within the Skeena watershed, the geographic distribution of kokanee is believed to be a result of landform or drainage changes that have isolated populations of formerly anadromous sockeye salmon (Ricker 1940, Foerster 1968).

Juvenile kokanee and sockeye are difficult to distinguish, and most of the understanding concerning kokanee developmental biology, fry behaviour, and juvenile ecology is based on studies for sockeye (Foerster 1968). Kokanee eggs are deposited in gravels of nursery lake inlet streams or on lake beach gravels. Eggs develop over fall and winter, and fry emerge in the spring. As juveniles, they move offshore, feeding primarily on zooplankton, although

they also feed on benthic invertebrate. Juvenile kokanee growth and survival rates are relatively variable and are determined by lake productivity and the intensity of feeding competition. Adults usually mature sexually at age two to four and migrate to spawning grounds in the early fall to deposit their eggs. As with sockeye salmon, kokanee die after spawning.

Kokanee are an important sport fish mostly caught by trolling in many of the larger lakes within the Skeena system. The deep red flesh is frequently considered by many to be the tastiest and finest eating fish in the watershed.

Whitefish

There are three or possibly four species of whitefish—salmonids in the subfamily Coregoninae—in the Skeena watershed. These are the Rocky Mountain whitefish *(Prosopium williamsoni),* pygmy whitefish *(Prosopium coulteri),* and lake whitefish *(Coregonus clupeaformis).* The fourth whitefish, the giant pygmy whitefish *(Prosopium* spp.), is probably not separable from the more common pygmy whitefish (Rankin 1999).

Rocky Mountain whitefish is one of the six species in the genus Prosopium. Rocky Mountain whitefish, also commonly called mountain whitefish, are the most widely distributed of the Skeena watershed fishes occurring in streams and lakes throughout the Skeena system (Godfrey 1955). They have been found, generally in fair abundance, in more than 20 sockeye rearing lakes in the system that vary from deep, cold, and opaque bodies of water to small, shallow, and warm ponds. Godfrey reported that abundance appears to be highest in higher nutrient status lakes, such as Lakelse Lake.

Mountain whitefish use a wide range of habitats for spawning and do not construct redds. Mainstem river resident- and lake-dwelling populations move into tributary streams in the late fall to spawn (Northcote and Ennis 1994); however, McPhail and Lindsey (1970) reported some cases of spawning occurring within lakes. Clearly, the habitat used for spawning should be determined for local populations. Mountain whitefish are generally nocturnal spawners (McPhail and Lindsey 1970). The eggs hatch in early spring, usually

Table 1. Lakes in the Skeena watershed with kokanee salmon

Watershed	Lake/Stream
Babine	Augier Lake
Babine	Babine Lake
Babine	Morrison Lake
Babine	Tahlo Lake
Babine	Nilkitkwa Lake
Babine	Taltapin Lake
Sustut	Bear Lake
Sustut	Sustut Lake
Lakelse	Clearwater Lakes
Lakelse	Onion Lake
Zymoetz	Burnie Lakes
Kispiox	Swan Lake
Kispiox	Stephens Lake
Bulkley	Goosly Lake
Bulkley	Toboggan Lake
Morice	Shea Lake
Morice	Morice Lake
Kitwanga	Kitwanga Lake
Skeena	Khtada Lake
Skeena	Kleanza Lake
Skeena	Slamgeesh Lake

at the time of ice break-up. Underyearlings generally leave near-shore habitat during the summer. There appears to be relatively little specific information in regards to yearling and subadult feeding, migration, and habitat. Adults occupy shallow portions of lakes and feed on aquatic insects and some small clams and snails (Godfrey 1955).

Although this whitefish has attracted moderate attention from anglers, there are surprising gaps in knowledge of it life history and biology. Information gaps exist in relation to stock recognition and impacts from forestry or other causes of water quality and habitat change. Significant winter sport fisheries for mountain whitefish have developed in B.C., particularly in the Similkimeen River, Elk River, and Okanogan and Kootenay Lakes (Northcote and Ennis 1994).

Pygmy whitefish *(Prosopium coulteri)* are commonly misidentified as juvenile Rocky Mountain whitefish. Pygmy whitefish are found in the Peace, Fraser, and Skeena River systems, and are thought to have spread into B.C. after the Ice Age from a refugium in the Columbia River basin (Lindsey and Franzin 1972). This species prefers deep-water habitats (McCart 1965) with a variable diet, showing the fish to be opportunistic benthic feeders. Piscivorous fishes such as trout and char prey upon pygmy whitefish. Two morphological types have been described by McCart (1970); all known pygmy whitefish populations in the Skeena system are categorized as the "low-raker" form (Rankin 1999). There is no direct evidence of spawning locations or timing, though Scott and Crossman (1973) suggested that it takes place in October to November depending on location. Much of the basic life history and ecological attributes of pygmy whitefish are unknown.

Lake whitefish *(Coregonus clupeaformis)*, also called the common whitefish, were rated by Scott and Crossman (1973) as the most important commercial freshwater fish species in Canada. Godfrey (1955) reported that lake whitefish have been found in only four lakes in the Skeena drainage—Babine, Morrison, Bear, and Azuklotz Lakes—all of which have oligotrophic characteristics and are relatively deep, cold bodies of water. In general, lake whitefish are pelagic and restricted to cool, well-oxygenated regions of lakes in close association with the bottom. Spawning usually occurs during October through December with eggs incubating over the winter and fry emerging in April or May. Lake whitefish fry remain in shallow inshore waters where they feed on planktonic and benthic organisms. They then move into deeper waters as water temperature increases and gradually adopt the benthic feeding habits typical of adult whitefish. Age of maturity varies widely, but is typically four to nine years. The main predators of lake whitefish are lake trout and burbot adults (Scott and Crossman 1973).

Giant pygmy whitefish *(Prosopium* spp.) occur in Tyhee Lake and possibly in Touhy Lake, both of which are within the Bulkley River drainage. The Tyhee Lake population has been red listed as a threatened species due to the rare occurrence of the giant pygmy whitefish and the eutrophication of the lake. Rankin (1999) concluded that Tyhee Lake giant pygmy whitefish show a distinct size at age curve and are clearly larger than pygmy whitefish, but the fish are not phylogenetically distinct from other pygmy whitefish and should remain *Prosopium coulteri*. Research is currently proceeding on the Touhy Lake population.

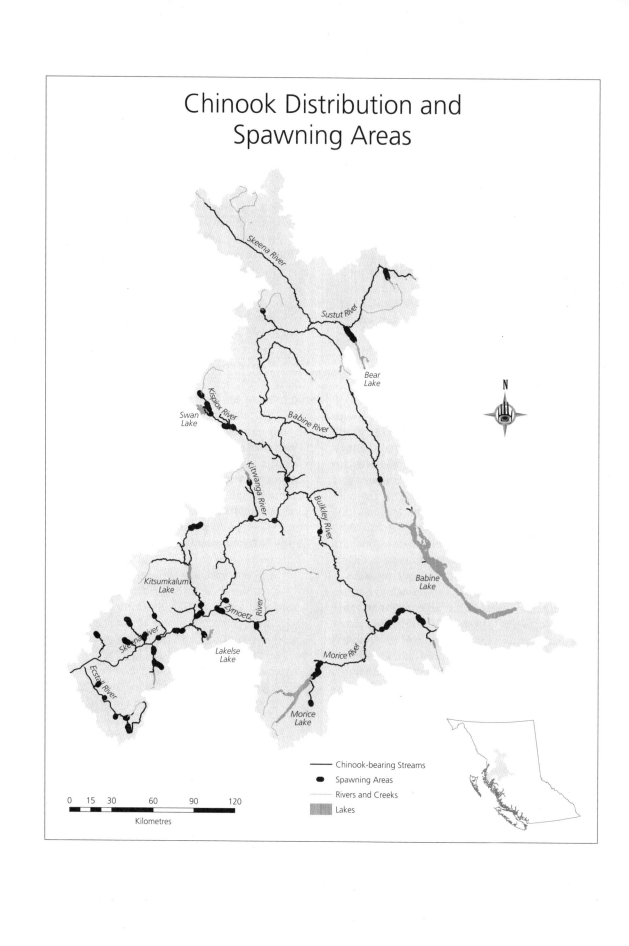

Chinook Distribution and Spawning Areas

Skeena River

Sustut River

Bear Lake

Kispiox River

Swan Lake

Babine River

Kitwanga River

Bulkley River

Babine Lake

Kitsumkalum Lake

Zymoetz River

Skeena River

Lakelse Lake

Morice River

Ecstall River

Morice Lake

—— Chinook-bearing Streams

● Spawning Areas

—— Rivers and Creeks

Lakes

0 15 30 60 90 120

Kilometres

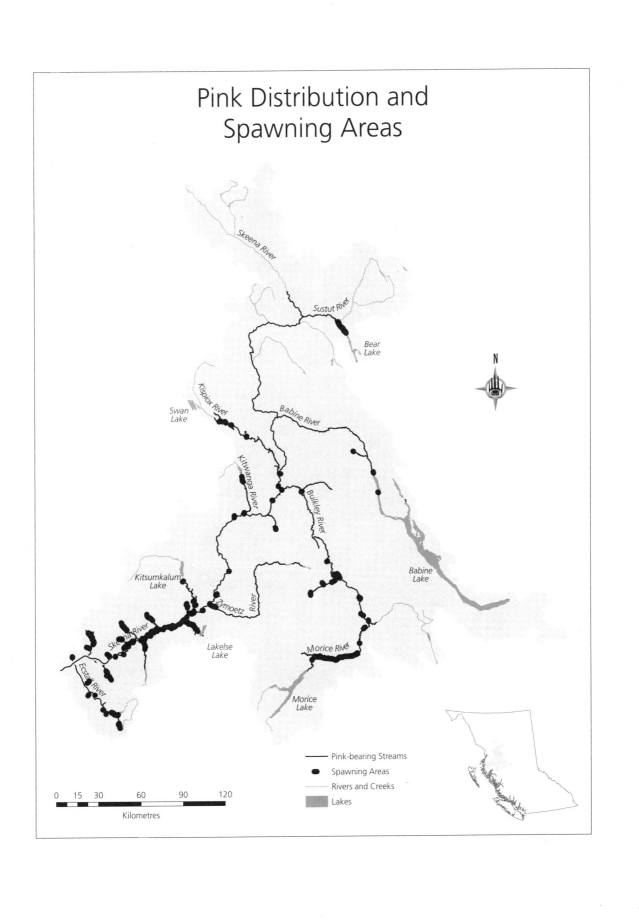

Pink Distribution and Spawning Areas

Skeena River

Sustut River

Bear Lake

Kispiox River

Swan Lake

Babine River

Kitwanga River

Bulkley River

Babine Lake

Kitsumkalum Lake

Zymoetz River

Skeena River

Lakelse Lake

Morice River

Ecstall River

Morice Lake

N

—— Pink-bearing Streams
● Spawning Areas
—— Rivers and Creeks
▨ Lakes

0   15   30   60   90   120

Kilometres

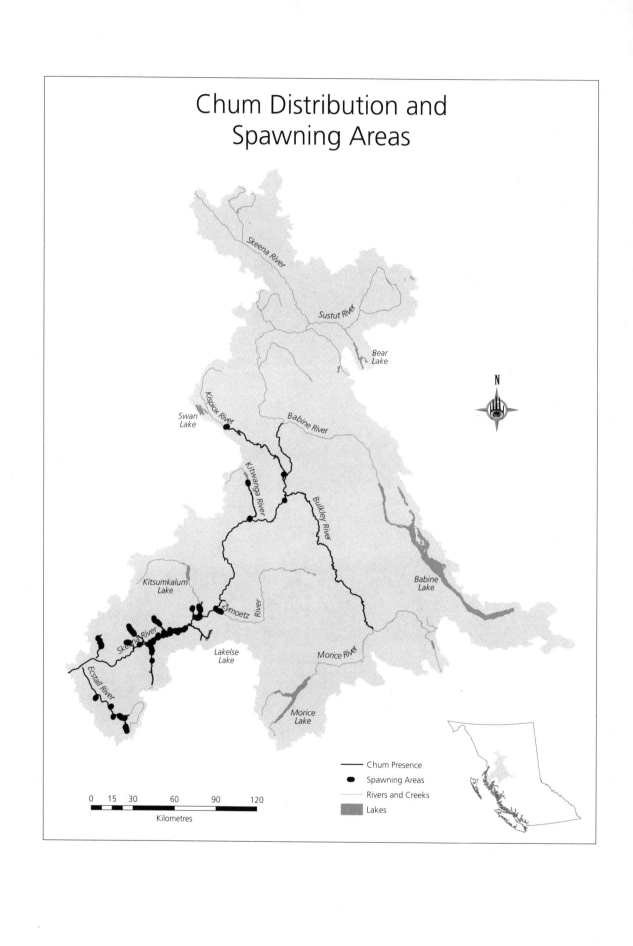

Chum Distribution and Spawning Areas

Skeena River

Sustut River

Bear Lake

Swan Lake

Kispiox River

Babine River

Kitwanga River

Bulkley River

Kitsumkalum Lake

Babine Lake

Zymoetz River

Skeena River

Lakelse Lake

Morice River

Ecstall River

Morice Lake

Chum Presence
Spawning Areas
Rivers and Creeks
Lakes

0   15   30   60   90   120

Kilometres

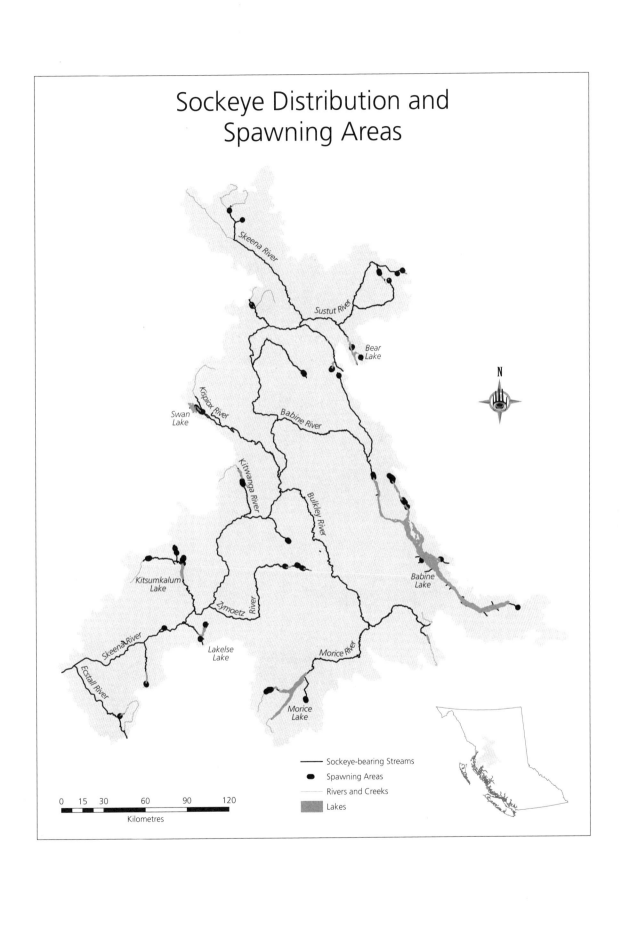

Sockeye Distribution and Spawning Areas

Skeena River

Sustut River

Bear Lake

Kispiox River

Swan Lake

Babine River

Kitwanga River

Bulkley River

Kitsumkalum Lake

Babine Lake

Zymoetz River

Skeena River

Lakelse Lake

Morice River

Ecstall River

Morice Lake

N

—— Sockeye-bearing Streams
● Spawning Areas
—— Rivers and Creeks
Lakes

0   15   30   60   90   120
Kilometres

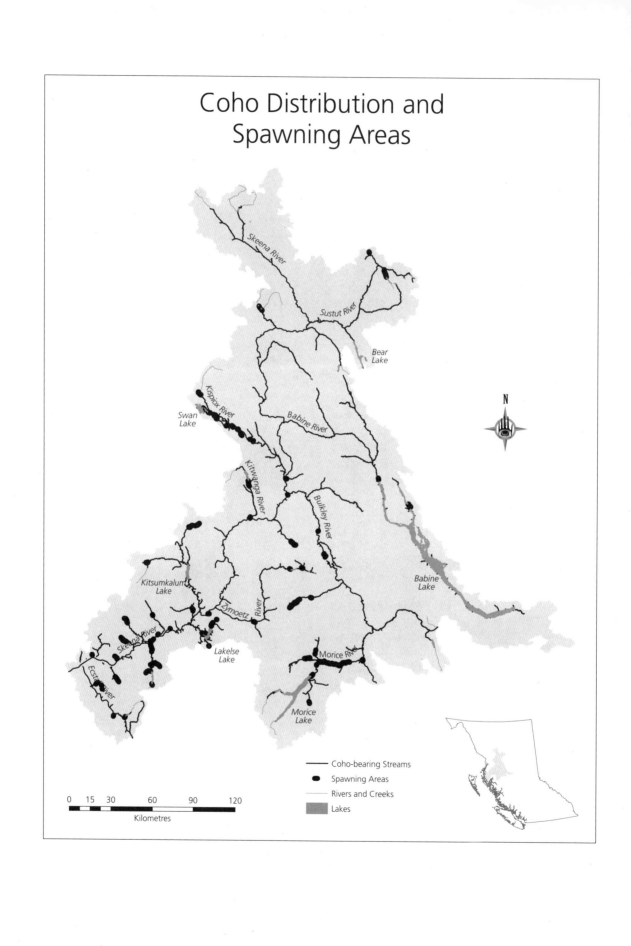

Coho Distribution and Spawning Areas

Skeena River

Sustut River

Bear Lake

Kispiox River

Swan Lake

Babine River

Kitwanga River

Bulkley River

Babine Lake

Kitsumkalum Lake

Zymoetz River

Skeena River

Lakelse Lake

Morice River

Ecstall River

Morice Lake

N

—— Coho-bearing Streams
● Spawning Areas
—— Rivers and Creeks
■ Lakes

0   15   30   60   90   120
Kilometres

3
Skeena Watershed Fisheries

First Nations Fisheries

The Coast Tsimshian, the Canyon Tsimshian, Gitxsan, Wet'suwet'en, and Ned'u'ten peoples occupy the Skeena watershed. The 10 Coast Tsimshian territories are located on the lower Skeena River downstream of the Zymacord River. The two Canyon Tsimshian First Nations, Kitselas and Kitsumkalum, utilize traditional territories upstream of the Zymacord River. Gitxsan occupy the middle and upper Skeena to the headwaters, upstream of Kitselas territories. Wet'suwet'en occupy the vast majority of the Bulkley drainage. Ned'u'ten occupy the upper Babine drainage; that is the country upstream of Shahnagh Creek on the Babine River and the watersheds that drain into Babine Lake.

The aboriginal salmon fishery formed the principal foundation of the economy. Arrangements for management of the fishery were deeply interconnected and woven into the fabric of society. Bodies of laws governing the fish resource generally and fishing specifically were based on values from a conceptual reality founded on thousands of years of interacting with the social, subsistence, and local environment dynamics. The majority of relevant fishing regulations were self-enforcing since they were founded on accepted community values shared by all its members.

Anadromous chinook, coho, sockeye, chum, pink, and steelhead stocks were typically harvested and processed close by their spawning grounds. This large-scale utilization of the abundant and predictable salmon stocks formed the foundation of the economy. Hereditary chiefs exercised authority for management and decision making. Principal management tools included ownership of specific sites, access allocation, control of harvest techniques and timing, and harvesting limitations imposed by processing capacity (Morrell 1985).

The harvest of surplus to conservation needs on a stock-by-stock basis allowed for optimal utilization of the salmon resource that was the core of the economy. These modes of management enabled the fishery system to adapt to changing natural situations and conditions, and facilitated allocation and regulation in managing the fishery, while encouraging habitat protection. In assessing the results of traditional fish management, it is a matter of record that First Nations management of the salmon fisheries left a fish resource that was vigorous, diverse, and healthy at the advent and incursion of the commercial Skeena fisheries in the late 19th century.

Salmon Fishery Management

Skeena First Nations salmon fisheries were governed by fundamental conservation elements, and waste was forbidden. Processing capacity was limited by smokehouse infrastructure, particularly the amount of space available on the lower poles, where fish were hung in the first stages of the drying process, and by the number of fish that could be dressed in the available time. When the daily processing limit was reached, the traps were removed from the water, and the salmon were allowed to proceed upstream. The predominant use of live-capture gear enabled the people to selectively harvest desired species, with the remainder released unharmed (Morrell 1985).

All fishing sites were considered property of the House — the basic social level of organization and territorial and resource owning unit — with particular sites being more or less delegated to individual chiefs or subchiefs within the House. The chief typically decided who would be fishing at specific sites and at which time. However, several Houses from various clans might share in the harvest distribution from productive weir and trap sites at villages that were strategically located to access the fishery. It was and is the responsibility of the chiefs to oversee the processing and distribution of the fish, so that all members of the House received sufficient amounts, even if they could not provide for themselves directly because of age, disability, or other circumstances.

Harvest and Processing

The abundant and predictable salmon runs provided the opportunity for the people to harvest and preserve a high-quality staple food in a few months of intensive effort. In June, the majority of House groups would congregate in their seasonal villages to prepare fishing gear and firewood and generally get ready for the salmon fishery.

The first salmon, the chinook or spring, usually reached the lower Skeena watershed area in early to mid June, marking the start of the fishery. This was the occasion for celebration and thanksgiving with the First Salmon Ceremony, in which the salmon were ritually prepared to ensure and herald an abundant harvest. At many Skeena River fishing sites, springs were readily caught in season, as the strong river currents during the snowmelt season concentrated them at particular points.

Following the spring salmon came the sockeye runs. Upriver sockeye were the most desired fish owing to a fat content that facilitated smoke-drying. They were fished heavily until the pink salmon showed up, by which time sockeye needs were usually met. Pink salmon are generally much less desirable, and if sockeye were still needed, fishing would take place in the deeper and swifter waters that the pinks avoided. The peak of the pink salmon marked the end of the sockeye season and signaled the beginning of berry picking and high country hunting.

Coho and steelhead would migrate into the lower Skeena watershed in early to mid-August and be harvested to a certain degree, but the main coho fishery would occur later in the many smaller, though important, tributary streams on the territories. The coho were especially useful to the people who did not go to the mainstem, but stayed out at their villages or camps on the remote territories. Due to their widely dispersed nature throughout the watershed, coho were often harvested and processed in headwater locations. In the same way, lake and stream fish such as rainbow trout, steelhead, Dolly Varden char, bull trout, and whitefish were also fished and processed in their respective habitats.

Salmon were eaten fresh during the summer, but the major directed fishing effort was focused on the preserving of salmon for use during the rest of the year. The salmon was split and hard-dried over slow, smoky fires in smokehouses, then stored in bark-lined excavated storage pits and covered over with the excavated dirt. These pits, often called cache pits, were usually located in drier (sandy or gravelly) soil types close to the village, winter camps, or other home places.

At Skeena River canyon or rock outcrop locations, where the salmon tended to be concentrated by strong currents, large woven baskets and/or lashed wooden strip traps were ingeniously made, some incorporating delivery chutes that moved the trapped fish to a waiting fisher who transferred it to the shore. Trap sizes varied, with larger ones being lowered and raised with stout poles operated by a strong and frisky crew. The various traps and dip net gear used depended on site location, fish quantities needed, and the number of people available to fish the gear and provide processing capacity.

At a few locations on the Skeena River mainstem, and on many of its tributaries, salmon and steelhead were traditionally caught with weirs (t'in) inset with a variety of large woven cylindrical or barrel basket (moohl) traps. Undoubtedly the most productive and ingenious of fishing gear, these weirs were built either right across smaller streams, or on the mainstems out on an angle to guide the migrating fish into shore-side traps. The wide variety of weirs and contiguous traps used were matched to the species, environment, placement, and building materials available.

Smaller tributaries often were fished with weir placements just upstream of the confluence with the mainstem, while larger tributaries had weirs strategically positioned close to lake outlets. These two types of sites are hydrologically suited for weirs because they are relatively protected from sudden floods following intense rainstorms. Gear types suited to single fish harvest included specialized dip nets with a closable mouth (bana) and spears. Spears were utilized in shallow, clear tributary streams where fish were readily visible.

Post-contact Cultural Context

Skeena watershed First Nations managed the coho, sockeye, chinook, pink, chum, and steelhead fisheries of their territories up to the mid 1870s. At this point, the incursion of Euro-Canadians with their colonial society concepts and the establishment of coastal industrial fisheries at the mouth of the Skeena River initiated a period of transition.

Early industrial development on the British Columbia coast saw the development of many new canneries, including in 1877, the first commercial salmon cannery on the Skeena River. Thirty years later, as markets were developed and investors looked for a certain return on their capital, 14 canneries supported by a fleet of 870 wind- and people-powered boats were in operation. This period was characterized by steady growth in both the number and size of the canneries, competition for sockeye, and the move to begin canning other species besides sockeye. In 1907, the canned pack totaled just over 159,000 cases (48 pounds each) of which two-thirds were sockeye, requiring a catch of approximately 1.6 million. The Nass River also supported a substantial commercial fishery with canneries first built in 1881. Fishing operations were restricted to the Nass estuary until the late 1920s, with gear increasing from approximately 100 sailboats in 1900, to 300 boats by 1930.

At the turn of the century, a campaign was initiated by cannery operators who wanted a larger share of the fish and a guarantee of harvesters and plant workers, both which could be accomplished by prohibiting the use of weirs by aboriginal fishermen. Legal action was taken prohibiting weir use by aboriginal fishermen and the sale of fresh and processed fish throughout the Skeena watershed. It was enforced by the Department of Marine and Fisheries (Newell 1993).

The legal action focused on the weirs at Wud'at, also known as Tsa Tesli (where the lake ends), which was the principal salmon season village on Babine Lake (Helgerson 1906). Located primarily on the right bank of the Babine River, it is for the most part currently overlaid by DFO's counting weir and camp. Salmon fishing was conducted as a cooperative clan endeavor with the fish caught in weirs across Nilkitkwa Lake and the upper Babine River. The dispute was somewhat settled with the Barricade Agreement of 1906; however, to this day, there are bitter feelings remaining with the Lake Babine Nation.

At the turn of the 20th century and continuing up to 1990, as fish stocks steadily declined, federal fishery administrators directed pressure against native fishermen. Enforcement of the Fisheries Act, which prohibited weir utilization from 1905 to 1913, made it difficult for traditional fisheries to continue as in the past. To replace the banned weirs and traps, gillnets were introduced.

The pressure to relocate the main Skeena River salmon fishery had many effects. The impact on native fishermen encountering another culture socially and politically was considerable. Over time, a shift occurred from many dispersed subsistence fisheries that were locally managed and closer to the spawning grounds, to a coastal, industrial, mixed-stock fishery with highly efficient capture methods. This shift had negative cultural and economic impacts on native fishermen. In general, government fisheries policies in the upper Skeena watershed between 1880 and 1980 resulted in a legacy of over-fished stocks, conflict, and marginalization of aboriginal people. In particular, the enhancement of two already productive Babine sockeye stocks led to excessive fishing pressure on less productive wild stocks of coho, steelhead, chinook, and sockeye, and resulted in the decline in both abundance and diversity of many.

Currently, salmon harvested for food, societal, and ceremonial use (FSC), as well as part of the Excess Salmon to Spawning Requirement (ESSR) fishery, are harvested with a range of gear types such as fish wheels, dip nets, gillnets, and beach seines. The importance of the salmon fishery, which is a blend of food resource, trade capital, cultural expression, and connection to ancestral practices, cannot be overstated.

Recreational Fisheries

The Skeena watershed is one of the last large Pacific drainages where wild salmon and steelhead can be easily angled in pristine wilderness settings. Many anglers testify that no other area in North America combines the superb angling fishery attributes in the manner that the Skeena system does. Freshwater fishing is the best known tourism activity in the watershed and is of international significance. This recreational fishery supports numerous tourism businesses such as fishing guides, fish camps and lodges, transportation services, and sport fishing retail services. The fishery also provides an important recreational opportunity for local residents. A recent conservative estimate of the value of the fishery, not including economic multiplier effects, is $110 million (IBM 2006).

The streams and rivers attract anglers who target a variety of species, particularly chinook, coho, rainbow trout, and steelhead. Pink and sockeye are also targeted in the Skeena. Other species sought include cutthroat trout, Dolly Varden, lake trout, and to a lesser degree, kokanee, whitefish, and burbot (ling). The activity occurs from the shoreline of rivers, with the aid of floatation devices, and in drift or jet boats. Jet boats are an important means of access for many of the larger rivers.

There is a diversity of anglers and angler preference for physical setting, type of gear, targeted species, and size of fish. Two broad types of angling exist—guided and unguided—with all anglers requiring a license. Freshwater fishing openings and closures by specific area are regulated for anadromous salmon by DFO and for steelhead and other freshwater species by the province. Angling is of two general types: catch and release fish or harvest. Most steelhead fishing is the former.

The current trend is increasing freshwater and coastal sport fishing activity. Though there are no official numbers arising from creel surveys, anecdotal estimates point to annual growth rates up to 10%. Demand for angler access to and overcrowding on classified waters led the province to initiate the Quality Waters Strategy planning. Tidal sport fisheries situated in Queen Charlotte Islands that target chinook have increased more than 10% annually since 1996. Winther (2006) reported that the total sport harvest estimate from Area 4 (marine) in 2004 and 2005 was approximately 2,700 chinook each year. The total in-river sport harvest estimate for Skeena chinook in 2004 and 2005 was 6,280 and 5,000, respectively (Winther 2006).

Recreational resident and nonresident anglers share an interest in protecting their experiences and the resource for future generations. The foundation of the Skeena sport fishery lies on a strong and vibrantly healthy fish resource. This in turn requires healthy freshwater and saltwater habitats and government management regimes that put fish first. It also necessitates sufficient information, of which there is currently a lack. Freshwater angling in the Skeena basin is not documented with an up-to-date profile that describes trends, economic outline, and species encountered and harvested.

Enhancement of Salmon Production and Habitat

Habitat enhancement has been ongoing since people first inhabited the watershed. Since the turn of the 20th century, DFO has taken the lead role. Habitat enhancement includes major projects such as spawning channels, as well as relatively small, ongoing activities such as stream clearance projects. Many of the salmon caught that are the progeny of Skeena River stocks are produced with human assistance in hatcheries and artificial spawning channels. Other Skeena River salmon stocks benefit from fishways built around barriers to their upstream migration, improvements to spawning beds, and occasionally, fertilization of the lakes where they spend the first period of their lives. Enhanced stocks now account for about one out of every two salmon caught; these are mostly sockeye from the Fulton River and Pinkut Creek spawning channels.

Fish enhancement projects have operated sporadically since the establishment of the first sockeye hatchery at Lakelse Lake in 1901, which was operational until 1937. The Morrison River Hatchery operated from 1907 to 1936, after which, most B.C. hatchery operations shut down due to the Depression and the negative report on hatchery effectiveness by R.L. Foerster (Taylor 1999). The early 1960s saw a resurgence of hatchery production within the watershed. The Nanika River Hatchery, in operation from 1960 to 1965, was not successful, most likely due to the use of transplant stock from Pinkut Creek. The Kleanza Creek pink salmon hatchery operated for several years until it was destroyed by fire in 1960. Scully Creek and Williams Creek, tributaries to Lakelse Lake, were the sites of hatcheries from 1962 to 1967.

Sockeye salmon studies in the late 1950s and early 1960s concluded that sockeye salmon production from Babine Lake was limited by the availability of suitable spawning habitat, and secondly, that the main basin of the lake was underutilized and could support additional sockeye fry. These results led directly to the Babine Lake Development Project (BLDP). Construction of this approximately $10 million project, consisting of artificial spawning channels and dams to provide for water flow regulation, took place at Pinkut Creek and Fulton River, tributaries of Babine Lake.

The first channel was completed on the Fulton River in 1965; the second on Pinkut Creek in 1968; a third channel was completed on the Fulton River in 1971 (Ginetz 1977). Sedimentation problems in the Pinkut Creek spawning channel led to a major rehabilitation program in 1976. The Fulton River spawning channel has capacity for approximately 630,000 sockeye, while the Pinkut spawning channel holds 160,000 sockeye. At Pinkut Creek, since 1973, an annual average of approximately 37,000 adult sockeye have been airlifted over the falls to allow utilization of a 6 km spawning area on the upper creek. In addition to the spawning channels, a small-scale hatchery at Fulton River has been in operation on a discontinuous basis since the late 1970s.

In 1977, the Department of Fisheries and Oceans announced the Salmonid Enhancement Program (SEP), with the primary goal of doubling salmon production. Within the Skeena watershed, five hatcheries under the Community Economic Development Program (CEDP) were established: Fort Babine, Kispiox, Kitsumkalum, Deep Creek, and Toboggan Creek. In addition, several minor hatcheries were established.

These are classified as Public Involvement Program projects, partially or fully funded by SEP, and include Chicago Creek and Eby Street. Numerous small projects such as incubation boxes, bioengineering investigations, biophysical studies, and habitat inventories have been conducted throughout the watershed under the auspices of SEP.

A pilot hatchery project was initiated at Kispiox in 1977; however, water quantity and quality problems were not resolved until 1983. Three wells were developed that supplied water of stable quality and constant temperature, at which time the present small hatchery was constructed. It is located close to the Kispiox-Skeena confluence and was initially operated by the Kispiox Band. This CEDP project under the Salmonid Enhancement Program was designed to increase the severely depressed Kispiox River chinook and coho stocks (DFO and MoE 1984). The hatchery continued to operate until 1995, when it was closed due to SEP program review budget cuts. Re-opened in 1997 under the auspices of the Gitxsan Watershed Authorities (GWA), with funding from a variety of sources, the hatchery allows for a flexible fish culture program.

Since 1983, the Fort Babine Hatchery, a CEDP project, has enhanced the Babine coho and chinook populations. Chinook brood stock is obtained from adults held at the counting fence, while coho brood stock is seined in the Babine River upstream of the fence. The eggs are incubated and reared with an approximate and variable release strategy of 50/50% fry and smolts. On an annual basis, enhanced coho typically number 150,000–200,000, while 80,000–100,000 chinook are raised.

In the Terrace area, a pilot facility was operated on Dry Creek from 1981 to 1982 to investigate the incubating and rearing of chinook salmon on that creek; however, the project was found to be unfeasible. Since 1984, Deep Creek Hatchery has been operated by the Terrace Salmon Enhancement Society (TSES), and was established as a chinook facility to augment chinook populations in the Zymoetz, Lakelse, and Kitsumkalum River systems (Tredger 1983b). Since the late 1980s, the hatchery has only supported Kitsumkalum River chinook.

Other small enhancement facilities include a small groundwater facility for the incubation and rearing of coho and chum, operated by the Kitsumkalum Band since the late 1980s, and a spawning channel for sockeye at the northeast end of Kitsumkalum Lake that enhances the Clear Creek and Cedar River stocks.

Toboggan Creek Hatchery is located on a Bulkley River tributary north of Smithers, B.C. This facility was built in 1984 and designed to raise steelhead fry and chinook and coho smolts. Adult coho returns have dramatically increased Toboggan Creek stocks. Upper Bulkley and Morice River coho were also enhanced with local stocks raised at this hatchery. Streams other than Toboggan Creek have not been effectively evaluated with respect to increases in stock strength.

Through the Ministry of Environment Fish and Wildlife Branch (FWB), the provincial government focused their SEP efforts on Skeena River summer-run steelhead. FWB sought to determine baseline biological data on juvenile steelhead distribution and densities, and survival rates of fry and parr. FWB reared and released steelhead, cutthroat trout, rainbow trout, and brook trout in many lakes, and to a lesser extent, in streams within the Skeena watershed. Hatchery steelhead fry were released primarily in the Suskwa, Bulkley River, Kitsumkalum, Kispiox, Zymoetz, and Morice Rivers.

The first major habitat enhancement project occurred in 1907. It involved blasting the 1891 rockslide in the lower Zymoetz River. In 1951, fishways were constructed at Moricetown Falls on the Bulkley River, and in 1958, rocks were blasted away in Hagwilget Canyon. The 1951 Babine slide was cleared effectively by 1952. In the late 1970s, the Fish and Wildlife Branch initiated a series of habitat modifications that included blasting the Harold Price falls and straightening the lower reach of Serb Creek. Morice Lake was one of the few lakes in the watershed fertilized; it was fertilized by aerial tankers in 1981 and 1985.

In the mid-1990s, the Watershed Restoration Program (WRP) was established by the B.C. government to accelerate the natural restoration of watersheds impacted by logging. The majority of the work in the first seven years of the program consisted of instream/riparian assessments and upslope/road assessment and deactivation. In the eastern portion of the Skeena watershed, a major impact of logging activities is the restriction of fish access to upstream waters, often the result of road crossings that were originally constructed as, or have become, barriers to fish migration.

Impacts are more complex in the western portion of the watershed with its naturally unstable slopes and high-energy stream systems. Identified potential restoration activities include road deactivation and pullback to prevent erosion and landslides, off-channel habitat and riparian zone restoration to stabilize channels and diversify habitat, and stabilization of highly mobile stream channels and gravel bars often associated with logged alluvial fans (MoF 2001). Millions of dollars were spent in the watershed on such efforts; however, few attempts have been made to evaluate the response of fish populations to habitat manipulations. Without such studies, it is difficult to assess the effectiveness of various habitat modifications.

SEP has committed substantial resources to education and public involvement activities. This education material is highly regarded among educators, partly because of the support they receive from community advisors and hatchery staff, and also because students and community members have an opportunity to learn "hands on" about salmon and salmon habitat (Donas 2002). The public involvement program coordinated by community advisors and supplemented by watershed stewardship coordinators and habitat stewards has supported scores of small enhancement projects – mostly incubation boxes for coho and chum, stream rehabilitation projects, and fish fence counts. The benefits of these projects are difficult to assess, but the public appears to consider them worthwhile.

Presently, the Skeena watershed has considerable enhancement capability for sockeye, chinook, and coho that was established both before and after SEP. The Babine Lake Development Project (BLDP), the largest spawning channel complex in the world, dramatically increased overall sockeye production and greatly increased the commercial sockeye catch. A serious concern about large-scale enhancement of Babine sockeye is the indirect effects on stocks of wild salmon as suggested by West and Mason (1987). This large-scale enhancement supports high catch rates in interception fisheries in Alaska and the mixed-stock fisheries close to the mouth of the Skeena River (Area 4), as well as First Nation in-river FSC and ESSR fisheries. These fisheries have impacted many of the wild sockeye and some other salmon stocks whose run timing is coincident with enhanced Babine stocks.

Compounding this problem is the fact that there is little reliable information available on the status of some wild stocks within the watershed. In general, there is widespread agreement that exploitation rates are too high for wild stocks and many wild stocks are well below desired optimum abundance and/or productive capacity levels. One solution to our present depressed wild stocks/enhanced stock dilemma is based on moving the enhanced Babine sockeye fishery inland to the Babine watershed, thus relaxing fishing pressure on wild stocks.

Enhancement of salmon stocks can damage wild stocks; this has been a longstanding concern not only here in the Skeena system, but in other countries with large-scale salmon enhancement programs. In recent years, the impact of adult hatchery fish spawning naturally in streams occupied by wild fish has become an issue of increased interest with respect to their impact on the biological health of wild populations (Waples 1991, Busack and Currens 1995, Campton 1995). Various perspectives challenge conventional views about the value of fish hatcheries and augmenting fish stocks.

Hilborn (1992) argued that hatchery programs that attempt to add additional fish to existing healthy wild stocks are ill advised and highly dangerous. Hilborn stated that large-scale salmonid hatchery programs have mostly failed to provide the anticipated benefits. Rather than benefiting the salmon populations, these programs may pose the greatest single threat to the long-term maintenance of salmonids.

4

Environmental Overview

Internal Environmental Issues

Anthropogenic environmental damage is widespread in the Skeena watershed, though not serious enough to interfere with normal functioning of most ecosystems. The habitat in general is in relatively good condition, although there are localized areas of high impacts. It is widely recognized that land management practices have permitted the degradation of forestlands, water quality, and fish habitat in the watershed. However, insufficient data exist for most sites to quantify cumulative and nonpoint impacts. Linear development and settlement have also affected areas of fish habitat.

Land-use practices in the Skeena watershed such as forestry, grazing, agriculture, urbanization, linear development, and mining disrupt aquatic ecosystems by altering watershed processes that ultimately influence the attributes of streams, rivers, lakes, and estuaries (Chamberlin et al. 1991). Most effects of human activities on watershed processes result from changes in vegetation and soil characteristics, which in turn, affect the rate of delivery of water, sediments, nutrients, and other dissolved materials from upland areas to stream channels.

Land-use activities within riparian zones can alter the amount of solar radiation reaching the stream surface, affect the delivery of coarse and organic materials to streams, and modify fluvial processes that affect bank and channel stability, sediment transport, seasonal streamflow patterns, and flood dynamics. Disconnecting streams from their floodplains further alters hydrologic processes, nutrient cycles, and vegetation characteristics (Beschta et al. 1995). Other human activities that influence salmonids and their habitats include large-scale harvesting of salmon and introduction of non-native species and hatchery salmonids.

Forest Development Activities

Damage from forest development activities is found in most of the Skeena tributary watersheds except in some of the northern subbasins where development has not occurred. Most watersheds are recovering well from logging impacts from 20 to 40 years ago that were concentrated in the lower portions of the watersheds—those most likely to be inhabited by anadromous fish. Logging in many areas is now focusing on headwater hill slopes that are frequently mountainous. Generally, early logging sites that have regenerated forests have not yet grown sufficient volume to be harvested as second growth.

Within the watershed, forest development activities alter natural fish habitats and ecology mainly through changing vegetative cover, watershed hydrology, sedimentation and erosion rates, and to a small degree, through application of silviculture chemicals.

Removal of riparian stands by forest harvesting has produced a troublesome legacy. Stream bank stability, which controls overall channel integrity, is a function of channel and stream bank components that are largely influenced by riparian vegetation (Swanson et al. 1982). In the past, destabilized stream banks due to logging practices often resulted in increased bedload movement, which causes widening and shallowing, as well as stream channel simplification. This in turn has led to habitat loss and declining fish productivity. In addition, removal of riparian forests results in warming of summer flows due to the loss of shading, often to the extent that salmonids are stressed by high temperatures. Riparian vegetation also keeps streams warmer in the winter and helps to limit the amount of streambed freezing. Riparian vegetation provides the bulk of the energy of the food web in small streams. Alteration of the amount and type of riparian vegetation can change the composition of stream algae and stream invertebrate communities. Stream invertebrates are important dietary staples of stream dwelling fish.

Forest development activities and clearing of utility corridors substantially reduced riparian habitat within the lower Zymoetz River floodplain, leading to a 25% to 30% loss of off-channel habitat (Lewis and Buchanan 1998). These activities also led to a large, highly mobile volume of bedload that will continue to pulse down channel and destabilize the floodplain for many years to come (Weiland and Schwab 1996).

Forest road construction can substantially increase delivery of sediments to streams through both surface erosion and slides. The effects of forest practices on sediment transport are greatly increased in the steeper topography and wetter climate in the western and middle portions of the watershed. These effects are dependent on a number of local site conditions including moisture, vegetation, topography, and soil type, as well as on specific aspects of the activity and proximity to the stream channel.

Within the Skeena watershed, the art of assessing watershed impacts of logging is in a state of growth. Critical threshold methods popular in the 1980s set a percentage of watershed cut restrictions that evolved into an expert system such as the Watershed Workbook (Wilford 1987). Watershed Assessment Procedures (WAP) 1 and 2 are indicator models that use point scores of measured watershed characteristics or land-use patterns to score the overall health or impact of logging on the watershed (BC. MoF and B.C. MoE 1999). Watershed management involves making decisions in a complex system with many variables. Generalizations are difficult to make and challenge geomorphological knowledge in establishing cause-and-effect links between management and hydrologic outcomes (Carver 2000).

Urbanization

Urbanization concerns in the Skeena watershed are relatively minimal at the present time. Settlement patterns in the watershed reflect the major river trunk system, with major urban settings at Terrace, the Hazeltons, Smithers, and Houston. Gitanyow, Gitwangak, Kitwanga, Kispiox, Moricetown, and Telkwa are smaller population centers. All these settlements are located on the mainstem Skeena or Bulkley Rivers, with the exception of Gitanyow, which is on the Kitwanga River. Future projections suggest that urban areas will occupy an increasing, but still relatively small fraction of the landscape.

These 10 population centers rely on facultative or aerated lagoons, and then discharge treated sewage into the adjacent rivers (Remington 1996). Remington notes that there is considerable variation in the amount and type of monitoring data collected for each of these sewage facilities. These community sewage discharges threaten their receiving waters with oxygen depletion, solids settling, enrichment or eutrophication of streams, and toxicity. These 10 communities utilize solid waste disposal landfill sites operated by the Bulkley-Nechako and Kitimat-Stikine Regional Districts. Landfill leachate has the potential to affect ground water and surface runoff down gradient of the landfill. The potential effect is dependent on leachate quantity and quality.

Urban development may change to the pattern of flooding on a floodplain and result in significant losses to floodplain habitat. In the Skeena basin, this is not a large concern; however, residential encroachment can create pressure for flood control. This is an issue in Dohler Flats located on the Bulkley River close to Smithers and Dutch Valley, north of Terrace.

The Lakelse watershed retains very high fish values centered at Lakelse Lake. There are moderate to strong concerns regarding settlement and development adjacent to and on the lakeshore. Lakelse Lake supports a relatively high number of seasonal and full-time residences enjoying a variety of rural, high-quality lifestyles. Mount Layton Hot Springs Resort operates pools, water slides, a restaurant, and a motel on the east shore of Lakelse Lake. Also located on the east side of the lake, as well as the northeast corner, are two provincial parks, which are popular stopping-off points for local and nonlocal water-based recreation, picnic, and camping activities.

The east and west sides of the lake are largely developed. These developments, with their associated septic systems and occasional stream diversions, affect fish and fish habitat. With any development at or impinging on Lakelse Lake, there is a standing concern about phosphate additions to the lake from sewage. The lake has been described as a phosphorous-limited system that is literally in danger of becoming mesotrophic, or even eutrophic (Remington 1996).

Transportation

The significant impacts along transportation corridors of the Skeena watershed are blocked fish access and habitat alteration. In many places, fish passage is restricted by roads and railroads because of poor design and construction of culverts and other drainage structures. The two most common problems with drainage structures are culverts with large outfall drops that establish migration barriers and culverts installed with excessive slope, which cause water velocity barriers. The highway and railway corridors passing through the Skeena River and Bulkley River valleys have, in the past, created access problems for returning spawners and restricted access to juvenile freshwater rearing habitat.

On the lower Skeena and in the upper Bulkley floodplains, the CN Rail and Highway 16 alignments cut off many back channels and sloughs. This has resulted in difficult or impossible fish passage and eliminates productive fish habitat, particularly at low flows. An example of altered habitat is the 70% loss of the floodplain downstream of the Highway 16 bridge crossing the Zymoetz River. This was caused by channelization that directed the flow under the bridge, which effectively moved the apex of the alluvial fan significantly downstream from its natural position (Pollard 1996). Other transportation related problems deal with the use of herbicides on electric transmission lines and placement and servicing of natural gas lines.

Agriculture

Agriculture within the Skeena watershed is concentrated along the wide valley bottoms of the Bulkley and Kispiox Rivers. Overall, agriculture has a relatively small impact in the watershed. Impacts to fisheries values are highest in the upper Bulkley valley and less significant in the lower Bulkley valley and the Kispiox valley. This is similar to the level of impact reported by Paish and Associates (1983).

One of the most important concerns is the clearing of agricultural land on floodplains, near streams, or on active fans, and the removal of riparian vegetation. Livestock grazing and watering along tributaries tends to break down stream banks leading to an increase in sedimentation. The production of nutrient laden runoff from farmyards and cattle feeding lots can be a local problem. Gaherty et al. (1996) noted that winter feedlots may contaminate water bodies due to manure runoff into nearby streams; high nutrient loads may result as documented by Remington and Donas (1999, 2001).

The use of water for irrigation, particularly from smaller streams, can contribute to increased water temperatures and inadequate streamflows for fish. Licensed water withdrawals in the upper Bulkley drainage for irrigation, domestic, stock watering, and water delivery is approximately 2.4 times the currently available average supply during summertime low streamflows.

Industrial Pollution

Industrial pollution in the watershed is relatively minor, mostly due to the lack of industry. The Port Edward sulphite pulp mill was constructed in 1950. Impacts to water quality were severe with near sterilization of Wainwright Basin. In the 1970s, the mill was renovated to a Kraft process mill, which greatly decreased pollution problems. The Port Edward pulp mill is currently mothballed, further reducing effluent pollution.

Mining

Parts of the Skeena watershed contain high-value mineral resources, some of which have been exploited in the past; others are likely to be developed in the future. There are no mines currently operating in the watershed; however, three mines are close to the environmental assessment stage of development. There are no outstanding major pollution problems at this time. The Duthie Mine, located in the upper Zymoetz drainage, has had tailings leachate problems over the last 50 years; however, remediation and mitigation of the minor acid rock drainage (ARD) has recently been completed. The Silver Queen mining property, located immediately east of Owen Lake, has shown associated elevated levels of zinc and copper in the lake. Remediation efforts consisting of wetland treatment and contaminated drainage improvements occurred throughout the 1990s (Remington 1996).

Three large open pit mines, which operated in the watershed and shut down in the 1990s, are known to have generated ARD: the Equity Silver, Bell Copper, and Granisle Copper Mines. These mines now operate elaborate monitoring and treatment programs to safeguard the receiving environment from ARD. In short, ARD is a serious environmental concern at these three mines; however, current toxicity and deleterious discharges are relatively minor and in control.

Oil and Gas

The Bowser and Sustut basins potentially represent the largest petroleum exploration target area within the intermontane region as shown in Figure 15. This area has received renewed interest in the last few years as a result of new thermal maturation data indicating that large portions of these basins may be within the oil and gas window (Evenchick et al. 2002). Prior to this, much of this area, particularly the Bowser basin, was considered to be overmature with respect to oil and in the upper end of the gas window (Hannigan et. al. 1995). These new data suggests the potential for hydrocarbon resources beyond those described in the report by Hannigan et al. (1995).

This suggestion has spurred the B.C. Ministry of Energy, Mines and Petroleum Resources and the Geological Survey of Canada to embark on a five-year program that ran till 2007. Program objectives were to better quantify potential oil and gas resources through the acquisition of new geoscience information. Current fieldwork and examination of surface and subsurface samples show oil staining and confirm the new thermal model generated for these areas (Hayes et al. 2004).

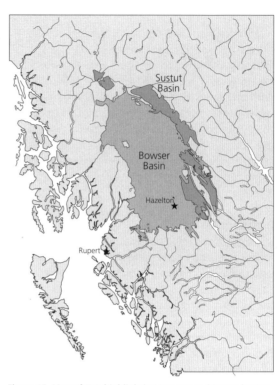

Figure 15. Map of B.C. highlighting Bowser and Sustut basins

The B.C. Ministry of Energy, Mines and Petroleum Resources is presenting areas within the Bowser basin with high-potential coal bed methane (CBM) resources as part of their service plan objectives to increase oil and gas production and stimulate investment in the province. Portions of the Klappan and Groundhog coal fields located in the Skeena headwaters have been leased to Shell Oil to explore for CBM with exploratory drilling that began in 2004. Threats to fish habitat from CBM development and the potential release of saline formation water have not been addressed to date by Shell Oil or the province. Rights to develop coal bed methane near Telkwa were awarded to Albertan energy promoters. The Wet'suwet'en and the local community are providing opposition to the development, which they believe poses unreasonable risk, hazard, and uncertainty to fish, water quality, and aspects of community health.

Natural Environmental Disturbance

In the Skeena watershed, natural environmental disturbance commonly occurs as large floods and changes in streams with unstable channel configuration. Coastal areas are likely to experience intense fall rainstorms that produce large floods. The prevalent steep slopes have many shallow landslides that carry sediment into the valley bottom and often into streams. These two factors typically produce high discharge and coarse sediment dominated systems with rapid changes to channel configuration.

Indirect impacts from logging may result in simplification of channels and conversion of small streams from pool and riffle to plane form. Disturbance of floodplain and riparian stands is associated with dramatic changes in rivers such as the Kitsumkalum and the Zymoetz Rivers. Streams are much quieter on the Nechako Plateau in the interior and do not seem to respond much to logging disturbance. In these areas, large stream alterations occur from extraordinary floods, and then mostly on steeper gradient reaches (Beaudry and Gottesfeld 2001). Streams in the transition zone, such as the Kitwanga and Kispiox Rivers, are intermediate in their responses.

The likelihood of stream channel change is also dependent on the geological setting. Streams with bedrock-bounded channels cannot move and change is limited to variation in the thickness of the gravel bed. Stream stability is also related to the availability of sediment. Watersheds with abundant loose gravel material are likely to have streams with wide and shallow channels easily modified by floods. On the other hand, streams that are inset into glacial lake deposits, a common situation in the eastern portion of the Skeena watershed, tend to have very stable conformations.

External Environmental and Anthropogenic Factors

The Skeena watershed is influenced by environmental changes and changes in human activities outside the watershed. The effects of some of these changes are key watershed issues and priorities. The most important issues are ocean survival of anadromous fish, global climate change, and coastal economic growth programs such as oil and gas development, marine interception fisheries, and fish farms.

Ocean Survival and Productivity

The physical, chemical, and biological state of the northeast Pacific Ocean environment influences the growth, survival, and distribution of Skeena anadromous fish. Understanding ocean processes and the portion of the salmon life cycle in the ocean environment is essential for Skeena salmon conservation. For many years, the potential for salmon production in the ocean was considered unlimited. Observations over the past two decades have clearly demonstrated that the ocean can constrain salmon production and that the capacity of the ocean to produce salmon is not constant over time and locations. Multi-year changes in the state of the ocean contribute directly to variations in salmonid growth, returning spawners, and recruitment.

Beamish et al. (1999) noted that the climate and ocean systems of the North Pacific are dynamic, and in addition to interannual variability, there are evident decadal-scale states. Climate-ocean systems experience these steady states, or regimes, that last for decades and also abruptly shift from one state to another within one year. The shift between the two states of the North Pacific climate is called the Pacific Decadal Oscillation (PDO). Regime shifts occurred in 1925, 1947, 1977, 1989, and late 1998 (DFO 2004). The productivity of British Columbia fisheries appears to respond to these abrupt changes in climate-ocean states.

A general decline in salmon abundance after 1989, as exhibited in fish catch, was evident in key stocks (PFRCC 2001). By the mid to late 1990s, coho abundance in the Skeena, particularly in the Babine system, was so low that unprecedented fishing closures were implemented to protect the at-risk populations, while production of sockeye, pink, and chum salmon was moderately depressed (DFO 2001). Other than coho, Skeena River salmon stocks did not experience the dramatic reductions that were evident in many Central and South Coast stocks in recent years. It has been concluded that there was very poor ocean survival in these stocks, probably in the early ocean life when juvenile salmon are inshore and subject to high predation and adverse habitat conditions (PFRCC 2001).

El Niño events such as that of 1991–1992, 1997–1998, and the somewhat weaker event in 2002–2003 bring warmer water to the B.C. coast. There is a consequent expansion northward of ocean fish such as hake and mackerel, which can be major predators of salmon. There is also a decline in deep water upwelling along the coast, which decreases seawater nutrients and hence plankton productivity. The decline of ocean survival and the reduced growth rate of first-year salmon are correlated with this ocean warming (DFO 2000).

In 2002, Mueter et al. emphasize the importance of regional summer sea surface temperature (SST) in the near-shore areas that are used by salmon for several months after they reach salt water. They posit that southern salmon stocks have decreased recruitment in times of high summer SST, while northern stocks have increased survival rates. The transition between the southern and northern pattern is in southeast Alaska for chum and pink salmon and in northern B.C. for sockeye. The difference in response to higher in-shore temperatures by these three salmon species may reflect differences in their patterns of early dispersal in the North Pacific.

Climate Change

One of the effects of the ongoing global climate warming is an increase in the occurrence and intensity of El Niño conditions that correlate with ocean warming along our coast. The Pacific Decadal Oscillation (PDO) index is another measure of the warming trend in the eastern North Pacific. It has been in a positive phase (warming) since 1978 except for a few brief periods, one of which was 1999–2001 (DFO 2000, 2001). In general, times of positive PDO index show decreased ocean survival of first-year salmon. It seems likely that ocean survival problems in northern B.C. will persist and perhaps become worse with continuing global warming over the next century.

Another correlate of global warming is an increase in drought in the northern interior of B.C. This area includes the portion of the eastern Skeena watershed subbasins on the Nechako Plateau. Climate modeling suggests that late summer drought conditions that result in low streamflows—that will potentially limit fish access in the upper Bulkley and the Babine Lake

area—will persist and, on the average, worsen with time (Boer et al. 2000, Price et al. 2001).

Marine Interception Fisheries

Adverse effects of harvest on salmonids are particularly difficult to control in mixed-stock fisheries where multiple species, stocks, and age classes are harvested together. Mixed-stock fisheries occur before stocks segregate into discrete spawning runs. Strong and weak stocks are harvested at comparable rates, as are wild and hatchery or enhanced salmon. In mixed-stock fisheries, there is strong pressure to maximize the catch of abundant productive stocks, with the result that weak stocks are overfished and decline.

With naturally small populations and populations that have been depressed by human activities, escapement through the mixed-stock fishery may be insufficient to maintain genetic diversity and offset the probability of undesirable declines. In addition to reducing the total escapement of salmonids, harvesting alters the population age and size structure. Ricker (1981) showed that the mean sizes of Pacific salmon harvested in B.C. waters has decreased over the past 30 to 60 years, which Ricker attributed in part to cumulative genetic effects caused by selective removal of larger individuals in troll and gillnet fisheries.

Fish returning to the Skeena River are subjected to a number of sequential mixed-stock fisheries. These include commercial seine, troll, and gillnet fisheries in Alaska and B.C. Since the modification of the Pacific Salmon Treaty in 1999, catches of B.C.-bound salmon have decreased in Alaska. However, coho and chinook interception remains a problem, especially since Skeena coho are caught as incidental catch in the very large Alaska coho fishery.

Mixed-stock fisheries in B.C. include seine and gillnet fisheries in Areas 1, 3, 4, and 5. Reduction in exploitation rates has been achieved for coho, chinook, steelhead, and early-run sockeye stocks. Still, overall exploitation rates remain high for chinook, steelhead, and non-Babine sockeye, and some stocks remain well below their productive capacity. There is a need for establishment of management goals targeting realistic exploitation rates that take conservation of wild stocks into consideration.

Recent development of selective fishing methods to harvest Babine sockeye (the primary Skeena target stocks) without harvesting nontarget steelhead and coho has shown modest success. While progress has been made in selective seine fisheries, there had been only limited results in selective gillnet fisheries until 2001, when mortality rates of released coho were significantly reduced from about 60% to well below 40% (PFRCC 2001). In the Skeena watershed, opportunities are now being pursued to harvest sockeye in up-river terminal fisheries that avoid most of the mixed-stock problem.

Fish Farms

In early 2003, the B.C. government proposed to double the number of existing net cage aquaculture facilities commonly known as salmon farms. This involves expanding salmon farm operations to the North Coast. Pan Fish (now Marine Harvest) has proposed at least 10 salmon farms for the southern, outer portion of the Skeena estuary. At the time of publication, two of the salmon farms have received government approvals and a third has been approved by the DFO but not the province. Consultation with First Nations is on-going. Other sites are at much earlier stages of processing. The impacts of these aquaculture facilities will vary depending on the local site characteristics, but it appears difficult to find sites that have adequate currents to disperse fish farm waste and to avoid negative interactions with passing wild salmon stocks.

Ecosystem baseline studies are minimal, yet the province and federal regulatory agencies streamlined the approval processes in 2003. Initial studies conducted by the Skeena Fisheries Commission in 2004 and 2005 investigated natural background levels of sea lice on wild salmon fry and smolts in the Skeena River estuarine waters of Chatham Sound and Ogden Channel. Results show that sea lice distribution is not geographically uniform and occur in relatively high-density level zones along the outer margins of the Skeena and Nass estuaries. These high-prevalence zones feature high salinity and long residence times by smolts

(Gottesfeld et al. 2004, 2005, 2006). Most of the infection of wild pink and chum salmon happens in these zones, which are also areas where returning adult chinook and coho linger.

Concerns about distant effects of fish farms revolve around two issues: disease transmission to wild fish populations and escapees of farmed salmon. Existing B.C. fish farms have had many outbreaks of sea lice *(Lepeophtheirus salmonis)* and diseases such as IHN (infectious hematopoietic necrosis, Saksida 2003), furunculosis *(Aeromonas salmonicida)*, and IPN (infectious pancreatic necrosis) as described by Ellis (1996).

IHN is an especially serious concern considering existing problems with this disease in Babine Lake sockeye and other sockeye stocks. The epidemiology of many of these diseases is poorly understood, but there are well-documented instances of transmission from one fish farm to farms several kilometres away. Laboratory and field studies suggest that diseases can pass to and from salmon farms and wild sockeye, chinook, and other native fishes. Fish farms may serve as a reservoir of some of these diseases. Although transmission patterns are complex, in general the more infected fish there are on the coast, the higher the likelihood is of the disease spreading to wild fish (Tully and Whelan 1993, Heuch and Mo 2001).

There have been severe outbreaks of salmon lice in the Broughton Archipelago east of Queen Charlotte Strait and Johnston Strait since 2002. This area has the highest density of salmon farms in B.C. and very high wild pink salmon populations. It is clear that these salmon farms are the source of large numbers of sea lice (Krkošek et al. 2005, Orr 2007), that the sea lice infect wild salmon (Morton et al. 2005), and that pink salmon die from these lice if they are infested at an early age (Morton and Routledge 2005, Krkošek et al. 2006). There are several strategies and treatments available to contain sea lice outbreaks and the next few years will hopefully see effective control of this disease problem.

The fish farm experience in Europe serves as a warning. Sea lice outbreaks originating from salmon farms have become so common that many populations of sea trout *(Salmo trutta)*, an anadromous form of brown trout, are in serious danger of extermination and no longer available for sports fishing. *Gyrodactylus salaris,* a parasitic flatworm, was spread by transfer of fish to and among salmon farms and has nearly wiped out many wild Atlantic salmon stocks in Norway. Efforts to control the spread of this parasite included poisoning all fish in streams with infested fish. Furunculosis, a virulent bacterial disease, spread from fish farms to many natural streams (Johnsen and Jensen 1994).

It is likely that most or all of the proposed salmon farms will raise Atlantic salmon. In southern B.C., hundreds of thousands of Atlantic salmon have escaped from salmon farms. McKinnell and Thomson (1997) showed that they have successfully matured and entered at least 29 streams on the South Coast with spawning taking place in some of these streams. First- and second-year parr are present in at least one stream (Volpe et al. 2000). Presumably, these fish could spread into other streams where they will compete with native species such as steelhead. The spread of an invading species is facilitated if native populations are not filling habitat to capacity, but the potential effects of the introduction of a new salmonid species to B.C. are unknown.

In 2007, a B.C. legislative committee recommended a ban on fish farming in all coastal waters north of Vancouver Island. In March 2008, the provincial government announced that they will not allow any fish farm applications or issue any licenses for coastal waters north of Klemtu, which is located north of Bella Bella. Most of the salmon farm operations in B.C. are concentrated on northern and western Vancouver Island and the Broughton Archipelago. These salmon farms remain controversial largely because of the potential impact on wild salmon from amplified amounts of parasitic sea lice found in salmon farms.

Cumulative Effects

The decline of various salmon stocks in the watershed has resulted from the cumulative effects of water and land-use practices, fish harvest management strategies, misapplied enhancement practices, and natural fluctuations in environmental conditions. Because of the longitudinal nature of river and stream ecosystems, the accrual of effects is significant along both spatial and temporal dimensions. Activities that take place in headwater streams influence habitats in downstream reaches, and may affect the response of ecosystem components to additional stresses. Similarly, activities that have occurred in the past may influence current habitat conditions through residual effects on landscape vegetation and aquatic biota.

Accumulation of localized or small impacts can result in cumulative watershed level changes to fisheries. Accumulations of effects, often from unrelated human activities, pose a serious threat to fisheries (Burns 1991). The effects of increased sedimentation on spawning gravels will be the same, whether the sediment resulted from livestock grazing, logging, road building, mining, or a pipeline crossing. The same is true of other variables such as water temperature, dissolved oxygen concentrations, channel morphology, and quantity and distribution of instream cover (Remington 1996). Loss of habitat elements such as large woody debris can have effects lasting from 80 to 160 years (Sedell and Swanson 1984). Cumulative losses of one element of fish habitat may result in long-term compound problems.

Within the context of conserving and restoring fish populations and their habitats in the Skeena watershed, the concept of cumulative effects is important. Individual actions that are by themselves relatively minor may be damaging when coupled with other actions that have occurred or may occur elsewhere in Skeena watershed subbasins. Historical and current patterns of land-use activities and practices, particularly forest development, have a significant bearing on how salmonid populations will respond to further anthropogenic disturbances. Within the Skeena watershed, management strategies that rely on site-specific analysis without regard for other activities that have occurred or may occur within the watershed will generally fail to protect salmonid populations against cumulative effects. This concept underlies the development of watershed and ecosystem approaches to land and resource management (Spence et al. 1996).

Secondly, declines in the Skeena watershed salmonid populations are the product of numerous incremental changes in the environment and fish populations. Recovery and conservation of salmonid populations may proceed in a similar way – through incremental improvements in habitat conditions, the use of alternative management strategies in relation to fish harvesting, and viewing the watershed as connected in regard to land-use activities and practices. This means that individuals can and must play an active role in salmonid conservation and restoration even if tangible efforts are slow to manifest.

To define cumulative impacts to fish habitats in the Skeena watershed, short-term and long-term datasets that measure water quality are critical. Few adequate datasets exist within the watershed, except perhaps for studies conducted in Babine Lake. Remington's (1996) review of water quality identifies datasets critical to the assessment of long-term cumulative impacts in the Skeena watershed. Environmental parameters not usually included in water quality sampling are also important to stream health. These include measures of the integrity of riparian areas, measures of snowmelt rates in nonforested areas (hydrological recovery after logging or other land disturbance), measures of runoff rates in urban and industrial zones, measures of stream channel form and rates of change, and monitoring of climatic changes.

Information constraints prevent many parameters from being articulated that define cumulative impacts to the tributary watersheds and the Skeena watershed as a whole. The lack of detailed information about fish populations and their habitats, currently and in the past, limits knowledge that relates to conservation goals. The first and most critical step in resolving land and resource use problems is recognizing that problems exist; then their nature and magnitude can be defined. Programs that address this lack of basic information concerning the status of fish populations and habitats will facilitate strategic and operational conservation solutions.

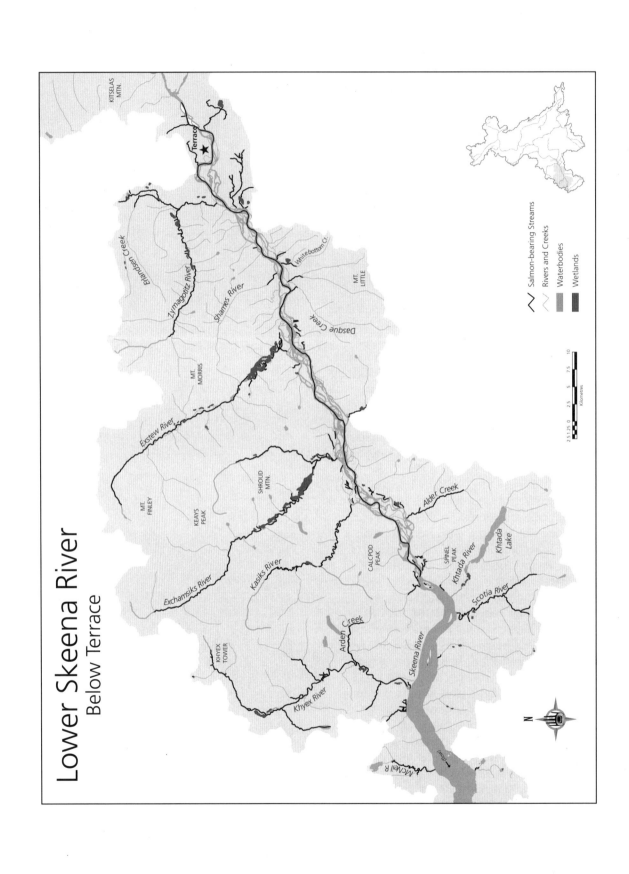

Lower Skeena River
Below Terrace

KITSELAS MTN.

★ Terrace

Erlandsen Creek

Zymagotiz River

Shames River

Whitebottom Cr.

MT. LITTLE

Dasque Creek

MT. MORRIS

Exstew River

MT. FINLEY

KEAYS PEAK

SHROUD MTN.

Alder Creek

Exchamsiks River

Kasiks River

CALCPOD PEAK

SPINEL PEAK

Khtada River

Khtada Lake

Scotia River

Arden Creek

KHYEX TOWER

Khyex River

Skeena River

McNeil R

N

Salmon-bearing Streams
Rivers and Creeks
Waterbodies
Wetlands

2.5 1.25 0   2.5   5   7.5   10
Kilometres

5
Lower Skeena River

Environmental Setting

The lower Skeena River area is defined here as encompassing the portion of river and tributaries between the mouth and the Kitsumkalum River. This section includes the mainstem and all tributaries other than the Ecstall, Gitnadoix, Lakelse, and Kitsumkalum watersheds, which are discussed in following sections.

Location

The lower Skeena River is located in northwest British Columbia and extends from the mouth at 54° N, just south of the port city of Prince Rupert, upstream 116 km to Terrace. This section of river is bounded to the north and the south by the Kitimat Ranges, to the east by the Hazelton Mountains, and to the west by Chatham Sound.

Climate

The lower Skeena River has a modified maritime climate that is controlled by the large-scale, mid-latitude weather frontal pattern and by the topography, particularly the Coast Mountains. There is considerable variation of temperature and precipitation from the Skeena River mouth eastward to Terrace. On the coast, there is abundant precipitation with cool summers and mild winters with average temperatures near 0° C. Usually, precipitation reaches a maximum in the fall and early winter, generally in October and November (Environment Canada 1993) when intense cyclonic storms from the North Pacific move across the coast on an almost daily basis.

The Coast Mountains, which trend northwesterly and parallel the coastline, cause uplift of the moisture-laden winds and augment the precipitation. The Prince Rupert Airport weather station, located in the lower Skeena estuary, receives 2,450 mm of total precipitation per year, and significantly higher amounts occur in the mountains. The meteorological station at Prince Rupert Regional Park, which is at the base of Mt. Hays at 91 m elevation, receives over 3,000 mm of precipitation. At Terrace Airport, the average total annual precipitation is 1,322 mm, about half that at the river mouth. The coastal portions and higher elevations of the lower Skeena drainage lack a dry season. Tributary valleys receive between 2,000 and 4,250 mm annual precipitation, of which approximately 80% occurs between October and April.

Hydrology

The hydrology of the lower Skeena River is complex due to three main factors: the tidal regime that exerts its influence from the mouth approximately 55 km upstream to the Kasiks River; the hydrological characteristics exerted by seasonal flows that are in part dependent on delayed runoff from precipitation stored in the form of snow and ice throughout the basin; and a climatological regime that is characterized by substantial winter precipitation.

These three factors combine to provide a unique hydrological environment in the lower Skeena River and its estuary. This hydrological situation influences the nontidal surface currents and salinity in Chatham Sound, the eastern portion of Dixon Entrance, and southern Clarence Strait, particularly in the late spring and early summer (Thomson 1981). The Skeena River estuary is defined by Pritchard (1967) as: "a semi-enclosed body of water which has a free connection with the open sea and the sea water within is measurably diluted with fresh water from land drainage." At peak nival flood flows, areas such as southern Chatham Sound, Telegraph Passage, and Ogden Channel may have salinities less than one half of undiluted sea water.

The upstream effects of tidal influences and the salinity regime for the Skeena River are not well documented. The tidal effects within the Skeena River estuary decrease from the mouth upstream to the Kasiks River, with tidal flooding in the lower reaches of the Ecstall, Khyex, Khtada, Scotia, and Kwinitsa Rivers, as well as several smaller tributary streams. Salt wedges – dense sea water under lighter freshwater – have been documented off the Ecstall River, though overall, data are generally limited (Trites 1956).

The Skeena River discharge in the estuary complicates the highly dynamic tidal regime. Skeena River freshwater introduces a seaward current component into the waters of the estuary. However, the effects of the run-off are largely masked by the strong tidal currents, other than at times of high snowmelt-related discharges in the spring and early summer (Hoos 1976). Tidal currents average three knots, though four-knot currents have been measured in deeper channels of Telegraph and Marcus Passages. When flood tides move into the Skeena River and encounter strong river discharge, violent riptides can result, particularly during the larger tidal ranges. The rip-tide effect also occurs when the wind blows in an opposite direction from the predominant flow of water.

The tidal range at the mouth of the Skeena River is among the largest on the B.C. coast and is semidiurnal; that is, it has two low and two high tides daily with a large difference between subsequent lows and highs. The mouth of the Skeena is defined for this discussion as Parry Point at the south end of DeHorsey Island to Hegan Point. Coastal charts show that the peak ebb-tide current seldom flows in the same place as the flood current. This mostly reflects the topographic features and their effects and is pronounced where short channels connect tidal course passages. Strong tidal currents contribute

to the relatively small amounts of Skeena River sediment deposition in many of the estuarine channels.

Wave action is dampened by Porcher and Stephens Islands, which form the western perimeter of Chatham Sound, and further by the islands, shoals, and banks in front of the Skeena River mouth. Shoals, submerged most of the time, allow varying amplified wave heights. River discharge may undergo significant year-to-year changes related to large-scale climatological factors, which in turn affect salinity, turbidity, and sediment deposition in the estuary.

Determining the size and seasonal variation in the streamflow on the lower Skeena River is difficult. The furthest downstream hydrologic station with a long record (08EF001) is located at Usk, 18 km northeast of Terrace. The Skeena River above Usk has a drainage area of 42,200 km^2 (Remington 1996). The typical annual hydrograph shows an extended peak flow in May and June, and one or more smaller storm flows in the fall. The size of the peak floods for each year is shown in Figure 16. Nearly all of the annual extreme floods occurred in the spring snowmelt season. The peak flows occurred between May 14 and July 2, except in 1974, 1978, and 1991 when fall rainstorms in October or early November produced the highest floods.

Considerable runoff is added to the Skeena River downstream of the Kitsumkalum River from large tributary streams, particularly the Exstew, Exchamsiks, Ecstall, Gitnadoix, and the Zymagotitz. Hydrometric stations have been operated in the past on the Exchamsiks River (station 08G012), the Zymagotitz River (station 08G011), and the Khtada River (station 08EG003). Peak flows on lower Skeena tributaries usually occur during the autumn

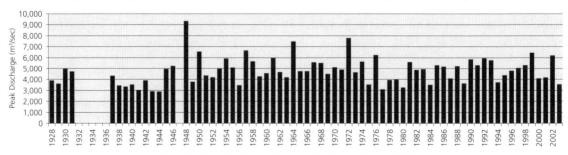

Figure 16. Annual extreme floods on the Skeena River at Usk

or early winter. In the 38 years of record (1962–1999) of the Exchamsiks River, all but one of the record annual floods (1970) were rainstorm flows. A typical annual hydrograph is shown in Figure 1.

Local climate conditions determine the hydrology of the tributary drainages and streams of the lower Skeena. Climatic characteristics such as amount of snowfall, seasonal distribution of rainfall, soil type, and topography are the primary factors influencing these systems. Water retention in terms of water storage in lakes and wetlands is small; Khtada and Gitnadoix Lakes are the only significant lakes.

Climate change could lead to potential streamflow changes as seasonal amounts and timing of precipitation in the drainage undergo change. Presently, due to a lack of baseline weather and streamflow data, there is no clear historical benchmark that could be useful in relation to the potential web of consequences stemming from climate change.

Stream Channels

The estuary formed at the mouth of the Skeena River is unusual for a large river on the Pacific, as it does not occur at the head of an inlet or open into a large body of water. Instead, the river discharges through a series of channels which separate several islands, including Smith, Kennedy, DeHorsey, and Porcher (Hoos 1976). Inverness Passage separates Smith and DeHorsey Islands from the mainland to the north; Marcus Passage separates Smith and DeHorsey Islands from Kennedy Island; and Telegraph Passage separates Kennedy Island from the mainland in the south.

There are no extensive or continuous tidal flats forming a distinct Skeena River delta. Rather, there is a series of pockets of shallow water where sediments have built up to form shoals and sandbanks (Hoos 1976). Major deltaic deposits include Robertson Banks, Flora Bank, the Base Sand, and Davies Bank.

The lower Skeena River mainstem is divided into two distinct reaches (Resource Analysis Branch 1975–79). Reach 1 passes from the mouth upstream to Kwinitsa Creek, with a regular or continuous profile and an average gradient of 0.0 to 0.1%. Channel movement

is limited by the valley walls near the edge of the floodplain. Upstream of Veitch Point, there are extensive banks including Skeena Banks, Raspberry Bank between Raspberry Islands and Raspberry Bluff, and the shoal between Raspberry Bluff and Windsor Point.

On the right bank of the Skeena mainstem, major shoals and banks include Tyee Bank and Khyex Bank. Mid-channel banks include Skeena Banks north of Hotspring Point, the large Ayton Bank seaward of Ayton Island, Walsh Bank that is northwest of Telegraph Point, and Carnation Bank, located upstream and downstream of Carnation Island. These floodplain and deltaic deposits are composed of silts and fine sands, except for some mid-channel deposits that are relatively coarser with scattered boulders. The channels between Ridley Island and Gibson Island are floored by medium to coarse sand, which accumulates and forms extensive bars that limit navigation (Anonymous 1977).

Reach 2 extends from above Kwinitsa Creek to just above Ferry Island adjacent to Terrace. This reach has an average gradient of 0.05%, or 0.4 to 0.5 m/km, and is bounded by the valley walls. This reach is an immense fluvial deposition zone. Numerous divided channels, backwaters, and swampy areas resulting from shifting islands and gravel bars characterize the lower Skeena floodplain, which averages 2.5 km in width. In this section of the Skeena River, islands range in size up to 300 ha and continuously undergo substantial change; this is the most dynamic section of the entire river. Hogan and Schwab (1989) reviewed the historical sequence of air photos from 1937 to 1988, which indicate that there has been significant change in stream channel morphometry over the last 50 years.

The Skeena River channel is characterized as a wandering gravel bed river with a straight and sinuous channel pattern. There are two or more main flow branches separated by islands (Hogan and Schwab 1989). Most of the river channels have complex histories of abandonment, filling, re-excavation, and reuse. The majority of sloughs, common between the Kasiks River and the Shames River, were created as a result of the road and rail routes blocking channels. The river slope decreases uniformly from Terrace to the estuary, as does the size of gravel in the river deposits. Gravels at Terrace are cobbles about 10 cm in diameter, while most sediment below Kwinitsa Creek is sand-sized.

Unique and conspicuous features of this river section are the large logjams, which range from a few tree lengths to logs piled massively and continuously for up to a kilometre (Bustard 2002). The larger logjams exceed 20,000 m³ in volume. There is no other B.C. river entering the coast with such a large extent of logjams and log piles. These log piles are important structural features that promote sediment deposition when located on or adjacent to a mid-channel bar; if positioned at the mouth of a back channel, they stabilize and train the river channel. Unlike logjams elsewhere in the Skeena watershed, these large logjams provide juvenile habitat at all river stages. Side channels regulated by logjams do not get scoured and provide important habitat for coho and chinook juveniles. A singular feature of these side and back channels are the very large amounts of stonefly larvae; these provide high-quality salmon food and are indicators of a healthy stream (Drewes 2002).

The Grand Trunk Pacific Railway (GTP) was constructed between 1910 and 1914 on the north bank of the Skeena River. Construction practices (riprapping) effectively reduced the lateral movement of the river in some areas and blocked off some of the floodplain and many side and back channels. Recent railroad upgrading and highway construction has further restricted river movement and prevents the rejuvenation of channels behind the transportation corridor. Coho use of these back channels is less than that of similar nearby areas that are more accessible to fish (Bustard 1991). Williams and Gordon (2001) reported on stream crossings along the CN Rail that may be obstructing fish passage or are problem stream crossings. Rabnett (2006b) reported on obstructed fish passage along Highway 16 and the CN Rail. The results indicate back-channel habitat east of the Exchamsiks River could be easily restored.

Major tributaries entering the lower Skeena River from the mouth upstream include the Ecstall, Khyex, Scotia, Kasiks, Exchamsiks,

Gitnadoix, Exstew, Lakelse, Zymagotitz, and Kitsumkalum Rivers. The majority of these single channel tributaries have been carved by glacial action into U-shaped valleys with steep sidewalls and floors in-filled with sediment.

Khyex River

The Khyex River flows into the right bank of the Skeena mainstem, approximately 25 km upstream of the mouth. Reach 1 of the Khyex River mainstem channel is characterized by a regular, sinuous pattern with an average gradient of 0.0% and a tidal influence that ranges at least 10 km upstream from the mouth. The floodplain ranges from 0.5 to 1.2 km in width and is mostly composed of alluvial sand and gravel with a widespread organic veneer. The lower 6 km of the mainstem channel substrate consists of fine silts. Extensive wetlands are prevalent. Reach 2, which starts at 10 km, shows a regular, sinuous, longitudinal profile that lies in a bounded floodplain 0.3 to 1 km wide (Resource Analysis Branch 1975–79). Arden Creek, a main tributary entering the left bank, has an average gradient of 2.5% and is essentially bounded by its valley walls.

On November 28, 2003, a low gradient landslide severed the Pacific Northern Gas (PNG) natural gas pipeline. The landslide is located on the west bank of Khyex River 6.8 km upstream from its confluence with the Skeena River. A steep main scarp characterizes the landslide that consists of glacio-marine sediments mantled by rubbly colluvium on and against smooth bedrock of the valley wall. The landslide covers an area of 32 ha and displaced about 4.7 million m³ of material. Displaced material flowed up and down river over a distance of 1.7 km, blocked the river and caused flooding upstream for a distance of 10 km. The landslide is classified as an extremely rapid, retrogressive liquefaction earth flow or a clay flow-slide and is similar to others in the region, such as the Lakelse slides in 1962 (Schwab et al. 2004).

Reach 3 of the Khyex River is characterized by a 2 km long canyon with falls and chutes, which acts as a barrier to anadromous fish passage. Reach 4 is regular in profile with an average gradient of 0.8% and is unbounded in an aggraded floodplain. The PNG natural gas pipeline and right-of-way follows the floodplain, making eight crossings across the lower Khyex and Arden Creek in order to utilize the drier and more stable portions of the floodplain. Khyex River is navigable to the 26 km cascades (Eastwood 1981).

Scotia River

The Scotia River is a left-bank tributary of the Skeena River approximately 36.5 km upstream of the mouth. The lower mountainsides throughout much of the watershed have been logged; the mainstem and the west fork are aggraded with lateral channel movement and sediment wedges (Jrykkanen 1997). Reach 1 is 3.4 km in length with an average gradient of 0.1%. The reach is comprised of two oxbows, and a strong tidal influence dominates with sedge marsh and mud flats in the riparian fringes (Jrykkanen 1997). Elevated mid-channel bars are prevalent with fines mostly making up the substrate. The floodplain is broad and the riparian habitat is mostly mature forest.

Reach 2 extends upstream 8 km to the west fork. A single, irregularly meandering channel characterizes the reach, which has an average gradient of 0.3% and a considerable number of elevated mid-channel bars. The relatively broad floodplain ranges between 500 m and 750 m in width with limited side- and off-channel habitat.

Reach 3 extends 5.4 km upstream to a waterfall that limits anadromous passage. The reach is typified by continuous meandering with an average gradient of 0.9%. The channel is bounded by a 100–170 m wide floodplain with a mostly gravel substrate. The west fork of Scotia River occupies a U-shaped valley possessing high-value fish habitat. The reach upstream to the forks is slightly meandering with an average gradient of less than 1%. Fines and small gravel dominate the substrate.

Kasiks River

The Kasiks River flows into the Skeena River right bank approximately 55 km upstream of the mouth. Reach 1 of the Kasiks River is characterized by a regular sinuous profile with an average channel gradient of near 0%. The highest Skeena tides influence the complete 9.5 km reach length. The floodplain averages 750

Upper portion of Skeena estuary

Lower Skeena islands and channels, with a view west to the Exstew River

m in width and is composed predominantly of sand, with minor amounts of gravel (Resource Analysis Branch 1975–79) overlain with organic deposits. Shallow side channels are typical, and high tides influence the channel and river flows.

The Kasiks back channel, 7.4 km in length with wetted widths exceeding 100 m and water depths to 5 m, enters the Kasiks left bank 2.25 km upstream of the Skeena confluence. This channel is a relic Skeena River channel cut off from the Skeena River flows by railroad and highway development at its upstream end (Bustard 1984d). Most of this channel is typified by short sections of flow between large ponded areas.

Reach 2 of the Kasiks River, approximately 6.5 km long, presents a regular sinuous profile and has an average gradient of 0.1% (Resource Analysis Branch 1975–79). The channel has well-developed bars and a mostly sandy substrate. Its floodplain averages 600 m in width. Gravel pockets occur at tributary mouths and adjacent to active avalanche chutes. The nearly vertical valley walls rise to over 1,200 m. Reach 3 is a 1 km long section with falls and cascades. Reach 4 occupies a narrow floodplain with a regular channel profile and an average gradient of 0.4%. The reach is aggraded with significant amounts of sediment and debris reflecting drainage from adjacent glaciers and avalanche deposition.

The PNG natural gas pipeline passes up the Kasiks valley, then west up Huckleberry Creek to enter the Khyex drainage (Bustard 1994c). Once upstream of their fans, most tributaries are too steep to be fish bearing. The Kasiks River is navigable for 18.5 km (Eastwood 1981).

Exchamsiks River
The Exchamsiks River flows into the Skeena River right bank approximately 63 km upstream of the Skeena River mouth and 53 km downstream of Terrace. The river trends southeasterly for some 48 km, draining steep mountainsides and many small glaciers. Reach 1 of Exchamsiks River is 4 km in length and is characterized by a regular, sinuous channel profile. The average gradient is 0.05%. The channel meanders through a sand and gravel floodplain, overlain with organics (Clague

1983). The floodplain width averages 700 m and is controlled by the valley walls. Tributaries include a large left-bank stream approximately 25 km in length that enters at 4 km.

Reach 2 is 23.6 km in length with an average gradient of 0.1%. The channel shows unbounded, irregular, and tortuous meanders amid extensive off-channel wetlands and scattered relic channels, on a floodplain that ranges in width from 200 m to 1,400 m. The streambed substrate is primarily sand with small amounts of gravel (Resource Analysis Branch 1975–79). Reach 3 is approximately 3 km in length with an average gradient of 0.4% and is confined by the steep valley walls. Reach 4 is 12.7 km in length with an average gradient of 0.8%. The channel is braided and aggrading across a narrow floodplain. It receives sediment from adjacent glaciers and debris from avalanches that deposit directly in the channel.

Exstew River

The Exstew River flows into the Skeena River right bank 36 km downstream of Terrace. Reach 1 is 14.5 km in length with an average gradient of 0.1%. The channel is a stable riffle-pool type with gravel and fine substrates dominant. The reach is occasionally confined and characterized by tortuous meanders that wander over the 900–1,300 m wide floodplain (Pollard and Buchanan 2000a). Considerable off-channel wetlands, oxbow lakes, and low gradient back channels occur on the floodplain. The main forestry access road follows the east bank of the mainstem for 8 km before crossing the river, then follows the west bank for 6 km. Other than the right bank stream, which flows into the Exstew River at 13 km, the tributaries are relatively steep once upstream of the floodplain.

Reach 2 is 18 km in length with an average gradient of 0.7%. Initially the reach is a confined single channel, which then opens up into a highly braided, aggrading channel. The floodplain ranges from 500 m to 700 m in width and is controlled by the valley walls. Reach 3 is predominantly a single channel, stepped, cascade-riffle type, approximately 10 km in length and headed by a glacial lake. The substrate is dominated by boulders and coarse sediment reflecting the glacial activity and avalanche deposition into the channel. High-velocity cascades at the lower end of reach 3 likely limit fish passage.

Shames River

Shames River is a Skeena River right-bank tributary approximately 18.2 km in length, 23 km downstream of Terrace. Reach 1 is a 1.8 km long active alluvial fan that has an average gradient of 1.5% (Resource Analysis Branch 1975–79). Channels are unstable except where trained by groins upstream of the Highway 16 and the CN Rail bridge crossings. Abandoned channels are evident upstream of the channelized segment where the riparian forests in reach 1 have been logged (Pollard and Buchanan 2000b).

Reach 2 is 4.9 km in length and is a single channel confined by steep banks and side slopes that occasionally fail as the toe is eroded. The average gradient is 4% with relatively high velocity flows through reach 2 cascades. The channel morphology is an aggrading cascade-pool type with a boulder substrate (Pollard and Buchanan 2000b). The main road accessing forestry and recreational skiing activities crosses the Shames River three times within reach 2.

Reach 3 is 3.5 km in length with a mix of single and braided channels, mid-channel bars, and islands. The average gradient is 3.2%, with side and off channels as well as intermittent wetlands occurring on the approximately 100 m wide floodplain. Channel morphology is an aggrading riffle-pool type with a gravel and boulder substrate (Pollard and Buchanan 2000b). Riparian forests along reach 3 have been extensively harvested with only 20% retention.

Khyex slide, November 2003

Reach 4 is approximately 6 km in length with an average gradient of 5.5%, in a relatively uniform, single-channel profile. The channel is a cascade-pool type with a predominantly boulder substrate.

Zymagotitz River

The Zymagotitz River, locally known as the Zymacord River, is a Skeena River right-bank tributary. The Zymagotitz River is approximately 32 km in length and flows east and south to enter the Skeena at New Remo, 10 km west of Terrace. The watershed encompasses approximately 600 km², and includes two primary tributaries, Erlandsen and Molybdenum Creeks, 21 km and 10 km in length, respectively. The Zymagotitz valley is typical of many lower Skeena tributaries: it is U-shaped as a result of glacial carving and gouging with a valley bottom fill of sediment beginning with thick glacio-marine deposits. The valley is characterized by a relatively broad floodplain dominated by nonforested wetlands, steep valley walls, and numerous alluvial fans created by the steep tributary streams (Gordon 1999).

Reach 1 of the Zymagotitz River passes from the Skeena River confluence upstream 7 km to the confluence of Erlandsen Creek. The reach is stable and slightly meandering with an average gradient of 0.4%. The stream channel is confined on the west by the valley wall and is unbounded to the east. Low-gradient tributary channels have been heavily altered by logging activity, and two side channels were cut off from the Skeena River when the railway was constructed.

Reach 2 is 2.3 km in length with an average gradient of 0.2%. A bouldery cascade is located 500 m upstream from the Erlandsen Creek confluence and is passable to fish. Two major slumps (in 1985 and 1989), located at 8.2 km on the left bank, were triggered by a combination of natural and logging-related causes (Golder 1990, Gordon 1999). The combined failure has been a chronic source of fine sediment to the lower reaches. The bridge that crossed the mainstem at 8.3 km has been removed.

Reach 3 is 9.2 km in length, and is characterized by a continuous meandering profile with an average gradient of 0.4%. The stream is stable, with a gravel substrate, though in low-gradient sections fine glacial sediments occur. Extensive off-channel habitat is found on both sides of the floodplain, which ranges between 300 and 800 m in width. Gordon (1999) noted that the off-channel habitats are likely critical overwintering areas for juvenile coho. Reaches 4, 5, 6, and 7 of the Zymagotitz River extend from 18.5 km to 27 km and have average gradients from 1.2% to 1.8%. These reaches consist of rapids and multiple braided channels, and offer good main-channel, side-channel, and off-channel habitat (Gordon 1999). Fish use above reach 7 is unknown and doubtful.

Reach 1 of Erlandsen Creek extends 3 km upstream to a rock waterfall that is an anadromous fish barrier. The reach is characterized by an aggrading, slightly sinuous and/or braided channel with an average gradient of 0.9% (Resource Analysis Branch 1975–79). Off-channel wetland habitat on the right bank is relatively moderate in size but of importance.

Water Quality

The lower Skeena River mainstem and estuary waters are very turbid, particularly during spring snowmelt and autumn rains. This is due to glacial flour and silt picked up by the river and its tributaries from the headwaters to the sea. Long-term data critical to the review and understanding of water quality are essential to detect changes or trends in water quality. Unfortunately, there are no long-term monitoring stations within the lower Skeena drainage.

Since 1953, water quality data, particularly dissolved oxygen levels and water temperature, have been studied in the estuary and the waterways adjacent to the pulp mill on Watson Island (Stokes 1953). Waldichuk (1962) and Goyette et al. (1970) showed that average dissolved oxygen values had deteriorated and were lower than the necessary 5 mg/l. Since 1970, there have been approximately 45 studies of water quality in the estuary, particularly Porpoise Harbour, Porpoise Channel, Flora Bank, Inverness Passage, and Marcus Passage (Williams 1991). The majority of these studies have been conducted in relation to the pollution from the pulp mills on Watson Island.

Geography

The Skeena River and the major tributaries cut through the heavily glaciated Kitimat Ranges of the Coast Mountains in distinct U-shaped valleys. The valley walls are composed of steep bedrock, which is locally covered with veneers of colluvium and till, with interspersed tributary stream alluvial fans and debris cones (Clague 1984). The bedrock is granodiorite and high grade metamorphic rocks.

During the retreat of Ice Age glaciers, sea levels underwent a complex series of fluctuations. For a brief period 10,000 to 11,000 years ago, the Skeena valley and low-elevation, tributary stream valleys were fiords. They were subsequently isostatically uplifted and partially filled with fluvial and marine sediments (Gottesfeld 1985). Slopes have colluvial and morainal blankets whose surface expression conforms to the underlying bedrock structure. Glacio-fluvial mantles and glacio-fluvial fans are the dominant surficial features of the valley bottoms (Clague 1984). The steep-sided former main fiord has been filled with hundreds of metres of sediment carried by the river since deglaciation, 11,000 to 10,300 years ago. Skeena River sediment deposition has formed a wide low-lying valley flat containing many islands and gravel bars, sizeable logjams, and extensive back channels and wetland areas.

The distinctly rugged landforms of the major tributary valleys are mostly northwest trending fault lineaments, typical of northwest British Columbia. The mountains are bold, impressive, massive, and largely comprised of granodiorites of the early Tertiary Coast Range Intrusives, emplaced into older highly metamorphosed suites. For the most part, the mountains are below 1,800 m in elevation. The heavily glaciated north and northeastern sides of the peaks have remnant small glaciers and ice fields (Holland 1976). Along the length of the steep slopes, water cascades over falls and down rock faces that are often swept clean of vegetation.

Forests

Other than the mountainous high country dominated by rock and alpine tundra, the majority of the landscape is covered with dense, coniferous forests. There are considerable

Mouth of the Exstew River

deciduous seral stands in the lower valley and on the floodplain, particularly on the Skeena islands between Kwinitsa Creek and Amsbury Creek. Forests are characterized by two Biogeoclimatic Ecosystem Classification (BEC) zones: Coastal Western Hemlock (CWH) zone and Mountain Hemlock (MH) zone.

The CWH zone represents the vegetation in the valley bottoms and on slopes up to 400–600 m elevation. Old-growth conifer stands of western hemlock, western red cedar, Sitka spruce, and amabilis fir dominate the forested landscape. Large spruce, hemlock, and deciduous trees with salmonberry, ferns and devil's club comprising the principal understory vegetation, characterize floodplains and alluvial sites. The CWH zone passes upward into the MH zone, which is dominated by mountain hemlock and amabilis fir with occasional spruce, red cedar, and yellow cedar. A well-developed shrub layer is mostly composed of Alaska blueberry, false azalea, and red and black huckleberries (Banner et al. 1993).

Fish Values

The lower Skeena River and its tributary streams have high fish values; the Skeena River as a whole is the second most important salmon producer in B.C., with annual escapements of over two million fish. Hoos (1976) noted that 122 fish species have been identified in the Skeena River system and estuary.

The Skeena River mainstem and adjacent channels from the Kasiks River upstream to the Shames is called Skeena River west. All species of salmon, as well as steelhead, cutthroat trout, and Dolly Varden char, are present in this section of river for some period of their life histories. Some species, such as sockeye, mostly migrate through the area, while pink and chum salmon spawn here in large numbers. Tributaries support many life stages for anadromous and freshwater species. Adult enumerations tend to be difficult due to turbid water conditions that can persist from May through early winter.

Juvenile fish migrating downstream to saltwater feed and rear along the extensive network of logjams, side channels, and wetlands, as well as the diverse and fertile estuary. Depending on the species, anadromous juveniles spend time ranging from days to months in the estuary, undergoing the physiological adaptation from their natal stream life to sea life. The varied salinity regime in the estuary facilitates this adaptation. Conditions are varying or challenging in other ways, as juveniles also encounter differences in topography, turbidity, sea temperatures, tides and currents, food abundance, and predator populations (Burgner 1991). Eulachon *(Thaleichthys pacificus),* one of four species of smelts known to inhabit the Skeena River, spawn in the river and its lower tributaries in early spring, some years in great abundance.

Despite the high fisheries values and ease of access to the lower Skeena River and its tributaries, fish abundance and use are relatively poorly understood. The basin is productive due to the outstanding spawning and rearing habitat that yields anadromous and resident fish in abundance; however, the fish populations and quality habitat are sensitive to ecosystem disturbance and high exploitation.

Chinook Salmon

Chinook spawning in the lower Skeena River area occurs in the mainstem and adjacent channels: Khyex River, Kasiks River, Exchamsiks River, Exstew River, and Zymacord River. All populations are relatively small and essentially have the same run timing.

Skeena River Mainstem and Adjacent Channels
SEDS data (DFO 2005) for the last 30-year period show an annual mean escapement of 935 chinook with a range of 200 to 3,500 fish for the lower Skeena River mainstem. Chinook salmon spawn primarily from mid-August through September in the lower Skeena River, with chinook ages four and five dominating the escapement (Godfrey 1968). Chinook salmon spawning occurs in the Skeena River mainstem in scattered pockets.

Fisheries studies conducted for BC Hydro (1983) inventory two chinook spawning sites: the north channel gravel bar, west of Hudson Bay Flats, and the main channel gravel bar, south of the mouth of the Shames River. Bustard (1991) reported anecdotal evidence of chinook spawning on the outside edge of Gravel Island during some years. The DFO (1991a) documented chinook spawning on the north bank of the Skeena for approximately 8 km downstream of Shames River confluence and a 500 m patch on the outside of the island south of the confluence. Chinook spawning also occurs upstream of New Remo to the Kitsumkalum River confluence on both mainstem banks.

Juvenile chinook were the predominant juvenile fish captured within the lower Skeena River during two of the three periods sampled by Bustard (1991). During April, chinook juveniles comprised 49% of the fish captured, rising to 64% of the total catch in late summer and declining to 23% during the early winter catch. Tredger (1984) reported that chinook fry dominated the boat shocking catch sampling in the lower Skeena in 1983. Tredger's estimates indicate 522 chinook fry per km of channel in this section of the Skeena during August. Side-channel estimates were about 40% higher than in the main channel.

Bustard (1991) reported finding no chinook juveniles present during the sampling in back channels upstream of Exchamsiks River that are blocked to adult movement by riprap. Shepherd (1978) reported chinook fry catches having between 1.5 and 18 fish per trap during sampling in September 1976. Juvenile chinook in the lower Skeena River are most likely a combination of migrants from upstream spawning locations and progeny from nearby spawning areas.

Information is limited in the Skeena River system in regards to chinook fry utilization of the estuary. Upon reaching the estuary, fry tend to delay movement to the sea, taking up residence in the estuary shallows for the summer

(Hoos 1976). Higgins and Schouwenburg (1973) reported finding chinook fry in their earliest May samples and in their last samples in August. The timing of abundance varied in different estuarine areas, though overall peak numbers occurred in mid-June, coincidental with the Skeena peak discharge.

Khyex River

Khyex River chinook escapement records date back to 1959, with intermittent reports since then. Escapement ranges from a high of 750 chinook in 1960 to 2 in 1992. Decadal escapements declined during the 1970s to an average of 44 spawners, then increased to an average of 180 in the 1980s, followed by 117 in the 1990s. Due to funding constraints escapement surveys have not been conducted since 2000.

Peacock et al. (1997) reported that run timing for Khyex River chinook starts in late July and continues through mid-October with the peak in mid-September. Bustard (1994c) noeted that spawning was finished by late August. The principal spawning grounds are located 0.5 km to 1.5 km downstream of the reach 3 falls (DFO 1991a). Chinook have occasionally been observed in lower Arden Creek, and Bustard (1994c) reported juveniles were trapped approximately 5 km upstream from the confluence.

Kasiks River

Kasiks River chinook escapement records date back to 1960 and are fairly continuous. Escapement ranges from a steady high of 400 to 25 (DFO 2005). Peacock et al. (1997) reported that run timing starts in mid-August and continues through mid-September with peak spawning in late August. The main spawning ground is the 2 km section downstream of Huckleberry Creek with scattered spawning occurring downstream (Bustard 1994c).

Exchamsiks River

Exchamsiks River chinook escapement records date back to 1964 and are fairly consistent to the present. Escapement ranges from a high of 400 in 1998 to many years of 20-odd chinook spawners. Peacock et al. (1997) reported that run timing starts early in August and continues through early September with peak spawning in mid-August. Principal spawning occurs over a 12 km section on reach 2, starting approximately 14 km upstream of the mouth. The DFO (1991b) documented chinook spawning at 30 km in the mainstem. The large left-bank tributary flowing into the Exchamsiks River at 4 km supports chinook spawning in the lower 3 km reach.

Exstew River

Exstew River chinook escapement records are consistent from the early 1970s to the present with spawners ranging from 200 to 25. Peacock et al. (1997) reported that run timing starts in early August and continues through mid-September with peak spawning in late August. The principal chinook spawning ground is situated in the upper section of reach 1.

Zymagotitz River

Zymagotitz River and Erlandsen Creek chinook escapement records date from the mid-1960s up to the present. Zymagotitz River chinook escapement ranges from 0 to 200, while Erlandsen Creek escapement ranges from 25 to 30. Peacock et al. (1997) reported that Zymagotitz River and Erlandsen Creek run timing starts in mid-August and continues through mid-September with peak spawning in late August. Chinook have been observed spawning up to 31 km in reach 8 with the only specific location recorded at 20.5 km in reach 5. The lower 3 km reach of Erlandsen Creek supports chinook spawners.

Pink Salmon

Pink salmon spawning is widespread throughout the mainstem and subchannel gravel bars including back channels. Pink salmon are the most abundant species of salmon spawning in the Skeena River west area, centered between the Kasiks and Shames Rivers. It is probable that spawning locations change on an annual basis depending on discharge levels and the changing river channels (Bustard 1991). Typically, pink spawning peaks the first week of September and is usually finished by late September to early October, though some stream runs occur slightly later.

Skeena Mainstem and Side Channels

Skeena River side channels downstream of Terrace are extensive, diverse, and constantly changing due to deposition and erosion activities. Winter low flows can freeze eggs, and siltation and scouring are common problems. Pink salmon escapements in the Skeena mainstem and side channels increased greatly in the late 1980s, and pinks became the second largest stock in the Skeena watershed. The odd-year cohort was dominant through the 1980s and 1990s. The mean annual escapement in the 1980s was 102,000 for even years and 334,000 fish for odd years. For the 1990s, an average of 68,000 pinks returned annually in the even years, while the mean annual escapement for the odd years was 405,000 pinks. Escapement estimates since 1970 range from 3,500 in 1982, to over 750,000 in 1989 and 1991 (DFO 2005). Spawning locations are numerous and the reader is referred to the maps included in the FHIIP Stream Summary – Subdistrict 4B (DFO 1991b).

McNeil River, Khyex River

McNeil River supports a modest-size run of pink spawners below Gamble Creek. There are five years of escapement records with an annual average of 195 spawners. Khyex River escapement records are relatively complete and date back to the early 1950s. Spawner abundance has ranged from 200 to 220,000 with no clear trends (DFO 2005). Principal spawning grounds are located in good gravel sections from 8 km to the cascades at 26 km. The first right-bank tributary entering the Khyex opposite IR 64 supports spawners for the lower 2.5 km. Spawners have been observed in Arden Creek upstream to the small chute at 3.3 km that appears to restrict pink salmon (Bustard 1994c). The first upstream left-bank tributary on Arden Creek has observations of spawners in the lower 400 m (Bustard 1994c). Whitewater Creek, which flows into the Khyex at Ksagwisgwas (IR62), supports spawners for the lower 4 km (DFO 1991a).

Ayton Creek, Scotia River

On the left bank of the Skeena River, Ayton Creek provides pink salmon spawning gravels for the lower 1.5 km; spawner abundance is unknown. Scotia River pink salmon escapement records are relatively complete from the mid-1960s; they show a range of 200 to 100,000 with no discernible trend. In Scotia River, pink spawning occurs to 18 km, though the principal mainstem spawning is noted for 4 km upstream and downstream of the west fork. This west fork supports pink spawning in the lower 1.5 km.

Kwinitsa Creek, Kasiks River

Kwinitsa Creek, which flows into the Skeena River right bank has a small pink run spawning in the lower 3 km. Escapement estimates for the last 25 years are not available, but from mid-1950 to the 1970s, spawners ranged from 25 to 1,200. Kasiks River has a close to complete pink escapement record from the mid-1960s to the present. Average even- and odd-year escapements more than doubled in the 1990s with 3,660 and 9,200 in the even and odd respective years (DFO 2005). Bustard (1994c) reported that the principal Kasiks mainstem spawning beds are scattered from 6 km upstream to the barrier falls at 22.5 km, and in gravel deposits at tributary mouths. Gravel derived from Huckleberry Creek and the upper Kasiks above the falls accumulates in a 2 km section below the falls, providing highly productive salmon and steelhead spawning.

Exchamsiks River

Exchamsiks River pink salmon escapement is well documented back to the early 1960s. The escapement ranges from 25 to 7,500 with no distinguishable trends. The principal spawning occurs from 7 km upstream to 17.5 km in gravelly patches, although pink salmon have been observed spawning up to 24 km.

Exstew River

Exstew River pink salmon escapement records are complete back to the early 1970s, with spawners ranging from 25 to 2,000 (DFO 2005). Spawning is scattered along the mainstem between 9 and 15 km in reach 1. Spawning may also occur in tributary streams on the floodplain.

Dasque, Whitebottom, and Alwyn Creeks

Dasque Creek, which flows north into the Skeena River left bank, supports pink spawners on the lower 300 m of its alluvial fan. Pinks spawn in Whitebottom Creek and its tributaries in the lower 1.5 km. Alwyn Creek supports pink spawning in the lower 0.5 km.

Shames River and Shames Slough

Shames River pink escapement records are not complete; average pink salmon escapement for six years between 1985 and 1994 was 870 with a range of 50 to 3,000. Spawning occurs upstream to 1 km on the alluvial fan. Escapement records for Shames Slough are available for 30 years since the early 1960s. Escapement has been highly variable and ranges from 6 to 87,500, with spawning principally in the lower 0.8 km.

Zymagotitz River

Pink salmon escapement records for the Zymagotitz River are complete from the early 1950s, with a range of 25 to 10,000, though most years' records range between 25 and 1,000. Spawning occurs in the lower 7 km and is especially concentrated downstream of Erlandsen Creek for 2 km. Pink salmon escapement in Erlandsen Creek ranges from 36 to 1,500 with spawning principally in the lower 1.3 km. Molybdenum Creek has two years of pink escapement, 1987 with 1,500 fish and 1988 with 75 fish, and spawning is located in the lower 1 km.

Chum Salmon

Chum are relatively common in the lower Skeena River and its tributaries. The most important spawning areas are located in the multi-channeled reach of the Skeena River below Terrace known as Skeena west, and the Ecstall River. These two areas contain the only abundant stocks in the Skeena watershed, accounting for approximately 80% of the total chum spawners. Escapements to these areas averaged between 8,700 in the 1950s and 8,500 in the 1990s to 20,000 in the 1980s. Lower Skeena tributaries supporting small spawning populations include the Khyex, Kasiks, Exchamsiks, Exstew, Shames, and Zymagotitz Rivers.

Skeena River Mainstem

The mean annual escapement to the Skeena River west (Kasiks to Shames) over a 30-year period, 1970 to 1990, was 2,750 chum salmon. This ranged from a low of 75 chum in 1982, to 20,000 fish in 1988. Escapement levels have fluctuated from year to year, although decadal averages have changed relatively little. The DFO (1991b) FHIIP maps show 22 spawning locations in the mainstem, sloughs, and side channels. In addition, eight tributaries support chum spawning. It is widely thought that the Skeena mainstem chum populations below Terrace are under-reported because of turbid water conditions at spawning time (Kofoed 2001, Bustard 2002, J. Culp 2002, Whelpley 2002).

BC Hydro (1983) described mainstem spawning sites as being cut-off channels or channels fed by groundwater seepage. Bustard (1991, 1993c) documented significant spawning populations in Andesite, Shames, S-Bend, Twin Islands, and Exstew side channels, as well as associated island complexes, which together are estimated to comprise 20% to 30% of the total Skeena River west escapement.

Bustard (1991) confirmed the importance of upwelling groundwater for chum spawning and notes the presence of warm groundwater inflows that kept some side channels open for 300 chum spawners while the majority of the lower Skeena was iced over.

Khyex River

Khyex River chum salmon escapement records are mostly continuous since 1950, ranging from 2 to 1,000 fish. The DFO (1991a) reported that mainstem spawning occurs for 4 km upstream of Arden Creek, as well as in the lower reaches of Whitewater Creek and Arden Creek. Bustard (1994c) noted chum spawning in the 9th and 10th tributary fans on the mainstem right bank.

Kasiks River

Kasiks River chum salmon escapement records are continuous since 1950 ranging from 25 to 1,000 with a mean of 250 fish. The meandering 3 km upper section of reach 1 (9–12 km) is the principal spawning ground (DFO 1991b).

Exchamsiks River

Exchamsiks River chum salmon escapement records are moderately complete since 1950, with ranges from 25 to 200 fish counted. The principal spawning occurs from 7 km upstream to 17.5 km in gravelly patches, though chum have been observed spawning upstream to 24 km (DFO 1991).

Exstew River, Shames River

Exstew River chum salmon escapement records consist of only four records since 1950, ranging from 25 to 200 fish. Spawning is documented from 9 km to 15 km in reach 1. Shames River has only four records since 1950, with a range of 0 to 200 chum spawners with spawning occurring on the fan for 0.5 km upstream.

Zymagotitz River

Zymagotitz River chum salmon escapement records are fairly complete since 1950 with ranges of 0 to 400 fish. Spawning occurs in the lower 7 km, and is especially concentrated downstream of Erlandsen Creek for 2 km. Chum salmon escapement in Erlandsen Creek ranges from 0 to 200 with spawning principally in the lower 1.3 km. Molybdenum Creek supports spawning in the lower 1 km, but there are no escapement records.

Sockeye Salmon

A small population of sockeye spawn in Esker Slough. Recorded escapements from the 1980s were 10 to 25 fish. This stock is noteworthy because it is one of the rare river type stocks in the Skeena watershed. Sockeye with river type life histories are better known in the Nass and Stikine Rivers where some successful stocks rear in habitat similar to the lower Skeena River back channels. Sockeye spawners have been recorded from one to several times in the Kasiks, Exchamsiks, and Exstew drainages, but these occurrences are not typical.

Coho Salmon

Coho are widely distributed throughout the lower Skeena River area, though data concerning escapement, distribution, and habitat are scant. Coho do not generally spawn in the Skeena River mainstem channel; however, Bustard

(1991, 1993c) reported observation of limited spawning in the mainstem. Kasiks Channel, Shames Slough, Exstew Slough, Esker Slough, Andesite back channel, and many unnamed back channels are important coho spawning grounds.

Coho are fish of small streams and are often dependent on off-channel habitat such as beaver ponds, back channels, and seasonally flooded areas. After emergence, coho fry disperse downstream along the margins of the Skeena River and move into back channels and wetlands that are accessible during spring and fall high water. Fry presumably leave after one year, during high water events, and only a few fish remain for a second year (Bustard 1993a).

Twenty-one tributaries support coho spawning in this area. Spawning occurs from the end of September through December, late spawning is especially common in the more coastal areas. The majority of coho in the Skeena River system mature at age three or four, after rearing for either one or two years in freshwater and slightly more than one year in the ocean.

Coho are known to have a relatively longer estuary inner residence than sockeye or pink salmon. Juveniles usually reach the estuary in early June and remain in the shallow waters of the sandbanks for several weeks to two months before moving out to sea (Higgins and Schouwenburg 1973).

McNeil River to Khyex River

The coho escapement records for the McNeil River are discontinuous, with ranges from 0 to 250. The DFO (1991a) reported observations of coho to 10 km above the mouth with no specific spawning locations. The Skeena River right-bank tributaries—Aberdeen Creek, Inver Creek, Antigonish Creek, and Ekumsekum Creek, which are influenced to a certain degree by the tidal flows—have small coho spawning populations in their lower reaches although there are no formal escapement records.

Khyex River

Khyex River coho salmon escapement records are discontinuous since 1950, with ranges from 0 to 7,500 fish. The record does not allow for interpretation of population trends. Specific spawning locales on the mainstem are unknown, though spawning also occurs at up to 5 km on the unnamed right-bank tributary at Khyex IR 64. In the Whitewater Creek drainage, coho spawn up to 4 km. It is likely that coho spawn in the lower sections of the many small tributaries and that these are important contributors to overall coho production. Bustard (1994c) noted that coho juveniles were present in many of the small tributaries.

In the lower reach of Arden Creek, spawning has been observed up to 6.5 km, though principal spawning grounds are in the lower 1 km and from 5.2 km to 6.5 km (Bustard 1994c). Bustard reported that results of minnow trap catch suggest that coho juveniles are present throughout Arden Creek below the impassable falls at 7.6 km.

Scotia River and Ayton Creek

Scotia River coho escapement records are few, with four counts averaging 450 fish. In the Scotia River mainstem, coho spawning is noted below 18 km, though principally occurs for 4 km upstream of the west fork. This west fork also supports coho spawning in the lower 3.4 km reach. Jyrkkanen (1997) suggested that the fines/gravel proportions on the mainstem may limit coho spawning.

Ayton Creek, located on the left bank of the Skeena River approximately 7.5 km downstream of Scotia River, provides coho salmon spawning gravels for the lower 1.5 km (DFO 1991b); there is no indication of spawner abundance. Triton (1998a) reported coho presence in low-gradient streams tributary to Ayton Creek.

Kwinitsa, Feak, and Alder Creeks

Kwinitsa Creek supports a small population of coho spawners. Escapement records since the early 1970s are discontinuous but show a range of 0 to 75 spawners using the lower 3 km. Feak Creek and Alder Creek support small numbers of coho spawners in their lower reaches with no escapement estimates available.

Kasiks River

The Kasiks River has a relatively large coho run with fairly continuous coho escapement records since the early 1950s. Escapement ranges from 300 to 7,500 with a decadal mean

of approximately 2,000 fish. Bustard (1994c) reported that the principal mainstem spawning beds are scattered throughout the system, with the coho relying on the lower sections of the many small tributaries in the valley bottom for both spawning and rearing. Bustard's observations also suggest that concentrated coho spawning occurs in the second reach of the mainstem below the falls. Gravel recruited from Huckleberry Creek and the upper Kasiks above the falls accumulates in a 2 km section below the falls, providing highly productive salmon and steelhead spawning habitat.

Coho juveniles remain for up to two years before migrating out of the system (Bustard 1994c). The 1984 fish inventory of Kasiks Channel, a large channel north of the rail and highway corridor, suggests that a substantial juvenile coho population is present and that recruitment is from fry moving out of the Kasiks River (Bustard 1984d). In 1998, three juvenile salmon population index sites were established in the Kasiks drainage on Kasiks Channel, Cabin Creek, and Blaze Creek; however, in 1999, the Kasiks drainage sites were dropped.

Exchamsiks River

Exchamsiks River coho salmon escapement has been fairly well documented since 1950. The escapement ranges from 25 to 6,000 with a 1990s decadal mean of 1,985 coho. The principal spawning occurs over a 12 km section in reach 2, approximately 14 km upstream of the mouth. Most likely, the lower sections of the many small tributaries on the floodplain also support spawners. The DFO (1991b) also documented coho spawning up to 30 km in the mainstem. The large left-bank tributary flowing into the Exchamsiks River at 4 km supports coho spawning in the lower 5 km reach.

Exstew River

Exstew River coho salmon escapement records are fairly complete back to 1966, with spawners ranging from 75 to 3,500, with an annual mean in the 1990s of 1,853 coho (DFO 2005). Spawning is scattered along the mainstem, particularly from 12 to 14.5 km in the upper section of reach 1, and spawning occurs in tributary streams on the floodplain. Sinkewicz

(1999) reported on redds and spawning adults on tributaries located at 5.5 km and 7.5 km.

In 1998, three juvenile salmon population index sites were established in the Exstew drainage (Biolith 1999a). In 1999, the sites were dropped and two new sites were established (Sinkewicz 2000). The purpose of establishing the index sites was to determine trends in fish populations in the Terrace area.

Shames River, Dasque Creek, Whitebottom Creek, and Alwyn Creek

Shames River appears to have one escapement record (1999) with enumeration of 165 coho spawners (Culp 2000). Spawning occurs largely in side-channel habitats and to a lesser extent in the mainstem, while in reach 2, coho were utilizing pockets of gravel found at pool tail-outs and behind boulders.

Dasque Creek historical escapement information is not available. On a creek walk in 1999, Culp (2000) reported 15 redds and 19 coho spawners, upstream of the lower 0.5 km, which is scoured and braided. Whitebottom Creek historical escapement records are not available. Coho have been reported spawning in the Whitebottom mainstem and its tributaries situated on the Skeena River valley flats. Alwyn Creek escapement records show fairly consistent enumerations between 1970 and 1986 with a range of 25 to 100. Spawning occurs on the mainstem to 2.5 km.

Zymagotitz River

Coho salmon escapement records for the Zymagotitz River are mostly complete from the early 1950s, ranging from 0 to 3,500; however, most years have ranged between 250 and 750 fish (DFO 1991b). The annual mean in the 1960s was 1,560; in the 1990s it was 911 coho. Spawning has been observed up to 28 km. Spawning is concentrated from 5 km to 7 km, or downstream of Erlandsen Creek for 2 km.

Coho salmon escapement has rarely been estimated in Erlandsen Creek. In 1984 and 1986, 20 and 300 coho were enumerated respectively. In 2000, Culp estimated 120 coho spawners from the mouth upstream to the waterfall at 4 km. Coho are also present in Molybdenum Creek, at least in some years. Culp (2000) estimated 90

coho from the Erlandsen confluence to 0.5 km upstream.

Steelhead

Skeena River Mainstem

Summer- and winter-run steelhead move through the lower Skeena River, though there is no documented evidence of spawning in the side or mainstem channels. Lough (1981) noeted that 14% of the 95 summer-run steelhead tagged in the lower Skeena River during August were still present in the lower river six months later. Lough suspected that most of these fish subsequently entered lower Skeena tributaries such as the Kitsumkalum, Gitnadoix, Lakelse, and Zymoetz Rivers to spawn. No steelhead were captured during driftnet surveys at seven possible spawning locations sampled in mid-May (Bustard 1991).

Juvenile steelhead use of the lower Skeena mainstem area is widespread. The catch composition of steelhead juveniles sampled by Bustard (1984d, 1991) and Tredger (1984) showed that the percentage of parr in the total catch was 16.6% in April and 5% in late summer/fall. Parr estimates for this area of the Skeena were 50 parr/km of channel. This compares to estimates of up to 200 parr/km in the lower Bulkley River, 83 parr/km in the Kispiox River, and 4 to 10 parr/km in the Morice River. Steelhead parr estimates were 2.5 times higher in side channels than in mainstem sites (Tredger 1984).

Juvenile steelhead may spend up to five years in freshwater, although most spend two to four years in tributary streams and the lower Skeena River before smolting (Whately et al. 1977). Bustard's (1991) and Tredger's (1984) juvenile steelhead samples indicate that 97% were age 1+ and 2+ parr. Steelhead juveniles frequented logjams and riprap sites most often in the spring, with riprap and cobble sites preferred through the summer and early winter. Bustard (1991) reported finding that side-channel sites received little use compared to mainstem areas, while Tredger (1984) reported having found that catches of steelhead parr were 2.5 times higher in side channels compared to mainstem sites. Nothing is known about the length of steelhead residence in the Skeena Estuary prior to movement to the open sea.

Skeena River Tributaries

The DFO (1991a) reported that steelhead are present in the Khyex River to the falls at 26 km. The principal spawning grounds are located 0.5 to 1.5 km downstream of the reach 3 falls. In Arden Creek, steelhead have been observed to 1.5 km, and Bustard (1994c) reported juveniles up to the falls at 7.6 km.

Steelhead are present in the Scotia River, though the only recorded spawning location is at 10 km on the west fork (Jyrkkanen 1997). Khtada River supports steelhead spawners below the cascade at 0.5 km (DFO 1991b). Steelhead have been observed spawning in the lower 1.2 km section in Kwinitsa Creek.

The Kasiks River supports steelhead spawners in the lower reaches of the first major right-bank tributary (downstream of IR 29) and Huckleberry Creek (Bustard 1994c, DFO 1991b). The DFO (1991b) reported steelhead present in the Exchamsiks River; besides that report, there is a lack of information. Steelhead have been observed at 31 km in the Exstew River; however, no specific spawning locations are documented. Steelhead are reported in Erlandsen Creek and are most likely present in the upper Zymagotitz (DFO 1991b).

Eulachon

Four smelt species are found in the Skeena estuary, the best known being the eulachon, sometimes spelled oolichan. Eulachons are small (<25cm) anadromous fish that spawn in the lower main channel and tidal portions of coastal tributaries of the Skeena River. Spawning occurs in the spring, usually between mid-March and mid-May (Hart 1973), from the Kasiks River downstream to the Ecstall and Khyex Rivers. The eggs adhere to the surface of the sand or fine gravel substrate for six weeks before hatching in May–June. The larvae average 4 mm in length and move to the sea immediately upon hatching (Lewis 1998).

Eulachon are a culturally significant fish. Oral histories and historical records are replete with commentaries concerning spring eulachon fishing and processing. The purpose of catching the fish in great quantities was to extract the rich and nutritious oil commonly called grease. The grease was in great demand and traded eastward throughout the Pacific slope. An introduction and overview of eulachon importance and value are presented in Drake and Wilson (1991).

The size of the population, the timing of migration, the site of spawning and egg incubation, and the preferred habitat of eulachon in the lower Skeena are unreported, though considerable anecdotal information is held by Tsimshian people. Pendray (1993) reported on larvae sampling in the lower Skeena, and Lewis (1998) analyzed data and described fieldwork conducted by the Tsimshian Tribal Council (TTC) in 1997. Ryan (2003) reported on continued TTC eulachon studies.

Eulachon spawning runs have declined over their range from California to southeast Alaska since approximately 1980 and especially since the mid-1990s. The causes of the decline are uncertain. Hay and McCarter (2000) review the growing awareness of the decline of eulachons and suggested that climate changes affecting ocean conditions, habitat alteration including subtle changes in streamflow discharge and timing, and offshore shrimp trawling bycatch are causes of the general decline. The decline and status of eulachons are a concern to Tsimshian and other upstream First Nation peoples, for whom the eulachon is of major cultural significance.

Fisheries
First Nations Traditional Use

The lower Skeena River area has long been the traditional homeland of coastal Tsimshian peoples. First Nations traditional use and occupancy of the lower Skeena areas was extensive and is well documented by oral history and early Euro-Canadian visitors. The Tsimshian culture has integral connections to the environment they inhabit. The social structure is composed of a matrilineal kinship society and exogamous clans divided into houses, with crests, oral histories, and a land tenure system of territories managed through a public forum process called the Feast.

Nine distinct groups held territories from Lakelse River downstream to the Skeena mouth. Tsimshian groups within the area included Gitziis, Gitwilgiots, Ginaxangiik, Gitandoh, Gitlan, Gitlutzau, Gitzaklalth (Gitzaxlaal), Gitnadoix, and the Gitwilksabe (thought to be extinct). These groups were centered in villages typically located close to the Skeena River (Figure 17) and used traditional territories owned by various lineage heads (Inglis and MacDonald 1979).

Gitziis territories were located in the Kasiks River drainage, with a village close to the mouth. Gitwilgiots and Ginaxangiik (Ginaxangits) occupied a village site at the mouth of the Exchamsiks River and used the rich resource territories in the Exchamsiks and Exstew drainages on the north side of the Skeena River, as well as several small tributary streams opposite the mouth of the Exchamsiks River. The Gitnadoix people used the Gitnadoix drainage with villages located at the mouth and by the Magar Creek confluence. In addition to the village site located close to the Shames River mouth, Gitando ancestral territories included the Shames River drainage (Inglis and MacDonald 1979).

After establishment of Fort Simpson in 1834, many of the lower Skeena First Nation groups moved to the area surrounding the fort and are now located in Lax Kw'alaams. Other Tsimshian reside in the Kitsumkalum village and continue to use the lower Skeena River area for traditional activities. Salmon are an integral part of the Tsimshian culture and one of the main food sources. The abundant chum salmon is a particularly important food resource in the lower Skeena River area. Because the fat content is relatively low and the dried product less likely to spoil, they are smoke-dried in great quantities.

Recreational Fisheries

The sports fishery in the lower Skeena River area provides important recreation and aesthetic values for the public. Salmon fishing draws local residents from northwest B.C. and nonresident tourists into the lower Skeena area with its extraordinary angling opportunities. Easy access to gravel bars and boat launch sites is afforded from Highway 16 or the railroad grade. The most popular boat launch sites are located at the mouth of the Kitsumkalum River, Andesite side channel, near the mouth of the Exchamsiks River, close to Kwinitsa, and near the mouth of the Exstew River.

The guided and unguided sport fishery targets chinook, coho, pink, steelhead, and chum salmon. Popular fishing locales are at Chicken

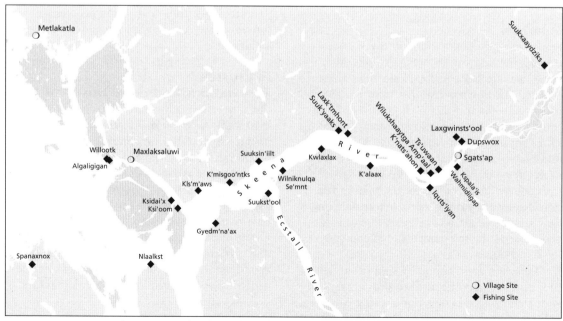

Figure 17. Map of traditional fishing and camp sites on the lower Skeena River

bar, Canoe bar, East and West Café bar, and Andesite bar. Other popular sites include China bar, Shames bar, Polymar bar, Gitnadoix bar, Chainsaw bar, Electric bar, Seventeen, Eighteen, and Nineteen Mile bars, Kraut bar, and Alberta bar (Thomas and Associates 1999). The one commercial fishing lodge on the lower Skeena is located close to the Shames River and has accommodations for 35 angling enthusiasts.

Anecdotal information suggests that intense angling pressure may be detrimental to chinook salmon populations at the mouth of the Kasiks, Exchamsiks, and Exstew Rivers (Kofoed 2002). Creel surveys were conducted on the lower Skeena for a number of years (Thomas and Associates 1995, 1999). Thomas and Associates (1999) reported on gear used, number of fish caught and released, fish health when released, and estimated mortality, which for chinook was more than 50%. The Skeena River mainstem from 1.5 km above the Kitsumkalum River downstream to the Exchamsiks River is designated Class II waters, July 1 to September 30, with a steelhead stamp required for steelhead angling (B.C. Fisheries 2005).

Enhancement Activities
The only known enhancement activity has occurred at the Andesite side channel, where a small groundwater-fed channel for chum salmon enhancement is operated by the DFO and Kitsumkalum Hatchery. Kitsumkalum Hatchery also incubates and rears coho that are marked with coded wire tags and released into the Zymacord River; but since 1998, the coded wire tag (CWT) program has been discontinued.

Foreshore fish habitat replacement projects have occurred in conjunction with three transportation projects. White (1988, 1990) reported on habitat compensation due to realignment and widening of Highway 16 at the Khyex to Tyee section. Miles and White (1996) proposed habitat replacement, and White (1997) discussed the establishment of foreshore vegetation following the construction of the Ayton mainline on the south bank of the Skeena River between Scotia and Ayton Creeks. Bustard (1993e) reported on the monitoring of water quality conditions and fish populations at the reclaimed Esker gravel pit.

Shames River sediment accumulated under the CN Railway bridge

Development Activities
The dominant development activities in the lower Skeena River area revolve around logging, transportation, and utilities. Mineral exploration and development have been slight and peripheral to the area of interest. Population and settlement other than by First Nations in the past have been very limited, except in the estuary (Figure 17) where seasonal settlements were associated with the fish canneries.

Forest Resource Development
The rich forest resources of the lower Skeena area have always been important to the native Tsimshian and in more recent times to Euro-Canadians. The early Tsimshian culture made extensive use of red and yellow cedar for their longhouses, canoes, poles and crests, cookware, and clothing. Other wood, such as spruce, alder, birch, and juniper, was carved or woven.

Commercial forest exploitation did not begin until the early 1890s, when large quantities of firewood were cut for steam-driven paddle wheelers, which regularly traveled the Skeena River. The construction of the railway between 1910 and 1914 necessitated clearing of the right-of-way, which was subsequently widened during World War II to complete the highway from Terrace to Prince Rupert (Kerby 1984).

Selective logging of spruce occurred in the 1940s. In the early 1950s, clearcut logging of the high-value Sitka spruce stands commenced on the lower Skeena River floodplain sites and islands. By the late 1960s, the floodplain Sitka spruce stands were nearly eliminated.

Enormous Sitka spruce logs were skidded with cats and dumped into the river where they were boomed and towed to the Port Edward pulp mill (Waldie and van Heek 1969). Some of the skid trails, which were often deeply engraved into the floodplain, are currently used by coho for rearing habitat.

In the late 1950s and early 1960s, forest harvesting activities began in the lower reaches of the Shames, Exstew, Exchamsiks, and Kasiks River valleys; most logging occurred on the floodplain or the toe of the slopes. At this time, the lower reaches of the Zymagotitz River and the lower Erlandsen were partially logged and clearcut. In the early to mid-1970s, extensive clearcutting occurred in the Zymagotitz drainage along the steep mountainsides and over alluvial fans.

In the mid-1950s through until the early 1960s, Columbia Cellulose (later called Canadian Cellulose and then Skeena Cellulose) dumped logs into the Skeena River from approximately half of its Terrace operations, as well as from logging operations downstream from Terrace. These logs were boomed, held in storage grounds, and towed down the Skeena to the pulp mill at Port Edward. In the late 1960s and into the 1970s, various areas in the lower Skeena including the McNeil River, Little Windsor, and Big Windsor saw the initiation of large-scale industrial logging.

Logging in Kwinitsa Creek was initiated in 1978 and continued into 1991 with approximately 64% of the operable area logged. In the Scotia River drainage, forest development started in 1980 and has continued into the present. Recent logging, particularly in the Shames, Exstew, and Scotia drainages, has been restricted to upland areas with both conventional and helicopter logging.

Transportation and Utilities

The Skeena River has been used for thousands of years as a transportation waterway, primarily by canoes. In the early 1890s, steam-powered paddle wheelers began regular, seasonal use that continued until the Grand Trunk Pacific (GTP) Railway was completed in 1914. Paddle wheel use in this section of the Skeena River did not require any rock removal or channel alterations.

At present, there is intensive linear development on the northern bank of the Skeena River with CN Rail and Highway 16 situated on the floodplain for much of their length. Railway construction practices caused many instances of blocked access to fish. The completion of Highway 16 linking Prince Rupert and Terrace in 1944 involved similar practices that restricted fish passage. Numerous back channels and side channels were cut off to returning adults, and abutments of bridges across tributary streams aggravated natural sediment deposition in fans, often requiring frequent channel dredging to avoid washouts.

Primary access to the lower Skeena River area is easily accommodated by Highway 16, which follows the north bank of the Skeena River from Terrace to Tyee, located 12 km upstream of the mouth. There is access off the highway onto resource roads reaching into side drainages, including the McNeil (Green) River and the Lachmach River road to Work Channel north of Skeena. Logging roads access the lower and mid portions of the Exstew, Shames, and Zymagotitz drainages. South of the Skeena River, resource road access from Old Remo extends west through Whitebottom past Dasque Creek. Highway 16 access provides boat launch options at Kwinitsa, the Andesite side channel, and at the mouth of the Kitsumkalum, Exstew, Exchamsiks, and Kasiks Rivers.

On the south side of the Skeena River, the floodplain hosts linear development in the form of the BC Hydro 287 kV transmission line and the PNG pipeline to Prince Rupert. This utility corridor is serviced by road to a point opposite Salvus, where the natural gas pipeline crosses the Skeena River and passes up the Kasiks River valley. The PNG pipeline was built in the 1960s with construction practices that impacted fish and fish habitat. In 1993, the pipeline across Skeena River broke due to river scour and was re-installed under the river by directional drilling (Powell 1995). Maintenance problems on the BC Hydro transmission line have been minimal, other than erosion along tower bases which were consequently protected with riprap.

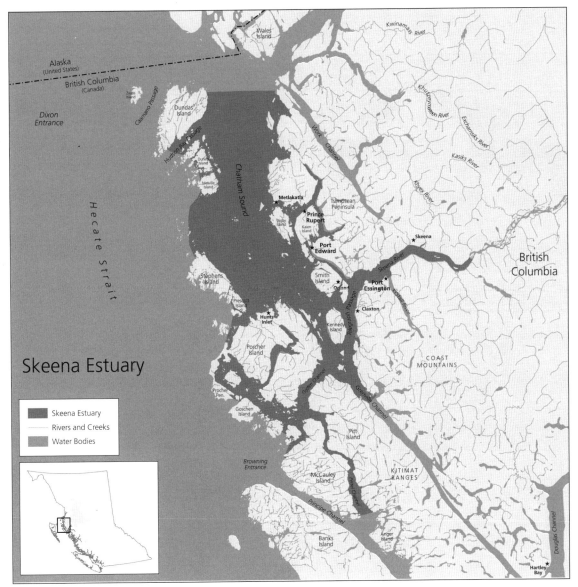

Skeena Estuary

Legend:
- Skeena Estuary
- Rivers and Creeks
- Water Bodies

Figure 18. Skeena estuary

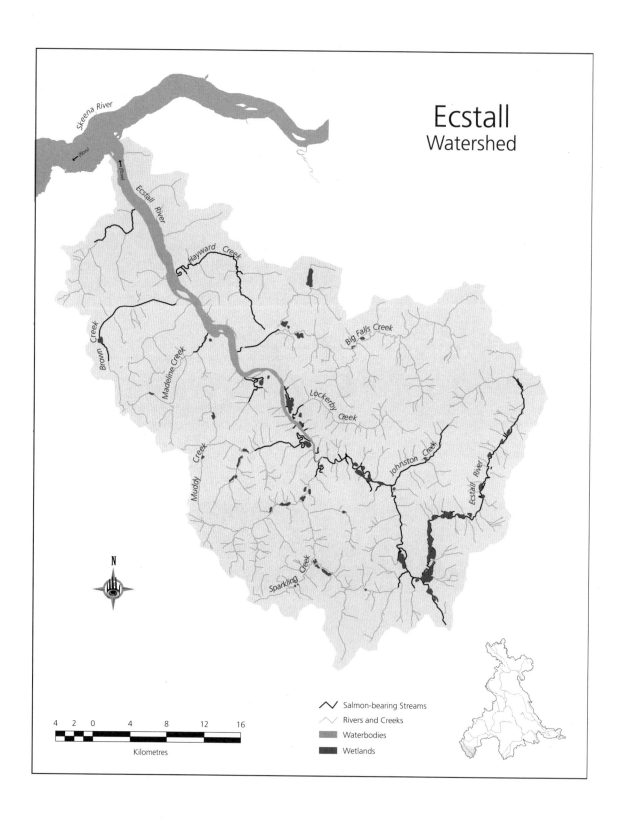

Ecstall
Watershed

Skeena River

(flow)

(flow)

Ecstall River

Hayward Creek

Brown Creek

Madeline Creek

Big Falls Creek

Lockerby Creek

Muddy Creek

Johnston Creek

Ecstall River

Sparkling Creek

N

4 2 0   4   8   12   16

Kilometres

⋀ Salmon-bearing Streams
⋀ Rivers and Creeks
   Waterbodies
   Wetlands

6

Ecstall Watershed

Environmental Setting

Location

The Ecstall (often pronounced "Oxstall") watershed is located in the southwest corner of the Skeena watershed. The watershed is a relatively small tributary subbasin of the Skeena watershed, situated 30 km southeast of Prince Rupert and approximately 10 km upstream from the mouth of the Skeena River. The watershed is bounded to the north by the Skeena River floodplain, to the west and south by tributary basins draining to Telegraph Passage, Grenville Channel, and Douglas Channel, and to the east by the Scotia River and the Giltoyees drainage.

Hydrology

The Ecstall River watershed is a fourth-order system with a catchment area of 1,485 km². Elevations range from 1,935 m to 0 m at the Ecstall-Skeena confluence. The Ecstall River flows generally northwesterly for 94 km from its headwater lake to discharge into the Skeena River estuary. There is an annual May to early June discharge peak due to high elevation snowmelt. The larger floods are rainstorm and rain-on-snow flows that can occur from September through the winter, but are most common in October and November. Total mean annual precipitation is 3,683 mm at the Falls River, located in the lower watershed. Climate conditions are characterized by high precipitation and generally moderate to cool conditions.

Because of the steep terrain within the watershed, the Ecstall and most of its tributaries are subject to high water flow fluctuations immediately following moderate and heavy rain events. Summer low flows are typically four to eight times greater than winter low streamflows and are sustained mainly by high-elevation snowmelt, while winter low flows are derived from lakes, groundwater, and unfrozen wetlands. Snowfall can occur anytime in the period between November and April, with an increasing proportion of the annual precipitation occurring as snowfall at higher elevations. Historic streamflow data exist for the Falls River and Brown Creek; both have been dammed to develop hydroelectric power.

Eleven medium- and large-size lakes contribute a substantial amount of water storage within the watershed. Johnston Lake, Lower Lake, and Ecstall Lake are significant within the Ecstall system for high-quality rearing habitat and for the moderating influence on their outlet streams' water flows.

The Ecstall River is the dominant hydrological feature in the watershed. Brown Creek, Sparkling Creek, Muddy Creek, and Madeline Creek are the major tributaries flowing into the left bank. Major right-bank tributaries are Hayward Creek, Big Falls Creek, and Johnston Creek.

Of the approximately 80 second-order or larger tributaries, 10 are of major consequence; however, most of these tributaries have gradients too steep for fish passage. These streams drain glacial headwaters, deliver moderate amounts of sandy sediment, and in the summer season have glacially derived silty and variable flows.

Stream Channels

Tidal influence extends approximately 42 km upstream of the Ecstall River mouth to Sparkling Creek. This reach of the river is a single, confined channel with relatively little floodplain and is mainly comprised of inside meander bends and tributary mouths. The lower section of the reach 1 exhibits a continuous channel, while the upper section possesses a sinuous, irregular pattern. At low water, the braided nature of the channel with its periodically changing shoals and bars is evident. The gradient is near 0%, but

View up the Ecstall River

the tide accelerates the river flow on the ebb tide and retards it on the flood.

Reach 2, which extends approximately 18 km from Sparkling Creek to Lower Lake Creek (also known as Ecstall Creek), is a single-thread, irregularly sinuous channel, with an average floodplain width of 1 km. A stable channel amid relic channels and extensive wetlands characterizes reach 3 with a floodplain of similar width. The upper reach, or reach 4, is braided and heavily aggraded, reflecting the drainage of huge quantities of sediments from the many adjacent glaciers and ice fields. Alluvial fans and avalanche deposition from the steep slopes dominate the 500 m wide floodplain. The majority of the tributary streams in this reach rapidly increase in gradient once off the floodplain.

Access

Access to and within the Ecstall watershed is limited to either air or boat travel. Road travel is possible to Hayward, Big Falls, and Carthew Creeks, and to an unnamed creek south of Big Falls via the logging road network that originates in the Scotia River drainage.

Riverboats travel a considerable distance up the Ecstall mainstem; however, some local knowledge is needed for successful navigation.

Geography

The distinctly rugged landforms of the Ecstall watershed are situated within the Kitimat Ranges of the Coast Mountains. The mountains are bold, impressive, massive, and largely comprised of granitic rocks of the early Tertiary Ecstall Pluton, one of the Coast Range Intrusives. For the most part, the mountains are below 1,800 m in elevation. The heavily glaciated north and northeastern sides of the peaks have remnant small glaciers and ice fields (Holland 1976).

The Ecstall valley is a northwest trending fault lineament that extends through the Quall River to Douglas Channel. During the retreat of Ice Age glaciers, sea levels underwent a complex series of fluctuations. The Ecstall valley was briefly a fiord 10,000 to 11,000 years ago, and was subsequently isostatically uplifted and partially filled with fluvial and marine sediments (Gottesfeld 1985). Modern landforms have colluvial and morainal blankets whose surface expression conforms to the underlying bedrock

ECSTALL WATERSHED

structure. The valley bottoms have glaciofluvial mantles and glaciofluvial fans as the dominant surficial features (Clague 1984). Typically, the steep, U-shaped valley slopes rise from the valley to the mountain summits. Along their length, water cascades over falls and down rock faces that are often swept clean of vegetation.

The coastal climate is reflected in the major ecological zones. The Coastal Western Hemlock (CWH) zone represents the vegetation in the valley bottoms and slopes to 400 m. Old-growth conifer stands of western hemlock, western red cedar, Sitka spruce, and amabilis fir dominate the forested landscape. Large spruce, hemlock, and deciduous trees, with salmonberry, ferns, and devil's club comprising the principal understory vegetation, characterize floodplains and alluvial sites. The floodplain upstream from Muddy Creek is dominated by large spruce and is undoubtedly one of the richest and most diverse ecosystems on the North Coast (Liepens 2002). The CWH zone passes upward into the Mountain Hemlock (MH) zone, which is dominated by mountain hemlock and amabilis fir with occasional spruce, red cedar, and yellow cedar. A well-developed shrub layer is mostly composed of Alaska blueberry, false azalea, and red and black huckleberries (Banner et al. 1993).

Fish Values

The Ecstall River watershed is a relatively small but biologically productive river system that has diverse high-value fish habitats. Fish species utilizing its habitats include sockeye, coho, chum, pink, and chinook salmon; steelhead and rainbow trout; Dolly Varden char; and mountain whitefish. Eulachon spawn in the tidal portions. The fish community contributes to the ecology, nutrient regime, and structural diversity of the drainage. It also provides strong cultural, economic, and symbolic linkages, particularly for First Nation peoples, and supports recreational and commercial fishing.

Chinook Salmon

The aggregate Ecstall River chinook stock is one of the top producers in the Skeena watershed. Ecstall River chinook were not as heavily harvested by the mixed-stock fishery as were many other Skeena chinook subpopulations.

Following the closure of directed-net chinook fisheries in 1983, the mainstem run appears to have rebounded until 1990. Escapements since then have been poorly sampled but appear to be depressed. The Johnston Creek chinook run, which averaged 2,900 adult spawners in the 1950s and 1960s, declined precipitously in 1970 to 300 returning adults, and has remained depressed to the present time (DFO 2005). The lower reach of Sparkling Creek and the outlet tail pond of Big Falls Creek also support chinook spawning; however, escapement data are scant and too limited to assess population trends over time.

The critical chinook spawning ground in the Ecstall system is the lower section of the 1.4 km reach of Johnston Creek between the Ecstall River and Johnston Lake. This low-gradient reach possesses excellent gravels. Adults return in early to mid-August and spawn shortly thereafter. Other significant spawning grounds on the Ecstall River occur in scattered pockets from the upper tidal limit, with higher densities between Johnston Creek upstream past Lower Lake Creek for 6 km (Ginetz 1976). Spawning is also prevalent, to a lesser degree, in the lower reach of Sparkling and Big Falls Creeks. The Ecstall River chinook start arriving in early July, spawning notably earlier than the Johnston Creek run (Jantz et al. 1989). The Johnston Creek run may remain suppressed because of inadvertent harvest in pink salmon seine catches (Winther 2002).

Ecstall chinook age structure is believed to be similar to that of the nearby Skeena test fishery, which does not sample the Ecstall River. It is thought that chinook fry emigrate from mid-April to early June, as is typical of Skeena chinook, and utilize the lower Ecstall and lower Skeena Rivers for rearing habitats. The Ecstall chinook are genetically separable from all other Skeena stocks and seem more related to North Coast populations (Winther 2002).

Pink Salmon

The Ecstall system has a relatively small pink population that spawns in the mainstem, Hayward Creek, Big Falls Creek outlet, Johnston Creek, Lockerby Creek, Madeline Creek, Muddy Creek and Sparkling Creek. Ecstall River pink salmon were abundant for four decades until pink adult abundance decreased in the 1960s. Populations rebounded in the early 1980s. Since 1994, odd years have been dominant with record high returns. Up to the late 1930s, Johnston Creek had relatively strong runs of pink salmon (Fisheries Research Board 1948). Johnston Creek and the other six creeks within the watershed that support pink spawners have few data to assess escapement trends.

Adult pink salmon typically arrive in the Ecstall River in early August, with peak spawning from mid-August through September. Principal spawning in the mainstem tributaries is generally in their lowest reaches, while mainstem spawning occurs in scattered pockets above the tidal range. Significant mainstem spawning beds are located in the 15 km stretch upstream of Sparkling Creek, extending to lower Lake Creek. Fry typically emerge in late March to mid-April and emigrate downstream to the estuary.

Chum Salmon

The Ecstall River receives the largest escapements of chum salmon in Area 4, averaging 65% of the total over the 40-year period between 1950 and 1990. Ecstall chum escapements fluctuate highly from year to year, with the range of 75,000 to 500 adult spawners since 1950; however, decadal averages are generally similar. The proportion of Ecstall spawners relative to the total chum escapement for Area 4 has increased from approximately 20% in the 1950s to as high as 85% in the 1990s. However, much of the recent predominance of Ecstall chum is due to declining effort in enumerating upstream Skeena chum stocks. In contrast to the upper Ecstall tributary, Johnston Creek has seen a steady decline from 1,500 spawners in the 1950s to less than 200 annually since the early 1980s.

Although escapement data are not available prior to the 1940s, Skeena fish catch information from 1900 to 1930 indicates significantly larger Skeena chum populations than at present. Annual catches in the 100,000–200,000 range were commonplace, with the majority of fishing occurring in the Skeena and Ecstall Rivers.

The main run of chum salmon into the Ecstall system occurs throughout August, with peak arrivals in mid-August. The main spawning grounds are along the mainstem above the tidal range, from Sparkling Creek to lower Lake Creek. Tributaries supporting chum spawning are Madeline Creek, Big Falls Creek, Lockerby Creek, Johnston Creek, and Sparkling Creeks – all of which currently have few spawners, or else have not been observed and counted in recent years.

Sockeye Salmon

Within the Ecstall watershed, sockeye return to spawn in the Johnston Creek and Lake system and the Ecstall Creek and Lake system. Brett (1952) estimated that average escapement to Johnston Lake was 1% of the total Skeena escapement. Johnston Lake escapement counts have been conducted regularly since the mid-1920s; escapement decreased rapidly in the 1950s and stayed depressed (<1,000) until the early 1970s. A peak return of 7,500 sockeye was recorded in 1972; this return probably reflects the lack of fishing due to the 1972 strike. Returning spawners then dropped off to less than 1,000 fish until 1995 when the historic abundance levels of the 1940s were reached and maintained until the present. A lack of sockeye escapement data for Ecstall Creek limits population trends discussion.

Sockeye run timing into the Ecstall River is reported by the DFO (1991d) to occur from June through August, with spawning from mid-September to mid-October. Sockeye have been observed in mid-June in Johnston Lake. In the Johnston system the principal sockeye spawning grounds are in the 600 m reach downstream of Johnston Lake's outlet and beaches adjacent to the inlet end. Ecstall Creek, located between Ecstall Lake and Lower Lake, is the primary spawning bed for Ecstall and End Lakes. Upon emergence from the gravel, fry rear in the stream-headed lakes – Johnston Lake, Ecstall Lake, and Lower Lake – for one year before emigrating to saltwater. Juvenile sockeye abundance and distribution were sampled in Johnston and Ecstall Lakes in 2005. Hume and Shortreed (2006) reported that Johnston Lake had an exceptionally high density of age-0 sockeye fry at 6,000/ha, while Ecstall Lake had a very low density of 71/ha.

Coho Salmon

Coho from the Ecstall drainage contributes a substantial proportion to the overall Skeena coho population. Along with the Ecstall River mainstem, Lockerby Creek, Johnston Creek, Madeline Creek, Hayward Creek, and several unnamed creeks support coho spawners. Semi-regular escapement counts only exist on the Ecstall River. Escapement in 1984 and 1990 reached 10,000 spawners, with the 18-year annual average escapement to the present indicating approximately 4,300 coho adults. Coho populations did not decline in the 1970s and 1980s as they did in the upstream portions of the Skeena watershed.

Coho typically arrive in the Ecstall River in late September and head to scattered spawning pockets along the mainstem and tributary mouths. The start, peak, and end of coho spawning periods are not known. Pre-spawning movement into back channels and tributary streams is often delayed due to low water flows. Coho spawn in the lower tributary reaches and in the lower section of Johnston Creek. In the mainstem, coho have been observed spawning upstream as far as 92 km. Most of the recent escapement to the Ecstall watershed has spawned in the mainstem from 5 km above Ecstall Creek to the headwaters (Wagner 2002). Recent DFO helicopter counts have observed strong numbers of adult spawners in Johnston Creek tributaries, the Ecstall Lake system, and generally the entire mainstem above the tidal range (Finnegan 2002a). Coho juveniles spend one or two years in freshwater before emigrating to saltwater.

Steelhead

Winter-run steelhead are known to be present in the watershed, but information about their life history and populations is scant. Steelhead have been observed in the Johnston system as early as mid-February, though most appear in March and April (Chudyk 1974). The DFO (1991d) reported spawning in the lower section of Johnston Creek.

Fisheries

It is important to note that before 1936, gill netting was permitted in the Ecstall River (Brett 1952). This historical commercial fishing practice could have significantly reduced the escapement of sockeye, chinook, chum, pink, and coho to the Ecstall River.

First Nations Traditional Use

First Nations traditional use of the Ecstall watershed was extensive and varied with village sites, home places, and fish houses or stations. The watershed is within the territory of the Tsimshian, who were divided into fifteen local groups that were politically autonomous but shared cultural and linguistic links. The Gitzaxlaal held the watershed territories with two main seasonal villages: Spiksuut, located at the Ecstall-Skeena confluence, and Txalmisoo', located at Big Falls Creek (Halpin and Seguin 1990).

The very abundant and predictable sockeye salmon stocks provided the Gitzaxlaal with opportunity to harvest and preserve a large amount of high-quality food in a relatively short time of intensive effort. The sockeye run was the major focus, as it provided the majority of high-quality dried fish needed to sustain the Gitzaxlaal over the year and to produce a trade item. Following the passage of the bulk of the sockeye, coho were available until early winter, providing both fresh and dried fish.

Salmon were primarily caught in weirs or by spear and then processed by smoke drying. Herring, various groundfish, marine invertebrates, and sea mammals were harvested; however, the eulachon was probably the most

Port Essington at the mouth of the Ecstall River

important nonsalmonid resource fished. A major aboriginal trail extended from the Ecstall valley to Douglas Channel. The trail ran parallel to the Ecstall River, passed Ecstall Lake, crossed the divide, and continued downstream along Quall River to Kitkiata Inlet.

Recreational Fisheries

Sport fishing in the watershed is limited due to access considerations; however, riverboat access provides transportation to the popular chinook fishery on Johnston Creek. In flood flows, a cascade is created at the canyon 1.2 km upstream of the Ecstall River. "Johnston Falls" generally attracts regional chinook angling enthusiasts (Neilson 2002). The Ecstall River and its tributaries are managed as Class II waters year-round, and a Classified Waters License is required (B.C. Fisheries 2005).

Development Activities

The principal development activities within the Ecstall watershed revolve around forestry and utilities. In the recent past, there was also mineral development and town site development.

Forest Resource Development

In 1872, Port Essington was established as a trading post; it then grew into a substantial and thriving community by 1890 with the high demand for lumber and steamboat cordwood. By 1900, the five canneries built at the mouth of the Ecstall River were a hub of activity, and the demand for lumber remained high. By 1906, water-powered Brown's Mill was supplying lumber and box wood for the many canneries on the Skeena River (Large 1996, Harris 1990).

Along the Ecstall mainstem, small-scale logging by hand, boat, and steam power was the preferred method for many years. Early logging occurred primarily within the tidal range and was directed to easily accessible spruce trees, singly and in clumps. Logging these trees by hand, boat, and steam power initially from the river, led to A-frame logging that was utilized into the 1960s to reach further back and up on to the side hills. Brown's Mill continued to operate until the late 1980s.

Large-scale industrial logging first occurred in the watershed in the late 1980s, when the logging road network from the Scotia River was extended through the low-elevation pass and down Carthew Creek. Logging also occurred along the Ecstall mainstem in the mid and late 1980s. There is currently extensive logging in the valley bottoms of Hayward Creek, Carthew Creek, Big Falls Creek, and Brown Lake. Jyrkkanen (1996) discussed logging-related impacts to these subbasin drainages assessed under the auspices of the Watershed Restoration Program (WRP). Progressive clearcuts are proposed for Lockerby Creek and many of the remaining forested stands in the above-mentioned creeks. The logging road network from Kumealon Inlet extends into Brown Lake, where, as of this publication, extensive logging is proposed.

The Ecstall watershed is within the Ministry of Forests, North Coast Forest District. International Forest Products is the major licensee operating in the area. The future trend will be a continuance of the present: logging the highly valuable and easily accessible valley bottom stands, with the priority effort directed to the most valuable stands in order to cover the high road development costs.

Mineral Resource Development

Mineral resource development within the watershed has focused on three polymetallic, volcanogenic massive sulphide deposits located near the Johnston-Ecstall confluence. The zone has been the object of many extensive drilling programs, and in 1999 the B.C. Geological Survey launched a mineral deposit evaluation, mapping, and sampling study of this district. Until economic conditions change substantially, this property will likely not progress to mineral production.

Transportation and Utilities

Two small hydroelectric developments are located within the Ecstall watershed: the Falls River Project and Brown Lake Project. The Falls River Project was built in 1930 for the Northern B.C. Power Co. and acquired by BC Hydro in 1964. The project consists of a dam, two penstocks, and a 7 MW nominal power plant with 27 km of transmission line running down the east side of the Ecstall River. The transmission line continues across the Skeena River and onward to Prince Rupert, providing electric power to the city and the larger electric grid (BCFWRP 2001).

The Brown Lake Project was built in 1997 for Synex Energy Resources Ltd. and consists of a 1.5 m high dam near the outlet of Brown Lake, with a 600 m tunnel to a powerhouse near sea level, generating 6 MW. The power is delivered via submarine cable to the BC Hydro grid and an interconnection with the 69 kV transmission line on the east bank of the Ecstall River (Sigma 1993).

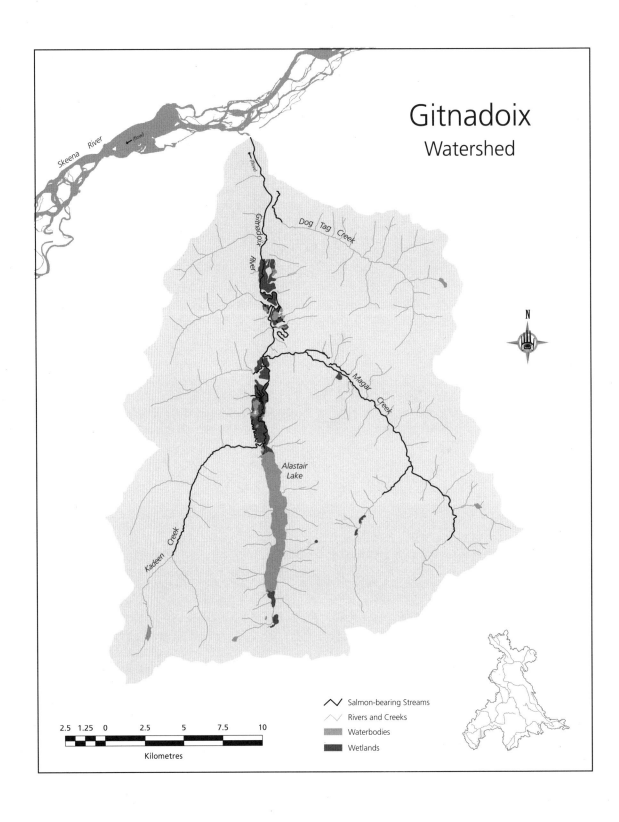

Gitnadoix
Watershed

Skeena River ←(flow)

Gitnadoix River (flow)

Dog Tag Creek

Magar Creek

Kadeen Creek

Alastair Lake

N

Salmon-bearing Streams
Rivers and Creeks
Waterbodies
Wetlands

2.5 1.25 0   2.5   5   7.5   10

Kilometres

7

Gitnadoix Watershed

Environmental Setting

Location

The Gitnadoix watershed is located in the Kitimat Ranges approximately 50 km west-southwest of Terrace. The Gitnadoix River flows north 18 km into the Skeena River left bank. The watershed is bounded in the south by the Giltoyees Creek drainage, to the west by the Khtada drainage, to the east by the Wedeene River, and to the north by the Skeena River floodplain.

Hydrology

The Gitnadoix River is a third-order system that drains a watershed area of approximately 546 km^2. Elevations range from approximately 24–2,180 m, with the watershed bounded by highly dramatic alpine rock and pocket glacial features. Mean annual precipitation for three years of measurement at Salvus, located close to Kasiks River, was 1,783 mm (Beaudry et al. 1990). Because of the strong coastal influence, it is likely that the highest discharge peaks of the Gitnadoix River occur during late autumn, rather than during snowmelt in May and June. Owing to the nature of the steep terrain in the watershed, the tributaries into the Gitnadoix mainstem are often subject to high water flows following heavy rain events.

The Gitnadoix River flows from the headwater lake, Alastair Lake. The major tributaries downstream are Kadeen, Magar, Dogtag, and Clay Creeks, which deliver a substantial volume of the total flow to the river mainstem. Alastair Lake dampens and buffers high water flows from the headwater tributaries. The Gitnadoix flows out of the north end of Alastair Lake with an average width of 45 m and an average gradient of less than 0.3%. The 6.9 km^2 Alastair Lake has an elevation of 30 m, an average depth of 24 m, a summer seasonal average thermocline depth of 10.4 m, and a cool hypolimnion. The deep water (>50 m) in the southern end of the lake is a very hospitable environment for juvenile sockeye. The lake is very productive and has one of the three highest photosynthetic rates ever recorded for B.C. sockeye lakes (Shortreed et al. 1998).

Stream Channels

The Gitnadoix River channel wanders and meanders through its approximately 1 km wide floodplain for the upper three-quarters of its length. The typically steep-sided valley walls limit lateral movement on the upper two reaches. The lower reach has a slightly higher grade and distinct lateral bar development. For the most part, the tributaries present either a stepped or continuous longitudinal profile with low to moderate gradients steepening in the headwaters.

Geography

The bedrock geology of the Gitnadoix watershed is characterized by the Coast Mountain Intrusives, consisting of largely eroded, dramatically bold, impressive monolithic granite (Holland 1976). The Gitnadoix valley is a small north-south lineament that originally was a fiord. Since the end of the Pleistocene, it has isostatically uplifted and, over time, filled with fluvial sediment at least in the northern section. The lower valley and hill slopes are occupied by the moist maritime Coastal Western Hemlock (CWH) zone, which is dominated by hemlock and spruce, with patchy stands of red cedar. The Mountain Hemlock (MH) Zone is reached at approximately 600 m elevation. The mountaintops have alpine vegetation. Annual avalanches, many of which reach the valley floor, sweep much of the mountain area. The upper Gitnadoix River floodplain is primarily swampy with single and dispersed cottonwood and patchy coniferous stands.

Fish Values

High fisheries values are generally prevalent throughout the low-gradient portions of the drainage. All five Pacific salmon species and steelhead are present, as well as rainbow and cutthroat trout, Dolly Varden and bull trout, Rocky Mountain whitefish, and large-scale sucker, peamouth chub, three spine stickleback, and prickly sculpin (Northcote and Taylor 1973). Bustard (2004) described the relative abundance and life-history strategies of char in the watershed from studies conducted in 2003.

Chinook Salmon

Although Gitnadoix River chinook escapement estimates have been recorded since 1967, escapement data are scant for Magar Creek, Dogtag Creek, and Kadeen Creek. These three tributaries each support 20 to 100 spawning chinook. Escapement on the Gitnadoix River was estimated at 750 chinook for 1967 and 1968. The 1970s average annual escapement was 213 chinook, ranging from 400 to 25 fish. The 1980s average annual escapement was 65 chinook, ranging from 200 in 1980 and 1981 to a low of two chinook in 1983. The 1990s average annual escapement was 38 chinook, ranging from 85 to 10 fish. Since 2000, escapement has been similar to the 1990s. However, since 1998, Magar Creek escapement has increased; the annual average for the last eight years has been 155 chinook. Overall, Gitnadoix chinook abundance has decreased and fluctuates at low levels.

The small run of chinook enters the Gitnadoix River in early to mid-August with spawning occurring from late August until mid-September. Knowledge of Gitnadoix chinook life histories is limited. However, critical spawning areas observed include the large, turbulent, lowest reach of Kadeen Creek; the lowest reach of Magar Creek; and a small population (15–20) in the lowest reach of Dogtag Creek (Graham and Masse 1975). Spawning beds exist on the Gitnadoix River mainstem near the mouth for approximately 400 m; on a 300 m section 1 km downstream of Magar Creek; and in dispersed patches throughout the mainstem channels (Pinsent and Chudyk 1973, Hancock et al. 1983, DFO 1991c). Chinook juveniles utilize the mainstem river channels and the lower reaches of accessible tributaries.

Pink Salmon

The Gitnadoix pink salmon run is relatively large for a lower Skeena River tributary, but small in comparison with the total Skeena River pink salmon escapement. Escapement estimates have been consistently recorded since the early 1960s and show an overall moderate increase in escapement. Average annual escapement in the 1960s was 770 for even-year and 3,125 for odd-year pink salmon. In the 1990s, it was 2,300 for odd-year and 7,750 for even-year pink salmon. Reported escapements since 2000 are 1,500 fish or less. Pink generally enter the Gitnadoix River in early to mid-August, primarily spawning in the lower 7 km of the river mainstem.

Magar Creek, Kadeen Creek, and Dogtag Creek also support small pink spawning populations. The few existing escapement estimates indicate that the average annual escapement of the odd-year pink runs in the tributaries is about one-third of the total Gitnadoix River escapement. In late April, pink

Gitnadoix River drains Alastair Lake

View down Gitnadoix River toward the Skeena River

fry emerge and migrate downstream to the Skeena estuary.

Chum Salmon

Chum salmon escapement in the Gitnadoix watershed has been surveyed for all but one year (1999) since the mid-1960s. Average annual escapement in the 1970s was 1,090 chum, with a decrease in the 1980s to 462 chum. The average annual escapement in the 1990s was increased to 614 chum. Chum escapement data for Dogtag Creek show average annual escapements of 42 chum in the 1970s, 575 chum in the 1980s, and 448 chum in the 1990s. Spawning also occurs in Magar and Kadeen Creeks, but escapement estimates are not available.

Chum typically enter the river in mid to late August, spawning in the lower reach of the mainstem in specific patches contiguous to seepage or groundwater flows. Most of the spawning occurs at the mouth of Clay Creek (Kofoed 2001) and in tributary lower reaches. Chum fry emerge and migrate directly to the sea in the late spring.

Sockeye Salmon

From 1950 to the present, sockeye escapement estimates have been recorded for Southend Creek and Alastair Lake. Westside Creek estimates were only recorded in the 1950s, with an average annual escapement of 1,442 sockeye, ranging from 3,500 to 400 spawners. Alastair Lake had an average annual escapement in the 1950s of 5,100 sockeye, which then abruptly decreased and remained moderately steady through the next three decades, with an average annual escapement of 1,691 sockeye. The average annual escapement estimate for the 1990s was 3,300 sockeye, ranging from 5,000 to 500 fish.

Southend Creek is the main producer of sockeye in the Gitnadoix system (Hancock et al. 1983). The average annual escapement estimate for the 1950s was 17,200 sockeye, ranging from 35,000 to 8,000 fish. Sockeye escapement then declined in the 1960s to less than 6,500 sockeye, and declined again in the 1970s to an average annual escapement of 2,740 sockeye. In the 1980s and 1990s, average annual escapement estimates were 4,600 sockeye. Since 2000, escapement has declined to an annual average

escapement of 3,670 sockeye. Gitnadoix sockeye escapement has been consistently depressed below the 1950s levels.

Gitnadoix sockeye are unique in that their spawning colours are atypical of other Skeena sockeye stocks. When mature, these sockeye possess a silvery body colour with only a dull red stripe, rather than the typical mature sockeye colors of a bright red body and a green head. Peak run timing as recorded by the Tyee Test Fishery is June 24–30 +/- 1.5 weeks. Sockeye typically enter the Gitnadoix drainage mid-August to mid-September, with an early run to Southend Creek and a late run to Alastair Lake (Whelpley 2002). They then move rapidly upstream into Alastair Lake, occasionally holding at the outlet and the south end of the lake while waiting for water conditions and ripening (Smith and Lucop 1966). Brett (1952) noted that early and late runs spawn in the east and west forks of Southend Creek, respectively. Westside Creek's very small alluvial fan is heavily utilized for 150–200 m, as is the small alluvial fan of the creek on the east shore, closest to the outlet.

Alastair Lake is the key nursery area for sockeye fry. Shortreed et al. (1998) noted that juvenile sockeye were distributed throughout the lake with high densities (6,200/ha) of limnetic fish (underestimated due to echo counting limitations at these high densities), of which two-thirds were stickleback and the remainder sockeye. Age-0 sockeye growth rates were low with an average of 1.7 g in October, most likely because of the intense competition from the large numbers of stickleback. Sockeye stomachs were only 20% full and contained nearly 100% *Bosmina,* a lower quality food item. In an earlier study, Simpson et al. (1981) also reported that fall fry from Alastair Lakes average less than 2 g. One-half to one-third of sockeye fry spend two years in Alastair Lake. This is an unusual pattern for a low-elevation coastal lake and is presumably an adaptation to the low food supply.

Coho Salmon

The Gitnadoix coho escapement record is relatively complete since the 1950s. The Gitnadoix River average annual escapement for the 1950s was 11,333 coho. There has been a fairly steady decline since that time. Escapement trends of tributary spawning streams closely follow the pattern indicated for the Gitnadoix River except for the lower portion of Magar Creek, which had increased escapement in the 1990s and received most of the coho spawners of the Gitnadoix watershed in 1999. Overall, coho escapement increased markedly in 1999 and 2000 after a decade or more of low escapements. Reported escapement estimates are only available for 2003, when there were 2,000 coho enumerated.

Coho have been observed to move into the drainage from late August to mid-October and generally spawn throughout the system. The most productive spawning grounds are spread along the mainstem of the river. The lowest reaches on Southend Creek, Kadeen Creek, and Magar Creek are the principal tributary spawning grounds (Kofoed 2001). Juveniles rear throughout the system utilizing off-channel habitat, particularly the extensive pond complex in the upper river.

Steelhead

The relatively small steelhead run in the Gitnadoix River is suspected to be a few hundred winter-run fish. There is no known evidence of a summer- or fall-run component (Hooton and Hipp, cited in Paish 1990). Triton (1998e) noted that steelhead migrate upriver between January and May and hold in large pools before spawning in tributaries in spring. Although knowledge of spawning sites is limited, it is thought that the principal spawning grounds are in dispersed gravel patches in the upper reach of the river and in the lower ends of the main tributaries. Rearing and holding areas are abundant and ideal due to the productive aquatic environments of the back channels and oxbow lakes of the upper reaches of the river (Pinsent and Chudyk 1973). Knowledge about winter parr habitat, smolts, sea life duration, and repeat spawning is limited.

Fisheries
First Nations Traditional Use

First Nations traditional use of the Gitnadoix watershed was extensive and varied with village sites, home places, and fish houses or stations. Gitandoix people, a poorly known group whose main village was located at the Gitnadoix-Skeena confluence, held the watershed territories. Dawson (1881) observed house remains standing at this traditional village that secured particularly rich hunting, fishing, and other resources in the drainage's interior territory. Currently, eight Indian Reserves within the watershed represent some of the important fishery and home-place locations.

Recreational Fisheries

The Gitnadoix watershed is now a provincial park and, as such, protects the natural and cultural heritage and recreational values located within it. The major recreational activity on the Gitnadoix River is sport fishing for salmon and freshwater species, principally Dolly Varden and cutthroat trout. There has been a steady increase in sport fishing activity on the Gitnadoix system for as far back as official and anecdotal records are available. This increase parallels a steady increase in other parts of the Skeena watershed.

Gitnadoix anglers are guided and unguided, with both generally targeting the spring steelhead and then the September to October coho fishery. The river once supported a chinook sports fishery, but is now closed to angling due to concerns about low escapement (DeGisi 1997a). There is a concern about the indiscriminate use of large, powerful boats, which detracts from the combined experience of angling and wilderness recreation and causes erosion on the riverbed and banks.

Development Activities

Gitnadoix Provincial Park is a visually scenic, pristine valley that has had no past development; currently, forestry, mineral, and hydroelectric development are prohibited. There is a natural gas pipeline corridor (Milepost 305) passing through the lower watershed approximately 400 m upstream of the Skeena mainstem confluence. Vehicles and machinery occasionally ford the river when replacement or specific maintenance is needed. There are no residences in the watershed, though two cabins support seasonal trapping efforts.

The Gitnadoix watershed has never been subjected to any type of industrial activity that would in any way disrupt or change the natural habitat and natural productive capability of the system. Therefore, the Gitnadoix could serve as a baseline for identification of habitat change and integrity. The principal influences on the biological and productive capability of the anadromous fish stocks are the exploitation of those stocks in the commercial and recreational fisheries on the Skeena, and changes in the ocean environment.

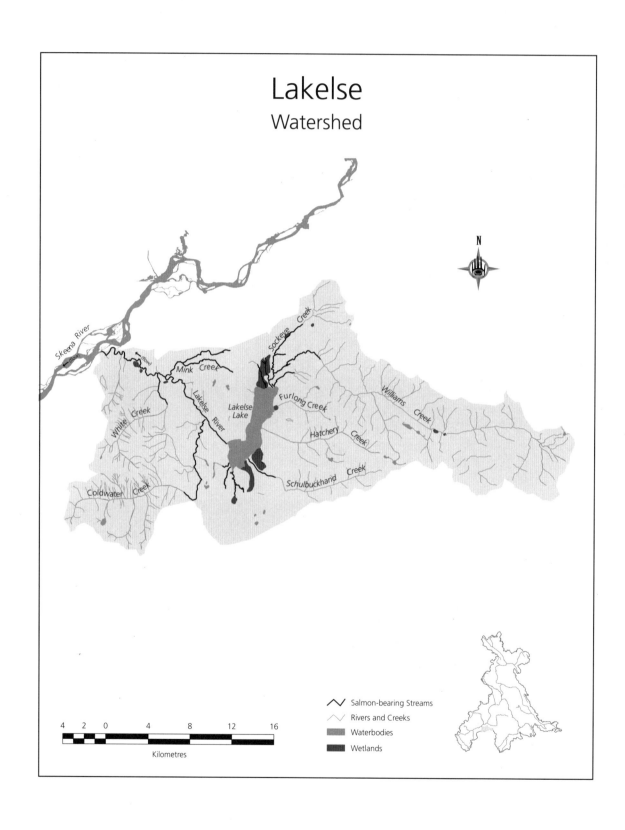

Lakelse
Watershed

N

Skeena River
(flow)
(flow)
Mink Creek
Sockeye Creek
Williams Creek
Lakelse River
Lakelse Lake
Furlong Creek
White Creek
Hatchery Creek
Coldwater Creek
Schulbuckhand Creek

⋀ Salmon-bearing Streams
⋀ Rivers and Creeks
  Waterbodies
  Wetlands

4 2 0   4   8   12   16
Kilometres

8
Lakelse Watershed

Environmental Setting
Location
The Lakelse watershed is located in northwest British Columbia, 20 km south of Terrace. The watershed is bounded to the east and west by the steep mountain slopes of the Kitimat Ranges and to the north by the Skeena River floodplain, while an ancient ice-contact terrace establishes the southern boundary within the Kitimat valley.

Hydrology
The Lakelse River drains Lakelse Lake at the southwest corner, in a northwesterly direction for approximately 15 km to reach the Skeena River left bank, about 17 km downstream of Terrace. The Lakelse River is a fifth-order system that drains a watershed area of 589 km². Elevation ranges from approximately 62 m to 1,845 m. Rainfall records based on a 30-year observation period, 1950–1980, show an annual average precipitation of 1,313 mm, 70% falling as rain and 30% as snow. Rainfall is greatest during September, October, and November, while snowfall is greatest during December and January. Cleugh et al. (1978) estimated that the greatest discharge occurs in the summer months, with maximum discharge in June. Lakelse Lake's mean annual discharge into Lakelse River is 20 m³/second (Kerby 1984).

Lakelse Lake is the predominant feature of the upper watershed. It covers an area of 14.5 km² (14,516 ha), with the majority of tributary streams feeding directly into the lake. Williams, Hatchery, and Schulbuckhand (Scully) Creeks are major tributary streams and the biggest of the 13 tributaries feeding Lakelse Lake. It is thought that Lakelse Lake is the warmest lake in northern B.C. It has a maximum depth of 32 m (at the north end opposite Furlong Bay), but a large portion (42%) of the lake is littoral. This extensive littoral zone affects temperature, dissolved oxygen, aquatic plants, and overall productivity of the lake (Cleugh et al. 1978).

Temperature profiles indicate the absence of a stable thermocline, which is probably a consequence of relative shallowness combined with strong prevailing southwesterly winds (Abelson 1976). Due to the shallowness of the lake and large volumes of water from tributary streams, Lakelse Lake flushes on the average of once every 58 days (Kerby 1984, Remington 1996). A number of hot springs with temperatures up to 85° C occur on the eastern shore of Lakelse Lake.

Tributary drainages are characteristically of two types: a meandering channel at a low gradient or a steeply graded channel within a narrow confined V-shaped valley, becoming braided or meandering on lower-gradient fans. The majority of steeply graded creeks originate in the mountains to the east or west of Lakelse Lake. They are fed from the large area of the alpine mountain slopes, which make up approximately 19% of the total watershed area. The creeks north and south of the lake are meandering in nature.

Stream Channels
The Lakelse River mainstem is nearly 20 km long, with a very low gradient (0%–1%) and no obstructions to anadromous fish species over its entire length. Major inlet streams are Herman, Coldwater, Mink and White Creeks. Stream banks are vegetated and generally stable, though minor amounts of riprap have been placed to increase stability. Sediment loading is high at times; however, this is mostly associated with the Mink Creek earth flow. Bedload movement is minimal with streamflow levels buffered by the lake. The channel is a 40–120 m wide meandering mainstem, occasionally confined with many exposed sandbars at low flows.

Reach 1 of Hatchery Creek extends from the lake upstream to Highway 37. The reach was channelized to support flood control measures across its fan and consequently has concentrated flows and a high degree of bedload movement. The source of the bedload is natural; however, construction of dikes and channelization along the creek does not allow it to be distributed in an alluvial fan (Gordon et al. 1996). Hatchery Creek is an established community watershed, which is defined as the drainage area above the point of diversion on a stream that is licensed under the Water Act for human consumption.

Schulbuckhand Creek, flowing into the southeast end of Lakelse Lake, has received restoration work in reach 1. In reach 2, large quantities of bedload are being deposited from upstream failures, and the flow is sometimes subsurface. Logging and the "Cat Fire" above the Scully Creek fan apex have exacerbated problems associated with levels of sediment and bedload mobilization (Reese-Hansen 2002). Maxwell (2003) reported that 80% of the streamflow in Scully Creek has been diverted by gabion baskets placed at PNG's right of way; other historic channel modifications have also affected the streamflow. This diverted stream now flows through a constructed channel for the lowest kilometre before entering the lake.

Coldwater Creek flows northeast into the Lakelse River 3 km downstream of Lakelse Lake and has several notable tributary streams. Reach 2 acts as a depositional area with associated bank instability, while reach 3 has moderate bedload movement resulting from high sediment contributions from its tributaries.

Mink Creek flows into the Lakelse River from the east about 8 km from Lakelse Lake. It had a massive earth flow (43 ha) of glacio-marine silt and clays in the early 1990s (Geertsema and Schwab 1995). This has led to bank instability, very high sediment loading, and a significantly unstable channel with a sediment wedge moving down the Lakelse River.

White Creek, which flows north into the Lakelse River 9 km below Lakelse Lake, has unstable channel changes related to three major logging-related impacts in reaches 2 and 3. Powerline Creek has been extensively logged with the result of unstable banks and channel (Gordon et al. 1996).

Williams Creek and its three main tributaries, Sockeye, Myron, and Llewellyn Creeks, comprise approximately 25% of the total stream length in the Lakelse watershed. Williams Creek alluvial fan is composed of the lower 6 km of the creek and segmented into four reaches. Reach channels on the fan are unstable; particularly reach 2, which receives large amounts of sediments transported downstream from the unconfined reach 3 and bank erosion in reach 4. Beaver activity followed the 1960s logging on Williams Creek fan, but it wasn't until the 1978 flood that the channel changed dramatically when large volumes of stored sediments worked downstream. The fan is now characterized by avulsions, fan sediments being reworked, and continuation of relatively high sediment transport rates. The 2001 avulsion of Williams Creek into Sockeye Creek left 3 km of creek bed dry at times of low flow (C. Culp 2002).

In 2006, Weilland and Bird conducted sediment source mapping in upland portions of the Williams Creek watershed. The mapping was based on interpretation of four sequential air photo series ranging from 1949 to 2001. The study indicates that the sediment regime is dominated by natural sediment sources and the future sediment transfer of these sources from the upland basin to the fan is expected to fluctuate depending on storm and flood event flows (Weilland and Bird 2007).

Furlong Creek, Granite Creek, Eel Creek, Herman Creek, and Killutsal Creek have received channel disturbance, and in some cases are unstable due to logging and road building activities (J. Culp 200). Maxwell (2003) reported that construction of the highway required diversion of the five original Furlong Creek channels into one consolidated channel and that, in the recent past, major landslides have been observed in the upper reaches. Clearwater Creek, Andalas Creek, Ena Creek, North Ena Creek, Powerline Creek, and Junction Creek, as well as their tributaries, have essentially undisturbed or stable channels.

Water Quality

In 1982, the Waste Management Branch of the Ministry of Environment, Lands and Parks (MELP) initiated a five-year monitoring program on major drainages of the Skeena River watershed. Data were collected monthly from seven stations located on the upper Bulkley River, Morice River, Bulkley River at Quick, Telkwa River, Kispiox River, Skeena River at Usk, and Lakelse River (Wilkes and Lloyd 1990). The Lakelse River station is located near the mouth of the Lakelse River. The Lakelse River is exceptional in that it has a very low TSS (total suspended solids) as a result of being lake-headed (mean = 9.1 mg/L, range = 1.79 mg/L). Turbidity, an indicator of suspended sediments, is also low. The Lakelse River is moderately coloured (mean = 16.7 TCU), which is attributable to natural organic substances that are not harmful to human health. Lakelse pH was near neutral (mean = 7.1), with a range of 6.7 to 7.6. Alkalinity is low (mean = 21.1 mg $CaCO_3$/L) and very near the water quality criterion, indicating the waterbody would be sensitive to acidic inputs. Nutrients are generally low, with metal concentrations often less than detection limits.

Water quality objectives were prepared for Lakelse Lake by the Ministry of Environment, Lands and Parks (McKean 1986). The accompanying assessment states that the impact of forestry on water quality was of concern in the Lakelse watershed, particularly the siltation of spawning and rearing streams. There had, however, been no data collected to quantify the degree of siltation attributable to logging. A specific objective for turbidity was established for tributary streams to the lake to protect spawning areas possibly affected by logging. Monitoring recommendations included intensive turbidity sampling one year prior to and following a logging operation.

Lakelse Lake is considered oligotrophic because of its low phosphorous concentrations, the low oxygen depletion rates of its bottom waters, and low chlorophyll *a* concentrations. These attributes, together with the lake's high water quality, help to account for the recreational and fisheries importance of the lake. Physics (light, climate, and thermal regime) and

Falls on Scully Creek

chemistry levels (nitrogen and phosphorous) suggest that increased nutrient loading would quickly increase lake productivity and phytoplankton biomass in Lakelse Lake. Further, already low nitrogen/phosphorous ratios indicate that increases in phosphorous loading without concomitant increases in nitrogen loading could result in the development of undesirable blue-green algal blooms or eutrophication (Remington 1996).

There are concerns regarding the protection of drinking and recreational waters, particularly associated with the rural residential developments that are served by septic systems, most of which are located in soils with moderate suitability for septic tank tile fields. McKean (1986) reported about 30% of the residential development, or 72 houses, are located on poor landforms. Of these, 60 houses are located adjacent to the lakeshore. Water quality monitoring was carried out between 1988 and 1992. Preparation of a biological and water quality database entailed monthly monitoring (May to September) of drinking water intakes for fecal coliforms and turbidity. The results show that most water quality objectives were met.

In 2000, *Elodea canadensis*, an aquatic invasive plant, was observed colonizing the littoral zone waters of Lakelse Lake. Its growth in the lake over the last several years has reached levels that seasonally occlude beaches and shorelines. As this weed has the potential to severely change fish habitat, a management plan was developed. In 2001, the Lakelse Watershed Society conducted regular shore surveys and documented the rate of *Elodea* colonization. The growth currently occupies most of the volume of several shallow bays and patches of shorelines (Kokelj 2003).

This aquatic growth infestation has the potential to severely change fish habitat. Kokelj (2003) prepared a management plan that addresses the seven community water quality priorities with intent to preserve and protect the quality and health of the Lakelse watershed. This concept has been moved forward by concerned community stewardship groups such as the Lakelse Lake Watershed Society and the Terrace Salmonid Enhancement Society. Water quality and *Elodea* growth continue to be monitored at Lakelse through a partnership with the provincial government, the Lakelse Watershed Society, and the Regional District of Kitimat-Stikine.

In 2002, sediment core samples were obtained from the north and south basins in Lakelse Lake. These cores allow the reconstruction of lake productivity levels over time. Results from the core analysis showed that Lakelse Lake appears to have been oligotrophic to slightly mesotrophic throughout the past several hundred years (Cummings 2002). There has been little change in the diatom species over the last several hundred years, with only small increases in eutrophic species after 1957. These small changes were not large enough to change the estimated phosphorous levels. The sediment core analysis indicated that in the north basin,

sediment delivery rates began to increase in the 1950s and peaked in 1991. In the south basin, loading was highest from 1967 to 1972 and 1981 to 1984 (Cummings 2002).

Significant human activities in the watershed began to occur in the 1950s and included a sawmill operation on the north end of the lake, increased logging activity, highway construction, and creek diversions. These activities may be related to the observed increase in sediment delivery. The changes in sediment input to the lake between 1950 and 1990 may have contributed to the creation of favourable habitat for *Elodea canadensis* colonization (Kokelj 2003).

Geography

The Lakelse watershed is situated within the Kitimat-Kitsumkalum trough, a broad north-south trending depression in the Kitimat Ranges south of Terrace. The deep-seated nature of the eastern boundary fault along the Kitimat-Kitsumkalum trough is demonstrated by hot springs activity. The bedrock geology of the Lakelse watershed is described as late Cretaceous to early Tertiary, consisting of plutonic rocks that form part of the coastal mountain belt. At the eastern end of the watershed, the bedrock geology changes to early and middle Jurassic volcanic and local sedimentary rocks of the

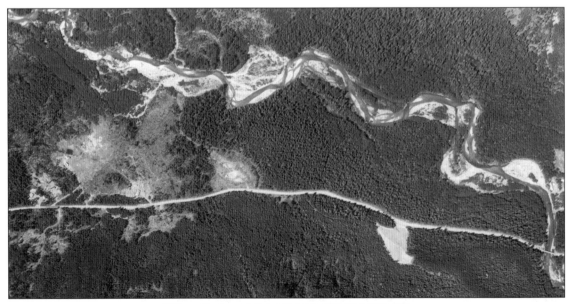

Williams Creek, reach 3, running downstream right to left

Hazelton Group. This group forms part of the larger Intermontane Belt (Clague 1984).

For a short period of time at the end of the Ice Age, between 10,000 and 10,600 years ago, the Lakelse valley was occupied by the sea (Clague 1984, Gottesfeld 1985). The maximum sea level was about 700 m. Due to postglacial erosion and burial beneath alluvium, glacio-marine sediments presently have a patchy surface distribution in the Kitsumkalum-Kitimat trough. The largest area of relic sea floor occurs between the Skeena River and Lakelse Lake. The area is bounded on the north by alluvium of the Skeena River floodplain and bordered on the east and west by alluvium (Williams Creek and White Creek fans) and the walls of the Kitsumkalum-Kitimat trough.

To the south, the relic sea floor is partially obscured by organic deposits but borders against the steep ice-contact face of the large deltaic platform south of Lakelse Lake. The west margin of the deltaic platform north of the lake is a relict foreset slope built into the sea by meltwater streams. The fore slope became inactive when the glacier flowing down the Skeena valley retreated back from the arcuate ice-contact face at the northeast (proximal) edge of the delta. Shortly thereafter, rapid isostatic rebound uplifted much of the glacio-marine terrain in this area above sea level. The continuity of this large area of glacio-marine deposits is broken by drift-veneered bedrock knobs and ridges such as Mount Herman, and by a large ice-contact delta north of Lakelse Lake (Clague 1984).

Certain portions of the low-lying landscape in the Kitsumkalum-Kitimat trough are riddled with earth flow landslide scars. These landslide scars include the two slides that occurred in May and June 1962, with the failures occurring in marine clay overlain in part by alluvial fan sediments (Clague 1978). The May slide between Furlong and Granite Creeks buried 540 m of the old road and the new highway and moved over a distance of 2.4 km on nearly level ground (Evans 1982).

The June slide buried 1.6 km of the old road and the new highway. The most significant example of unstable glacio-marine deposits occurred in December 1993 or January 1994, when 23 ha of glacio-marine sediment, located on nearly level ground, flowed and slid rapidly into Mink Creek, a tributary of the Lakelse River.

The predominant biogeoclimatic zone in the Lakelse watershed is the Coastal Western Hemlock (CWH) zone, which merges into the Mountain Hemlock (MH) zone at approximately 550–650 m. The historic, natural vegetation of the watershed was dominated by old-growth conifer stands (rainforests) of western hemlock, western red cedar, and amabilis fir. Sitka spruce is common but never dominant, and occurs mainly on alluvial soils. Seral stands were uncommon before clearcut logging began on a major scale, except for some south-facing slopes, where lodgepole pine, birch, and aspen were well established. Red alder and cottonwood occur mainly on floodplains and landslide scars where disturbance exposes mineral soil (Banner et al. 1993). The MH zone is distinguished by the presence of mountain hemlock and the lack of red cedar.

Fish Values

The Lakelse watershed possesses very high fisheries values and is a major producer of sockeye, coho, and pink salmon, which are fished commercially and recreationally, making it one of the premier watersheds of the Skeena system. Steelhead and the spring cutthroat runs support a major sport fishery. Resident species present in the system include rainbow trout, Dolly Varden, mountain whitefish, and the following resident fish: prickly sculpin, largescale sucker, redside shiner, northern pikeminnow, peamouth chub, and threespine stickleback. There are no fish species known to be at risk within the watershed.

Chinook Salmon

The chinook salmon population is relatively low in the Lakelse watershed and is one of the smaller Skeena populations. The decadal mean since 1950 is 183 chinook spawners, with a range from 293 in the 1970s to 91 in the 1990s. Chinook enter the Lakelse system in mid-August through early September. Chinook spawning principally occurs below the lake outlet, with limited spawning in the Lakelse River mainstem in a patchwork of small areas (Pinsent and

Chudyk 1973, MoE 1979). Historically, chinook have spawned in low numbers (20–30) in Coldwater Creek, White Creek, Sockeye Creek, and Williams Creek (Smith and Lucop 1966, DFO 2005, Kofoed 2001).

Pink Salmon

The Lakelse River is one of the major pink salmon producing rivers in the Skeena system, with pink escapement exceeding 1.5 million fish in some years. The mid-season pink salmon run typically averages 50% of Area 4 production (DFO 1985). Pink salmon escapement and catch were comparatively high from the early 1980s through the mid-1990s.

The Lakelse pink salmon run enters the Lakelse River in late August, peaks early to mid-September, and ends mid-September to mid-October. The odd-year pink salmon usually enter and spawn over a longer time period than even-year pinks. Pink salmon spawn virtually throughout the mainstem, with extremely heavy spawning taking place between Coldwater and Herman Creeks. This area often has the latest spawning timing in the Lakelse watershed. Wisley (1919) reported that on September 22, 1919, the Lakelse River, from the lake outlet to Coldwater Creek, was literally filled with a mass of spawning humpbacks. This was a common observation throughout the 1920s. The lower reaches of White Creek, Mink Creek, Coldwater Creek, Herman Creek, Scully Creek, Hatchery Creek, and Granite Creek are also occasionally utilized for spawning.

Chum Salmon

The chum salmon run into the Lakelse River is modest. Escapement data are scant, though current escapements are less than 5% of adult chum returns in the 1970s. Run timing typically starts in late August and peaks in mid-September, usually ending by mid-October. Hancock et al. (1983) showed patches of chum spawning grounds scattered sporadically from below Mink Creek upstream to the lake outlet. Chum have been observed spawning at 6 km in Coldwater Creek.

Sockeye Salmon

Sockeye are significant in Lakelse Lake, which, for its area, supported one of the largest sockeye runs in the Skeena. Sockeye spawning population abundance was moderate in the 1950s, increased through the 1960s, severely declined in the 1970s, regained strength in the 1980s and 1990s, and recently declined.

The Lakelse sockeye salmon run usually enters the system in June, holds in Lakelse Lake, and ascends the streams in August (Sword 1904, Whitwell 1906, Bams and Coburn 1962). Williams Creek and Schulbuckhand Creek are the two important spawning streams; however, spawning occurs in the lower reaches of many Lakelse Lake tributaries, including Andalas Creek, Clearwater Creek, Hatchery Creek, Granite Creek, Sockeye Creek, and Blackwater Creek. Williams Creek has excellent beds of medium-coarse gravel, and Sockeye Creek also provides good gravel. Sockeye rear in Lakelse Lake for one year before migrating to saltwater in late May and June. Adults return after two or three years at sea (Rutherford et al. 1999).

Comprehensive propagation and migrant survival studies of Lakelse sockeye were conducted as a component of the Skeena River Investigation (Brett 1952). The Skeena River Investigation occurred from 1939 to 1966 and was focused on increasing sockeye abundance in the watershed with research prioritized at Lakelse

Williams Creek sockeye spawners

and Babine Lakes. Sockeye studies on Lakelse Lake in the early 1960s followed the loss by fire of the pink hatchery facilities at Kleanza Creek. This prompted construction of a hatchery, fish fence, and ancillary facilities at Schulbuckhand Creek. Fish fences, holding ponds, and spawning facilities were also built on Williams Creek. Tow netting, trap netting, and lake-pond studies were used in an effort to observe and better understand sockeye behaviour.

Sockeye escapements to Lakelse Lake have been low in recent years and are depressed relative to historic levels. Based on the last 12 years of visual escapement surveys for Lakelse Lake (1992 – 2003), the Lakelse Lake sockeye stock has experienced a 92% decline over the last three cycles. Exploitation rates for Lakelse Lake sockeye have been low to modest since 1970, primarily because of the early timing of this stock through mixed-stock interception fisheries targeting enhanced Babine Lake sockeye. Fisheries exploitation is not believed to be the major factor affecting escapements and subsequent sockeye production from Lakelse Lake. Kokanee have also been observed in this lake.

Recent lake trophic studies indicate that Lakelse Lake provides a favorable rearing environment for juvenile sockeye. In 2003, juvenile sockeye densities in Lakelse Lake were less than 5% of estimated lake rearing capacity, representing the progeny from just 750 spawners. Lakelse Lake has the capacity to rear the progeny from approximately 29,000 spawners. Lakelse Lake is fry recruitment limited and is producing sockeye well below potential production. Degraded or limited tributary spawning habitat, relative to historic levels, is believed to be restricting spawner access and spawning success. Increasing fry recruitment by increasing escapements, combined with spawning habitat restoration and/or fry outplanting, has been suggested for improving sockeye production from Lakelse Lake.

The Lakelse Sockeye recovery process was initiated when a recovery committee was established in 2003. The committee is composed of the Lakelse Watershed Society, Terrace Salmon Enhancement Society, Kitselas Fisheries, the DFO, and the provincial Ministry of Environment and Ministry of Forests. The recovery strategy created the Lakelse Lake Sockeye Recovery Plan, which includes determining current conditions, setting recovery objectives, implementing recovery activities, and conducting monitoring and evaluation. Recovery activities to date include numerous fish passage improvements, improving several off-channel spawning sites, outplanting sockeye fry, and assessing sediment source and transport and erosion risks on Williams Creek.

Coho Salmon

The Lakelse coho aggregate stock remains one of the most productive coho stocks in the Skeena drainage. Coho escapement in the 1950s annually averaged 21,000 fish, with an increase in the 1960s to 34,000 annual spawners. The coho escapement declined severely by the mid-1970s to an annual average of 8,000 fish, with a similar trend in the 1980s. Declining escapement in the 1990s culminated with a low of 900 coho in 1997. The stock rebounded in 1998 and 1999 to 10,600 and 12,800, respectively. Since 2000, the average annual escapement has been 2,892 coho from a range of 270 to 7,574.

Lakelse system coho enter the Lakelse River in early to mid-September through early or mid-October; and by early December the run has tapered off. Approximately 75% of the spawning occurs in the Lakelse River below the lake outlet. Spawning has also been noted to occur in the lower reaches of White Creek, Mink Creek, Coldwater Creek, Herman Creek, Ena Creek, Andalas Creek, Clearwater Creek, Schulbuckhand Creek, Refuge Creek, Hatchery Creek, Granite Creek, Furlong Creek, Blackwater Creek, Williams Creek, and Sockeye Creek (Hancock et al. 1983, DFO 1991c). Coho use has been nil in the once productive Mink Creek since the large earth flow in the early 1990s (J. Culp 2002).

Dams and Bustard (1996) noted that in 1995, spawning in Clearwater Creek peaked during the last week of October and the first week in November, with completion by the middle of December. The spawning on Sockeye Creek peaked in mid-November to early December, with spawners still present in the last week of December. Coho juveniles are widespread throughout the accessible portions of the Lakelse system.

Steelhead

Information concerning Lakelse steelhead escapement and population trends is not available. Steelhead trout enter the Lakelse watershed in two distinct runs: a spring run from March until May, and a winter run from October until January – the latter being one of a few substantial winter-run steelhead populations in the Skeena River system. Anecdotal information suggests that a summer run of steelhead enters the river in September. Spawning takes place in a patchwork of small areas spread throughout the Lakelse River mainstem (Pinsent and Chudyk 1973). The major spawning area is in the river section immediately downstream of the lake outlet (DFO 1991c). The Lakelse River Project documented steelhead spawning at various sites in the mainstem, from the lake outlet to the Skeena confluence (Whelpley 1983, 1984).

Whelpley (1983) also observed steelhead spawning in the lower reach of Herman Creek, White Creek, and Williams Creek, and evidence of spawning in Coldwater Creek. J. Culp (2002) considered Coldwater Creek as one of the most important steelhead spawning streams in the Lakelse system. Local anglers note that winter-run steelhead generally use the upper river to spawn, while spring-run steelhead utilize the lower river (Whelpley 1984, Grieve and Webb 1997). Tagging records of 347 steelhead collected over a 30-year period showed that a majority of these fish spent three years in fresh water and two or three years in saltwater. These records also indicate that 15.6% of all spawners were repeat spawners.

Juvenile steelhead utilize the low-gradient streams throughout the watershed for rearing.

Lakelse steelhead

Juvenile steelhead or rainbow trout have been observed in the lower and middle reaches of White Creek, Clearwater Creek, Junction Creek, Coldwater Creek, Johnstone Creek, Eel Creek, Ena Lake, and Williams Creek. Although steelhead do not generally overwinter in the Lakelse River or its tributaries, they have been known to do so in the lake (Tetreau 1982, Whelpley 1983, 1984). First Nations anecdotal information also describes the netting of steelhead off the inlets of Andalas and Clearwater Creeks. It is likely that Lakelse system steelhead also overwinter in the Skeena River, both upstream and downstream of Lakelse River (Grieve and Webb 1997).

Fisheries
First Nations Traditional Use

First Nations traditional occupation and use of the Lakelse watershed is thought to be extensive, but information is limited. The English name is derived from Lax gyels, or "place of the mussels," which is a reference to the abundant freshwater mussels in the river. Pre-contact, the Lakelse River watershed was territory occupied by Gitlutzau (Barbeau 1917), considered one of the nine Allied Tribes of the Lax Hwalams (Garfield 1939). At the southern edge of the Lakelse watershed, Haisla territories occupied the flatlands. Gitlutzau, also called Killutsal, was an important settlement on the east bank of the Lakelse River at the Lakelse-Skeena confluence (Dawson 1881). Most of the occupants of this village moved to Port Simpson prior to 1900, and the village was largely abandoned at this time. There are three other documented village sites: Lakgeas, a village site along the Lakelse River approximately 1.5 km downstream of IR 25; a summer fishing village site located just downstream of the outlet of Lakelse Lake; and the fishing station (village) site at the mouth of Herman Creek (Kerby 1984).

Recreational Fisheries

The Lakelse River supports a strong recreational steelhead, coho, and cutthroat and rainbow trout fishery. The cutthroat trout sport fishery is described by Bilton and Shepard (1955), Imbleau (1978), Hatlevik et al. (1981), and DeLeeuw (1991). Due to a substantial winter steelhead run and easy access, there are potentially eight months of steelhead fishing. A large and popular coho fishery takes place in September, particularly on the lower half of the river. Proximity to Terrace and Kitimat and high aesthetic values also contribute to this popular angler destination. The recreational importance, use patterns, and economic values and opportunities were surveyed and documented by Sinclair (1974). Lakelse River steelhead was comprehensively reviewed by Grieve and Webb (1999).

The Lakelse River is designated Class II waters with specific regulations applicable to the river and its tributaries, including use of a single barbless hook, a bait ban, and, on a seasonal basis, catch and release and fly fishing only. The fall and winter fishery (October to January) is principally located from the CN Rail bridge crossing upstream to Herman Creek, with access from Beam Station Road. There is also fishing throughout the entire Lakelse mainstem with several hot spots, but the easiest access points receive most of the angling pressure.

Enhancement Activities

The Coldwater Creek–Lakelse River confluence was the site of the first hatchery in the Skeena system, which was constructed in 1901 and operated until 1920. Fish were trapped for egg take at the mouth of Sockeye River (presently Williams Creek) and taken to the hatchery, which had capacity for four million fry. Coldwater Creek was dammed for a water supply, but the dam failed on a regular basis in 1902, 1903, 1904, and 1905, flooding the hatchery (Sword 1903, Whitwell 1906). Due to cold water and flooding, the hatchery moved in 1920 to Granite Creek and operated until 1935 when it closed due to flood damage and lack of funding. Escapement of sockeye salmon to Lakelse Lake averaged 175,000 fish during operation of the hatchery (Kerby 1984).

From 1960 to 1962, the Fisheries Research Board operated counting fences on the lower Lakelse River, Scully Creek, and Williams Creek. Fish eggs from these fences were raised at an experimental hatchery operated on Scully Creek. Scully and Williams Creeks were the sites of

Alluvial fan of Williams Creek at Lakelse Lake

hatcheries from 1962 till an unknown date, possibly 1967 (Hancock et al. 1983).

Since the early 1900s, remedial work has periodically been implemented on lower Williams Creek to improve fish passage by countering aggradation effects on the alluvial fan. Sockeye Creek received channel improvements as well as logging debris cleanup in the mid-1960s. Various studies for enhancement opportunities were undertaken under the auspices of SEP, particularly reconnaissance for sites with good groundwater flow (Brown 1980). In the 1980s, a small volunteer facility at Howe Creek in Terrace called Eby Street Hatchery began enhancing many of the small streams flowing into the east shore of Lakelse Lake. This group consistently produced coho fry from broodstock collected on Clearwater Creek for at least eight years. In the late 1980s, Deep Creek Hatchery conducted chinook enhancement on Coldwater Creek.

B.C. Ministry of Highways has recently upgraded Highway 37S culverts to open bottom structures that have mitigated fish passage concerns. Lakelse Watershed Society recently conducted activities that have included one project for coho on Schulbuckhand Creek, improvements to and creating several off-channel spawning sites, outplanting sockeye fry, and assessing sediment source and erosion risks on Williams Creek.

Development Activities

Principal development activities in the Lakelse watershed are settlement and housing development, forest development activities, and

Lakelse River counting fence, 1960

transportation and utilities. There is no known mineral resource development in the Lakelse watershed, though there are active claims west of the Lakelse River.

Forest Resource Development

The conclusion of World War II brought a great demand for lumber, and small mills selectively logged portions of the most valuable timber stands. The Whitebottom area of the Lakelse watershed was awarded to Columbia Cellulose as Tree Farm License (TFL) #1 in 1948. In 1960, the area south of Lakelse Lake was awarded to Eurocan Pulp and Paper Co. as TFL #41.

The majority of roads in the watershed were built between 1964 and 1972, when logging was most active. Very few patches of accessible, viably commercial, mature timber were left standing. Over this period, the following areas were intensively logged: Herman Creek, along Beam Station Road; in the drainages of lower Coldwater and White Creeks; and north of Lakelse Lake in the Sockeye Creek and Blackwater Creek watersheds. South of Lakelse Lake there was logging development in the Andalas Creek area, the Ena Lake area of Coldwater Creek, the south end of the lake, parts of the Clearwater Creek watershed, and at Onion Lake flats.

The Lakelse WRP project (Triton 1996b) stated that of the 64 stream reaches rated for impacts to riparian habitats, 25% were rated as having very high impacts, 31% as having high impacts, 22% were given moderate impact ratings, 6% low impact ratings, and 16% had no riparian impacts. Results of the fisheries assessment noted a total of 63 reaches assessed, with 43 reaches being rated as very highly impacted (68%), and eight reaches as highly impacted (13%). Ten reaches were rated as moderate (16%); no reaches were rated with low impacts, while only two reaches had nil impact (3%).

In 1992, the Thunderbird Integrated Resource Management Plan was established with the recognition that future timber harvesting activities would be constrained, due to past practices and the high fisheries, wildlife, and recreation values in a portion of the Lakelse watershed. This plan has since been subsumed into the higher-level Kalum Land and Resource Management Plan (LRMP) with specific directions related to land use in the watershed. The Lakelse River Corridor (1 km width from each bank) has been designated a Special Resource Management Zone (SRMZ) with a conservation orientation to maintain the natural integrity of this highly productive and unique river (MoF 2001). The Lakelse River, Williams Creek, Hatchery Creek, Scully Creek, Furlong Creek, and Coldwater Creek are to be evaluated by the Coastal Watershed Assessment Procedure (CWAP); it is unknown whether the CWAPs have been implemented. The Lakelse Lake, Ena, End, and Clearwater Lakes are to be managed for water quality, fisheries, wildlife, recreation, and other uses. Hatchery Creek will be an established community watershed. Mount Herman and the Lakelse Lake wetlands at the south end of the lake are now established provincial parks.

Population and Settlement

The Lakelse watershed supports a relatively high number of seasonal and full-time residences providing a variety of rural, high-quality lifestyles. According to the Ministry of Environment, Lakelse Lake is believed to be the most heavily utilized recreational lake in the region. Mount Layton Hotsprings Resort, a family orientated facility, operates hot and cold pools, water slides, a restaurant, and a motel on the east shore of Lakelse Lake. The resort has proposed an 18-hole golf course and a convention center. There are two provincial parks, located on the east side and at the northeast corner of the lake, which are popular stopping-off points for local and nonlocal water-based recreation, picnics, and camping.

The east and west sides of the lake have many homes. These developments, with their associated septic systems and occasional stream diversions, may have fish and fish habitat impacts. Property owners have expressed interest in lowering the lake outlet to facilitate property drainage during spring and fall floods. This proposal would directly threaten critical spawning habitat at the lake outlet. Fertilizer use on residential lawns and future developments may also be an issue. There is concern associated with any housing development at or near Lakelse Lake because of the contribution of phosphorus to the watershed.

Federal, provincial, and First Nation governments, as well as community organizations, each representing differing values and interpretations, are working together to implement plans and regulations regarding settlement, water quality issues, recreation facilities, and other developments. The developments surrounding the lake appear to be closely monitored, and this will most likely continue for the foreseeable future.

Transportation and Utilities

A network of transportation and utility systems traverses the Lakelse watershed. Linear development includes Highway 25, a major north-south transportation route connecting Terrace and Highway 16 with Kitimat and tidewater to the south. Alongside the highway are PNG's natural gas pipeline and a BC Hydro major transmission line, which were built in 1957. The transmission line forks north of Lakelse Lake with a branch transmission line heading down the south side of the Skeena River.

Secondary roads through the watershed include two main north-south and two main east-west roads, with many secondary roads providing access to the two provincial parks, forest development activities, and residential developments. Access to the western portion includes the CN Rail branch line, built in 1955,

and the road to the west of it. The majority of roads in the watershed were built between 1964 and 1972, when logging was most active. Approximately 300 km of logging-related roads access most parts of the Lakelse watershed.

Known impacts from linear development within the watershed include degradation of riparian habitat, a reduction in stream channel complexity at stream crossings, channelization, and bank erosion. Current activities are directed towards deactivation of roads and an increasing awareness of the importance of fish, fish habitat, and riparian zones. Both commercial and private floatplanes utilize Lakelse Lake as a base.

Lakelse Lake looking south

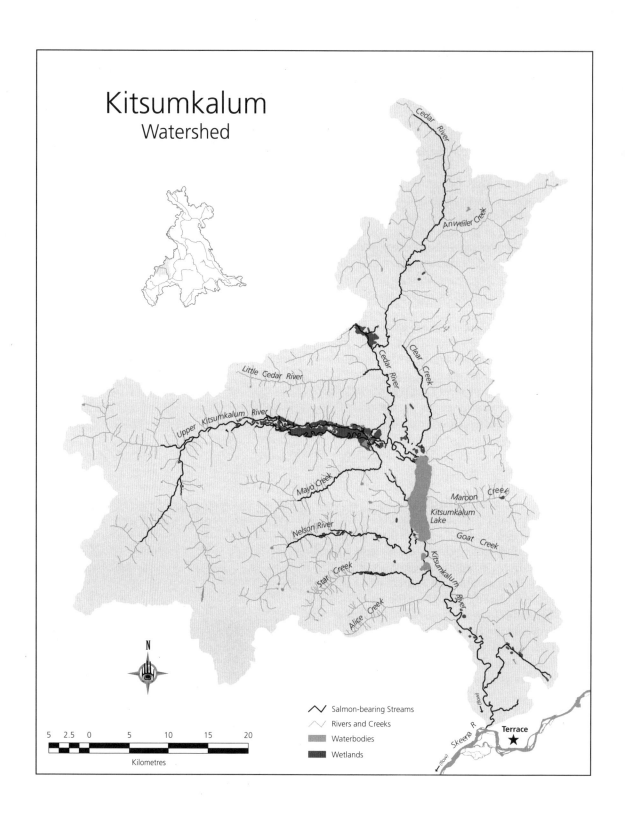

Kitsumkalum
Watershed

Cedar River

Anweiler Creek

Little Cedar River

Cedar River

Clear Creek

Upper Kitsumkalum River

Mayo Creek

Maroon Creek

Kitsumkalum Lake

Goat Creek

Nelson River

Star Creek

Kitsumkalum River

Alice Creek

N

Salmon-bearing Streams

Rivers and Creeks

Waterbodies

Wetlands

5   2.5   0   5   10   15   20

Kilometres

Skeena R.

(flow)

(flow)

Terrace ★

9
Kitsumkalum Watershed

Environmental Setting
Location

The Kitsumkalum River (also known as the Kalum River) flows southerly 30 km from Kitsumkalum Lake to join the Skeena River right bank, 2 km west of Terrace. The watershed is bounded on the south by the Skeena River floodplain, to the north by the Tseax River drainage flowing to the Nass River, to the west by northern Kitimat Ranges, and to the east by the Nass Range. Locally, the Kitsumkalum River upstream of Kitsumkalum Lake is known as the Beaver River, and the section of the Kitsumkalum River from the lake outlet to the canyon is known as the upper Kitsumkalum. From the canyon downstream to the Skeena River confluence, local usage denotes the river as the lower Kitsumkalum.

Hydrology

The Kitsumkalum watershed drains the northern portion of the broad low-lying Kitsumkalum-Kitimat trough. The Kitsumkalum is a fifth-order stream that drains an approximate area of 2,255 km². Elevation ranges from 55 m to just over 2,300 m with spectacular mountains to the west and east, rising rapidly from the valley floor to alpine rock and glacial features. The Kitsumkalum River has a late May to early June discharge peak, due to snowmelt, with subsequent peaks in most years due to fall rains, generally in October and November. Deep Creek, a tributary of the Kitsumkalum and a portion of the District of Terrace water supply, is the location of the only hydrometric station (08EG017) currently operating within the watershed. Terrace precipitation records show annual precipitation of 1,313 mm per year, with 70% falling as rain and 30% falling as snow. Precipitation increases in relation to the elevation gained in the watershed, especially in the Kitimat Ranges. Kitsumkalum Lake somewhat buffers the lower river from extreme flow conditions.

Kitsumkalum Lake, the canyons at 8 km and 11 km, and the Skeena River confluence control the slope of the river. The lower reaches of tributaries draining east into the Kitsumkalum system are generally low gradient (less than 2%), with middle and upper reaches steepening to their headwaters. The exceptions to this general profile are the upper Kitsumkalum River (Beaver River) and the Nelson River, which have relatively low gradients for 40 km and 22 km, respectively. Both streams carry significant amounts of glacial silt from glaciers in their headwaters. The tributaries draining westward into the Kitsumkalum basin trough have relatively low gradients in their lower, high-value, fish-bearing reaches, and then quickly steepen in their headwaters. Clear Creek and the Cedar River drain the generally low-gradient northern section of the trough.

Kitsumkalum Lake at 122 m elevation has a surface area of 18 km² and a mean depth of 81 m. The shoreline of the lake is predominantly rocky and slopes steeply into the lake, restricting productive littoral areas to the shallows at the mouths of tributary streams. The lake is glacially turbid, with an average euphotic zone depth of only 3.8 m, thus limiting primary productivity by light availability. Summer surface water temperature averages 12° C. The average thermocline depth in late summer is 25.5 m. Total dissolved solids of 28 mg/L represent low nutrient levels.

Zooplankton biomass is low and consists mainly of diaptomid copepods. As a result, Kitsumkalum Lake has the lowest productivity of any B.C. sockeye nursery lake studied (Shortreed et al. 1998). Primarily because of its high turbidity and its fast flushing rate, Kitsumkalum Lake is considered highly oligotrophic (Shortreed, cited in Remington 1996). Redsand

Lake, located 2 km downstream of Kitsumkalum Lake, has a maximum depth of 16 m, total dissolved solids of 33 mg/L, and an estimated littoral area (<6 m) of 55% of the total area (Grieve 1996). Treston (Mud) Lake, immediately below Redsand Lake, has a maximum depth of 37 m, total dissolved solids of 31 mg/L, and an estimated littoral area (<6 m) of 50% of the total area.

For much of the summer and fall, the Kitsumkalum River is turbid due to the significant glacial areas in the watershed. This opaque water hinders the observation and recording of fish escapement.

Stream Channels

The lower Kitsumkalum River from Kitsumkalum Lake to the Skeena River is a gravel-bed sinuous river meandering within a valley flat that is 1–2 km wide. In the last 10,000 years, the glacio-fluvial deposits of the valley were incised and terraced by the Kitsumkalum River (Clague and Hiscock 1976). Three canyons were cut into the Hazelton Group: the largest, north of Deep Creek (Kitsumkalum IR 2), was widened in the 1920s to allow cedar poles to be floated down the river from the Rosswood area (Kerby

1984); and the other two are between Lean-to and Canyon Creeks. Stream gradient in the Kitsumkalum River ranges from 0.1% below the lake to 0.5% downstream of the main canyon. The river generally flows free throughout the year, although shelf ice forms during colder winters.

The river channel has made notable shifts in its course since 1947, with log drives having an appreciable effect. Beginning in the 1940s through the mid-1950s, logs were dumped into Kitsumkalum Lake and at several other mainstem locations and driven down the river (McNicol 1999). Modifications to the river between 1954 and 1956 to facilitate log movement included placement of guard logs, log cribbing, placement of rock berms, armouring, and sheer sticks on the outside of river bends; dyking to close off back channels; construction of log storage sites; and the use of dynamite and heavy equipment to dislodge log jams (Grieve 1996, Paish 1975).

Historically, the outlets of both Kitsumkalum and Treston (Mud) Lakes were major spawning beds, until large amounts of gravel were removed to aid log driving (G. Llewellyn, cited in Grieve 1996). River drives of logs caused scouring and increased bedload movement in

Kitsumkalum River lower reach on the right, entering the Skeena River

the channel. The main channel has been more or less stabilized in the half-century since the log drives. The large amounts of habitat lost by closing off back channels has not been restored or reconnected. C. Culp (2002) reported that for several years substantial numbers of chinook spawners have been observed at the outlet of Kitsumkalum Lake; recently, more chinook have been spawning at the outlet of Treston Lake.

In their 1996 WRP overview assessments, Grieve (1996) and Triton (1996) noted many logging related impacts to Kitsumkalum tributary channels. These included changed channel morphology, increased bedload movement, bank failures, sediment loading, and debris accumulation. They note that several streams, particularly Star Creek and the Cedar River, showed heavy erosion and scour.

Water Quality

The city of Terrace landfill, located on the first gravel terrace immediately above the Kitsumkalum River floodplain, has an intermittent creek passing through the landfill and draining to the river. An initial investigation in 1992 (Gartner Lee 1993) showed relatively minor elevated leachate concentrations in the intermittent creek. The pollutants are much diluted in the Kitsumkalum River, where the estimated dilution is 61,500:1.

In 1964, the city of Terrace dammed Deep Creek in its headwaters and diverted water into Spring Creek (Hancock 1983). Further water withdrawals on Deep Creek include the seven licenses downstream. The lower 5 km of Spring Creek have been influenced by rural residential development along its banks, with the creek supporting water removal under 13 licenses. The city of Terrace has a dam for water impoundment with water removal of 0.289 m³/s.

Geography

The Kitsumkalum-Kitimat trough is the result of the deep-seated boundary fault between the Kitimat Ranges to the west and the Nass Range to the east. The Kitimat Ranges bedrock geology is described as late Cretaceous to early Tertiary plutonic rocks that form part of the coastal mountain belt. The Nass Range bedrock consists of early and middle Jurassic volcanic and minor sedimentary rocks of the Hazelton Group (Clague 1984).

The surficial geology of the Kitsumkalum valley bottom is a legacy of the last Pleistocene glaciation. As deglaciation proceeded, the rising sea advanced far up the depressed valleys. A huge valley glacier retreated up the Kitsumkalum valley with three standstills, forming large sand and gravel deposits. The present day Kitsumkalum Lake is impounded by one of these sandy glacial deposits. In the 10,000 years since deglaciation, the valley bottom was modified by stream erosion and minor deposition of sand, gravel, and silts (Gottesfeld 1985).

Vegetation in the wide, gently sloping valley and on the low-elevation slopes consists of forest stands of spruce, hemlock, and limited amounts of cedar representing the Coastal Western Hemlock (CWH) biogeoclimatic zone. As elevation is gained, the CWH is replaced by the Mountain Hemlock zone, while higher yet lies the Alpine Tundra (AT) zone (Banner et al. 1993).

Fish Values

High fisheries values are present throughout the Kitsumkalum watershed. Species present include all five Pacific salmon species and steelhead, as well as rainbow trout and cutthroat trout, Dolly Varden char, bull trout, Rocky Mountain whitefish, largescale sucker, peamouth chub, threespine stickleback, and sculpin (DFO 1991). These fish constitute an important resource to First Nations, recreational anglers, and commercial fishermen who depend on the Kitsumkalum watershed production.

Chinook Salmon

The Kitsumkalum River is one of the three major chinook producers in the Skeena watershed, along with the Bear and Morice Rivers. This stock is especially important, as it has consistently produced the largest-bodied chinook not only in the Skeena watershed, but on most of the Pacific Coast (DFO 1985). In 1984, the Kitsumkalum River summer chinook stock was chosen for monitoring under the chinook "key stream" program, which was initiated in response to objectives set out in the Canada-U.S. Pacific Salmon Treaty (McNicol 1999). The goal was to use escapement and exploitation information from this stock as an indicator of harvest and exploitation rates on B.C. North Coast chinook.

There are two distinct chinook runs in this system. A small early run spawns above Kitsumkalum Lake in the Cedar River and Clear Creek in late July through late August (Morgan 1985, Alexander and English 1996). An unknown number of chinook also spawns in the upper Kitsumkalum River above the lake. A much larger summer run spawns in the Kitsumkalum River, downstream of the lake in September.

Chinook escapement estimates have been recorded since 1961 for the lower Kitsumkalum River, Cedar River, and Clear Creek. Escapement to the lower Kitsumkalum River increased through the 1960s and 1970s, with a substantial peak in the late 1980s and early 1990s that averaged 20,000 chinook annually. From the early 1990s to 2000, there has been a slow decline to an annual average of approximately 10,000 chinook. From 2000 through 2005, the average annual escapement has been 21,000 chinook.

Cedar River and Clear Creek chinook enter the lower Kitsumkalum system relatively early, passing through the coastal zone before the intense commercial fishing activity begins and the run is lightly exploited. Cedar River escapement estimates indicate that approximately 80% of the upper Kitsumkalum chinook run spawns there. The average annual escapement estimate in the late 1950s was 1,125 chinook. The spawning population precipitously declined in the 1960s to an average annual escapement of 94 chinook, ranging from 25 to 200 fish. Throughout the 1970s, a steady increase led to an average annual escapement of 945 chinook, with an average of 890 chinook spawners in the 1980s. In the 1990s, the average annual escapement was 564 chinook. From 2000 through 2005, the average annual escapement has been 1,360 chinook. Cedar River has significantly productive spawning beds upstream to Sterling Creek and possibly further. Clear Creek supports extensive chinook spawning upstream to the second reach break.

Clear Creek escapement estimates decreased from the late 1950s and early 1960s, when the average annual escapement was 333 chinook, ranging from 400 to 200 fish. Between 1962 and 1985, it dropped to 140 chinook. Escapement dramatically increased to 400 chinook spawners in 1986 and 1987, and then fluctuated at low levels, annually averaging 120 spawners to 2005.

Critical spawning beds in the lower Kitsumkalum are concentrated in the lower 3–9 km mainstem section and in the 0.5 km reach downstream of the Kitsumkalum Lake outlet. Spring Creek and Deep Creek debouch into this lower Kitsumkalum spawning reach. Deep Creek spawning has dropped from the 400 recorded in 1965 to 10, 3, and 0 recorded in 1991, 1992, and 1993, respectively. Spring Creek had escapement of up to 75 chinook in the 1960s and 25 in the 1970s (Graham and Masse 1975).

A radio-tagging program was implemented in 1985 to study early-run chinook salmon stocks returning to the upper Kitsumkalum watershed (Alexander and English 1996). The Cedar River and Clear Creek chinook stocks are of concern due to low escapements. The study confirmed the timing separation between the early- and late-run Kitsumkalum chinook, established the early-run spawning proportions (Cedar River 82%, Clear Creek 18%), and provided an estimate of spawning population size through mark-recovery. The early upper Kitsumkalum chinook are smaller in size than the well-known lower Kitsumkalum fish (Triton 1996a).

Following fry emergence, chinook have been observed to rear in the lower reaches of the upper Kitsumkalum, throughout the three lakes, downstream to the Kitsumkalum mouth, and in the lower reach of Spring Creek, Deep Creek, Lean-to Creek, Glacier Creek, Clear Creek, Cedar River, Hadenschild Creek, and Mayo Creek. Juvenile out-migrant enumerations indicate that chinook juveniles migrate out of the Kitsumkalum River into the Skeena mainstem, primarily as 30- to 60-day-old fry, from mid-April to mid-May (Morgan 1985). A small number of 1+ smolts out-migrate, typically in mid-April.

Upper Kitsumkalum (Beaver) River flowing into Kitsumkalum Lake

Pink Salmon

Pink salmon escapement records date back to 1950 for Deep Creek and to 1961 for the lower Kitsumkalum River. The Kitsumkalum pink salmon run has been dominated by the odd-year cohort since the early 1960s. Overall, annual escapement was very low until 1977, when 100,000 pink spawned in the lower Kitsumkalum River. For 14 years until 1991, the variable odd-year escapement ranged from 40,000 to 150,000 pink salmon. Since then, in the few years of escapement observations to 2005, there have been less than 2,000 spawners.

The Kitsumkalum pink salmon run enters the river in early to mid-August and spawns in areas scattered throughout the mainstem up to Treston Lake and Star Creek. There is minor spawning occurring in the lower reaches of the following: Deep Creek, Lean-to Creek, Star Creek, and Spring Creek. Following emergence in the late spring from the spawning beds, fry migrate downstream to the estuary.

Chum Salmon

Chum salmon escapement has been recorded since 1950 for Deep Creek and since 1968 for the lower Kitsumkalum River. Deep Creek escapement estimates peaked in 1950 with 200 chum and have slowly declined to 25–75 spawners over the years. Lower Kitsumkalum River escapement has fluctuated widely since the 1960s, ranging from 10 chum in 1982 to 2,500 in 1988. Decadal average escapements are similar for these three decades at approximately 650 chum spawners. Since 1994, there are no recorded escapement estimates. Anecdotal information suggests that chum spawners are underestimated in the Kitsumkalum system due to glacial turbidity in the river.

The Kitsumkalum River chum run generally enters the river in August, with spawning occurring shortly thereafter on beds mainly from 1 km to 7 km above the mouth of the Kitsumkalum River. Chum have been observed in the extensive eastside oxbow lakes, south of Glacier Creek (Grieve 1996). Occasionally, chum salmon are observed spawning in the lower reach of the small creek just north of Lean-to Creek. The fry emerge from the gravel and migrate directly to saltwater.

Sockeye Salmon

Escapement estimates for the Kitsumkalum aggregate sockeye stock indicate that in the

1950s, average annual escapement was 3,435 sockeye. The average annual escapements of the 1960s and 1970s decreased to 2,650, with a further decrease in the 1980s to 1,430 spawners. The 1990s average annual escapement increased to 3,586 sockeye and further increased from 2000 through 2005, with an average annual escapement of 5,000 spawners. It is assumed that the increase in sockeye escapement is due to the establishment in the late 1980s of the spawning channel at the north end of Kitsumkalum Lake. Bocking and Gaboury (2002) estimated that the total spawning ground capacity is currently in the order of 10,000–11,000 spawners.

The Kitsumkalum sockeye run enters the Kitsumkalum River in August and passes through the lower river to Kitsumkalum Lake. The Kitsumkalum Lake sockeye spawning beds are principally located at the northeast end of Kitsumkalum Lake, where there are good gravel beds as well as seepage of clear water. Shore spawners have also been observed close to Hall Creek and Goat Creek. The Cedar River spawners are in scattered patches, though concentrated spawning takes place close to the mouth of Little Cedar River. Clear Creek spawning is patchy, with critical beds located from 1 km to 3 km above the mouth. The upper Kitsumkalum River is reported to have sockeye spawning to above Mayo Creek. Dry Creek, Mayo Creek, Wesach Creek, and Goat Creek all have spawners at least in some years in their lower reaches. After fry emergence, juvenile sockeye mainly rear in Kitsumkalum, Redsand, and Treston Lakes. Juvenile sockeye have been observed in the lower reaches of Mayo Creek, George Creek, Allard Creek, Cedar Creek, and Clear Creek.

Cedar River sockeye escapement was strong from 1963 to 1974, averaging 1,500 spawners. Since that time, escapement has fallen; recent observations have been few, and it is thought that average annual escapement is currently less than 200 spawners. The Clear Creek sockeye show a similar trend. Since the mid-1960s, the estimated average annual spawner population has decreased from 1,500 sockeye to less than 100 sockeye. For both the Cedar River and Clear Creek a lack of recent escapement surveys is disturbing. Dry Creek decreased from an average annual escapement in the 1950s of 200 sockeye

to 61 in the 1980s. The current status of Dry Creek sockeye is unknown.

Coho Salmon

Escapement records for the Kitsumkalum coho aggregate stock are complete from 1960 to the early 1990s. The average annual escapement estimate for the 1960s was 4,150 coho; for the 1970s, 4,548 coho; and for the 1980s, 3,895 coho. Mean escapement in the 1990s was 2,740 coho, ranging from a high in 1991 0f 7,300 to a low in 1997 of 15. Between 2000 and 2005 the annual mean was 1,100, ranging from 46 to 2,651. Escapement for the last two decades was highly variable.

Upper Kitsumkalum River coho had high escapements of 7,500 coho in 1955 and 1958. Since that time, average decadal escapement estimates have been generally under 1,000 coho. Lower Kitsumkalum River coho escapements were under 1,000 until the 1980s, when the average annual escapement was 2,675 coho, though this apparent increase in escapement reversed in 1992. The Cedar River coho average annual escapement was estimated at 713 during the 1960s; 1,170 coho during the 1970s; and 506 coho during the 1980s,. Average annual escapement estimates have generally stayed below 400 coho for Clear Creek. Escapement for Clear Creek ranges from 1,500 in 1964 to 75 coho over several years in the 1960s.

Coho return to the Kitsumkalum system in two main runs. The early run typically arrives in late July to early August, making its way up into Kitsumkalum Lake where they hold and mature. The late-run coho enter the system through September and into October and typically spread throughout the system to their respective spawning grounds. The main coho spawning production area is in the side channels and groundwater-fed beaver ponds of the lower Kitsumkalum River; however, coho also spawn upstream to 45 km (DFO 1991). Dry Creek had an average of 350 spawners between 1965 and 1974. Clear Creek supports coho spawning to the upper reaches, while coho in the Cedar River utilize spawning beds up to 25 km. Hadenschild Creek has coho spawning up to at least 5.6 km.

Historically, the lower reach of Spring Creek supported 300 spawners (Graham 1975) with

Deep Creek supporting 580 coho spawners. Graham (1975) reported 50 coho spawners in Glacier Creek. Goat Creek had a 10-year average (1965–1974) of 225 in the lower 2–3 km. Star Creek has coho spawning in the lower reach (Paish 1975), while the majority spawn in the marshes of upper Star Creek (C. Culp 2002). Mayo Creek has spawning in the lower 6 km. After emerging as fry, juvenile coho utilize all the forenamed streams for rearing. Other areas used for the freshwater rearing phase include the lower reaches of Luncheon Creek, Benoit Creek, Allard Creek, George Creek, Redsand Lake, and side channels throughout the system.

Steelhead

Information concerning Kitsumkalum steelhead escapement and population trends is not available. The Kitsumkalum watershed has three main timing return groups: one peaking in the late summer, another in the late fall, and a third in the spring (DFO 1991). Only summer-run steelhead were reported to spawn upstream of Kitsumkalum Lake. Radio telemetry studies (Lough and Whately 1984) in the 1980s and tagging records point to the spatial segregation of the runs. Summer fish move above the lake to overwinter; fall-run steelhead move into the middle portions of the mainstem and lakes to overwinter; and the spring arrivals remain in the lower river to spawn (Grieve and Webb 1999a). Major overwintering areas are the upper Kitsumkalum–Cedar confluence, the lakes, and slow, deep portions of the Kitsumkalum mainstem, particularly above and below the canyons.

Spawning sites recorded above Kitsumkalum Lake include the upper Kitsumkalum River, the Cedar River mainstem and its tributaries, Hadenschild Creek, and Clarence Creek, as well as the Clear Creek mainstem. C. Culp (2002) reported substantial numbers of steelhead spawners in Deep Creek, with lesser amounts in Lean-to Creek and Glacier Creek.

The Kitsumkalum Lake outlet provides high-quality and well-used spawning habitat that extends 750 m downstream. Further downstream, spawning occurs on the Kitsumkalum River mainstem to the Skeena confluence and occasionally in side channels from Alice Creek to Lean-to Creek. Spawning steelhead have also been reported in Deep Creek and Lean-to Creek. Repeat spawners are reported to be 7.5% of total spawners (Grieve and Webb 1999a). Lough and Whately (1984) reported most spawning during early and mid-May, with most kelts leaving the Kitsumkalum by June. Juvenile

View west at Terrace

steelhead are reported to be present throughout most of the watershed streams.

Fisheries
First Nations Traditional Use
The Kitsumkalum people are part of the Tsimshian Nation. They occupied the Kitsumkalum and Zymacord watershed territories, although the Cedar River watershed has stated Gitxsan interest. Traditionally, the main town was Dalk Gyilakyaw, which is usually translated as Robin Town (McDonald 2003). Dalk Gyilakyaw is situated at the main canyon on both sides of the Kitsumkalum River. Gitxondakl, another important town, was located between Dalk Gyilakyaw and Kitsumkalum Lake. Kitsumkalum village, the one from which the people take their name, is located at the mouth of the Kitsumkalum River (McDonald 2003). The strong economy was based around the summer salmon food fishery and mid-winter feasting, with dispersal into smaller family groups during the rest of the year to fish, hunt, and gather on the House territories.

Recreational Fisheries
A large sports fishery attracting local residents and nonresidents occurs within the Kitsumkalum watershed. Angling effort is primarily directed to coho, steelhead, and chinook. The river is Class II waters and is open between December 1 and May 31. It requires a steelhead stamp and the release of steelhead angled in the watershed above the signs below the lower canyon. The steelhead fishery begins with the arrival of summer steelhead in late August to early September, with angling distributed throughout the entire river. Winter steelhead fishing is concentrated in the sections from the Kitsumkalum Lake outlet downstream to Glacier Creek, with less effort directed to the upper Kitsumkalum and Cedar Rivers. Coho angling effort is concentrated at the north end of Kitsumkalum Lake in September and continues in the lower Kitsumkalum through the fall to early November.

Enhancement Activities
In 1982, a pilot hatchery was operated on Dry Creek to investigate the incubation and rearing of chinook salmon on that creek; however, the project was found to be unfeasible. The Deep Creek Hatchery has been operated by the Terrace Salmon Enhancement Society (TSES) since 1983, initially functioning as a chinook facility to augment chinook populations in the Zymoetz, Lakelse, and Kitsumkalum systems (Tredger 1983b).

Since 1986, the hatchery has augmented only Kitsumkalum chinook. TSES annually conducts a mark recapture program and a dead recovery program providing an estimate of abundance for Kitsumkalum chinook. The society also takes on average 250,000 chinook eggs to be raised in the hatchery, and, since 1998, assesses coho stocks on a variety of Skeena tributaries that include the Lakelse, Zymacord, Kitsumkalum, Zymoetz, and Exstew (C. Culp 2002). TSES also assesses local chinook and coho stocks.

Kitsumkalum Hatchery, which has been operating since 1986, focuses on incubating and rearing Kitsumkalum chinook and coho and Skeena chum. Recent enhancement of Kitsumkalum chinook and coho has been in partnership with Deep Creek Hatchery.

A spawning channel for sockeye at the northeast end of Kitsumkalum Lake was built in early 1984, upgraded in the late 1990s, and continues to function well (C. Culp 2002). From 1985 to 1988, TSES reared Kitsumkalum summer-run steelhead to fry at Deep Creek Hatchery, releasing close to 50,000 fry annually. Various programs over the years have investigated potential habitat enhancement possibilities, fish abundances and population sampling, salmon habitat surveys, and identification of groundwater upwelling areas.

Development Activities
The principal development in the Kitsumkalum watershed consists of forestry, linear, mineral resource, and settlement activities.

Forest Resource Development
Timber in the Kitsumkalum watershed is administered by Ministry of Forests, Kalum Forest District. At the conclusion of World War II, a great demand for lumber led to selective logging of the most valuable

timber stands, which were processed in small bush mills. Large-scale industrial logging started in 1950 with the award of TFL #1 to Columbia Cellulose. Over the last 50 years, the Kitsumkalum watershed has been extensively logged causing very high impacts to fish habitat. The effects of this long history of disturbance have, for the most part, been assessed by the Watershed Restoration Program (WRP).

Watershed health is improving, and hydrological recovery is underway with approximately 200 km of road deactivated, including pullback of steep fill slopes. Riparian, instream, and off-channel habitat restoration works are helping to re-establish more stable channels and diverse habitats in tributary streams (Ottens 2002). The WRP has completed a full Restoration Plan (RP) for the Clear Creek and Douglas Creek watersheds, while interim RPs have been composed for the remainder of the Kitsumkalum watershed, excluding the Cedar River. Watershed restoration activities in the Kitsumkalum have tapered off since 2001 when government ministries changed their business plans and the program was disbanded. In addition, the major licensee Skeena Cellulose Inc. went bankrupt.

The current health of the watershed is affected by past land-use practices that have produced cumulative impacts on fish habitat. The most important watershed management issue is dealing with the impact of past logging. The effects of increased erosion, sediment movement, and riparian destruction have declined in the past two decades, but will remain as important issues for the next generation. Future resource development will be guided under the direction statements contained within the recently completed Kalum Land and Resource Management Plan. Proper land management practices must be promoted to maintain the integrity of streams and riparian zones.

Mineral Resource Development

Within the watershed there are approximately 30 mineral occurrences including two formerly producing mines. Recent activity has been dominated by the Eagle Plains exploration program located 38 km northwest of Terrace. The prospecting and geological work on this zone of gold-silver mineralization has included airborne and geochemical surveys, geological mapping, and diamond drilling.

Population and Settlement

Relatively minor settlement has occurred in the watershed. Dutch Valley, at the mouth of Spring Creek, and Rosswood located north of Kitsumkalum Lake, provide scattered rural residences for approximately 300 people. The city of Terrace, located at the southeast edge of the watershed, along with its adjacent suburbs, has a population of approximately 19,650 people. The economy is dominated by the forest industry, although tourism is growing and contributing to the growth of sport fishing. It is estimated that 25% of the area's population are freshwater sports fishing enthusiasts, fishing approximately 40,000 days per year in the lower Skeena valley region. Terrace serves as the regional community and economic services center.

Transportation and Utilities

In the Kitsumkalum watershed, the two main roads – the west side road and the Nisga'a Highway – branch off Highway 16 and provide access to communities to the north in the Nass valley. These also access a network of secondary forestry roads that branch out into most tributary drainages. To an extent, fish access has been impeded due to culvert placement and rerouting of watercourses on the main north-south roads that cross all major tributaries in the watershed. Secondary forest development roads crossing unstable slopes, as well as roads crossing alluvial fans of high-energy systems, have had and will continue to have impacts to fish-bearing streams with high rehabilitation costs. Utilities corridors exist on both the east and west sides of the valley, generally paralleling the main roads. A BC Hydro transmission line passes north-south on the west side, while a BC Hydro transmission line and Telus telephone line run through on the east side. These utilities have had few impacts, which are mainly associated with riparian zones.

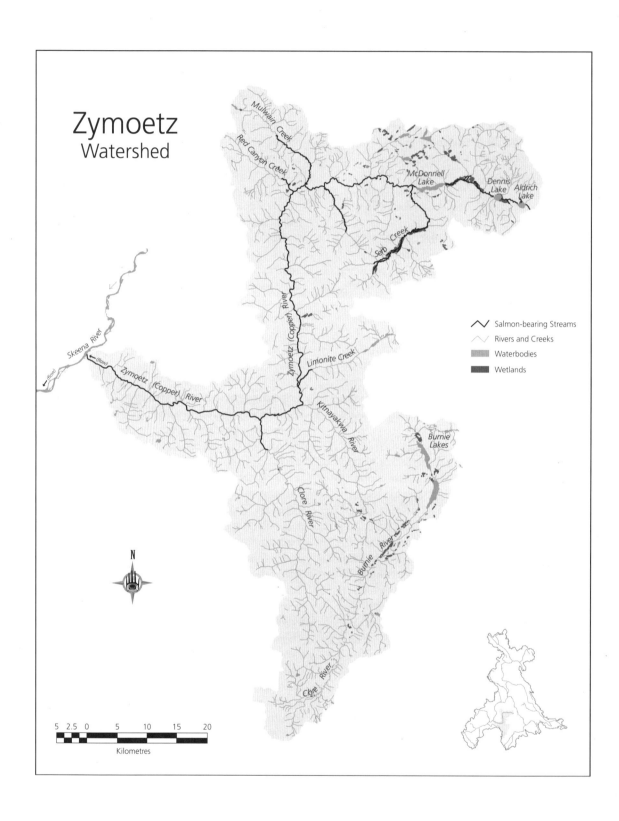

Zymoetz
Watershed

Mulwain Creek

Red Canyon Creek

McDonnell Lake

Dennis Lake

Aldrich Lake

Serb Creek

Skeena River

(flow)

(flow)

Zymoetz (Copper) River

Zymoetz (Copper) River

Limonite Creek

Kitnayakwa River

Burnie Lakes

Clore River

Burnie River

Clore River

N

Salmon-bearing Streams

Rivers and Creeks

Waterbodies

Wetlands

5 2.5 0   5   10   15   20

Kilometres

10
Zymoetz Watershed

Environmental Setting
Location
The Zymoetz watershed is located in northwest British Columbia, in the south central portion of the Skeena watershed. The Zymoetz River, locally known as the Copper River, flows generally westerly into the Skeena River left bank, approximately 8 km northeast of Terrace. The headwater lakes in the north are approximately 20 km southwest of Smithers.

Hydrology
The Zymoetz River is a sixth order system that drains a watershed area of approximately 3,028 km². It is a major tributary of the Skeena River, contributing approximately 10% of the flow (Kerby 1984). The Zymoetz River drains a portion of the Bulkley Ranges of the Hazelton Mountains. It has approximately 20 salmon-bearing streams, each with tributaries. Elevation ranges from 120 m at the Skeena-Zymoetz confluence to 2,740 m in the Howson Range. McDonell Lake is situated at 830 m elevation.

Snowmelt controls the hydrology with a mean annual discharge for the system of 105 m³/s. Monthly mean discharge ranges from a low in March of 25.7 m³/s to a high in June of 358 m³/s (WSC gauging station 08EF005 1963–1993). The river shows a prolonged late-May/early-June discharge peak due to snowmelt, and in most years, one or more fall floods following rainstorms. In about 40% of the years, the annual peak flood is a fall storm event.

The headwaters of the Zymoetz watershed include Hudson Bay Mountain (2,250 m) near Smithers, the rugged and heavily glaciated Howson Range, the Burnie Lakes basin, and the northern slopes of Atna Peak at the head of the Clore River. Important tributaries are the Clore River with a drainage area of 625 km², Kitnayakwa Creek with a drainage area of 274 km², and Limonite Creek with a drainage area of 83 km². Hydrological characteristics, landforms, and stream processes vary greatly throughout the drainage due to its large size and transitional climate.

Stream Channels
The Zymoetz River mainstem arises from a chain of headwater lakes – Aldrich, Dennis, and McDonell Lakes – and flows approximately 120 km to the confluence with the Skeena River.

The Zymoetz River forms a wandering gravel bed channel for about 6 km above its mouth; it is then confined within the lower canyon for 3 km and in the upper canyon for 2 km. There is a 10 km stretch of unconfined river between the canyons. The river then widens out again into a multi-channel, wandering gravel bed river for about 20 km up to the Clore River confluence. Typical elements of a wandering gravel bed configuration include a wide continuous floodplain with a mostly unconfined channel, periodic channel avulsions, and several flood channels. The Zymoetz River above the Clore River is confined by numerous bedrock obstructions.

The Zymoetz River is a volatile system subject to extreme discharges and channel changes. The river experienced extreme fall flood events in 1935, 1936, 1945, 1951, 1961, 1962, 1964, 1966, 1974, 1978, 1987, 1988, 1991, 1992, and 1993. Extreme spring flood events occurred in 1936 and 1950, with a 1954 mid-winter thaw flood event (Septer and Schwab 1995). Dramatic changes in channel morphometry occur frequently.

Subsequently, Weilland and Schwab (1996) conducted a study to provide geomorphologic information on the floodplain in order to document the history of channel changes and provide management interpretations. The study assesses historic channel changes from 1949 to 1992, noting that until 1973, the channel

Lower Zymoetz River mouth, Skeena River in the foreground

was characterized as relatively stable. Heavy autumn rainstorms and related extreme peak flows are considered the primary reason for the onset of lateral channel instability and bank erosion commencing in 1974. The period of accelerated floodplain erosion and frequent large flood events coincided with road and pipeline development and clearcut harvesting on the floodplain. During the 100-year flood in 1978, bank erosion, rapid meander bend migration, and channel avulsions mobilized a large bedload volume. This large volume of highly mobile bedload will continue to pulse down-channel and destabilize the floodplain for many years to come.

The distribution of large organic debris (LOD) within the lower mainstem is similar to most coastal systems where the majority of LOD is ephemeral, and only piles deposited during extreme high waters will be retained for any length of time. Typically these piles will be left high above normal water levels and are consequently of little value to fish habitat. Approximately 70% of the LOD piles in this system are ephemeral (Pollard 1996).

Riparian habitat has been substantially reduced within the lower watershed; first by fire, later through logging, and then by pipeline and electric transmission corridor clearing activities. Lewis and Buchanan (1998) indicated that the loss of 25% to 30% of the off-channel habitat was due to road construction practices.

Geography

Regional bedrock consists mainly of early Jurassic to Cretaceous age sedimentary and volcanic rocks with small intrusions of late Cretaceous to early Tertiary granitic rocks (Hutchison et al. 1979). Red volcanic rocks of the Telkwa Formation are abundant in the central part of the watershed, lending a characteristic reddish colour to turbid floodwaters. As valley glaciers melted following deglaciation, silty-gravelly kame deposits were laid down against the lower valley sides, where they now form terraces. The long, comparatively narrow valley bottom was filled with glacio-fluvial sediments that were dissected and terraced when the river down cut (Clague 1984).

The forested valleys and hill slopes are mainly covered by the Coastal Western Hemlock (CWH) biogeoclimatic zone that changes into the Mountain Hemlock (MH) zone at higher elevations. The upper Zymoetz River watershed is characterized by its broad valley that makes a gradual transition into the Nechako Plateau, where it is dominated by the Sub-boreal Spruce zone with Engelmann Spruce–Subalpine Fir (ESSR) zone at higher elevations. Engelmann Spruce-Subalpine Fir dominates forested areas in the upper Clore and Burnie Rivers.

Fish Values

High fisheries values are prevalent throughout the Zymoetz watershed with use by all species of Pacific salmon, steelhead, rainbow trout, burbot, cutthroat trout, Dolly Varden char, Rocky Mountain whitefish, sculpins, longnose dace, resident sockeye (kokanee), longnose sucker, bull trout, and peamouth chub (DFO 1991b, DFO 1991c, Triton 1999). Burbot and kokanee are present only in Burnie Lakes (B.C. MoE). Cutthroat and rainbow trout, bull trout, mountain whitefish, and Dolly Varden char are found throughout the river and accessible portions of most tributaries. The Zymoetz River is relatively steep. Two canyons located 6.4 km and 19.6 km upstream of the Skeena confluence obstruct pink and chum salmon passage.

Chinook Salmon

Chinook escapement information is limited to the lower Zymoetz River, with estimates available from 1968 until 1994. The average annual escapement was less than 300 chinook in the 1970s, increasing to an average of 700 chinook spawners in the 1980s. Between 1990 and 1994, an average of 375 adult chinook returned annually (DFO 2005). There were no recorded escapement surveys from 1995 through 2005.

Chinook enter the Zymoetz River in late June, with spawning occurring from the end of August to the end of September. The presence of chinook salmon is documented in the Zymoetz River from the second canyon upstream to Limonite Creek, although no barriers exist that would definitely prevent them from migrating farther up the system (DFO 1991c). C. Culp (2002) reported that chinook salmon have been consistently observed for 10 years at Corner Creek, close to Red Canyon Creek. Bustard (1993b) reported scattered chinook spawning in a side-channel site 2 km downstream from Red Canyon Creek.

Critical spawning habitat occurs in patches throughout the mainstem, the lower 3 km reach of Limonite Creek, and in the lower reach of Thomas Creek, a tributary of the Clore River (DFO 1991c). Bustard (1996c) reported that chinook are present in the Clore River to at least 27 km and they are expected to extend farther upstream. The lower reach of Salmon Run Creek has a high concentration of chinook spawners (Kofoed 2001). The lower portion of Cole Creek has not been accessible to chinook for several years due to beaver dams (C. Culp 2002). Simpson Creek supports on average 30 chinook spawners from six years of estimates. Although enhanced chinook have been released at McDonell Lake, no returning adult chinook have been observed.

Pink Salmon

Pink salmon escapement in the Zymoetz system has been recorded since the early 1970s and is relatively small. The last two decades averaged approximately 2,000 pinks annually, with annual ranges from 75 to 35,000 fish. Pink salmon enter the Zymoetz River in August and spawn in September and October within the largely unconfined first reach below the canyon. Upon emergence from the gravel, pink fry migrate downstream to the ocean (DFO 1991c).

Chum Salmon

Chum salmon escapement into the Zymoetz River was recorded from 1960 to 1989. The 1960s annual average escapement was 50 chum salmon; the 1970s annual average escapement was 322 chum; and the 1980s annual average escapement was 50 chum. The highest escapement estimate was in 1978, with 750 adult chum returns (DFO 2005). Chum salmon enter the Zymoetz River in August and spawn in September and October within the largely unconfined first reach below the canyon (DFO 1991c). It is likely that habitat loss due to the repositioning of the Highway 16 bridge and to channelization efforts below the bridge in the 1970s and early 1980s has contributed to low chum returns.

Sockeye Salmon

Sockeye escapement records for the Zymoetz River indicate moderate fluctuations of abundance in the last 50 years. Average annual escapement in the 1950s was 2,550 sockeye, ranging from 5,000 to 750 fish. The annual average escapements of the 1960s and 1970s were under 1,500 fish, while the average annual escapement of the 1980s was 1,860 fish. The 1990s escapement data are incomplete; however, the 1990 to 1994 average annual escapement was 3,650 sockeye, with a high of 7,500 in 1993 (DFO 1991b, DFO 2005). Four years of surveys from 2000 to 2005 averaged 3,670 spawners annually.

Sockeye enter the Zymoetz River in July, spawning primarily during the months of August and September in the upper watershed. Critical spawning areas are in the Zymoetz River mainstem from Serb Creek to McDonell Lake,

Debris flow (landslide) deposit in the lower Zymoetz River, 2003

and the reaches upstream of McDonell Lake to Aldrich Lake. Upstream of McDonell Lake, the meandering low-gradient reaches and the lakes themselves, are stable with moderated flow and temperature regimes. This area supports the majority of the spawning. Several inlet streams to McDonell, Dennis, and Aldrich Lakes, particularly lower Silvern Creek, are also reported to be used for spawning (DFO 1991b).

Coho Salmon

Anecdotal information indicates that there are distinct run components for upper and lower Zymoetz River coho. Lower Zymoetz River escapement averaged 820 coho in the 1960s, 1,925 coho in the 1970s, 6,700 coho in the 1980s, and 500 coho in the early 1990s. There have not been any recorded escapements since 1992.

Lower Zymoetz River coho enter the river in early August, and spawn throughout the months of September to December. Critical spawning habitat in the lower Zymoetz River includes the lower 8.5 km reach of Clore River and the lower reach of Salmon Run and Thomas Creeks. Bustard (1996c) reported that coho are present in the Clore River to 17 km and are expected to extend farther upstream during certain years of peak abundance. Moderate concentrations of spawning occur in the lower mainstem side channels.

Upper Zymoetz River shows a long-term decline in coho from about 1960 to 1990. Escapements averaged 1,861 in the 1950s and 622 in the 1970s. The three counts made in the 1980s and 1990s averaged only 7 coho. Coho

escapements in the lower portion of the Zymoetz River have also declined since 1972. The average was 1,925 coho in the 1970s and 325 in the 1990s. The upper watershed experienced a recovery in coho escapement in 1990, when the escapement was 776. Escapement has more recently increased, with the four years of surveys since 2000 showing annual averages of 1,870 coho.

Coho juvenile rearing and overwintering areas are in accessible low-gradient habitat. Critical rearing and overwintering habitat for coho include the off-channel habitats adjacent to the mainstem between McDonell and Aldrich Lakes, and the lakes themselves (Triton 1999). In the upper Zymoetz River, spawning occurs in the mainstem at the outflow of McDonell Lake and in the following tributaries: Coal Creek, Sandstone Creek, Willow Creek, Passby Creek, Silvern Creek, and Serb Creek. In addition, the lower ends of many of the small tributaries flowing into the upper Zymoetz mainstem between McDonell and Aldrich Lakes support coho spawners (DFO 1991b, DFO 1991c).

Steelhead

Steelhead trout are the most intensely documented fish species within the Zymoetz River watershed, most likely due to the high value of the recreational steelhead fishery. However, escapement estimates are not available. Since 1999, the Terrace Salmonid Enhancement Society (TSES) has conducted steelhead enumerations in the Zymoetz and Clore Rivers (C. Culp 2002).

Adult steelhead enter the system from July to November and spawn the subsequent year in late May and early June. Zymoetz River steelhead are believed to be a mixture of summer-run and winter-run fish, though they are predominately summer-run (Tetreau 1997, cited in Lewis and Buchanan 1998). Local anglers report that a spring run of steelhead enters the system in March and April (Lewis and Buchanan 1998). This has not been clearly documented to date but is clearly feasible considering the proximity of Kitsumkalum and Lakelse Rivers, both of which have strong spring runs of steelhead. Repeat spawners compose

Lower Zymoetz River canyon

View west over McDonell Lake

16% of Zymoetz River steelhead, similar to the proportion (17.9%) found within the Kispiox River (Whately 1977).

Most summers, steelhead spawn in the upper 20 km of the river, with approximately 15% of the fish spawning at the outlet of McDonell Lake. Approximately 30% of the steelhead appear to spawn in tributary streams including Serb Creek, Willow Creek, and Coal Creek, and in the mainstem between Coal and Sandstone Creeks (Lewis and Buchanan 1998). Anecdotal information suggests that the Clore River also has a significant run, though there is little information available on it. Bustard (1996c) reported steelhead presence in the Clore to the canyon at approximately 38 km.

Steelhead overwinter in McDonell Lake and in areas of the mainstem between Limonite Creek and the Clore River. Radio telemetry studies conducted on Zymoetz River steelhead in 1978 and 1979 indicate that steelhead winter in the Skeena River and move into the Zymoetz system in the spring.

Fisheries
First Nations Traditional Uses

Traditional use of the upper Zymoetz River watershed by Gitxsan and Wet'suwet'en people was extensive, and village sites, home places, and fish houses existed throughout. A major aboriginal trail provided connection from the upper to the lower Zymoetz River and to the Skeena River. Another major trail went through Limonite Creek and passed down the Telkwa River. Major trails to the upper Copper River and McDonell Lake connected the villages at Gitsegukla and Moricetown Canyon.

The aboriginal fishery primarily utilized a weir at the outlet of McDonell Lake and spears at the Six Mile flats close to Dennis Lake (Naziel 1997, Rabnett et al. 2001). Information is unknown about aboriginal fishery activities on the lower Zymoetz River, although the mouth of the river is within IR 5, and it is assumed that a fishery was operated at the mouth and possibly the first canyon.

Logging, roads and power line development on the lower Zymoetz River

Recreational Fisheries

The Zymoetz River is considered one of the top-ten steelhead rivers in B.C. (Bustard 1975). Since the advent of logging and consequent easy access, angling pressure has increased over the past 35 years, particularly in the lower and upper reaches. Angler success rates appear to be dropping from their high point in 1975, though reasons are difficult to clearly define. The upper reaches of the watershed have been designated as a special management area, and special restrictions apply. Estimated annual steelhead catch is 1,700 fish, which includes guided angling effort (Lewis and Buchanan 1998).

Creel census data were collected in five studies conducted from 1974 to 1990. Almost all of the anglers (94%) were nonguided and had greatest catch success September 15–30, a reflection of when river conditions were best for angling. Angling restrictions regulate steelhead fishing times and methods in the Zymoetz River. The river is a "classified water" from September 1 to October 31, with Class I licensing upstream of Limonite Creek and Class II licensing requirements downstream. This classification requires that a license be purchased by anglers over and above the basic angling license. For the past several years, a kill ban has been instituted for the entire Skeena River watershed to protect steelhead runs from harvest. Overwintering protection is afforded by angling ban regulations between January 1 and June 15, from McDonell Lake downstream 3 km, between signs in Zymoetz Canyon, and above the signs at the transmission line crossing (Anonymous 2001a).

Enhancement Activities

In 1907, the Department of Marine and Fisheries blasted a rockslide (in one or the other of the lower canyons) that had partially obstructed migration since 1891. In 1968, a rockslide was removed by blasting approximately 10 km upstream from the mouth. In 1973, a fishway was blasted around a series of difficult falls located 7 km upstream from the confluence.

In 1980, the lower reach of Serb Creek was diverted to extend and improve spawning and rearing habitat, and gravel was placed to rehabilitate the outlet of McDonell Lake. An incubation box was placed at Fossil Creek, and was utilized from 1981 to 1983. Brood stock collection and hatchery operations were undertaken from 1981 to 1985, with steelhead fry releases in 1981, 1983, and 1985 and chinook fry in 1984 and 1985. All fry releases were in the upper Zymoetz River watershed (B.C. MoE).

Development Activities

The primary human activities in the Zymoetz River watershed are forest harvesting, mineral resource, and linear developments including electric transmission lines and gas pipelines. Population and settlement impacts are very light and limited to a segment of floodplain close to the river mouth.

Forest Resource Development

The lower and middle portions of the watershed are located in the Kalum Timber Supply Area (TSA), with administration from Terrace. The portion of watershed north and east of Red Canyon Creek is located in the Skeena Stikine Forest District and administered from Smithers. The Burnie River upstream of the Clore River is located in the Nadina Forest District and administered from Burns Lake. All planning efforts originate with the Integrated Land Management Bureau in Smithers. For the last 45 years, forest harvesting has proceeded at a steady pace but has slowed down due to exhaustion of the high and moderate value timber.

Forest harvesting began in the late 1950s under Canadian Cellulose Ltd.'s Tree Farm License (TFL) #1 with the construction of a mainline forestry road up the lower river. By 1965, the road had reached Limonite Creek. Since that time, road systems have been developed into all the major tributaries and most of the minor tributaries. In 1985, the Kleanza road was extended into the Zymoetz drainage north of Nogold Creek and, in 1995, approached the TFL boundary. Currently, total area clearcut logged is just under 10,000 ha. Harvesting and road building techniques have heavily impacted the fish producing potential and habitat in the lower watershed (Pollard 1996).

Timber harvesting and road construction activities have been ongoing in the upper

Zymoetz River watershed from 1968 to the present. Early harvesting began in the upper reaches; however, with construction of the Copper Forest Service Road (FSR) north of McDonell Lake in approximately 1980, access and development of Hankin, Willow, Passby, Sandstone, Coal, and Mulwain Creeks have been relatively rapid. Areas west of Mulwain and Lee Creeks proposed for development have been identified as potentially highly hazardous terrain due to erodible soils and the presence of natural failures. Timber harvesting has not occurred in the Burnie River drainage to date, though Canadian Forest Products Ltd.'s current forest development plans (FDPs) do have major proposed developments for the area southeast of lower Burnie Lake.

The Watershed Restoration Program (WRP) was initiated in 1995 to restore the terrestrial and aquatic productivity of watersheds negatively impacted by forest development and harvesting. The upper Zymoetz watershed was assessed at the fish and riparian overview level in 1999; results indicate that forestry-related impacts to fisheries values were generally low (Triton 1999). A total of 131 potential impacted sites were identified within the study area, with the majority (104 or 80%) related to road crossings of streams. Fifteen sites were rated as high priority, 41 sites as moderate priority, and 75 sites as low priority.

The lower Zymoetz watershed has received numerous assessments under the WRP. The 1995 Fisheries Assessment Overview identified 45 sites requiring restoration or further assessment, with the majority of impacts acute and occurring 20 to 30 years in the past (Pollard 1996). Most of these sites were regenerating naturally or were irreparable, and while the majority of sedimentation sources were natural, sediment transport have been accelerated by forestry operations. The most significant forestry impact on fisheries values is the restriction or loss of critical rearing and overwintering habitat due to road construction practices. Recommendations were made for more in-depth fisheries assessments and analysis.

Riparian habitat, especially off-channel, in the lower Zymoetz valley, has been substantially reduced. Changes in channel morphology have been accelerated with harvesting of floodplain forests and road building. While it is difficult to quantify exactly how much change is related to forestry, the result is a decrease in fisheries habitat.

Since 1996, 12 studies have examined the lower watershed that included detailed aquatic, riparian, and upslope site impact assessments along with several reports concerning aquatic habitat restoration projects. Various rehabilitative site works have been completed, as well as approximately 50 km of road deactivated (Ottens 2002). The WRP has completed an interim Restoration Plan (2002–2006) for the lower Zymoetz. Proposed activities include approximately 10 ha of landslide treatments, 30 ha of riparian treatments, and access and habitat improvements to at least 10 ha of salmonid rearing areas. Since the demise of Forest Renewal BC, which funded the WRP, and the declaration of bankruptcy by Skeena Cellulose Inc., watershed restoration has come to a standstill.

Mineral Resource Development

Approximately 50 known mineral occurrences are dispersed throughout the watershed. Mineral resource development in regards to developed prospects and past producers is primarily centered in the upper watershed. Precious and base metal occurrences predominate, although coal and copper-molybdenum bulk tonnage is noteworthy. Many mineral prospects have been worked on from 1910 up to the present, but only the Duthie Mine advanced to production.

The Duthie Mine, located 12 km west of Smithers on the McDonell Lake road, operated from 1922 to its closing in the 1930s, due to poor markets. It operated again from 1950 to 1953; sporadic operations at a small scale continued into the mid-1980s. During its productive life, 47,000 tonnes of tailings flowed down to a swampy area below the mill where they were impounded in settling ponds.

Field investigations from 1982 to 1983 found that the dyke around the old tailings was failing, causing contaminated run-off to Henderson Creek, subsequently to Glacial Creek and then Aldrich Lake (Maclean 1983). The tailings were found to have a high concentration of arsenic, iron, aluminum, and lead. Metal levels were found in Glacial Creek and Aldrich Lake that exceeded water quality criteria for cadmium, copper, lead, and zinc. Arsenic concentrations in the lake were found to exceed maximum allowable water drinking standards. Lake bottom sediments had elevated metals, but there was no corresponding evidence of metal contamination in fish. Subsequent covering of the tailings and diversion of leachate were carried out, and in 1987 the tailings were contained behind a berm. Sampling in 1993 found that the tailings were still generating acid rock drainage (ARD). A remediation plan and mitigation of the ARD were implemented, and remedial work has been carried out in the past several years.

Transportation and Utilities

Logging roads that leave Highway 16 provide access to both the upper and lower portions of the Zymoetz River watershed. From Smithers, the McDonell Forest Service Road (FSR) accesses all the major tributaries north and south of the Zymoetz, as far west as the mid reaches of Mulwain Creek. From Terrace, access to the lower watershed is by the Copper River FSR, which is located on the west side of the river and extends to the fossil beds approximately 50 km upstream.

In the past, the Copper River FSR provided access upstream on the Zymoetz River past Limonite Creek; however, this road has been washed out. After departing the river, the road followed the Pacific Northern Gas (PNG) gas pipeline route through the Limonite Creek drainage and then over the Telkwa Pass. Access to the middle watershed is now via the Kleanza Mainline, which crosses the headwaters of Nogold Creek and descends to the Zymoetz River 58 km upstream of the Skeena confluence. This road proceeds up the Zymoetz River to within 2 km of Red Canyon Creek.

Highway 16 crosses the Zymoetz River near its mouth. In 1964, the Highway 16 bridge was relocated from the fan apex to its present location. The span placement was threatened by the extreme high water event of 1978. Subsequently, work was carried out to stabilize the upstream banks of the fan channel approaching the bridge. This work resulted in moving the fan apex significantly downstream from its natural location, and consequently 70% of the floodplain was cut off (Pollard 1996).

Following severe bank erosion during the 1978 flood event, riprap channel protection was placed in reaches 5 and 6a. In reach 6a, the road and the pipeline were routed on opposite riverbanks with the riprap forcing the river channel into a straight flume. The 1978 flood caused several new channel crossings to be installed for the pipeline, as well as extensive road relocation between Kitnayakwa and Limonite Creeks, where the road was destroyed. The Copper FSR was also relocated at the Clore Junction. After additional washouts, the natural gas pipeline was relocated to a position on the southern hillslope over most of the route.

From the 1980s to 1991, repeated road washouts required repair work and minor rerouting at the Clore Junction (km 34) and reaches 3, 5, and 6, along with riprap maintenance in reach 6. In 1991 and 1992, the second and third major road relocations downstream of the Clore junction were necessary after flooding. Road access and the pipeline were cut by a debris flow in early June 2003. The gas pipeline connection was restored within a week but the road reconstruction was not complete for more than a year.

Pacific Northern Gas (PNG) and BC Hydro both utilize the east-west corridor that passes through the Telkwa River, over the Telkwa Pass, then downstream on Limonite Creek and Zymoetz River to the Skeena River. Between 1967 and 1975, the natural gas pipeline was constructed, with large sections of the original pipeline built alongside the Copper River FSR. The road itself was built on the floodplain, often near the channel and in several locations, the pipeline crossed the river. By 1975, the BC Hydro power line, located midslope along the north side of the valley, was near completion. The right-of-way encroaches on the riparian zone of the Zymoetz River in several places; however, in general, the loss of riparian habitat is not significant.

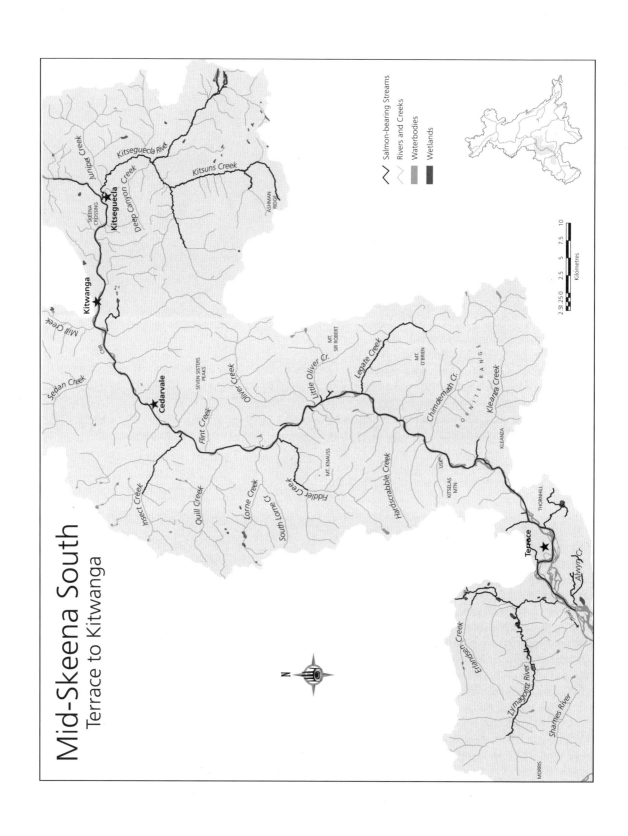

Mid-Skeena South
Terrace to Kitwanga

Salmon-bearing Streams
Rivers and Creeks
Waterbodies
Wetlands

2.5 1.25 0   2.5   5   7.5   10
Kilometres

N

Juniper Creek
Kitseguecla River
Kitsuns Creek
Deep Canyon Creek
Skeena Crossing
Kitseguecla
Kitwanga
Mill Creek
CNR
Sedan Creek
ASHMAN RIDGE
SEVEN SISTERS PEAKS
Oliver Creek
Little Oliver Cr.
MT. SIR ROBERT
Legate Creek
MT. O'BRIEN
Chimdemash Cr.
BORNITE RANGE
Kleanza Creek
KLEANZA
Cedarvale
Flint Creek
Insect Creek
Quill Creek
Lorne Creek
South Lorne Cr.
Fiddler Creek
MT. KNAUSS
Hardscrabble Creek
USK
KITSELAS MTN
THORNHILL
Terrace
Alwyn Cr.
Erlandsen Creek
MORRIS
Zymagotitz River
Shames River

11

Mid-Skeena River South:
Terrace to Kitwanga

Environmental Setting

The mid-Skeena River south area is defined as the area encompassing the Skeena River and its tributaries from the Kitsumkalum River upstream to Kitwanga River. This section includes the mainstem and all tributary drainages other than the Kitsumkalum, Zymoetz, and Kitwanga watersheds.

Location

The mid-Skeena River south area is located in west-central British Columbia. The City of Terrace is located on the western boundary, while Gitwangak is located on the eastern bounds. The mid-Skeena River south area is composed of the Hazelton Mountains; the Bulkley Ranges on the east, and the Nass Ranges on the west.

Hydrology

The drainage area is for the most part mountainous with high relief. Elevation ranges from 2,756 m in the Seven Sisters to 56 m at the Skeena-Kitsumkalum confluence. The Skeena River essentially bisects the mid-Skeena River south area on a rough northeast-southwest line. Skeena River peak discharges occur in May and June due to spring snowmelt throughout the upper Skeena watershed east of the Coast Ranges. In almost all years, the peak spring freshets are the largest floods in the lower river. Water levels in the main river channels and back channels fluctuate seasonally; typically they are high from May to early July, drop for the summer months, rise to intermediate levels in the fall, and reach their annual low levels late in the winter season. Monthly mean discharge for the Skeena River at Usk, station 08EF001, is 160 m³/s in March and 2,830 m³/s in June.

Major tributaries draining into this reach of the Skeena River from the north and west are the Kitwanga River, Sedan Creek, Insect Creek, Quill Creek, Lorne Creek, Fiddler Creek, and the Kitsumkalum River. From the south and east, major tributaries draining into the Skeena River are Price Creek, Boulder Creek, Oliver Creek, Little Oliver Creek, Legate Creek, Kleanza Creek, and the Zymoetz River.

The peak discharges of the major tributaries are typically in May and June due to the majority of tributary drainage lying at high elevations. Discharges then decrease until September when fall rains and early snowmelt increase streamflows through October. High streamflows and floods can also occur due to rainstorms in the fall (late September to early November). These high streamflows are typically of short duration.

Streamflows decrease through November and December when precipitation falls as snow. Discharge continues to decrease until late in the winter, with the annual minimum flow generally occurring in January through March. Summer low flows are typically four to eight times greater than winter streamflows and are principally sustained by high-elevation snowmelt, while winter low flows are derived from groundwater, lakes, and unfrozen wetlands. The surrounding glaciated mountains help to maintain moderate summer streamflows. Originating from glaciers, these streams produce moderate amounts of natural sediment that contribute to the wash load eventually deposited in the Skeena estuary.

The Hazelton Mountains, which are composed of the Nass and Bulkley Ranges, exert the major hydrological influences, and tributary streamflows have a moderately high and rapid response from rainfall due to the high gradients of the major tributaries. The wide Nass valley and Nass basin to the north and Douglas Channel to the south allow coastal weather systems to easily enter the area, leading to heavier snow packs in the mountains of

View across the Skeena River showing forests and remnant glaciers of the Seven Sisters

the northwestern portion of the drainage. The general climate of the watershed is transitional between a temperate, maritime coastal climate and the colder, continental climate that characterizes the interior of the province. Summers are warm and fairly moist, with a relatively long growing season. In the Skeena valley, snowpacks are light to moderate (up to 1 m) and the valley is prone to cold air ponding as well as to summer droughts.

Stream Channels

Between the Kitsumkalum River and the Kitwanga River, the Skeena River is approximately 90 km long. This is the lower half of reach 3, which extends from Terrace to the Bulkley-Skeena confluence (Resource Analysis Branch 1975–79). Overall, the Skeena River in this section is a single-thread, slightly sinuous channel incised into the valley bottom, with a moderately narrow valley floor. From Gitwangak to Cedarvale, the river channel is, for the most part, comparatively wide. Downstream of Cedarvale, the river carves a generally confined and stable channel to Kitselas Canyon. From Kitselas Canyon downstream to the Kitsumkalum River, there are multiple channels and moderate bar development. The average gradient of the river within reach 3 is 0.001 with no obstructions to anadromous fish.

Valley bottom deposits are largely the product of rapid deposition of fluvio-glacial material during the melting of the last ice sheet about 10,000 years ago. This abundant coarse sediment filled most of the valley to at least several hundred meters. Wide benches (terraces) were established as the Skeena River cut its way down to the present position, which it reached at least 5,000 years ago. Highway 16 is often located on the remnants of the incised terraces. Thin soils, colluvium, and rock outcrops characterize the mountainsides sloping up from the Skeena River and tributaries. On many middle slope positions, a morainal veneer is found while upper areas are dominated by rock and colluvium.

Stream channel change on the Skeena mainstem has been relatively slight since the end of the Little Ice Age (approximately 1850), though the Skeena River bed has probably degraded slightly in this period. There have been minor channel changes from meander movement and bank disturbances from high water events. Channel modifications may also have come about to a minor degree due to rock removal that was carried out to facilitate upriver passage of sternwheel riverboats and log driving activities.

Over the years, major storms, floods, and torrents have caused significant changes to tributary streambeds. Every few decades, spring or fall floods cut new channel sections or mobilize sediment wedges and channel bedload, causing significant erosional and depositional features. The spring floods in 1936, 1948 and 1972 and the fall flood in 1978 produced some of these effects (Septer and Schwab 1995). For example, following the 1978 flood, fish passage from the Skeena River was restricted into many of the tributary streams including Chimdemash Creek, Legate Creek, Oliver Creek, Coyote Creek and Price Creek, due to the steep-faced gravel fans that were established at their mouths.

The tributaries flowing into the Skeena River are generally short, high-energy, and steep-gradient streams that level out only in the last kilometre or less before entering the Skeena. Most of the streams possess stable channels throughout much of their length. However, occasional debris torrents during rainstorm floods may move hundreds of thousands of cubic meters of gravel onto their alluvial fans and into the Skeena River. The trend since the 1970s to place bridges, rather than culverts, over the streams, has facilitated coarse sediment passage and reduced the impacts of the crossings.

Water Quality

Water quality data for the Skeena River was collected from 1966–1990 at Usk. Remington (1996) reported that waters are soft and have a mean pH of 7.4 (neutral to slightly alkaline) with coloration similar to most upstream sites (mean=20 TCU). Filterable residue is low. The increased turbidity is the result of the much higher suspended sediment loading received from the steep tributary drainages downstream of Kitwanga River.

Geography

Within the Kitsumkalum River to Kitwanga River area, there is a diverse assemblage of rugged mountains and short, steep valleys. Within the drainage, the highest peaks are the lofty Seven Sisters, Mount Sir Robert, and Mount Quinlan, all of which possess small glaciers. The latter two mountains are separated from the Seven Sisters by the deeply cut Oliver Creek valley that trends east-west. Although in retreat, the glaciers and snowfields are still massive. The Nass and Bulkley Ranges were the source of large, coalescing ice sheets that covered the entire watershed during the Fraser Glaciation, which lasted until approximately 10,000 years ago.

The Skeena River valley was significantly altered by ice moving down it, and subsequently by glacial debris and alluvial deposits. During melting of the ice sheet, kame terraces formed between the ice and the valley; subsequently, many were modified by slumping (Gottesfeld 1985). Within this reach, these kame terraces are discontinuous features best developed adjacent to the tributaries flowing into the Skeena valley.

Thin soils, colluvium, and rock outcrops characterize the mountainsides sloping up from the Skeena River and tributaries. On many middle slope positions, a morainal veneer is often found, while upper areas are dominated by rock and colluvium. On the highest mountains, above 1,500 m, there was alpine glacial activity during the last few thousand years that significantly modified the highest peaks.

The mid and low elevations of the mid-Skeena River south area are covered with dense coniferous and deciduous forests. Within the drainage area, three principal forest types are recognized: the Coastal Western Hemlock (CWH) zone, the Mountain Hemlock (MH) zone, and the Interior Cedar Hemlock (ICH) zone (Pojar et al. 1988). The Engelmann Spruce–Subalpine Fir (ESSF) zone is marginally represented at middle elevations of the eastern section of the watershed.

In the downstream portion of the mid-Skeena River south area, the Coastal Western Hemlock (CWH) zone characterizes the low- to mid-elevation forests in the Skeena valley. The CWH zone lies in the valley floor and on the hillsides, and protrudes into the tributary valleys to approximately 900 m elevation. Upstream of Pacific, B.C., the CWH zone lies above the ICH zone in elevation and below the Mountain Hemlock zone. The CWH is dominated by old-growth coniferous (rainforest) stands of western hemlock, amabilis fir, and subalpine fir, though hybrid spruce is common. The CWH zone at 900–1000 m elevation passes into forest cover that is dominated by mountain hemlock and subalpine fir, and the absence of deciduous and seral conifers such as pine is typical. The Mountain Hemlock (MH) zone covers more of the forested ground than the other zones, mirroring the mountainous nature of the drainage.

Upstream of Pacific, the climate becomes drier and the Interior Cedar Hemlock (ICH) zone makes up most of the lowland forests. The major tree species are hemlock, spruce, subalpine fir, and, to a certain extent, red cedar, though there are extensive seral forests of lodgepole pine, trembling aspen, and, to a limited degree, birch. Due to the pervasive aboriginal and natural fire history, lodgepole pine and deciduous seral stands on upland sites are extensive, particularly along stream terraces and on southern aspect slopes. Black cottonwood prevails in floodplain stands, and is typically mixed with immature spruce and cedar.

Geology

The mid-Skeena River south area is underlain mostly by rocks formed in the Mesozoic Era. The important bedrock units are the volcanic and sedimentary rocks of the Hazelton Group of mostly early Jurassic age and sedimentary rocks of the middle to late Jurassic Bowser Lake Group. These were followed by volcanic rocks from the late Cretaceous Skeena Group and then by the Bulkley Intrusives with minor amounts of the early Tertiary Babine Intrusions.

As part of the western edge of North America, the geologic history of this area is complex. About 225 million years ago (ma) off the west coast of the North American continent, there existed a chain of islands, called an island arc system, which was similar to present day Indonesia. These islands originated from volcanic activity caused by plate tectonics. This activity produced great quantities of volcanic

flow rocks such as basalt and andesite, as well as ash and mud that flowed into the sea and formed sedimentary deposits of sandstone, siltstone, and shale. This process was continuous during the Jurassic period, more than 160 ma.

At about 157 million years ago, this island arc was pushed into and accreted with the North American continent forming a shallow sea along the west coast. This resulted in the Bowser Lake Group of rocks being formed 157–136 ma in the Jurassic age. The majority of rocks in this area, approximately 70%, belong to the Bowser Lake Group. In the mid-Skeena River south area, they comprise a series of mudstones and sandstones formed in nearshore marine and nonmarine environments (Gottesfeld 1985).

About 136 ma, a new period of volcanism formed the Skeena Group, made up of a series of lava flows interspersed with sediments formed in a shallow marine environment and terrestrial areas, including large swampy areas where peat and plant life accumulated to form coal beds. The Skeena Group represents approximately 25% of the rock underlying the area.

From 100 to 65 ma, uplifting, volcanic, and plutonic activity created the Bulkley and Babine Intrusives that are chiefly composed of porphyritic rocks with minor amounts of granitic rocks. These intrusions into the sedimentary and volcanic rocks occurred at depth, and provided heat to produce alteration zones and local mineral deposits including metallic ore bodies. These alteration zones are in relatively close proximity and relationship to the major faults that occur through the watershed. These intrusives account for approximately 5% of the rocks in the watershed.

Numerous faults and contacts prevail with a strong regional pattern oriented at 340°. The geological structure is dominated by block faulting. These linear breaks now control the location of the major mountain valley systems, as well as the many rock suites and mineral deposits (Richards 1990). Overall, metamorphism is light, aside from the contact effects near intrusive bodies and mechanical metamorphism along the major fault zones.

Fish Values

The mid-Skeena River south area has high fish values with the presence of chinook, pink, sockeye, coho, and chum salmon, and steelhead trout. Kokanee are resident in Kleanza Lake. Rainbow and cutthroat trout, Dolly Varden, bull trout, kokanee, mountain whitefish, slimy and prickly sculpin, largescale sucker, threespine stickleback, and peamouth chub are also present in the drainage system (DFO 1991b, 1991c).

Compared to other areas in the Skeena, such as Kispiox or Babine, the mid-Skeena River south area is a relatively moderate producer of salmon. In general, the most widely dispersed salmon species is coho, while Dolly Varden, rainbow trout, and mountain whitefish are found in most fish-bearing waters.

Chinook Salmon

Within the Skeena watershed, there are no known documented chinook spawning sites on the mainstem, but turbid summer flows might obscure some sites. The Skeena mainstem provides rearing and migration habitat for some of the Skeena system chinook populations. Tributaries with known, current chinook spawning grounds are Fiddler Creek and Kleanza Creek. Chinook spawning in Fiddler Creek is documented up to 4 km. Escapement was recorded from 1966 to1980; abundance ranged from 25 to 200, with a mean of 125 (DFO 2005).

Kleanza Creek chinook escapement has been documented with nine records since 1966. Spawners range from 2 to 25 fish, with spawning located below the falls at 3.5 km. Hancock et al. (1983) reported that no chinook were counted through the Fisheries Research Board fence between 1957 and 1961. Chinook were observed spawning in Boulder Creek in the late 1920s and early 1930s in the lower 0.2 km of the fan. Chinook salmon have been observed at 1.5 km in Insect Creek (Smith and Lucop 1966).

Pink Salmon

Skeena River Mainstem and Tributaries
In years of high returns to the Skeena, pink salmon spawn in the lower reach of many streams tributary to the Skeena mainstem. Pink salmon spawners were recorded at Mill Creek in the early 1930s; since then, there have been no observations. Price Creek has had 14 historic observations of pink spawners up to 3 km since 1953, with no recent observations.

Pinks utilize the lower 1 km of Sedan Creek, with spawning observed to 2.5 km upstream. The few recorded escapements are mostly zero, but 1983 had 150 pinks. Price Creek had an average of 2,500 pink spawners in the 1950s but low abundance since – perhaps from habitat change accompanying the Highway 16 bridge construction. Pink salmon were observed spawning in Boulder Creek in the late 1920s and early 1930s; however, there have been no documented observations since then (Smith and Lucop 1966, DFO 2005).

Pink salmon spawn in the lower 0.2 km of Wilson Creek, with escapement ranging from 75 to 400 spawners. Few observations have been made since the 1960s. Pink salmon were observed spawning in Insect Creek in the late 1920s and early 1930s; however, there have been no documented observations since then (Smith and Lucop 1966, DFO 2005). Pink salmon have been observed spawning at 0.7 km in Lorne Creek, although spawner abundance is not documented. Little Oliver Creek supports pink spawning in the lower reach (DFO 1991), though there are no escapement records. On Fiddler Creek pink salmon spawn up to 2 km, with the heaviest spawning below the CN Rail bridge. Escapement has been discontinuously recorded since 1943 (DFO 2005), but the recorded escapement of pinks from 1968 to 1972 was 1,500.

Legate Creek supports an unknown number of pink spawners in the lower 0.6 km. Shannon Creek escapement records document 200 spawners in 1947 (Hancock et al. 1983). Chimdemash Creek has supported 0 to 40 pink spawners on the lower 200 m on a discontinuous basis. Lowrie Creek supports pink spawners that range from 0 to 750 fish on the lower portion of its fan. Pink salmon in Hankin Creek have been observed spawning on the lower 100 m of the fan. Kleanza Creek has fairly consistent escapement records to the present that show pink salmon spawning in the lower 3.5 km, with a range from 10 to 10,000. Singlehurst Creek had 200 – 400 spawners in the 1980s and 1990s in the lower 1.8 km. Between 0 and 250 pinks spawners have been recorded on the lower 1 km section of Thornhill Creek.

Chum Salmon

Chum salmon arrive in the Skeena watershed from late July to early September. In general, one can conclude that chum are probably the Skeena watershed salmon species in the greatest danger of significant spawning stock and genetic diversity loss.

Field observations suggest that chum are highly specialized in their selection of spawning sites. Several of the Skeena River spawning sites used every year are less than a few hundred metres long. Chum continue to use these patches of gravel even when channel reorganization separates them from their former source of flow. At the Coyote Creek alluvial fan, spawners use a patch of gravel that was at the mouth of the creek until the flood of 1978 but is now several hundred metres upstream of the mouth.

There is little documentation of chum spawning sites on the Skeena mainstem; however, it is quite likely that spawners use areas of freshwater seepage and upwelling especially in side channels a few kilometres below Kitwanga and, in the past at least, just upstream of Little Canyon. Kleanza Creek has chum spawning in the lower 3.5 km below the falls, with a range of 4 to 400. On Fiddler Creek, chum salmon spawn up to 2 km, with the heaviest spawning below the CN Rail bridge. Escapement has been recorded discontinuously since 1943, with a range from 0 to 200 fish.

Both Insect and Boulder Creeks had documented chum spawning in the late 1920s until the early 1930s; however, there are no documented observations since then (Smith and Lucop 1966, DFO 2005). Price Creek has one record of chum spawners: 25 in 1966. Mill Creek had 300 to 500 chum spawners recorded from 1931 to 1935 (Smith and Lucop 1966). The small population at Coyote Creek has declined to only a few spawners or no spawners in the past few years.

Coho Salmon

Coho salmon are widely dispersed throughout the mid-Skeena River south. Coho usually spend one to two winters in freshwater before migration to the ocean. Jack coho are found in the Skeena only as far upstream as the Kitwanga River (GWA, unpublished data).

Coho migrate through the study area throughout August, typically moving to tributary outlets to hold. Dependent on high water flow, coho will wait for fall rainstorm floods before moving into the tributaries to spawn from late September to late November. Coho spawning populations within the watershed are relatively small and dispersed, though notably important.

In the late 1920s and early 1930s, Mill Creek had recorded coho escapements ranging from 300 to 500 fish spawning in the lower 0.5 km (Smith and Lucop 1966); current status is unknown. Price Creek had small runs recorded in the late 1950s until 1979, with a range from 6 to 25 coho using the lower 5 km of the creek (Hancock et al. 1983); Price Creek coho have not been observed recently. Boulder Creek had recorded coho escapements (light) from 1929 until 1934; however, coho have not been observed recently. Wilson Creek supports coho spawners that number less than 100 in the lower 0.5 km; coho escapement has not been recorded since the mid-1970s (DFO 2005).

Coho spawning in Insect Creek has been observed to approximately 12 km upstream (DFO 1991c). In the late 1920s, escapement was listed as light. Quill Creek has documented coho presence 1.5 km upstream to the 5 m falls and likely supports coho spawning. Coho spawn in Lorne Creek up to the falls at 1.5 km (DFO 1991). The only documented escapements are from the late 1920s and early 1930s, ranging from medium to heavy (Smith and Lucop 1966).

Coho spawners have been observed at 0.7 km in Lorne Creek. Escapement estimates from the late 1920s to early 1930s were medium to heavy; current escapement is unknown. Coho spawners have been observed at 0.1 km on the Gosling Creek fan. Little Oliver Creek supports an unknown number of coho spawners up to 0.4 km on the fan. Culp (2000) reported an estimate of five spawners up to the falls 1 km upstream of the Highway 16 crossing. Legate Creek escapement records show 25 coho in 1986, with spawning up to the cascades at 5.5 km. Culp

(2000) reported an estimate of 12 spawners up to the falls in 1999.

Fiddler Creek is the principal coho producing stream between the Zymoetz River and the Kitwanga River and is located about half way between these two larger systems. Fiddler Creek supports moderate amounts of coho spawning and rearing habitat for approximately 14 km upstream and in 12 tributaries (Smith and Lucop 1966). Escapement was recorded from the mid-1940s to the mid-1980s, with spawners ranging from 50 to 750 fish (DFO 2005).

Recent escapements in Fiddler Creek have been very low. Culp (2000) reported an estimate of 100 spawners in 1999. In 2001, a relatively good year, the escapement was 75 (GWA unpublished data). This coho population is important to the Gitxsan; however the Lax Skiik have not harvested Fiddler Creek coho in recent years because of low numbers.

Coho spawners have been observed at 1 km in Shannon Creek, though abundance is unknown. Culp (2000) reported an estimate of 10 spawners in 1999 in Hardscrabble Creek. Chimdemash Creek supports coho spawners up to the falls at 1.5 km with abundance unknown; Culp (2000) reported an estimate of 25 spawners up to the falls in 1999. In Lowrie Creek, coho spawn in the lower 1.5 km. Escapement is unknown, though in the late 1920s and early 1930s, it ranged from medium to heavy (Smith and Lucop 1966). Singlehurst Creek escapement records start in 1970 and are discontinuous to the present. Coho spawners have been observed up to 2 km with a range from 0 to 100 fish. Culp (2000) reported an estimate of 300 spawners up to the beaver dam upstream of the Highway 16 crossing.

Kleanza Creek supports coho spawners up to 8 km. Escapement records show medium and heavy spawning from 1923 to 1930. From the early 1950s to the present, fairly consistent escapement records show a range of 25 to 750 spawners. Culp (2000) reported an estimate of 70 spawners up to the falls in the canyon area.

Gossen Creek has a small coho population in its limited gravels; Culp (2000) reported an estimate of 14 spawners in 1999. Thornhill Creek supports coho spawning in the lower 1 km below the falls, with a range from 0 to 1,100 fish. Juvenile coho have been released above the falls.

Steelhead

The mid-Skeena River south supports summer-run steelhead populations that arrive in the area beginning in August and continuing into autumn. Steelhead spawning occurs from March through May, coinciding with warming water temperatures and an increase in streamflows. Steelhead fry emerge between mid-August and mid-September and are widespread throughout smaller tributaries that offer suitable refuge. There are few good data to record steelhead escapements at individual streams.

The Skeena mainstem is used as a migration route for upstream Skeena River and Bulkley River steelhead stocks. Steelhead overwinter in the Skeena River at the mouth of the Kitwanga River. Steelhead have been observed spawning in the lower 1 km below the falls in Thornhill Creek. Steelhead have been recorded at 12 km in Kleanza Creek and presumably spawn in the system. In Singlehurst Creek, steelhead have been observed spawning at 2.2 km. Steelhead have been observed upstream to 14 km on Fiddler Creek. Steelhead have been observed at 7 km in Lorne Creek and in the lower reach of Insect Creek.

Resident Freshwater Fish

There are 12 known species of freshwater fish in the study area drainage system (McPhail and Carveth 1992). In comparison to salmon, information is sparse on resident, freshwater fish in both river and lake habitats of the mid-Skeena River south area.

Known freshwater species and documented populations inhabiting the mid-Skeena River south area include rainbow trout *(Oncorhynchus mykiss)*, cutthroat trout *(Oncorhynchus clarki clarki)*, bull trout *(Salvelinus confluentis)*, Dolly Varden char *(Salvelinus malma)*, mountain whitefish *(Prosopium williamsoni)*, peamouth chub *(Mylocheilus caurinus)*, largescale sucker *(Catostomus macrocheilus)*, threespine stickleback *(Gasterosteus aculeatus)*, river lamprey *(Lampetra ayresi)*, kokanee *(Oncorhynchus nerka)*, Coastrange sculpin *(Cottus aleuticus)*, and prickly sculpin *(Cottus asper)*. It is possible, but not known, whether the range of green sturgeon *(Acipencer medirostrus)* extends into the mid-Skeena area.

Within the mid-Skeena River south drainage, rainbow trout have been documented in the Skeena River mainstem and the following streams: Price, Boulder, Sedan, Wilson, Insect, Coyote, Flint, Lorne, Fiddler, Oliver, Little Oliver, Legate, and Kleanza Creeks.

The cutthroat trout in the Skeena watershed belong to the coastal cutthroat trout subspecies *(O. clarki clarki)*. Within the mid-Skeena River south area, cutthroat trout have been documented in the Skeena River mainstem and many first- and second-order tributaries, as well as the following systems: Kitwanga, Wilson, Legate, Chimdemash, Singlehurst, and Kleanza Creeks. Cutthroat trout are also abundant in lower-elevation lakes such as Watson Lakes.

Dolly Varden char *(Salvelinus malma)* are blue listed by the B.C. CDC as a species of concern. Dolly Varden are the most common freshwater fish in the mid-Skeena River south area, particularly in the upper reaches of small streams throughout the watershed. Beyond knowledge of distribution and general life history, Dolly Varden char have not received extensive management or biological study in the mid-Skeena area. Cedarvale constitutes the regulatory upper limit of Dolly Varden anadromy; the remaining populations are assumed to have fluvial or adfluvial life histories. Dolly Varden char are only targeted as sport fish in the lower Skeena and its coastal tributaries, due primarily to their small size in upper watershed drainages. Current daily catch quotas allow anglers to keep five Dolly Varden.

Dolly Varden have been documented in the Skeena mainstem as well as the following systems: Mill, Sedan, Price, Coyote, Insect, Quill, Lorne, Fiddler, Oliver, Little Oliver, Legate, Chimdemash, Singlehurst, and Kleanza Creeks. Dolly Varden are also found in many unnamed first- and second-order Skeena tributaries.

Bull trout *(Salvelinus confluentus)* are actually a char that are blue listed as a species of concern due primarily to limited global distribution. Within the mid-Skeena River south area, bull trout have been documented in the Skeena mainstem and in Fiddler, Kleanza, and Lorne Creeks. This apparent lack of fish is most likely due to past inventory and classification efforts that possibly did not identify the difference between Dolly Varden and bull trout. Bull trout are also common in Skeena tributaries upstream of Cedarvale.

Fisheries

The mid-Skeena River south has long had a large First Nations harvest fishery based on returning salmon and steelhead. Prior to European contact fishing took place mostly in rockbound canyons or rockcrop areas and in smaller tributary streams. The past 100 years has seen aboriginal fishing effort focused more on the mainstem Skeena. There are also high-value sport fisheries that are conducted on the Skeena River.

First Nations Traditional Use

The Tsimshian and Gitxsan salmon fisheries within the mid-Skeena River south were concentrated along the Skeena mainstem and dispersed on tributary streams. The food fishery, which is a blend of food resource, trade capital, cultural expression, and connection to ancestral practices, is important. The fishery continues into the present, mainly harvesting Babine sockeye and, to a small extent, the other five anadromous salmon species that spawn in or migrate through the watershed.

At Skeena River canyon or rock outcrop locations, such as Kitselas Canyon or Gwax Tseelixit (Klootch Canyon), where the salmon tended to be concentrated by strong currents, large woven baskets and/or lashed wooden strip

Singlehurst Creek culvert outlet

traps were employed. Trap sizes varied, with larger ones being lowered and raised with stout poles operated by a strong and agile crew. The various traps and dip net gear used depended on site location, fish quantities needed, and the number of people available to fish the gear and provide processing capacity.

Over the last decade, the Gitxsan Watershed Authorities and the Kitselas Band, as part of the Skeena Fisheries Commission and working with the support of the Aboriginal Fisheries Strategy, have been active in fishery management on the Skeena River and some of its tributaries. Food, social, and ceremonial (FSC) needs of the Gitxsan and Kitselas people are met with river gillnets, driftnets, fishwheels, beach seines, or dip nets in fisheries that target chinook, sockeye, and coho sequentially from late May to mid-October.

In the past decade, there have been a few special in-river commercial fisheries carried out under the auspices of Excess Salmon to Spawning Requirements (ESSR). These ESSR fisheries have been exclusively live capture selective fisheries targeting enhanced Babine Lake sockeye stocks that escape through the marine mixed-stock commercial fishery and are surplus to First Nations food, social, and ceremonial fisheries and conservation requirements. Within the mid-Skeena River south area, the Kitselas Band has used fishwheels for this fishery, while the Gitxsan use predominantly beach seines.

Recreational Fisheries

The mid-Skeena River south sport fishery supports seasonal recreational use from local-area residents and nonresident visitors. Resident anglers from Terrace, Kitwanga, Gitwangak, Gitanyow, Hazelton, and Smithers dominate the recreational angling. The most popular sport fishing locales are Ferry Island at Terrace, the mouth of Zymoetz River, and the mouth of the Kitwanga River.

The Ferry Island sports catch is directed towards chinook and pink salmon with coho, steelhead, and sockeye generally being released (Thomas and Associates 1999). The Skeena-Kitwanga confluence sport fishery is primarily angling directed to chinook, coho, and steelhead.

Creel surveys were conducted for the chinook and coho sport fisheries at the mouth of the Kitwanga River in 2000, 2001, and 2002. The sports fishing harvest recorded is small; 113 chinook in 2000 (Gottesfeld 2001), and 25 and 144 coho in 2001 and 2002, respectively (Hall and Gottesfeld 2002, 2003a).

Enhancement Activities

The mid-Skeena River south has seen limited enhancement effort. Kleanza Creek Hatchery operated from 1957 to 1961, incubating and rearing pink salmon fry for release into Kleanza Creek. Brood stock was taken from the Lakelse River (1957 and 1958), the Kitwanga River (1959), and Kleanza Creek in all years. Kleanza Creek Hatchery burned in early 1962 and was not rebuilt.

Various projects to restore fish passage have occurred on productive streams since Highway 16 was constructed. A typical example is Singlehurst Creek, one of the most productive coho streams in the mid-Skeena watershed (Culp 2000). From 1965, when Highway 16 was rerouted, until 1978, the Highway 16 culvert across Singlehurst Creek had a 1.7 m outlet drop that prevented fish migration. The DFO installed two timber weirs that created pools and effectively backwatered the culvert (Southgate 1978, Hancock et al. 1983). Williams and Gordon (2001) assessed potential problem stream crossings and provided recommendations to CN Rail, which subsequently applied habitat and fish passage restoration efforts to Newton Creek and MP 127.

Development Activities

Land-use and development activities in the mid-Skeena River south area are relatively moderate overall. The principal development activities are forestry, transportation, mining, and settlement.

Forest Resource Development

The mid-Skeena River south area is located within the Ministry of Forests Skeena Stikine Forest District east of Fiddler Creek, while the western portion is within the Kalum Forest District. Currently, the Ministry of Forests provides land-use management zoning, objectives, and strategies. Forestry development is concentrated in the low- and mid-elevation forested slopes arising from the Skeena River and in major tributary drainages where timber is accessible.

Forest development activity began with cutting cordwood needed for the steam-powered sternwheel boats plying the Skeena River in the 1890s. Beginning in 1909, the construction of the Grand Trunk Pacific Railway required clearing of the right-of-way and wood for ties. Ties continued to be produced on a steadily diminishing basis until the depression. The railway facilitated growth of the forest industry, providing shipping for timber and specialty products such as spruce boom sticks (20 ft), straight-grained spruce for airplanes, cedar poles, and cedar shingles. Small-scale lumbering led to dispersed, small bush mills, while the post-World War II economic boom generally increased the demand for lumber. Independent cedar pole loggers also saw a demand for poles at this time.

In 1947, the B.C. Forest Act was significantly amended to provide for "sustained yield units." In 1948, Columbia Cellulose was granted TFL #1, which initiated the trend toward the centralization of license holding and milling capacity. In 1952, Columbia Cellulose began operations in Terrace. Since the mid-1960s, clearcut harvesting has been the silviculture system of choice. In 1963, consolidation of seven or eight small mills led to the establishment of Hobenshield's mill at its present location in Kitwanga. The Canadian Cellulose Ltd. mill in Kitwanga was constructed in the early 1970s.

In the late 1950s and early 1960s, industrial-scale logging was initiated in the Kleanza valley, and by 1975, progressive clearcuts were developed in the lower portions of the Chimdemash and Legate drainages. During the 1980s, clearcutting occurred in the Coyote, Gull, Cedarvale, Flint, Oliver, and Little Oliver drainages. Access was established westward across the Skeena River upstream of Hardscrabble Creek. In the late 1990s and early 2000, development was characterized by logging low-elevation forests in the Bonser flats and Gosling Creek areas, and in the Quill Creek, Insect Creek, Mill Creek, and Price Creek drainages.

Many of these forest development activities raised fish and fish habitat concerns with affected First Nations, local residents, and fish conservation interests. In the late 1980s, concerns came to a head and focused on two main points of forest management: the rate and quality of timber cut and forest practices. The unsustainable rate of logging, often called overcutting, occurs when old-growth forests are cut faster than the regenerated forest can replace them. There were concerns in regard to the effects of logging on the resilience of forest ecosystems within the watershed. In 1984, local residents presented a plan that laid the groundwork for establishment of the Seven Sisters Park area.

Overall, the past and current government forest management strategies, which have perpetuated overcutting and highgrading, have resulted in enormous economic, environmental, and social losses. These losses are primarily due to changes occurring to forestland, timber and nontimber values, and the ecosystem itself. Furthermore, these government management strategies have created social and economic hardships to people, businesses, and the communities. The lack of a transition strategy that would provide work opportunities as the forest resource was exhausted resulted in chronic unemployment.

Between 1995 and 2001, the Watershed Restoration Program (WRP) was involved in assessing the forestry-related impacts and upslope, sediment-producing areas in relation to fish and fish habitat within the mid-Skeena River south drainage. Few WRP stream

CN Rail grain units derailed into the Skeena River, 2004

assessment and remediation works have been completed in this area in comparison to adjacent watersheds. Due to the demise of Forest Renewal BC, minimal restoration activities occurred.

Mineral Resource Development

Exploration and utilization of rocks, crystals, and minerals within the watershed have been occurring since aboriginal people settled into the watershed many thousands of years ago. Euro-Canadian exploration for base and precious minerals has been ongoing since the late 1860s. By 1884, early European efforts were focused on placer gold deposits that were sought and worked along the Skeena River and on Lorne Creek and Fiddler Creek. This was followed by mineral exploration and development of mainly high-grade silver and gold-rich veins.

In 1917, large-scale hydraulic operations on Lorne Creek were attempted but were mostly unsuccessfully. Most exploration and mining endeavours slowed down with the onset of the Depression, though Lorne and Fiddler drainages saw placer exploration and workings into the 1970s. The early 1960s saw a resurgence of exploration interest focused on large porphyry copper-molybdenum-gold deposits. Mineral exploration interest since the mid-1990s has been limited but focused on high-grade gold and silver-vein deposits.

Occurrences in the Nass Ranges are principally between Maroon Mountain and Sedan Creek, while the Seven Sisters mainly host a variety of polymetallic vein-type showings containing silver, lead, zinc, and gold. Quartz and auriferous veins are the probable sources of placer gold along Quill, Lorne, and Fiddler Creeks.

The Fiddler property on Mt. Knauss was productive from 1914 to 1952 with a total of 621 tons shipped that produced 185 oz. gold, 729 oz. silver, 14,982 lbs. lead, and 4,817 lbs. zinc. In the late 1930s, 111 tons of marl (agricultural lime) were produced from the Buccaneer of the North marl deposit, which is located 1 km west of Ritchie. Roads developed to facilitate mineral exploration and mining properties include the road that accesses claims near Knauss Creek, Coyote Creek, Oliver Creek, and Lorne Creek.

Population and Settlement

The Skeena valley and the territories in the watershed have been home to Gitselasu and Gitxsan people for thousands of years. Euro-Canadian settlers arrived following completion of the railroad in 1914, mainly attracted by the agricultural possibilities. The mid-Skeena River south population base has slowly grown in the communities of Terrace, Thornhill, Usk, Kitwanga, and Kitselas area subdivisions. Scattered rural residents are located along Highway 16, in the Copper River area, and at Cedarvale.

Most area residents derive their income from the manufacturing, wholesale/retail trade, accommodation, logging and forestry, and government sectors. Many of the rural residents are engaged in full- and part-time employment, hobby farming endeavours, and living a rural lifestyle. However, severe job losses related to both the forestry and fishing sectors have caused massive unemployment problems. The population trend for the Gitselasu and Gitxsan First Nations communities points to slow, steady growth. Rural population levels are expected to remain steady. Terrace, as the regional shopping and service center, is expected to steadily grow over the long term.

Transportation and Utilities Development

The Skeena River has been used for thousands of years as a transportation waterway, primarily by canoes. In the early 1890s, steam-powered paddle wheelers began regular, seasonal use, which continued until the Grand Trunk Pacific (GTP) Railway was completed in 1914. Sternwheeler use required many rock removals in the Skeena River.

The existing road network in the watershed reflects 80 years of steady development based on First Nations trail infrastructure. Generally, trails were initially widened for packhorses and later improved for wagons, then further improved for vehicular traffic. Overall, the development pattern was first spurred by mineral exploration and mining activity, and then with the motive of extracting forest products.

Linear development is relatively moderate on the northwest bank of the Skeena River, with

CN Rail dominating the riverside for much of its length. Above the railway, gravel roads provide access to residents and logging traffic. Railway construction practices effectively caused, and continue to cause, many instances of blocked access to fish habitat. Numerous back channels and side channels were cut off to returning adults, and abutments of bridges across tributary streams aggravated natural sediment deposition in fans, often requiring frequent channel dredging to avoid bridge washouts.

On the southeast side of the Skeena River, the paved, all-weather Highway 16 links Hazelton and Highway 37N to Terrace. It was initially constructed in 1944, and upgraded in 1952 and 1975.

Secondary roads off the highway primarily serve residents and forest development. From Highway 16 at Gitwangak, Highway 37N runs northerly, following the Kitwanga River past Kitwanga Lake and across the divide into the Nass drainage. An airstrip at Woodcock, originally constructed in World War II, provides access for fixed-wing aircraft.

Secondary forest development roads provide access to drainages of Mill Creek, Price Creek, Sedan Creek, Boulder Creek, Wilson Creek, Coyote Creek, Insect Creek, Flint Creek, Quill Creek, Lorne Creek, Hell's Bells Creek, and Oliver Creek. Many dirt roads lead down to fishing sites along the Skeena River.

Fish access has been slightly impeded on nonbridged fish-bearing streams crossed by Highway 16. Secondary forest development roads crossing unstable slopes and roads crossing alluvial fans of high-energy systems have caused impacts to fish-bearing streams and could have a high rehabilitation cost. A utilities corridor generally parallels Highway 16 and consists of a BC Hydro power line and Telus telephone line. These utilities have had light impacts that are mainly associated with riparian concerns.

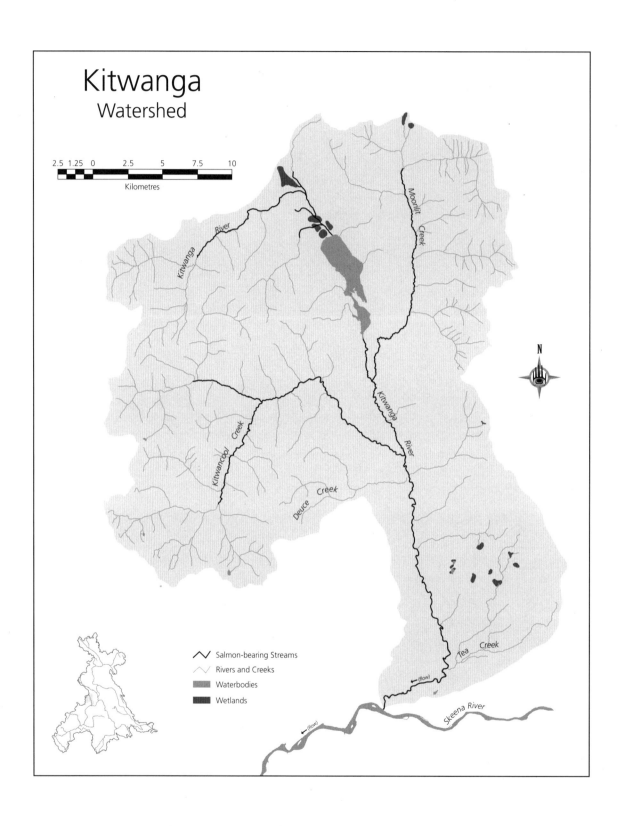

Kitwanga
Watershed

2.5 1.25 0   2.5   5   7.5   10
Kilometres

Kitwanga River

Moonlit Creek

Kitwanga River

Kitwancool Creek

Deuce Creek

Tea Creek

← (flow)

← (flow)

Skeena River

⋏ Salmon-bearing Streams
⋏ Rivers and Creeks
▨ Waterbodies
▩ Wetlands

N

12

Kitwanga Watershed

Environmental Setting
Location

The Kitwanga watershed is a tributary subbasin draining south into the right bank of the Skeena River about 250 km from the coast. The watershed is bounded to the west by the Nass Ranges, to the east by the Kispiox Range, to the north by the Cranberry River drainage, and to the south by the Skeena River. The watershed is located north of Gitwangak, which is 65 km west of Hazelton. Kitwanga Lake, also commonly called Kitwancool or Gitanyow Lake, lies in the middle of the Kitwanga watershed.

Hydrology

The Kitwanga watershed is a fifth-order system with a catchment area of approximately 833 km². Elevation ranges from 2,096 m in the Kispiox Range to 172 m at the Skeena River confluence. The Kitwanga River peak discharges typically occur in May and June due to spring snowmelt, then decrease until September when fall rains and early snowmelt increase streamflows through October. Streamflows decrease through November and December when precipitation falls as snow, with low discharges recorded from January through March.

Summer low flows are typically four to eight times greater than winter streamflows and are principally sustained by high-elevation snowmelt, while winter low flows are derived from groundwater, lakes, and unfrozen wetlands. Historic stream-flow data for the Kitwanga River are not available; however, Gitanyow Fisheries Authorities (GFA) has recently installed stream-gauging stations below Kitwanga Lake and close to the mouth of Kitwanga River.

The Hazelton Mountains to the west and the Nass basin to the north exert the major hydrological influences. The Kitwanga valley has a broad, low-gradient valley bottom with a low watershed divide to the Nass drainage

that allows coastal weather systems to enter the watershed. This generally leads to heavier snow packs in the mountains and the northern half of the drainage. The general climate of the watershed is transitional between temperate, maritime coastal climates and the colder, continental climates that characterize the interior of the province. The watershed as a whole has a moderately high response from water input due to the high gradients of the major tributaries.

Kitwanga Lake and the extensive wetlands at the northern end of the Kitwanga valley constitute the primary water storage. The upper Kitwanga River, Kitwancool Creek, and Deuce Creek, which drain the Nass Ranges, and Moonlit Creek, which drains the bulk of the Kispiox Range to the east, are the major tributaries flowing into Kitwanga Lake and the river mainstem. These tributaries contribute to the wide variations in water flows in the mainstem. They also transport moderate amounts of bedload in average flood flows and often carry large amounts of suspended sediments. The silt and clay are derived from mudstones of the early Cretaceous age, which were ground up by Ice Age glaciers and left behind as a mantle over the landscape. These sediments are easily mobilized by natural and man-induced stream instability and landslide failures.

Kitwanga Lake lies at an elevation of 376 m. It has a surface area of 7.8 km², and is relatively shallow with a mean depth of 5 m. The results of limnological sampling in 1995 showed a pronounced midsummer thermal stratification, with an average thermocline depth of 5.7 m, and epilimnetic temperatures exceeding 18° C. These extended to the lake bottom for up to 30% of the lake's total area during a four to six week period (Shortreed et al. 1998). The GFA has collected limnological data annually since 1999, and Cleveland (2001) suggested that this data show no obvious environmental constraints on sockeye fry. The lake is clear, with the euphotic zone encompassing the entire water column in most areas of the lake. Macrozooplankton biomass is relatively high, and *Daphnia* abundance makes up more than 60% of the total. With this high production and strong thermal stratification, summer oxygen concentrations may become very low in the lake's hypolimnion (Shortreed et al. 1998). Consequently, in long warm summers, there may be no deep cold-water refuge available for sockeye juveniles, potentially resulting in decreased growth and reduced ability to avoid predators.

Stream Channels

The Kitwanga River mainstem between Kitwanga Lake and the Skeena confluence is approximately 36 km in length. It is low gradient with no effective barriers to anadromous fish passage. The northern two-thirds is a wandering gravel-bed river. Because of the high-suspended sediment load, abandoned channel segments and backwaters fill up with mud more rapidly than in typical gravel-bed rivers. The lower 12 km is entrenched 30–100 m into a bedrock canyon. From the Skeena River upstream to the lake, mainstem gradients vary from 0.5% to 0.7% slope (MoE 1979).

Giesbrecht et al. (1998) noted that channel instability in reaches 3 and 5 of the mainstem is due to logging and clearing of the riparian area, as well as to past road construction. The major problems in the watershed include damaged riparian areas, bank erosion, channel instability, reductions in large woody debris (LWD), and barriers at stream crossings, particularly blocked and perched culverts. Forest development activities and poor placement of drainage structures along Highway 37N have adversely affected many tributary streams. Ten Link Creek is the community watershed for Gitanyow village. Currently, erosion generated from an access road compromises water quality.

Riparian logging and beaver damming adversely affect an area located north of Kitwanga Lake between the 26-Mile Forest Service Road (FSR) and the Webber FSR. Over 20 beaver dams block a 5 km section of the river to annual adult salmon migration (Cleveland 2002a). Cleveland also notes that logging has

View downstream of the lower Kitwanga River

negatively influenced most of the streams entering Kitwanga Lake. Recent investigations by the GFA have revealed that several of the larger, high fish-value streams are subterranean in their lower reach, due to bank de-stabilization from logging in upstream riparian areas and resulting downstream aggradation.

Geography

The Kitwanga watershed is comprised of two mountain masses, the Nass Ranges and the Kispiox Range, divided by a linear down-faulted trough. The Kitwanga River on the south end and the Cranberry River to the north, which is part of the Nass watershed, occupy the trough. The watershed is principally underlaid by bedrock composed of early Cretaceous, Skeena Group sedimentary rocks in the Nass basin, and Bowser Lake Group sedimentary and volcanic rocks with a minor granitic intrusion at Hazelton Peak (Gottesfeld 1985). The fluvial and surficial geomorphology of the watershed is strongly influenced by its recent glacial history. Ice from the Coast Ranges flowed southerly down the Nass basin, with a portion flowing through the Kitwanga valley into the Skeena. The scoured and grooved mountainsides, as well as Kitwanga Lake, are part of the glacial legacy.

The gentle divide (385 m) between the Cranberry and Kitwanga drainages lies north of Kitwanga Lake; it is occupied on the west by the compound alluvial fan of both the upper Kitwanga River and the Cranberry River. The east side of the divide is dominated by an extensive floodplain and wetland complex. Because of these indeterminate drainage features, it is easy to imagine fish passage from the Skeena to the Nass valley.

Within the Kitwanga River valley, inactive river terraces, higher and wider than the present floodplain, may have been deposited during the Little Ice Age cool-wet period between 200 and 500 years ago (Gottesfeld 1985). These terraces have been the preferred sites for farmland south of the Gitanyow village. The mountain valleys and side walls of the main valley are covered by thick blankets of glacial till. The surface expression conforms generally to the underlying bedrock surface, with bedrock exposure along deeply incised streams and on steep-sided hillocks.

The coastal/interior transition climate is reflected in the major ecological zones. Vegetation in the lower-elevation valley is represented by the Interior Cedar Hemlock (ICH) and Coastal Western Hemlock (CWH) zones, which are dominated by forest stands of hemlock, spruce, subalpine fir, and, to a certain extent, red cedar. In the higher elevations, the ICH and CWH zones pass into forest cover that is dominated by mature and overmature subalpine fir, represented by the Engelmann Spruce – Subalpine Fir (ESSF) biogeoclimatic zone (Pojar et al. 1988). Historically, frequent fire disturbance by Gitxsan people created successional stands of aspen and birch on upland sites. Black cottonwood prevails in floodplain stands, typically mixed with immature spruce and cedar.

Fish Values

The Kitwanga River watershed is a relatively small, but biologically rich river system that has significant high-value fish populations and habitat. Fish species utilizing this habitat include sockeye, coho, pink, chum, and chinook salmon, as well as steelhead, rainbow and cutthroat trout, Dolly Varden, kokanee, bull trout, and mountain whitefish. The depressed sockeye salmon abundance is a serious concern and the focus of rebuilding efforts.

Chinook Salmon

The chinook population in the Kitwanga River contributes a few percent of the total Skeena chinook escapement. The escapement trend from 1950, with 700 adults, to the mid-1980s, with less than 100, indicates a collapse of the Kitwanga chinook substock. Following the closure of directed net chinook fisheries in 1983 and the Pacific Salmon Treaty in 1985, the stock increased to record escapements with average escapement of over 1,500 in the 1990s. Mean escapements since 2000 have been slightly more than 2,000 chinook.

Chinook salmon typically enter the Kitwanga system in early August and head to their spawning grounds. The principal chinook spawning areas are the reach below the lake outlet, the lowest reach of the river near the Skeena, and the mainstem reach immediately downstream of the Kitwancool Creek confluence (DFO 1991a). Recent spawning assessment surveys conducted by the GFA showed that 40% of chinook spawning in the Kitwanga system occurs in the mainstem section 1 km below Moonlit Creek. The GFA also noted that approximately 20% of the total chinook spawners were evenly distributed over the lower 12 km of Kitwancool Creek (McCarthy et al. 2002). Rearing takes place throughout the mainstem and lower tributary reaches.

Pink Salmon

The Kitwanga River is one of the major pink salmon producing rivers in the Skeena system. The Kitwanga even- and odd-year pink runs do not have a well-developed dominance. Compared to other Skeena subbasin pink populations, Kitwanga pink abundance since 1950 shows a slight increase. Mean annual escapement since the 1950s to the present is 153,000 odd-year and 123,000 even-year pink salmon, with ranges from 400,000 in 1987 to 5,000 in 1988.

Pink salmon generally enter the Kitwanga River in early to mid-August, in two timing groups, though this is variable. Early spawners use the lower Kitwanga River below Kitwancool Creek; though, in years of high abundance they make use of the river up to and including Moonlit Creek. The later (7 to 10 days) spawning group uses the mainstem between Moonlit Creek and the lake outlet (Woloshyn 2002). Major concentrations of pink spawners have also been recorded by the GFA from Tea Creek downstream to the Skeena confluence (Cleveland 2002).

Chum Salmon

Adult chum salmon returning to spawning grounds in the Kitwanga River made up approximately 40% of the total reported Skeena system chum escapement in the early 1950s (DFO 2005). In 1957, chum adult populations abruptly declined until the mid-1980s, when returning adult numbers increased to an annual average over five years of 1,500 chum into the early 1990s. Since 1993, chum escapement has annually averaged less than 400 fish. Recorded escapement averages 1,855 chum since 2000.

Chum salmon typically return to their Kitwanga River spawning grounds in mid to late August. Principal chum spawning occurs in the lowest reach of the river, in the mainstem reach below the confluence of Kitwancool Creek, and in the section of river below the lake outlet. The GFA has recently observed chum spawners in the mainstem adjacent to Moonlit Creek and to 200 m upstream on Moonlit Creek (Cleveland 2002). Upon emergence as fry, the juveniles migrate directly to sea.

Sockeye Salmon

Escapement records maintained between 1919 and 1950 show Kitwanga River sockeye escapement ranged between 35,000 and 2,300 with an approximate mean of close to 10,000 spawners (Smith and Lucop 1966). Gitanyow elders indicated that declines in salmon returns began in the 1960s, with most sockeye fishing sites along the Kitwanga River abandoned by the early 1970s (Jacobs and Jones 1999).

Escapement since the 1960s has rarely exceeded several hundred spawners. The average escapement between 1998 and 2002 was 225 sockeye (Cleveland 2002). Since 2003, when a counting fence was installed on the mainstem, escapement has ranged between 937 and 5,146 with a mean of 2,681 sockeye. The abundance trend for Kitwanga sockeye over the last 50 years indicates escapement fluctuating at low levels, and presently the stock is only 5% of the system's potential production.

Fisheries exploitation rates on Kitwanga sockeye have been high (>50% in most years) since the inception of commercial fishing at the mouth of the Skeena River in 1877. Kitwanga sockeye run timing overlaps with the enhanced portion of Lake Babine sockeye. Overfishing has been a significant factor in reduced Kitwanga sockeye abundance.

Sockeye adults typically return to the Kitwanga River in early August and pass up the mainstem to their principal spawning grounds. Historically, spawning occurred in the mainstem below the lake outlet and above the lake inlet on the Kitwanga River, with extensive spawning along the northern and western lakeshore (Smith and Lucop 1966, Jacobs and Jones 1999). Spawner surveys in 1998 recorded sockeye spawners only from the lake outlet to just downstream of Moonlit Creek confluence. Recent GFA observations noted sockeye spawners in Kitwanga Lake only (Cleveland 2002).

Shortreed et al. (1998) described Kitwanga Lake as one of the most biologically productive lakes in the Skeena. The zooplankton is dominated by *Daphnia,* a preferred food for sockeye fry. Based on the trophic status alone, Kitwanga Lake could potentially produce about one million sockeye smolts annually, which would result from an optimum adult escapement that exceeds 18,000 sockeye. The potential production of this lake for sockeye is shared to some extent by kokanee, which are almost indistinguishable in the fry stage.

Changes to the freshwater ecosystem and sockeye spawning and rearing habitat in particular have occurred over the last 45 years. In the low- and mid-elevation portions of the landscape, logging has altered the ecological conditions. Critical habitat for Kitwanga sockeye spawning and rearing areas have been adversely affected by changes to the annual water flow pattern, water temperature regime, and increased sediment input, which in turn has increased macrophyte growth.

Overall, high fisheries exploitation rates and habitat modifications have led to reduced sockeye abundance. In response to this sockeye conservation concern, the GFA has initiated a sockeye rebuilding plan in cooperation with the DFO, the Skeena Fisheries Commission, and the province of B.C. An expert panel reviewed the available data and recommend options to rebuild Kitwanga sockeye. The main goal is to restore Kitwanga sockeye abundance while retaining the genetic endowment of the stock. The plan identifies potential limiting factors to production and recommends recovery actions, forming the basis of the Kitwanga Sockeye Salmon Recovery Strategy. The plan will be updated as stock rebuilding activities proceed and require amendment.

Increasing fry production by increasing adult escapements and by outplanting fry, combined with restoration of spawning habitat, are the essential elements to improve Kitwanga Lake sockeye production. Projects that received priority for 2006/07 include:

- The collection and culture of Kitwanga sockeye brood in a hatchery setting in order to release 100,000 fry into Kitwanga Lake in the spring of 2007. If successful, this will allow for a production boost in excess of 85% from the egg to fry stage, helping to increase adult recruitment in 2010 and 2011.

- Pilot small-scale restoration programs of known sockeye spawning areas within Kitwanga Lake including cleaning sediment deposits from spawning grounds, removal of macrophytes, and adding superior quality spawning gravel.

- Assessing Kitwanga sockeye smolt output during the spring of 2006 and 2007 and the enumerating escapement at the counting fence.

Coho Salmon

Coho are the most widespread salmon in the Kitwanga watershed. Records show average escapements of less than 300 coho for the 1960s, while the 1970s and 1980s show average escapements of approximately 600 fish annually (DFO 2005). Escapement records are incomplete for the 1990s; however, there were 2,500 coho in 1990. Cleveland (2002) reported mainstem stream walk enumerations of 2,300 coho salmon in 2000 and approximately 4,000 coho adults in 2001. Escapement enumerated through the counting weir was 2,000 and 7,100 in 2003 and 2005, respectively. There are no known escapement data for any of the tributaries.

Generally, coho migrate into the Kitwanga system from early September to mid-October. They hold in the mainstem and off tributary mouths until fall storm flows permit passage

Coho anglers at the Kitwanga-Skeena confluence

into smaller streams. Coho spawning grounds on the mainstem are concentrated from Kitwanga Lake downstream to 1 km below the Moonlit Creek confluence. Other principal spawning areas are the lowest reach of the river and the reach downstream from Kitwancool Creek. The major tributaries – the upper Kitwanga River, Moonlit Creek, Kitwancool Creek, Deuce Creek, and Tea Creek – support coho spawning in varying degrees. Coho juveniles are widespread throughout the tributary streams and mainstem. Migration to saltwater occurs one or two years after hatching, but the proportion of these two age classes is unknown. Coho adults return after about 16 months at sea.

Steelhead

The Kitwanga watershed supports one of the significant steelhead populations in the Skeena watershed. There are insufficient and uncertain data to adequately assess the current stock status, past escapements, or trends in abundance over time.

Kitwanga River steelhead enter the Skeena River as summer-run fish. They migrate into the Kitwanga River from August through December. Lough (1983) noted that adult steelhead overwinter in the lower 12 km of the Kitwanga River in the area largely composed of canyons, bedrock outcrops, and pools. It appears from resistivity counter data (Gitanyow Fisheries Authority unpublished data) that many or most steelhead overwinter in the Skeena River at or adjacent to the mouth of the Kitwanga River. Lough's (1983) radio telemetry study also found steelhead overwintering in the Skeena River, and although no steelhead were found in Kitwanga Lake, the lake may be an important overwintering locale. Lough (1983) reported that Kitwanga River steelhead moved in early May onto spawning sites located throughout the mainstem below Kitwancool Lake. Spawning activity peaked in mid-May, with most kelts departed by the end of May. The dominant steelhead spawning ground appears to be from Kitwanga Lake downstream to the Moonlit Creek confluence.

The Kitwanga River from Kitwanga Lake to the Skeena has been documented as juvenile rearing habitat (Bustard 1992a, 1993, Beere 1993). Over three years, the overall fry and parr combined density was 0.80 juvenile/m^2, significantly greater than that found in similar sampling programs in the Morice system (0.23/m^2) and in the Zymoetz system (0.27/m^2) (Beere 1993). Bustard (1992) reported that 64% of the juvenile fish captured during the 1991 study in the Kitwanga River mainstem were juvenile steelhead, while in lower Moonlit Creek they comprised only 9.4%. Scale aging data from Kitwanga River steelhead indicated that juveniles remained in freshwater for three to five years (mean=3.67), and that the majority of juveniles spent three years (44%) or four years (44%) in freshwater. The data further showed that the majority of steelhead (66.7%) spent two years in the ocean before their first spawning at age six (Grieve and Webb 1999b, Cleveland 2002b).

Fisheries
First Nations Traditional Use
Traditionally, Gitxsan from the Gitanyow and Gitwangak villages occupied and used the Kitwanga watershed. Two Gitxsan House groups from Gitwangak, Gaxsbgabaxs, and Sakxum Higookw, maintain territory in the lower portion of the drainage. Gwaas Hlaam and Gwinuu, House groups from Gitanyow, use and maintain the upper, major portion of the watershed as their home. The Gitanyow village is the only Gitxsan settlement removed from the Skeena or Babine River mainstems. Its location is along the "grease trail" from the Skeena to the Nass River (Derrick 1978).

The abundant and predictable sockeye salmon stocks provided the Gitxsan of the Kitwanga River with opportunity to harvest and preserve a large amount of high-quality food in a relatively short time of intensive effort. The sockeye run was the major focus, as it provided the majority of high-quality dried fish needed to sustain the Gitxsan over the year, and to produce a trade item. Following the passage of the bulk of the sockeye, coho were available well into autumn, providing both fresh and dried fish. Rainbow trout, steelhead, and Dolly Varden char were also fished in their respective habitats and then processed.

Weirs were the most intensive fishing effort on the Kitwanga River. The last remaining documented weir was located immediately below the Gitanyow village. This productive weir, built across the shallow river, supplied most of the salmon needs for the Gitanyow people. Posts pounded into the river bottom and then overlaid with panels of split cedar secured on the upstream side supported barrel-type traps. These traps were fitted with a movable panel through which fish could be dipped or gaffed out, or released, dependent on whether the species was desired. Multiple weirs were in operation at various locations on the river, such as the outlet of Kitwanga Lake, at sites downstream from the Kitwanga-Kitwancool confluence, at sites adjacent to Gitanyow, and a site approximately 8 km north of Kitwanga.

Recreational Fisheries
The Kitwanga River and Kitwanga Lake attract a moderate recreational fishery that is dominated by regional residents. Angling for chinook and steelhead is popular, particularly at the Skeena-Kitwanga confluence, though steelhead anglers utilize various holes along the mainstem. Kitwanga Lake is fished in all seasons for resident cutthroat and rainbow trout, and

Gitwangak, 1899

Weir on the Kitwanga River, 1918

is easily accessible. Since 1991, guiding on the Kitwanga River has not been permitted. Creel surveys for the chinook and coho fisheries at the mouth of the Kitwanga River were carried out in 2000, 2001, and 2002. The sports fishing harvest recorded is small and discussed in the mid-Skeena River south section. Kingston (2002) discussed the creel survey data collected for the river and lake. Current angling regulations designate the Kitwanga River as Class II waters year-round. A steelhead stamp is mandatory between September 1 and October 31 and a bait ban is applicable from September 1 to December 31 (B.C. Fisheries 2005).

Enhancement Activities

Habitat enhancement activities in the Kitwanga watershed have been ongoing for centuries according to Gitanyow elders, particularly in regard to beaver populations and fish access management. The Fisheries Research Board operated a counting fence on the lower river close to the Skeena confluence from 1957 to 1960 (DFO 1930–1960). A coho incubation box located at Gitanyow was in operation in the late 1990s and early 2000s to raise 10,000–12,000 coho eggs annually (Jacobs and Jones 1999, Kingston 2002). Since the mid-1990s, the GFA has been active in enhancement work primarily

directed to facilitating fish access, beaver dam mitigation, and salmonid fry salvage (Cleveland 2002). In 2003, the GFA installed a salmonid enumeration fence across the Kitwanga River.

Development Activities

The principal development activities involve forest development, settlement, and linear development. There are few mineral occurrences and no known mineral developments in the watershed.

Forest Resource Development

The Kitwanga watershed is located within the Ministry of Forests Skeena Stikine Forest District. Forest development activity began with agricultural clearing by settlers following completion of the Grand Trunk Pacific Railway in 1914. Small-scale lumbering led to small bush mills, and the post-World War II economic boom increased the demand for lumber. Independent cedar pole loggers also saw a demand for poles at this time. Over the years up to 1960, logging was selective with a moderate proportion of residual timber left standing, particularly in the southeastern portion of the watershed. Timber was processed by small, on-site sawmills, whose sawdust piles are still clearly visible from the air. By the mid-1960s, vast areas of accessible

Kitwanga River salmonid enumeration fence, 2003

timber both west and east of the Kitwanga River up to Kitwanga Lake were logged.

Since the mid-1960s, clearcut harvesting has been the preferred logging system. In 1963, consolidation of seven or eight small mills led to establishment of Hobenshield's mill at its present location in Kitwanga. From the mid-1960s to the 1970s, the valley bottomlands north of Kitwanga Lake began to be logged. The Canadian Cellulose sawmill was constructed in Kitwanga in the early 1970s. During the 1970s, most logging was in the lower, eastern portion of the watershed and in the low-lying country north of Kitwanga Lake, with minor development in the lower Moonlit Creek area. In the 1980s, the upper Kitwanga valley, along with the slopes to the east of Gitanyow and around Kitwanga Lake, saw extensive logging development. Forest development activities also occurred in Moonlit Creek and some mainstem tributaries draining from the east. Further development in the 1990s was concentrated in McKenzie, Manuel, Hanna,

and other headwater drainages of the upper Kitwanga River, with widespread and dispersed development elsewhere in the watershed.

In 1995, the Watershed Restoration Program (WRP) became involved in assessing the forestry-related impacts and upslope sediment-producing areas in relation to fish and fish habitat (Wildstone 1995). Watershed health has benefited from road deactivation, riparian, in-stream, and off-channel site works to a certain degree. Habitat restoration activities conducted under the WRP include culvert backwatering, placement of large woody debris, and riparian site works (McElhanney 2001). McElhanney (2001) summarized the 12 assessment and site works projects conducted in the watershed since 1995, and concluded that approximately $750,000 of logging-related, prioritized restorative work was still needed.

The large wetland complex drained by the Kitwanga River and located north of Kitwanga Lake, which was adversely affected by logging, remains an outstanding compound problem from a fisheries perspective (Cleveland 2002). This problem is due to beaver colonization following the spread of deciduous trees into clearcuts. The beaver dams have greatly dispersed streamflows from the upper Kitwanga River, blocked anadromous fish passage, and caused increases in stream water temperature.

Population and Settlement

The Kitwanga valley has been home to Gitxsan people for thousands of years. Many Euro-Canadian settlers arrived following completion of the railroad in 1914, attracted by the agricultural possibilities. The Kitwanga watershed population base has slowly grown to total 1,315 people (Statistics Canada 1996, SNDS 1998). Community populations are as follows: Gitanyow 403, Kitwanga 383, and Gitwangak 529. Severe job losses related to the depressed forestry and fishing sectors have caused massive unemployment problems. The population trend for the two First Nations communities points to slow, steady growth.

Transportation and Utilities

The existing transportation network in the watershed reflects years of steady improvement based on the First Nations trail infrastructure, particularly Highway 37, which follows the "grease trail" (Derrick 1978). The railroad provided the main transportation link to the mouth of the watershed up to 1950, when the Kitwanga backroad, a series of pole logging roads, was connected to roads from Hazelton (Hobenshield 2002). In 1975, improvements to Highway 37 included alignment, pavement, new drainage structures, and the Skeena River bridge crossing. Overall, the development pattern has been spurred by the motive to extract forest products.

Other than the paved highway, all other roads are gravel surface. Secondary roads branching off Highway 37 include the Kitwanga backroad, Tea Lakes FSR, Mill Lakes FSR, Ten Link Creek Road, Kitwancool FSR, 18 Mile Branch, Moonlit Branch, Rehab Branch, and Kitwanga Main. Utilities are limited within the watershed. Electricity is supplied by BC Hydro's provincial grid, with the transmission line closely following the highway and servicing the communities of Gitanyow, Kitwanga, and Gitwangak.

Collecting sockeye brood on Kitwanga Lake

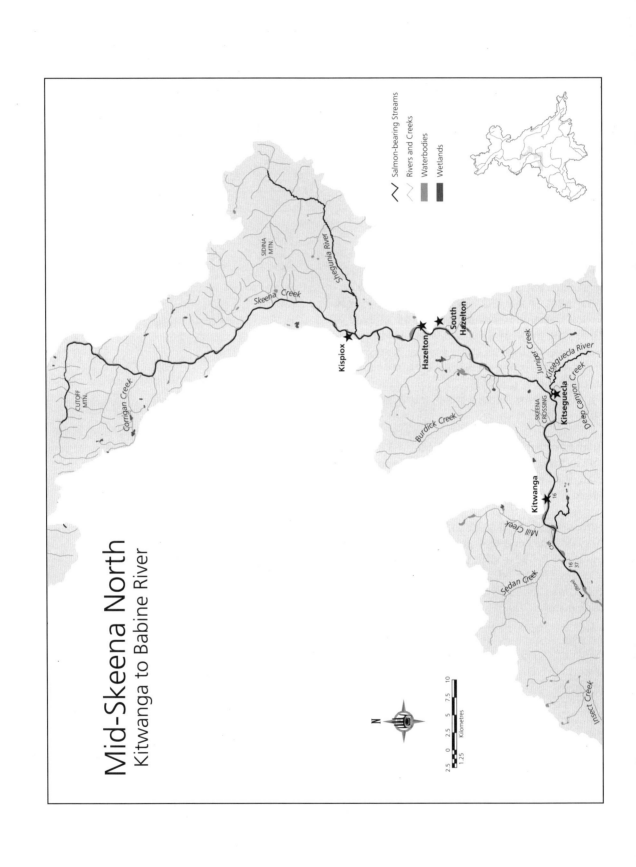

Mid-Skeena North
Kitwanga to Babine River

Salmon-bearing Streams
Rivers and Creeks
Waterbodies
Wetlands

SIDINA MTN.

Shegunia River

Skeena Creek

Kispiox

Hazelton

South Hazelton

Juniper Creek

Kitseguecla River

Deep Canyon Creek

Kitseguecla

SKEENA CROSSING

Burdick Creek

CUTOFF MTN.

Corrigan Creek

Kitwanga

16

Mill Creek

CNR

16 37

Sedan Creek

Insect Creek

N

2.5 0 2.5 5 7.5 10
1.25   Kilometres

13

Mid-Skeena River North: Kitwanga to Babine River

Environmental Setting

The mid-Skeena River north area is defined as the area encompassed by the Skeena River and its tributaries from the Kitwanga River upstream to the Babine River. This section includes the 99 km mainstem and all tributaries other than the Kitwanga watershed, Bulkley watershed, and Kispiox watershed. These three major watersheds are discussed elsewhere in this book.

Location

The mid-Skeena River north area is located in west-central British Columbia. Gitwangak is located on the western bounds, and Hazelton, which is at the confluence of the Bulkley River and Skeena River, is central to this area. The Kispiox Range and the Kispiox watershed bound the drainage to the west, and the Rocher Déboulé Range and the Babine Range form the east boundary. The Zymoetz watershed bounds the area to the south.

Hydrology

The mid-Skeena River north area is, for the most part, mountainous with high relief; elevations range from 2,421 m in the Rocher Déboulé Range to 172 m at the Skeena-Kitwanga confluence. The Skeena River essentially bisects this area on a roughly north-south lineament. Major streams within the watershed area include the Kitseguecla River and the Shegunia River, with inflow from the Bulkley, Kispiox, and Babine Rivers.

Skeena River peak discharges occur in May and June due to spring snowmelt throughout the upper Skeena watershed east of the Coast Ranges. In almost all years, the peak spring freshets in this zone are the largest floods in the major rivers. Water levels in the main river channels and back channels fluctuate seasonally; typically they are high from May to early July, drop for the summer months, rise to intermediate

levels in the fall, and reach their annual low levels late in the winter season.

Hydrometric station 08EB003 is located on the Skeena River at Glen Vowell, a few kilometres upstream of Hazelton. Monthly mean discharge for March is 97.3 m^3/s and increasing to 1,960 m^3/s for June. These flow values are about two-thirds of the flow at Usk, which is located another 100 km downstream and which receives the additional inflow of the Bulkley, Kitseguecla, and Kitwanga Rivers, as well as other smaller rivers.

Tributaries upstream from Kitwanga that flow into the Skeena River right bank include Burdick Creek, Hazelton Creek, the Kispiox River, Carrigan Creek, and Blackstock Creek. Skeena River left-bank tributaries include Shandilla Creek, the Kitseguecla River, the Bulkley River, the Shegunia River, Sediesh Creek, and Shewililba Creek. Most tributaries north of the Shegunia River are short and steep.

Similar to other interior portions of the Skeena watershed, the hydrology of the mid-Skeena River north is dominated by snowmelt. The peak discharges of the major tributaries are typically in May and June due to spring snowmelt. Discharges are moderate in size as melting occurs progressively at higher elevations with the seasonal rise in temperature. Discharges then decrease until September when fall rains and early snowmelt increase streamflows through October. High streamflows and floods can occur due to rainstorms in the fall (October–November). These high streamflows are typically of short duration but sometimes of high magnitude, and usually are due to intense rainfall or rainfall on snow. Streamflows decrease through November and December when precipitation falls as snow, with low discharges recorded January through March.

The Hazelton and Skeena Mountains exert the major hydrological influences on tributary drainages. Tributary streamflows have a moderately rapid response from storm-water input due to the high gradients of the major tributaries. The low watershed divide through the Kispiox valley to the Nass drainage allows coastal weather systems to easily enter the area. Rainfall records at New Hazelton based on a 39-year observation period show annual average precipitation of 509 mm per year, with approximately 77% falling as rain and 23% as snow.

The general climate of the watershed is transitional between a temperate, maritime coastal climate and the colder, continental climate that characterizes the interior of the province. Summers are warm and fairly moist with a relatively long growing season. In the Skeena valley, snowpacks are light to moderate (up to 1 m) and the valley is prone to cold air ponding as well as summer droughts. The wide Skeena valley allows warm, humid coastal weather systems to penetrate, resulting in a gradual loss of warmth and moisture as well as heavy snowpacks in the mountains.

Shandilla falls

Stream Channels
Skeena Mainstem
The Skeena River between the Kitwanga River and the Bulkley River is the upper part of reach 3 (Resource Analysis Branch 1975–79). It is approximately 42 km in length, with an average gradient of 0.1% and no obstructions to anadromous fish passage over its length. Reach 3 of the Skeena River is characterized by a single-thread, irregularly sinuous channel pattern controlled by occasional bedrock outcrops or incised into the bedrock, reworked till, and alluvial deposits on the moderately narrow valley floor. Boulders and cobble beds, with the occasional gravel bar, dominate the substrate. The Kitseguecla River delivers a moderate to heavy contribution of sediment in peak flows into the Skeena mainstem. Valley bottom depositions are largely a result of glacial sediment supply that filled in the wider portions of the valley bottom, creating the floodplain and river benches that favour settlement and farming (Gottesfeld 1985).

Reach 4 of the Skeena is the section between the Bulkley confluence and Pinenut Creek. In this reach, the Skeena River is slightly meandering in a single channel with an average gradient of 0.2%. The gradient is controlled by the incised bedrock of reach 5, the bedrock at Four Mile Canyon, and the confluence of the Bulkley River. Potential flood areas include alluvial plains, low terraces, and alluvial fans of tributary streams. Bank erosion is common at concave banks of certain bends.

Reach 5 extends from Pinenut Creek to Carrigan Creek and is characterized by the single-thread, confined channel incised into mostly bedrock with an average gradient of 0.2%. Though there are numerous falls, all are passable to anadromous fish. Reach 6 from Carrigan Creek to the Babine River is relatively wider and bounded within a narrow floodplain. The channel has a mostly cobble/gravel substrate, and an average gradient of 0.1%.

Kitseguecla River
The Kitseguecla River is a major left-bank tributary, 33 km in length to its headwaters in Kitseguecla Lake. Major tributaries draining into the right bank of the Kitseguecla River include

Juniper Creek and Laura Creek, while Kitsuns Creek, Deep Canyon, and Jack Mould Creek flow into the left bank.

Reach 1 of the Kitseguecla River extends from the mouth at the Skeena River upstream to the Kitsuns Creek confluence. This reach is characterized by a regularly confined, single-thread channel with an average gradient of 1.2%. There are six major bank slumps on this reach, which, for the most part, are failing fluvio-glacial terraces. The channel is incised and exhibits significant bedrock control, particularly at the 0.5 km canyon bisected by the mouth of Laura Creek and the crossing of the 400 Road.

Prior to major logging activity in the 1960s, air photos show the main channel as relatively stable. Since then, the main channel has become less stable, and large amounts of sediment have started moving down the river during flood flows (Gilchrist 1998). This decrease in channel stability is shown by the increase in internal sediment stored in the mainstem and the general increase in channel width. The increase in channel width in turn increases bank failures contributing additional sediment. Causes of the decline in channel stability include a series of major floods in the 1970s through the 1990s, and the consequences of logging in small tributaries where the equivalent clear-cut area is in excess of 30% of the tributary watershed area (Gilchrist 1998).

Reach 2 extends from Kitsuns Creek upstream 2.5 km. The channel becomes more entrenched with the gradient averaging 1.7%, though some sections are up to 6%. The channel is mainly riffle-glide type habitat with the substrate composed of boulders and cobbles. Reach 3 extends upstream to Jack Mould Creek, with an average gradient of 1.7% in a slightly irregular meandering channel, often confined by steep valley walls. Stream morphology is characterized by riffle-pool-glide type habitat, with the substrate composed of boulders, small cobbles, and the occasional pocket of gravels and fine sediments. Both sides of the channel appear to be unstable, with numerous natural failures. Reach 4 extends from Jack Mould Creek to Kitseguecla Lake. The upper portion is a wide, low-gradient wetland complex that is an

Skeena River, reach 5, downstream of the Babine River

extension of Kitseguecla Lake. The substrate is predominantly fine organic matter.

Juniper Creek

Juniper Creek flows west and enters the Kitseguecla River right bank approximately 2.8 km upstream from its confluence with the Skeena River. The Juniper subbasin is confined by steep-sloped terraces, bedrock, and hillsides throughout, except for the narrow, gentle sloping fan as it nears the Kitseguecla River. The drainage hydrology is considered flashy, and the mainstem often experiences a large amount of bedload movement. Reach 1 is characterized by the alluvial fan and is approximately 2.7 km in length with an average gradient of 4%. Reach 2 extends approximately 5.6 km upstream to the confluence of Brian Boru Creek, the major tributary. The reach is entrenched and confined within the valley walls. Natural mass wasting resulting from steep slopes and pockets of lacustrine soils characterizes this reach. The stream shows a stepped profile and the substrate is dominated by boulders and cobbles. The Red Rose Mine access road closely follows the right bank of the stream channel.

Kitsuns Creek

Reach 1 of Kitsuns Creek is approximately 5.2 km in length, extending to the West Kitsuns Creek confluence. The channel has an average gradient of 1.5% that ranges from 1% to 3% (Acer and Gitsegukla Band Council 2000). The reach is confined and entrenched throughout with very steep bedrock walls, rising directly from the edge of the active channel. The aggrading channel is characterized by long riffles (45%), deep glides (45%), and very few distinct pools (10%). Disturbance indicators include many failing banks, numerous debris jams, and elevated bars, which are due to periodic high-energy flows resulting in major movements of debris and bedload (Acer and Gitsegukla Band Council 2000).

Reach 2 of Kitsuns Creek passes upstream from the West Kitsuns Creek confluence for approximately 11.7 km in a stepped and confined profile with an average gradient of 1.5%. The channel receives high sediment loads from glacial deposits and bank failure sediments from the west fork draining Ashman Ridge.

West Kitsuns Creek

West Kitsuns Creek generally flows north and east, draining into Kitsuns Creek 5.6 km upstream of the confluence of Kitsuns Creek and the Kitseguecla River. The drainage is characterized by various-sized natural and logging-related landslides that often spill directly into West Kitsuns or tributary stream channels. Reach 1 is approximately 4 km in length, with an average channel gradient of 3%. The stream is deeply incised into bedrock and is generally bedrock controlled. The substrate is comprised mostly of boulders and cobbles with excellent riffle-pool habitat (Biolith 1999b). There is significant bank erosion and natural mass wasting from unstable soils that contribute to the high turbidity during flood events.

Reach 2 extends 1.6 km upstream to the bridge on the 700 Road and is characterized by confining bedrock sidewalls and fast-flowing water. The average channel gradient is 2.5%, with the substrate dominated by cobbles and boulders. Naturally unstable soils and road-related failures contribute to the mass wasting into the stream channel. The reach has been altered for about 0.5 km by the bridge-crossing site.

In this reach, there are multiple impacts that include loss of riparian vegetation; sediment deposition from the complex slide close to the intersection of the 700 and 400 Roads; sediment deposition from the overburden placed between the 700 Road and the stream near the northern end of the bridge; and bank erosion and slides between the road and stream channel for 400 m downstream of the bridge. A large area adjacent to the three bridges is heavily impacted and destabilized, with mass wasting on both sides of the stream from road building activities (Biolith 1999b).

Reach 3 of West Kitsuns Creek extends upstream for approximately 3 km to a 4 m falls over bedrock that is considered an upstream migration barrier to all salmonids at all flow levels. The channel is characterized by riffle-pool morphology with an average gradient of 2–3%. Cobbles and boulders dominate the substrate with gravels in pools. The stream channel is frequently bounded with bedrock walls on both banks, and there is evidence of natural bank instability. Except for one road-related slide into the channel, the reach is considered to be in its natural condition (Biolith 1999b).

Shegunia River

The Shegunia River flows approximately 37 km south and west to enter the left bank of the Skeena mainstem 1 km upstream of the Kispiox River confluence. West Fork and Duncan Creeks, which flow into the mainstem right bank, are the two major tributaries.

Reach 1 of the Shegunia River is 2.9 km in length and represents the alluvial fan; the gradient ranges up to 1.2% with a gravel-cobble substrate. Reach 2, which extends 3.6 km, is entrenched and sinuous with an average gradient of 1.7% and contains falls and cascades. Reach 3 is a canyon extending 8.7 km upstream, with the channel frequently bounded by bedrock walls. The average gradient is 1.8% with a boulder-cobble substrate and a riffle-pool sequence. Reach 4 is entrenched with a regular, sinuous profile and an average gradient of 1.1%. This reach extends approximately 10 km with a predominantly gravel-cobble substrate. The

Shandilla Creek culvert running beneath Highway 16

reach is bounded upstream by a series of falls and cascades at 25 km that limit anadromous fish passage.

Minor Skeena River Tributaries
Many of the smaller tributaries along the Skeena left bank have road- and/or railroad-related fish access problems. The Highway 16 culvert (4.5 m arch) at Shandilla Creek has an outlet drop of 1.5 m and is a full barrier to fish passage (Rabnett and Williams 2004). The Andimaul Creek Highway 16 culvert is a partial barrier to upstream fish migration. The Highway 16 crossings of Shandilla and Andimaul Creeks are velocity barriers (Rabnett and Williams 2004). Comeau Creek is subject to beaver activity and seasonal water fluctuations that effectively limit adult spawners to the 150 m reach upstream from the Skeena River (McCarthy 2000). CN Rail's Chicago Creek culvert crossing has a 1.5 m outfall drop and is generally impassable to fish. Seasonal water levels in both Chicago Creek and the Skeena River occasionally cause difficult spawner access.

On the less disturbed right bank, seasonal low water in Burdick Creek occasionally prevents fish from entering the creek. The fan at the mouth of Hazelton Creek periodically builds up to a steep gradient due to failing banks 1 km upstream. Glen Vowel Creek experiences years of seasonal low flow and is subject to extensive beaver damming.

Water Quality
The mid-Skeena River north generally has good water quality; however, during flood events or stream bank failures, water quality is compromised, particularly in Kitsuns Creek and West Kitsuns Creek. Cumulative effects from logging operations are not well understood or documented for the watershed. Major concerns include the integrity of small streams, hydrologic change, temperature change, sedimentation, and effects to the physical stream structure.

Water quality is of high concern to the communities of Glen Vowell and Gitsegukla. Glen Vowell Creek, also commonly called Sikedakh Creek, is a community watershed that supplies domestic water for Glen Vowell. Kits Creek, the designated community watershed for Gitsegukla, has experienced logging-related sediment problems. Channel restoration and erosion control work in 2001 mitigated the major potential detrimental consequences to the community water supply.

Geography
The principal landform of the mid-Skeena River north area is the Skeena River valley, set amid the fairly steep mountainous terrain of the Bulkley, Kispiox, and Babine Ranges. The Hazelton Mountains are composed of the Kispiox and Rocher Déboulé Ranges, and the Skeena Mountains are composed of the Babine Range. Major valleys in the area became established in early or middle Tertiary period during uplift of the Hazelton Mountains; however, it was the recurrent glacial erosion and deposition of the Pleistocene period that modified the area to its present-day form (Clague 1984).

The ice that covered and flowed down the Skeena valley during the last glacial period altered the mountain slopes and basins, leaving a legacy of glacial erosion and depositional features. Glaciers moving along the major valleys caused oversteepening of mountain slopes and contributed to the present-day U-shaped valley profiles, while the rounded summits are the result of ice moving across the mountain ridges. Above 1,500 m, alpine glacial activity significantly modified the high country. The presence of many ice-free cirques in the high mountains is the result of several centuries of warming.

Thick blankets of glacial till cover the main valley and mountain valleys, and extend up the valley sidewalls, though the surface expression conforms generally to the underlying bedrock surface. Thin soils, colluvium, and rock outcrops characterize the higher mountainsides. On many mid-slope positions, a morainal veneer is often found, while upper areas are dominated by rock and colluvium. Bedrock is exposed along deeply incised streams and on steep-sided hillslopes. The distinctive landform features are the strikingly displayed effects of glacial erosion and depositional features.

The Kitseguecla valley, a prominent north-south lineament that is relatively broad and opens into the Nechako Plateau to the south, sits between the Rocher Déboulé Range and Ashman Ridge. Ashman Ridge, which forms the glaciated headwaters of Kitsuns Creek, presents a striking visual appearance rising out of the Kitseguecla valley. This glaciated mountain possesses substantial pocket glaciers that are the remnants of former mountain glaciers.

Forests

Other than the mountainous high country of the Rocher Déboulé and Babine Ranges, the majority of the drainage is covered with dense, coniferous forests. Smaller but significant amounts of deciduous forests cover the low-elevation valley bottoms and lower mountain slopes along the Skeena and major tributary valleys.

Within the drainage, the principal forest types are represented by the three Biogeoclimatic Ecosystem Classification (BEC) zones: the Coastal Western Hemlock (CWH), the Interior Cedar Hemlock (ICH) zone, and the Engelmann Spruce–Subalpine Fir (ESSF) zone. A fourth type, the Mountain Hemlock (MH) zone characterizes minor amounts of subalpine forests along the western margin of the drainage.

The CWH zone characterizes a limited amount of the low- to mid-elevation forests in the Skeena valley, especially on northerly aspect slopes between the Kitwanga and Kitseguecla Rivers and the West Kitsuns and Kitsuns Creek drainages. The CWH is dominated by old-growth coniferous (rainforest) stands of western hemlock, amabilis fir, and subalpine fir, though hybrid spruce is common.

The ICH zone characterizes the low-elevation coniferous and deciduous forests in the Skeena valley and tributary valleys. As the dominant BEC zone in the mid-Skeena River north, the ICH occupies valley floors and mountain slopes with a transitional coast-interior climate. These forests of coniferous and deciduous stands reflect the close proximity of warmer, humid, maritime moisture. Major tree species in this zone are western hemlock, spruce, pine, and subalpine fir, with a mix of local stands of red cedar.

Both the CWH and the ICH zones merge into the ESSF zone at higher elevations that range from 900 to 1,300 m, depending on local topography, aspect, and climatic conditions. The ESSF zone possesses a shorter, cooler, and moister growing season, with continuous forests passing into subalpine parkland at its highest elevations. Subalpine fir is the dominant tree species, with small amounts of lodgepole pine and white spruce hybrids in drier slope positions or fire-influenced areas. Avalanche tracks are common in many ESSF forests with vegetation ranging from shrubby slide alder to cow parsnip and false hellebore.

Geology

The geology of the mid-Skeena River north is composed of three main types of rock units over three geologic time periods: volcanic and sedimentary rocks of the early to middle Jurassic Hazelton Group; sedimentary rocks of the middle to late Jurassic Bowser Lake Group; and primarily volcanic rocks from the late Cretaceous Skeena Group. These were followed by the Bulkley Intrusives along with minor amounts of early Tertiary Babine Intrusives. The final faulting into the valley and range structure now found in this area occurred within the last 15 ma. The geologic history is discussed in more detail in the chapter on the mid-Skeena River south area.

Fish Values

The mid-Skeena River north has high fish values with the presence of chinook, pink, sockeye, coho, and chum salmon, and steelhead trout. Rainbow and cutthroat trout, Dolly Varden, bull trout, mountain whitefish, slimy and prickly sculpin, largescale sucker, threespine stickleback, and peamouth chub, are also present in the drainage system (DFO 1991c).

Despite having high fish values, the area is a relatively moderate producer of salmon overall. Tributaries to the Skeena from the Shegunia River upstream to the Babine River are not known to have adult salmon presence; this is attributed to the steep gradient of streams flowing into the entrenched Skeena River. In general, the most widely dispersed salmon species is coho, while Dolly Varden and rainbow trout are found in most fish-bearing waters.

Chinook Salmon

Chinook salmon are present in the Skeena mainstem, which is mostly used for a migration route but also to an unknown extent as rearing habitat. Small populations spawn in the Kitseguecla and Shegunia Rivers. Escapement records note other sporadic chinook spawning in Burdick Creek (1929–36), Comeau Creek (1990), Chicago Creek (1929), and Hazelton Creek (1930) (DFO 2005).

Kitseguecla River

Kitseguecla River chinook escapement estimates have been recorded discontinuously since 1966; the only prior record is 1929. Escapements for the mainstem have ranged from 0 to 300 spawners; though most years the range is 25–75 fish (DFO 2005). There are no chinook escapement counts for Kitsuns Creek.

Adult chinook salmon begin their migration into the Skeena River usually in the third and fourth weeks of July, arriving at the Kitseguecla River throughout the month of August with the peak in mid-August. Spawning occurs on selected pockets of the mainstem up to 19 km. Chinook have also been observed spawning in Kitsuns Creek 1 km upstream from its confluence with the Kitseguecla River. It is not known if chinook juveniles leave the system and head into the Skeena mainstem shortly after hatching or rear in the Kitseguecla River.

Shegunia River

The name Shegunia is corrupted from the Gitxsan language name Xsu Gwin Ya'a, which translates as "river of chinook salmon." Shegunia River chinook escapement was first recorded in 1949 and has been recorded fairly continuously since the mid-1960s, with a range from 0 to 1,800 spawners. The annual mean for the 1990s was 81 fish. The 1980s annual mean was 327 spawners, with relatively high abundance in the mid and late 1980s. The principal spawning beds are located in the lower two reaches, though chinook have been observed at 15 km on the mainstem. Falls on the mainstem at 3 km limit fish passage at high and low flows. This is where the Salmon River Road crosses and in season, fish can often be seen holding for a change in water stage. Since the mid-1950s, the ease of human access to this canyon site has frequently caused chinook mortality.

Pink Salmon

Pink salmon spawners are present in the lower reach of most of the mid-Skeena River north tributaries from Shegunia River downstream. Pink salmon arrive at their spawning grounds in mid-August through September and spawn in gravel areas soon after ascending the river. Fry emerge in April and May and immediately head downstream to the estuary.

Skeena Tributaries

The lower 30 m of Shandilla Creek, essentially that portion within the Skeena high-water zone, supports pink spawners. Escapement was recorded from 1979 onward, with a range of 0–750 spawners. The average escapement for even and odd years in the 1980s was 140 and 500 fish, respectively, while for the even and odd years of the 1990s there were 270 and 381 pinks, respectively. The bulk of Shandilla Creek habitat is completely blocked by the Highway 16 culvert that has a 2.5 m outfall drop.

Comeau Creek sustains pink spawning in the lower 0.5 km with a range of 0–1,800 fish. Currently, there is a 2 m high beaver dam located 150 m upstream from the Skeena River that effectively limits adult spawners (McCarthy 2000). Burdick Creek supports pink spawners throughout the lower 1 km with a range of 0–6,000 fish. Typical escapement during the 1950s was approximately 500 for both even and odd years and for the 1980s was 94 and 164 for even and odd years, respectively.

Chicago Creek escapement was recorded inconsistently for over 70 years; it ranged from 0 to 3,500 fish, with most typical returns composed of 300 pinks utilizing the lower 0.8 km. Hazelton Creek escapement records show a range of 0–2,500 fish, with most enumerated years averaging 320 spawners using the lower 2 km reach. Glen Vowel Creek supports pink spawning in the lower reach in years of sufficient flow with records of 0–200 fish.

Kitseguecla River

Kitseguecla River pink spawner escapement has been recorded irregularly since 1954. Spawners have ranged from a high of 5,000 in 1991 to a low of 25 fish. From 1983 to 1993, the 10-year average of the even-year run was 1,020 fish; the odd-years run averaged 1,858 spawners.

Adult pink salmon arrive between mid-August and mid-September. The principal spawning ground is from the mouth to Kitsuns Creek (reach 1). Pink spawning has been observed in various gravel pockets upstream from Kitsuns Creek on the mainstem to 19 km and throughout the lower 1 km of Kitsuns Creek. Nearby Deep Canyon Creek had 350 spawners in 1989.

Shegunia River

Shegunia River pink salmon escapement has been highly variable, ranging from 0 to 40,000 fish. Typical escapement in the 1980s was 600 in the even years and 2,320 in the odd years. Most spawning occurs in the lower 1 km reach, which consists of the channel through the alluvial fan.

Chum Salmon

Chum salmon arrive in the mid-Skeena River north from late August to early September. Their migration coincides with the much larger runs of pink salmon, and they usually spawn in places that also have spawning pink salmon. Chum salmon are rare in the Skeena River above the Kispiox River confluence and in the Bulkley River.

Skeena River Mainstem and Tributaries

There are few documented chum spawning sites on the Skeena mainstem; however, it is quite likely that spawners use small areas of freshwater seepage and upwelling. There is rumour of a small chum population spawning immediately downstream of the Babine River confluence on the Skeena mainstem, but this spawning report has not been substantiated (Woloshyn 2003). The GWA has been collecting DNA from a small population spawning in a Skeena back channel 4 km upstream of Kispiox. Chum spawning occurs in the lower sections of several Skeena tributary streams and on their alluvial fans. Chicago Creek has scattered records of small numbers of chum salmon from 1931, 1984, 1989, and 1990 (DFO 1905–1949). The peak records are 10 chum in 1984 and 50 in 1989. Comeau Creek hosts a small number of chum spawners in the lower 50 m and in the Skeena River at the creek mouth.

Kitseguecla River

Chum salmon typically return to their Kitseguecla River spawning grounds in mid to late August. Spawning is scattered in the reach between the mouth and Kitsuns Creek; chum have not been observed spawning upstream of the mainstem bridge crossing at 10 km. Adult chum salmon returning to spawning grounds in the Kitseguecla River have been enumerated in only four years: 25 chum in 1986, 100 in 1987, and 25 chum in 1989 and 1990. The status of chum salmon in the Kitseguecla drainage is largely unknown.

Sockeye Salmon

Small sockeye populations have been reported from two mid-Skeena River north watersheds: Chicago Creek and the Kitseguecla River. Historically, Seeley Lake in the Chicago Creek system supported sockeye spawners. Fishery officer reports from 1929 to 1934 described sockeye that spawned at the outlet of Seeley Lake. These sockeye were apparently an early run and appeared on the spawning grounds in early July. Sockeye have not been documented at Seeley Lake since 1934.

Kitseguecla System

Two apparently distinct but little-known sockeye stocks spawn in the Kitseguecla watershed: stream-type spawners in West Kitsuns Creek and spawners in the upper Kitseguecla mainstem that are thought to rear in Kitseguecla Lake, though they could possibly be stream-type juveniles (Woloshyn 2003). Little is known about the sockeye that spawn for approximately 2 km downstream of Kitseguecla Lake. Escapement of 50 spawners was recorded in 1982, and their presence since then has been periodically noted by fishery officers (Woloshyn 2003). It is not known if sockeye fry, after hatching, migrate upstream to the lake or move out of the system. In 2004, Gitksan Watershed Authorities operated a fence downstream of Kitseguecla Lake and recorded no sockeye presence.

Spawning sockeye have been noted in West Kitsuns Creek in the upper portion of reach 2 and the lower portion of reach 3, close to the twin bridges crossing site (Woloshyn 2003). These spawners appear to be river type sockeye, which are relatively unusual in the Skeena system. In addition, West Kitsuns sockeye are unique in that their spawning colours are atypical of other Skeena sockeye stocks except Gitnadoix Lake sockeye. When mature, these sockeye possess a silver body colour with only a dull red stripe. Timing and enumeration data are scant.

Kitseguecla River

Coho Salmon

In general, most mid-Skeena River north tributary coho arrive in mid-September. They usually spawn in the fall, from the end of September through December in small tributary streams; there are no known coho spawning grounds in the Skeena mainstem.

Skeena Mainstem Tributaries
Four years of records since 1950 show that Comeau Creek sustains coho spawning in the lower 0.5 km with a range of 0–150 fish (DFO 2005). Comeau Creek is subject to beaver damming, seasonal water fluctuations, and inundation of the productive lower 1 km of channel by Skeena River floods.

Five years of records since 1950 show that Burdick Creek supports coho spawners throughout the lower 1.6 km with a range of 0–90 fish. Hancock et al. (1983) reported that low water in Burdick Creek occasionally prevents fish from entering the creek.

Chicago Creek coho spawner escapement from 20 years of inconsistent records range from 0 to 400 fish, with most typical returns composed of 25–75 coho utilizing the lower 1.5 km of channel. In addition to seasonal low water levels, CN Rail's perched culvert in Chicago Creek has contributed to depressed levels of spawners.

Hazelton Creek escapement records from the 1930s report coho use, but those are the only records. Glen Vowel Creek supports coho in scattered spawning locations in the lower reach (up to 3 km) in years of sufficient flow with records of 0–200 fish.

Kitseguecla System
Coho spawner escapement for the Kitseguecla system has been recorded discontinuously since 1965. The lack of consistent observations does not allow overall escapement trends to be noted. Spawners have ranged from a high of 1,500 in 1968 to averages in the low range of 25 fish. From 1980 to 1989, the longest consecutive record, the 10-year mean was 368 coho spawners.

Preferred documented spawning areas are the 3 km reach downstream of Kitseguecla Lake, on the Kitseguecla mainstem from the Kitsuns Creek confluence upstream to Jack Mould Creek, as well as in the lower reach of Jack Mould Creek, and the lower 2 km of Kitsuns Creek (Morris and Eccles 1975, DFO 1991c, Triton 1998). Biolith (1999b) reported coho spawners in the West Kitsuns mainstem upstream of the Kitsuns Creek confluence and close by the twin bridges. It is most likely that coho spawn throughout the system in all accessible, low-gradient habitats in the mainstem and tributaries, particularly in years of high escapement.

Shegunia River
Shegunia River coho escapement records are spotty; these data indicate that spawners range from 0 to 400 with most years averaging 150–200 fish. Principal spawning grounds are the lower 1 km of the river and the 1.4 km downstream of the canyon. Coho have been observed as far up as 14 km on the mainstem.

Steelhead

The mid-Skeena River north supports summer-run steelhead populations that enter the mouth of the Skeena from July through September and arrive in the area beginning in August. Steelhead spawn in this area in May and June, coinciding with warming water temperatures and an increase in streamflows. Within the area, the two known small steelhead populations are in the Kitseguecla and Shegunia Rivers. Chicago Creek steelhead were reported to have spawned in May for six years between 1929 and 1942. It is not known if steelhead still occur in this drainage.

Skeena River Mainstem and Tributaries
Steelhead overwinter in the Skeena River at or adjacent to the mouth of the Kitwanga, Kitseguecla, and Kispiox Rivers (GWA unpublished data). Lough's (1983) radio telemetry study found steelhead overwintering in the Skeena River. The Skeena mainstem is used as a migration and out-migration route for upstream Skeena River and Bulkley River steelhead stocks.

Kitseguecla System
Kitseguecla steelhead are summer-run fish that enter the Kitseguecla system in late August or early September; it appears that an unknown percentage of these steelhead overwinter in the lower reach of the mainstem or close to the mouth in the Skeena River. The population status of Kitseguecla steelhead is unknown.

Documented spawning is dispersed in side and main channels on the mainstem, particularly in the mid reach between the Kitsuns Creek confluence and the mouth of Jack Mould Creek, and in the lower reach of Kitsuns Creek. There are no data to assess the proportion of repeat spawners or the number of years spent in freshwater before smolting. Recent reconnaissance level (1:20,000) fish and fish habitat inventories in the Kitseguecla drainage show steelhead and/or rainbow trout well distributed in accessible, low-gradient habitat (Triton 1998a, Lorenz 1998a, Lorenz 1998b, Biolith 1999b).

Shegunia River
Shegunia River steelhead are summer-run fish that likely overwinter in the Shegunia, Skeena, and Kispiox Rivers. Lough (1983) reported observing one steelhead overwintering in Kispiox River and then moving into Shegunia River to spawn. Spawning locations are not documented; it is likely steelhead spawning is scattered and sporadic up to 15 km.

Resident Freshwater Fish

In comparison to salmon data, information is scant in regard to resident, nonanadromous or freshwater fish in both fluvial (river) and lacustrine (lake) habitats of the mid-Skeena River north. Dolly Varden, and, to a lesser extent, cutthroat and rainbow trout, are the most widely dispersed species and are present in most fish-bearing waters. Freshwater species and documented populations inhabiting the mid-Skeena include rainbow trout (*Oncorhynchus mykiss*), cutthroat trout (*Oncorhynchus clarki clarki*), bull trout (*Salvelinus confluentis*), Dolly Varden (*Salvelinus malma*), mountain whitefish (*Prosopium williamsoni*), northern pikeminnow (*Ptychocheilus oregonensis*), longnose dace (*Rhinichthys cataractae*), redside shiner

(Richardsonius balteatus), and prickly sculpin *(Cottus asper)*.

Fisheries
First Nations Traditional Use
In the past, Gitxsan salmon fisheries within the mid-Skeena River north were concentrated along the Skeena mainstem with smaller dispersed fisheries along the upper Kitseguecla River and at Seeley Lake. Presently, the majority of Gitxsan fishing sites are located on this section of the mainstem of the Skeena River, mainly due to the richness of the fish resource and the high rate of catch at traditional fishing sites.

Gitxsan fishing sites on the Skeena mainstem are numerous. Between Gitwangak and Burdick Creek there are 50 sites mapped (GWA 2003b). Between Burdick Creek and Kispiox there are 69 fishing sites mapped (GWA 2003c). Between the Shegunia River and the Babine River there are 37 fishing sites mapped (GWA 2003d). The highest densities of sites are clustered upstream of Gitsegukla, upstream of Hazelton in Four Mile Canyon, and upstream of the Shegunia River to Pinenut Creek. These 156 sites signify the importance of this mainstem section to the Gitxsan.

Gitxsan traditional fishing gear, including traps, dip nets, and weirs, is discussed by Holland and Starr (1999). The various traps and dip net gear used depended on site location, fish quantities needed, and the number of people available to fish the gear and provide processing capacity. At some Skeena River fishing stations, usually where the salmon tended to be concentrated by strong currents, large ingeniously woven baskets and/or lashed wooden strip traps were constructed; some incorporated delivery chutes that moved the trapped fish to a waiting fisher who transferred it to the shore.

Undoubtedly weirs were the most productive and ingenious of fishing gear. They were constructed either right across smaller streams or on the mainstem, built out on an angle to guide the migrating fish into shore-side traps. The wide variety of weirs and contiguous traps used were matched to the species, environment, placement, and building materials available. Wilson (1982) noted the fish weir installation

at Miinhltakwit, which is an ancient village site, located approximately 25 km upstream on Kitseguecla River close to Xsan Gokhl.

Currently, the vast majority of utilized fishing sites employ fixed gill nets set along eddies and points of the Skeena River. Drift gill nets are used for a minor portion of the catch. Gitxsan fisheries on the Skeena mainstem mainly harvest Babine sockeye. About 80% of the FSC catch is sockeye. For a few weeks early in the season, chinook salmon are targeted, and in years when limited fishing continues into September or October, coho are targeted. The other salmon species – pinks, chum, and steelhead – are mainly incidental catches.

For the Gitxsan, salmon represents the most important cultural foundation as well as the substantial, singular element of diet. The Gitxsan fishery maintained an understanding of the tools and techniques that allowed management for optimal utilization and escapement on a stock-by-stock basis. These modes of management enabled the fishery system to adapt to changing natural situations and conditions, and facilitated allocation and regulation in managing the fishery, while encouraging habitat protection.

Recreational Fisheries
The majority of sport fishing within the drainage occurs on the Skeena River. This fishery supports medium scale, seasonal recreational use from local-area residents and nonresident visitors. Angling pressure has increased over the last 45 years due primarily to access made available by development of a logging road network. The sports fishery, which is pursued by aboriginal and non-native residents as well as nonresident anglers, has steadily increased in popularity since the late 1960s, with the projected trend indicating further increase. Angling by lure and fly fishing is primarily directed towards chinook, coho, and steelhead, with mandatory release of steelhead from July 1 to December 31. Creel surveys conducted at the mouth of the Bulkley and Kitwanga are reported on by Hall and Gottesfeld (2002 and 2003a).

Enhancement Activities
Enhancement activities in the area have been modest. Chinook fry are raised at Kispiox

Hatchery, and Chicago Creek Hatchery has operated from 1993 into the present. Chinook releases into the Shegunia River between 1987 and 1995 totaled 175,000 fry and smolts. This outplanting from Kispiox Hatchery may have contributed to a short-term increase in escapement, but long-term recovery results are not apparent.

Chicago Creek Hatchery was started in 1993 to incubate and then feed coho fry for release into Station Creek. The intent was to mitigate the effects of the impassable Highway 16 culvert that was and still is a velocity and migration barrier to all fish passage. This small community facility continues to annually rear and release approximately 10,000 coho fry into Station Creek, as well as supplying local elementary schools with coho eggs for educational programs (Houlden and Donas 2002).

Development Activities

Over the last 40 years, land-use and development activities in the mid-Skeena River north have dramatically increased. The principal land-use activity is forestry, which has been concentrated in lower-elevation valleys. The early part of the 20th century saw extensive mineral exploration in the Rocher Déboulé Range and the upslope areas northeast of Hazelton. Linear development along the southern bank of the Skeena Rivers accommodates the CN Rail line, Highway 16, urban and rural settlements, and BC Hydro and Telus Communications utilities. Transportation has been the key to growth in land-use and resource development.

Forest Resource Development

The mid-Skeena River north is located within the Ministry of Forests Skeena Stikine Forest District. Currently, the Integrated Land Management Bureau provides land-use management, zoning, objectives, and strategies through the Kispiox Land and Resources Management Plan (LRMP). In the early 1960s to the mid-1990s, the area and volume logged increased from small-scale operations to clearcutting the easily accessible and high-value timber in most portions of the drainage.

Currently, the undeveloped portions of the drainage include forests that are either not economically feasible to log; are high-elevation, biologically unproductive forests; or are forests with environmental constraints such as unstable terrain or sensitive soils.

Until 1960, logging was selective with a moderate proportion of residual timber left standing, and was concentrated along the lower slopes of the Skeena valley and the lower Kitseguecla valley. However, as forest management became more sophisticated and

Bulkley and Skeena Rivers with New and Old Hazelton

the industry became more competitive and integrated, small mills were replaced by large, stationary sawmills, which in turn complemented the pulp-mill complex constructed at Port Edward.

Since the mid-1960s, clearcut harvesting has been the silviculture system of choice. Since then, the trend to the centralization of forest licenses and milling capacity has laid the foundation for the current situation. The history of overcutting and highgrading led to a collapse of logging activity in the late 1990s as the resource became depleted. In turn this has produced chronic unemployment and a general lack of opportunities. Past logging activities have foreclosed on most forestry development options for the near and mid-term future. Forestry opportunities are limited, for the most part, to forest health-related logging and nontimber forest products.

From 1995 to 2000, the Watershed Restoration Program (WRP) was involved in assessing logging-related disturbances within the Kitseguecla watershed. Within the mid-Skeena River north, Kitseguecla was the only area assessed in relation to fish, fish habitat, and upslope sediment-producing areas. The WRP has mostly succeeded in assessing the 25-year long history of disturbance, but restorative site works have been limited.

A large contribution of the WRP has been to increase the awareness in the forest sector of best management practices regarding water quality, fish, and fish habitat. Watershed health has benefited from the program principally through road deactivation. Hydrological recovery, a natural healing process, has mitigated many problems over the years. Currently, very little information exists in regard to cumulative effects on fish and fish habitat, and few attempts have been made to evaluate the response of fish populations to habitat manipulations.

Mineral Resource Development

Exploration and utilization of rocks, crystals, and minerals within the mid-Skeena River north area have been ongoing since the Gitxsan established themselves many thousands of years ago. The archaeological overview presented by Allbright (1987) describes various lithic tool types and their functional aspects, as well as a review of past lithic studies regarding Gitxsan territories.

Euro-Canadian explorations for base and precious minerals have been ongoing since the 19th century; however, focused efforts increased after the railway was completed in 1914. Early exploration efforts were focused mainly on high-grade silver and gold-rich veins, with most endeavours slowing down with the onset of the Great Depression. From 1942 to 1954, there was a resurgence of exploration and mining interest focused on tungsten deposits, primarily due to high demand and interruption of the Asian supply. In the early 1960s, mineral exploration interest increasingly focused on porphyry-copper-molybdenum.

Most known mineralization occurs in the Rocher Déboulé Range, with the greatest concentration in upper Juniper Creek and in the Nine Mile Mountain vicinity across the Bulkley River. These two localities have mineralized zones associated with the Bulkley and Babine Intrusives. The main producing mines in Juniper Creek drainage were the Red Rose Mine and Rocher Déboulé Mine, which were developed to supply tungsten. From 1915 to 1951, these mines produced 1,140 tons of tungsten.

The main producer close to Nine Mile Mountain was the Silver Standard Mine. From there and five other nearby locations, 232,180 tons of ore have produced 15,140 oz. of gold, 7,758,009 oz. of silver, 9,084 tons of lead, 13,698 tons of zinc, and 223 tons of copper from 1913 to the present.

Although the drainage is known for its variety of mineral deposits, the main types of mineral occurrences are polymetallic veins containing silver, lead, zinc, copper, gold, tungsten, nickel, and uranium. Current potential evaluations of metallic minerals are broad based and nonspecific. The area has been delineated into mineral tracts on the basis of geology and distribution of known mineral occurrences, with most of the region rated as having moderate mineral potential.

Population and Settlement

The mid-Skeena valley and its tributaries have been home to Gitxsan people for thousands of years. Euro-Canadian settlers arrived following completion of the railroad in 1914, attracted by agriculture, mining, and other opportunities. This population base remained relatively stable with small increments of growth until the early 1970s, when logging activities expanded and rural living became a more popular lifestyle.

Approximately 1,730 people currently reside in the mid-Skeena River north area (RDKS 2002). This includes the communities of Gitsegukla, South Hazelton, Hazelton, Glen Vowell, as well as one-half the total population of Kispiox and Gitanmaax (half of Kispiox and Gitanmaax reside on the Kispiox and lower Bulkley watersheds and are discussed elsewhere).

Historically, and up to the last few years, many people derived their income from the forestry sectors; however, severe job losses have curtailed this income. Most residents derive their income from public- and service-sector employment in the villages and Hazelton area or through seasonal forestry-related work. Nontimber forest products, especially pine mushrooms, have been an important part of the local economy. Mid-Skeena area residents experience unemployment rates considerably higher than those faced by the rest of the regional population.

There are currently an estimated 48 relatively small ranches or farmsteads in the drainage, with an unknown number of breeding cows that graze on Crown land. Land titles are typically small, though larger holdings (greater than 60 ha) have Agriculture Land Reserve restrictions. The population trend projects growth for aboriginal communities and stability for other residents. Recreation and tourism-based incomes are projected to grow over the next decade.

Transportation and Utilities

The existing transportation network in the watershed is based on the ancient aboriginal trail infrastructure with trails being initially widened for packhorses, later improved for wagons, and then further improved for vehicular traffic. The mountainous terrain limits the number of feasible transportation routes and largely determines settlement patterns.

Highway 16 is the major road providing access to the mid-Skeena drainage. Forestry road development throughout the drainage has recently tapered off, and future trends point toward off-highway road maintenance responsibilities being off-loaded to industry or dropped altogether. Linear development parallels the Skeena River, with Highway 16, CN Rail, BC Hydro, and Telus utilizing a corridor on the south side of the river and taking advantage of the geography and settlement patterns.

Over the last two decades, Highway 16, which is located on the south bank of the Skeena River, has assumed an increasingly important role for tourist travel, recreational vehicles, and commercial trucking operations. This reflects the great many changes affecting the highway transportation sector. Secondary roads include the well-established road network in the upper and lower Kitseguecla drainage, the Salmon River Road that parallels the Skeena River for 53 km to Babine River, and the Hazelton-Kitwanga backroad that connects those communities and runs on the north bank of the Skeena River.

CN Rail serves the region with connections to the rest of the country and to saltwater at Prince Rupert and Kitimat. The railroad provided the main transportation link accessing the watershed up to 1950, when the Kitwanga backroad, a series of pole logging roads, was connected to roads from Hazelton (Hobenshield 2002). Until the 1980s, the rail line was used to transport grain and manufactured goods produced outside the mid-Skeena area and wood products produced within the drainage. Since the mid-1980s, the vast majority of freight moves via unit trains carrying coal or grain to the Port of Prince Rupert. Passenger traffic is maintained at a low level; Via Rail schedules passenger trains twice a week.

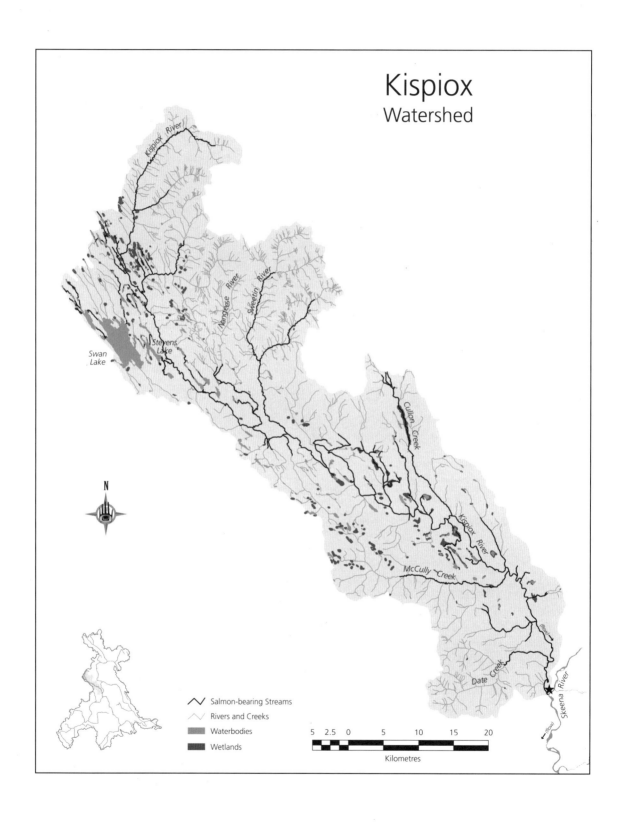

Kispiox
Watershed

Kispiox River

Nangeese River

Sweetin River

Swan Lake

Stevens Lake

Cullon Creek

Kispiox River

McCully Creek

Date Creek

Skeena River

N

Salmon-bearing Streams
Rivers and Creeks
Waterbodies
Wetlands

5   2.5   0   5   10   15   20

Kilometres

14

Kispiox Watershed

Environmental Setting

Location

The Kispiox River is a large tributary of the Skeena River. It flows 140 km southeast from its headwaters to the confluence with the Skeena River (right bank) at the Kispiox village, which is approximately 12 km north of Hazelton. The watershed is bounded in the north and the east by the southern Skeena Mountains, to the south predominantly by the Kispiox Range, and to the west by the low-relief Nass basin.

Hydrology

The Kispiox is a fifth-order stream with a catchment area of 2,088 km². Elevation ranges from approximately 200 m at the mouth to 2,090 m on Kispiox Mountain and 1,850 m in the Skeena Mountains. This major tributary contributes about 9% of the Skeena River flows (Remington 1996). Kispiox River peak discharges typically occur in May and June due to spring snowmelt, and then decrease through July and August. In September, fall rain and runoff from early snowmelt once again increases streamflows through to October. Streamflows decrease through November and December when precipitation falls as snow, with low discharges recorded January through March. Hydrometric station 08EB004, located downstream from the McCully Creek confluence, recorded a monthly mean discharge of 128 m³/s for June, while low flows in February average 7.8 m³/s over a 42-year observation period (1963–2005). Summer low flows are typically four to eight times greater than winter low streamflows and are principally sustained by high-elevation snowmelt draining from the Skeena Mountains. Winter low flows are derived from groundwater, lakes, and unfrozen wetlands (Wilford 1985).

Climatic information from the Murder Creek weather station (AES 2005), located in the lower Kispiox, shows mean annual precipitation of 631 mm over a 20-year period, of which rainfall accounts for 71% (Environment Canada 2005). Total annual precipitation (TAP) is much greater in the upper watershed, particularly at higher elevations. The Skeena Mountains to the north and the Nass basin, which broaches the northwest and western perimeter of the watershed, exert the major hydrological influences. The low-elevation watershed divide to the Nass drainage in the west allows coastal weather systems to enter the watershed, leading to heavy snow packs in the mountains and the upper half of the drainage.

The watershed as a whole has a moderately high response from water input due to the steepness of the upper Kispiox and many major tributaries. In general, the streams flowing from the west into the upper Kispiox River arise from a myriad of lakes and bogs; hence, they are relatively stable in flow, temperature, and water quality characteristics. The Kispiox Range to the southwest is principally drained by Date and McCully Creeks, both of which transport large amounts of bedload and suspended sediment originating from natural sources and have active alluvial fans at their confluences with the Kispiox River (Weiland 2000a). Tributaries to the upper Kispiox from the northeast, principally the East Kispiox River, Sweetin River, and, to a lesser extent, the Nangeese River, drain glacial headwaters and transport moderate amounts of sediment from natural sources. The wide variations in Kispiox mainstem water flows are primarily attributed to these tributaries. Through the summer season these streams generally have glacially turbid, unstable flows.

Swan and Stephens Lakes, the two most important high fish value lakes in the watershed, are located close to the Nass drainage divide in the upper watershed. These two clear-water lakes have ice cover for up to six months of the year, receive approximately 1,500 mm of total

View northwest across Swan Lake

annual precipitation, and lie at an elevation of 520 m (Stockner and Shortreed 1979). Compared to the other upper Kispiox tributaries, Swan and Sephens Lakes are warm in the spring and summer and contribute warm flows to the Kispiox River via Stephens Creek (Pinsent and Chudyk 1973).

Stephens Lake is small, relatively shallow with a mean depth of 11 m, and located 3 km downstream of Swan Lake. The results of limnological sampling showed a pronounced thermal stratification with a strong thermocline at a depth of 4.8 m and maximum surface temperatures not exceeding 18° C (Stockner and Shortreed 1979). Average euphotic zone depth was 13.1 m. Although there were no data on phosphorous levels and photosynthetic rates, the lake appears to be oligotrophic and is likely nutrient-limited (Shortreed et al. 2001). Macrozooplankton biomass is relatively high, as is *Daphnia* abundance (Rankin and Ashton 1980). Simpson et al. (1981) reported mean sockeye fry weight of 3.1 g for Stephens Lake, which is moderate to high in relation to other sockeye lakes in the Skeena system.

Swan Lake discharges through Club and Stephens Lakes into the Kispiox River. Swan Lake's physical environment is excellent for juvenile sockeye, with a 15.3 m euphotic zone depth, a stable cool epilimnion, and a large hypolimnion (Shortreed et al. 2001). *Daphnia* was abundant relative to other northern lakes (Rankin and Ashton 1980); however, despite a good physical environment, an abundant food supply, and low planktivore densities, Swan Lake fall fry averaged only 1 g (Simpson et al. 1981). Given the apparently good rearing conditions, it is unclear why Swan Lake sockeye fry do not exhibit higher growth rates. Swan Lake has limited spawning gravels, thus it is likely that the majority of sockeye fry in the Swan-Club-Stephens system rear in Stephens Lake.

Stream Channels

The Kispiox River is divided into three distinct reaches from the Skeena River upstream to Sweetin River. These lower three reaches are composed of a mix of pools, riffles, and runs, which offers holding, rearing, and spawning habitat. The mainstem channel presents a regular profile with a gradient of 0.3% slope or less (MoE 1979). Bedrock outcrops are infrequent and bank erosion is common. Minor amounts of sediment are received from most tributaries other than Date Creek and McCully Creek, which contribute comparatively large amounts of natural sediment. Low summer flows may compromise off-channel habitat rearing capacity.

Reach 4, from Sweetin River upstream to Gitangwalk Canyon is frequently confined by bedrock, which becomes more evident in reach 5 starting at the bottom end of Gitangwalk Canyon. Gitangwalk Canyon, defined as reach 5, is approximately 1 km in length with an average gradient of 0.6%. The lower end of the canyon presents a 200+ m long cascade with two 1–2 m drops that restrict pink and chum salmon access to the upper reaches of the river. In some years of low water flows, late running sockeye have been unable to ascend these falls and were observed spawning just below. Wadley and Gibson (1998) noted that the DFO carried out blasting in the cascade-falls section to facilitate fish passage.

Adjacent to this section of the river is the ancient village site of Gitangwalk and the river crossing for the grease trail (Rabnett et al. 2001). Above Gitangwalk Canyon, the river has gravel banks and a lower gradient. Reach 7 and 8 both have average gradients of 0.4%; however, upstream of the Williams Creek confluence, a series of falls limit anadromous fish passage.

Three major assessments of the mainstem and tributary channels were conducted between 1998 and 2001. The Kispiox River channel assessment in 1998 notes that of the 10 reaches surveyed, one was considered stable, six reaches were relatively stable, and three reaches indicated light disturbance (Nortec 1998). Floodplain stability mapping of the Kispiox River mainstem and selected tributaries was conducted with sequential air photos for the periods from 1950 to 1975 (period 1) and from 1975 to 1992 (period 2). Hudson (2001) reported that for 10 of the 11 reaches surveyed, maximum bank retreat numbers are greater during period 1; bank retreat is most dramatic in reach 1 for both periods, and the active channel area and width for most of the reaches expanded from 1975 to 1992.

Hudson's (2002) further analysis of the floodplain stability mapping indicates that channel impacts are greatest in the lower gradient reaches where sediment storage occurs. Direct effects of forestry-related peak flows are obscured by scale effects such as lake and channel storage and groundwater effects. Channels with high natural levels of overbank flooding, lateral migration, and avulsion (for example, reach 1 of the Sweetin River) can be expected to display the greatest degree of instability.

Weiland (2000b) noted that several tributaries in their lower reaches have very low gradient channels with very low sediment transport capability. Overall, most Kispiox River tributary channels downstream of Hodder Creek have received logging impacts ranging from moderate to severe when compared to natural conditions prior to large-scale industrial logging. Triton (2001) stated that forestry impacts to fish habitat are extensive through the low-gradient reaches of most tributaries due to obstructions to fish passage, logged riparian zones, surface erosion, reduced instream habitat complexity, subsurface flows, and degraded habitat quality.

Murder and McCully Creeks have avulsions in their lower reaches due to the combined effects of agricultural clearing of the floodplain and riparian zones, followed by high streamflows and resultant downstream sediment deposition.

Low-gradient reaches of many tributaries in the watershed contain large numbers of beavers. Nortec (1997) described 12 creeks where channel

Upper Kispiox River

changes, bank erosion, and decreases in riparian suitability for conifers were due to beavers and their dams. Riley and Lemieux (1998) found that beaver activity on Kispiox River tributaries created large areas of habitat that supported high densities of coho fry. They recommended no removal of beaver dams unless it could be demonstrated that beaver activity had resulted in negative effects on coho populations. They suggested that beaver dam removal contravened the DFO's No Net Loss policy.

Water Quality

In 1982, the MELP Waste Management Branch initiated a five-year water quality monitoring program with a station (Site 0400205) on the Kispiox River close to the Skeena confluence. The program concluded that the Kispiox River is a soft-water river, with neutral to slightly alkaline pH and clear, slightly tea-coloured waters for most of the year. This coloration is due to natural organic substances, such as humic acids, contributed by swamps and wetlands in the drainage. Alkalinity and calcium concentrations are in a range that would provide moderate buffering from acidic inputs. Total suspended solid loadings are much higher during freshets than the remainder of the year. Nutrient concentrations are low. Mean total levels of metals are generally very close to MELP criteria for the protection of aquatic life at the hardness levels present (Wilkes and Lloyd 1990).

Community watersheds located on Dale and Quinmas Creeks supply domestic water for the Kispiox village. Licensed water withdrawals within the Kispiox watershed are mainly from small tributaries presenting minor impacts on instream flows for fisheries. The effects of clearcut logging on increased peak flows in the Kispiox River have been a persistent issue raised by the public. Keeping logging debris out of small creeks, particularly in winter-logged areas, was reported to be a major difficulty (Remington 1996). Loedel and Beaudry (1993) noted that their investigation of interception and throughfall water at Date Creek was initiated by concerns from waters licensees, Native peoples, and others that clearcutting may increase peak flows, decrease low flows, and/or alter the timing of these flows.

The Kispiox WRP overview (Jyrkkanen et al. 1995) concluded that logging-related erosion, obstructions, sedimentation, gravel aggradation, and altered water yield are the primary sources of impacts to the Kispiox system. Concerns about turbidity and poor water quality from tributaries, as well as the impact on sports fishing, were also noted. Nortec (1997) reiterated that water quality was impacted from forest development activities. Weiland (2000b) conducted a reconnaissance sediment source mapping survey, which identified natural sources and activity in the watershed. This study reports that natural sources supply by far the most sediment in the Kispiox River and the mountainous subbasins, and that currently, sediment transport in the watershed appears to be in overall steady-state equilibrium.

Geography

Three physiographic units are present in the Kispiox watershed: the Nass basin, the Kispiox Range, and the Skeena Mountains to the north and northeast. The Nass basin is an area of low relief, which generally falls below 700 m and forms the valley floor. The Kispiox Range, which bounds the watershed to the southwest, is largely drained by Date and McCully Creeks. The southern Skeena Mountains form the headwaters of the Kispiox River and its major tributaries, the Sweetin and East Kispiox Rivers.

The broad northwest-southeast trending Kispiox valley, approximately 100 km long and averaging 20 km wide, resulted from a block faulted Basin and Range structure zone associated with plate tectonics on the west coast of North America. The down-faulted block fault lines control the break between the valley and the mountains. The relative uplift of the mountains has been on the order of several kilometres.

Folded and faulted Bowser basin marine sediments characterize the underlying bedrock in the Kispiox watershed; minor amounts of an intrusive granitic stock appear in the Kispiox Range. The ice that covered and flowed down the Kispiox valley during the last glacial period strongly glaciated the mountain slopes and the basin, leaving a legacy of drumlin fields, hundreds of small lakes, and a generally linear drainage pattern. Thick blankets of glacial till cover the main valley and mountain valleys and extend up the valley sidewalls.

On the lower elevations of the Kispiox valley south of Swan Road, alluvial terraces of gravel and sand, often separated by moderately steep scarp faces, dominate the surficial materials. This is primarily due to the river downcutting into outwash materials from the end of the Pleistocene glaciation. On the very lowest elevations, within the current Kispiox River floodplain, recent alluvial deposits range from sand and gravels to predominantly silty materials.

The soils of the watershed have been influenced by the sedimentary origins of the parent materials, with relatively fast weathering and a higher natural fertility than comparable till from granitic bedrock (Kerby 1997). The forest cover, climatic factors, topography, drainage, and elevation differentiate the six basic soil types found. Soil series identified in the valley bottom primarily include the Moricetown series, the Kispiox series, Barrett series, and the McCully series (Farstad and Laird 1954, RDKS 1991). The majority of soils of the higher river terraces north of Elizabeth Lake possess limitations to agriculture.

The coastal-interior transition climate is reflected in the major ecological zones. Vegetation in the wide, gently sloping valley below approximately 750 m is represented by the Interior Cedar Hemlock (ICH) biogeoclimatic zone, which is dominated by forest stands of hemlock, spruce, subalpine fir, and, in the southern half, red cedar. Before industrial logging, the majority of forest stands were mature hemlock and fir. These stands have been replaced with plantations of spruce and pine; while a major portion of the valley bottom has been replaced by deciduous forests. With increasing elevation, the ICH zone passes into a forest dominated by mature and overmature subalpine fir, representing the Engelmann Spruce–Subalpine Fir (ESSF) biogeoclimatic zone (Pojar et al. 1988).

Fish Values

The Kispiox River watershed is composed of approximately 100 km of mainstem and 300 km of tributary streams that are high-value fish habitat, provide a migration corridor, and support spawning and rearing (Nortec 1997). The numerous salmonids utilizing this habitat include sockeye, coho, pink, chum, chinook, and steelhead salmon, as well as rainbow, lake and cutthroat trout, Dolly Varden, bull trout char, and mountain whitefish. Lamprey and several other resident fish (*Cottidae* and *Cyprinidae*) are also found in the watershed. Gottesfeld et al. (2002) rated the Kispiox subbasin as the most productive in the Skeena watershed.

Chinook Salmon

Kispiox River chinook salmon are one of the large and important stocks in the Skeena watershed. Since the 1950s, there has been a long-term population decline due in part to the mixed-stock fishery with related incidental interception and to a targeted sports fishery. This has slowly turned around with escapement numbers recovering in the last two decades. Chinook escapement has varied widely in the past, from about 400 in many years of the 1970s and 1980s to 15,000 in 1957 and 1992 (DFO 2005). The 10-year mean escapement was 12,560 chinook for the 1950s, 2,801 for the 1980s, and 5,493 for the 1990s. Three years of surveys between 2000 and 2005 averaged 6,270 chinook annually, showing overall increased abundance.

Typically, chinook salmon enter the Kispiox system in June and July and disperse to their spawning areas where they spawn from late July through August. The bulk of the spawning is concentrated in the mainstem. Critical chinook spawning areas include portions of reach 3 upstream and downstream of Murder Creek, pockets south of Elizabeth Lake, and dispersed areas throughout the mainstem (DFO 1991c, Wadley and Gibson 1998). These are often just downstream of tributary outlets that provide sources of fresh sediment and increased hyporheic or shallow intragravel flow.

A variety of bedrock pools in reach 3 support holding areas for mature chinook, coho, and steelhead (Wadley and Gibson 1998). Reach 4 contains moderate to heavy spawning in suitable sections with good holding pools. Reach 5, which is essentially Gitangwalk Canyon, has no known reports of chinook spawning. Reach 6, located above the canyon upstream to the mouth of Stephens Creek, has excellent spawning beds in the upper section. Dispersed, heavily-used spawning areas exist in reach 7, especially in the upper portion.

Tributaries with noted chinook spawning in their lower reaches include Date Creek, McQueen Creek, Cullen Creek, Sweetin River, Nangeese River, and Stephens Creek (particularly near the mouth), lower Club Creek, and lower Williams Creek (Smith and Lucop 1966, Hancock et al. 1983, DFO 1991c).

Stuart (1981) conducted a biophysical assessment of the Kispiox mainstem and 13 of the major tributaries, reporting that chinook fry were present only in Date Creek and in the Kispiox mainstem. Recent juvenile trapping efforts by the GWA has yielded juvenile chinook in the Nangeese and Kispiox Rivers; other known chinook spawning streams have not been sampled.

Pink Salmon

The Kispiox River is one of the major pink salmon producing areas of the Skeena River system. Kispiox River pink salmon are distinguished by their early run timing, with no dominant cohort year. Pink escapement fluctuates widely from cycle to cycle. Relative to recent escapements in the Skeena, Kispiox River pink salmon have not experienced a dramatic increase in escapement. The 10-year mean escapement for the period of 1990–1999 shows 33,500 for the even-year mean and 56,800 for the odd-year mean (DFO 2005). The largest return was an odd-year escapement of 750,000 in 1959.

Typically, pink salmon enter the system in mid to late August and disperse to spawn throughout the mainstem and its lower tributaries. Gitangwalk Canyon is a barrier to upstream movement and is consequently the upriver limit of pink salmon spawning. The area of heaviest spawning occurs from Seventeen Mile Bridge upstream to Cullon Creek (Smith and Lucop 1966). Wadley and Gibson (1998) reported moderate mainstem pink spawning in areas of suitable substrate upstream from McQueen Creek; heavy pink spawning from McCully Creek up to Cullon Creek; and dispersed patchy

spawning in reach 4. Pink salmon also utilize the lower reaches of the following tributaries: Date Creek, McQueen Creek, McCully Creek, Murder Creek, Cullon Creek, Ironside Creek, Twin Creek, Corral Creek, Skunsnat Creek, Clifford Creek, and the Sweetin and Nangeese Rivers. Upon emerging from the gravel in spring, pink salmon fry migrate immediately to the saltwater.

Chum Salmon

Kispiox River chum are the farthest upstream of any large chum population spawning in the Skeena system, but the escapement has been severely depressed since the late 1950s. The 10-year mean escapement for the 1950s was 4,083; for the 1960s, 553; for the 1970s, 1,108; the period of 1980–1989 recorded 131 chum, while the 10-year mean escapement for the period of 1990–1999 was 400 spawners (DFO 2005). Since 2000, no chum escapement surveys have been recorded.

Generally chum move into the Kispiox system in August, spawning in selected sections of the mainstem, principally reaches 1 and 2. Chum spawners have also been observed scattered along the mainstem close to the mouths of Date Creek, McCully Creek, McQueen Creek, Murder Creek, Elizabeth Creek, Steep Canyon Creek, the Sweetin and Nangeese Rivers, and in the lower reach of Date Creek and the Nangeese River. Migration downstream to the saltwater begins immediately following fry emergence in the spring.

Sockeye Salmon

The Kispiox River watershed is among the eight most important sockeye producing watersheds in the Skeena system. Kispiox River sockeye are a unique population with spawning taking place primarily in streams tributary to Swan, Club, and Stephens Lakes.

Figure 19 shows the historical escapement estimates for the Swan Lake system. Estimates prior to 1992 were based primarily on Club Creek (upper and lower) counts. In 1992, the GWA included Jackson and Barnes Creeks, and in 2001 the GWA enumerated all known spawner areas in the Swan Lake system, including Club Creek. A counting weir was set up in 2001 to obtain accurate escapement for the Swan Lake

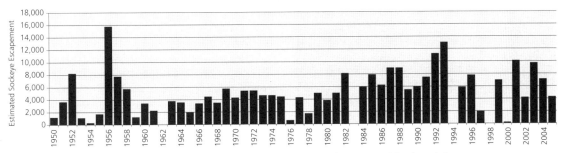

Figure 19. Historical sockeye escapement estimates for Swan and Stephens Lakes

drainage; 10,109 sockeye were enumerated. Mean annual Kispiox aggregate escapement between 2000 and 2005 was 7,010 sockeye.

Sockeye adults typically enter Stephens Creek in August and the beginning of September and migrate upstream to spawn. The major spawning grounds are located on upper and lower Club Creeks (Fisheries Research Board 1948, Sterritt and Gottesfeld 2002). The spawning habitat in Club Creek is exceptional for the Skeena system in that the spawning substrate is primarily boulder size. Other spawning areas are found on four creeks tributary to Swan Lake, of which Falls Creek is the most important. Some sockeye spawn in Swan Lake and possibly also Stephens Lake. Following emergence from the spawning beds, most juvenile sockeye (>95%) spend one year in Swan, Club, or Stephens Lakes before migrating to the sea (Rutherford et al. 1999).

Small numbers of sockeye regularly spawn in downstream sections of Ironside, Clifford, and Skunsnat Creeks and in the Nangeese River (GWA, unpublished data). It is presumed that these fish are river-type sockeye. River-type sockeye are rare in the Skeena watershed, although this life-history type is common in Asia and makes up a significant part of the total escapement of sockeye to the Stikine River and, to a lesser extent, the Nass River.

Coho Salmon

Coho salmon are widely distributed throughout the Kispiox River system with 28 recorded spawning stream localities. The greatest abundance is within the Kispiox mainstem where escapements as high as 35,000 (1958) have been recorded. Coho escapement within the watershed has decreased by an order of magnitude over the last few decades (Plate et al. 1999). Coho

synoptic and stock assessment studies were conducted by the Gitxsan Watershed Authorities (Plate et al. 1999, GWA 2000, GWA 2001, Sterritt 2001). Kispiox watershed coho escapement declined steadily from the 1950s until the last few years, when widespread fishing restrictions and improved ocean survival led to a marked rebound. Escapement records show the annual aggregate mean between 2000 and 2005 as 4,320 coho.

Generally, coho return to the Kispiox system throughout September, spawning from late September through December, usually dependent on water flow and levels. Coho spawning grounds are concentrated in the upper half of the mainstem; however, select areas adjacent to creek mouths in the lower portion are also used. The Nangeese River and the Stephens watershed have large escapements. Approximately 20 other streams support smaller spawning groups.

The majority of tributaries support coho rearing, while mainstem rearing principally occurs in side channels. Comprehensive mark and recapture sampling, as well as smolt out-migration fence counts conducted by the GWA, have resulted in determining wild smolt and enhanced smolt habitat distribution, density, and migration timing. These studies also identified underutilized stream sections that are potentially suitable for future coho hatchery releases (Plate et al. 1999, GWA 2000, GWA 2001, Sterritt 2001).

Steelhead

The world-renowned Kispiox River steelhead population is distinct from other Skeena River stocks due to the large average size of the returning adults. Though uncertainties exist as to steelhead escapement levels given the large estimated population size (Tautz et

Coho in Stephens Creek

al. 1992) and continuing high sports fishery catches, the steelhead population appears to be relatively stable. It was estimated that 4,027, and 2,514 steelhead returned in 1994 and 1995, respectively (Koski et al. 1995, Alexander and English 1996). Ward et al. (1993) calculated commercial harvest rates of Kispiox River steelhead for the period of 1986–1991, with the mean being 41.2% incidental harvest in the Area 4 commercial fishery. Efforts to decrease commercial catches of steelhead have been made since 1992. The intention is to keep commercial exploitation rates below 26%. However, most recent years have had a much lower exploitation rate.

Steelhead migrating up the Skeena River enter the Kispiox system in late August and September and overwinter in deep pools, mainly in the lower Kispiox River below Cullon Creek and in the mainstem Skeena below the confluence (Lough 1980, 1983). Steelhead have been observed spawning from mid-May through mid-June, primarily in mainstem side channels; though Stephens Creek and the Club Creek system are likely the most concentrated spawning grounds (Chudyk 1972b).

In 1979, a radio tagging study showed 80% of the radio tagged steelhead spawned in tributaries including Cullon Creek, Ironside Creek, Skunsnat Creek, and the Nangeese River (Lough 1980). Lough (1983) reported a small concentration of fish spawning in the mainstem between Date and McQueen Creeks and that two steelhead left the Kispiox and spawned in Skeena River side channels, while one steelhead moved over to the Shegunia River to spawn. Other tributaries known to support steelhead spawners

include the lower reaches of Williams Creek and the Sweetin River (DFO 1991c, Baxter 1997a).

Kelts leave the river in late May. In a detailed study of Kispiox River steelhead during 1975, Whately (1977) found that 12.1% of steelhead adults were repeat spawners (S1+) and 0.6% were second time repeat spawners (S1S1+). Most of the repeat spawners were females because males experience a higher mortality during spawning.

Steelhead juveniles remain in the Kispiox system for 1+ to 4+ years. Scale sample analysis of upstream migrating adults showed an average age of three years in the river before moving to the ocean (Whately 1977). Fry densities are generally lower in the mainstem than in sampled tributary sites; however, parr densities in both rearing areas are largely similar. Cullen Creek has the highest fry densities by a factor of five, in relation to other monitored or sampled sites (Stuart 1981, Tredger 1983a). Recent mark and recapture sampling conducted by the GWA reported generally lower fry densities that averaged $0.01–0.31/m^2$ (Gottesfeld et al. 2000). Their results suggest that the Kispiox watershed as a whole is underutilized. It is likely juvenile recruitment is low due to incidental catch from high exploitation rates in the mixed-stock fishery.

Resident Freshwater Fish

In comparison to salmon, information is sparse on resident, nonanadromous or freshwater fish in both fluvial and lacustrine habitats of the Skeena watershed; indeed, much of the watershed is poorly known and may contain populations of special interest or status that are presently unknown. Ecological and life-history information that permits good conservation planning is simply not available. Of the 21 known species of fish in the Kispiox watershed system, 15 are freshwater resident species (McPhail and Carveth 1993).

Known freshwater populations inhabiting the Kispiox watershed include rainbow trout *(Oncorhynchus mykiss)*, cutthroat trout *(Oncorhynchus clarki clarki)*, kokanee *(Oncorhynchus nerka)*, bull trout *(Salvelinus confluentus)*, lake trout *(Salvelinus namaycush)*, Dolly Varden char *(Salvelinus malma)*, mountain whitefish *(Prosopium williamsoni),* lake whitefish *(Coregonus clupeaformis),* northern pikeminnow *(Ptychocheilus oregonesis),* largescale sucker *(Catostomus macrocheilus),* longnose sucker *(Catostomus catostomus),* river lamprey *(Lampetra ayresi),* longnose dace *(Rhinichthys cataractae),* redside shiner *(Richardsonius balteatus),* and prickly sculpin *(Cottus asper).*

Fisheries

First Nations Traditional Use

For the Gitxsan, salmon are a very important cultural foundation, icon, and food source. Traditionally, sockeye followed by coho have been the most important species to First Nations groups harvesting Kispiox River fish stocks. The Kispiox village, also called Ans'payaxw, is one of seven main Gitxsan villages spread along the Skeena River and its tributaries. Gitangwalk and Lax Didax, both abandoned in the early 1900s, were villages logistically located to intercept the upstream migration of sockeye and coho salmon to upper Kispiox River spawning areas.

Many seasonal fish camps positioned along the mainstem were used to harvest fish. The "grease trail," which runs from the Kispiox village to the Nass, passed along the eastern side of the Kispiox River and provided access to many of the fish harvesting sites listed in Table 2, as well as many other resource gathering localities.

Currently, Kispiox watershed salmon harvested for personal, societal, and ceremonial use are primarily gillnetted and, to a small degree, dip-netted. The importance of the fishery, which is a blend of food resource, trade capital, cultural expression, and connection to ancestral practices, cannot be overstated.

Recreational Fisheries

The Kispiox watershed attracts a large sports fishery that includes local residents and nonresidents. Adult steelhead that return to the Kispiox River are among the largest in the world, and the river is an international destination for anglers. Generally, angler access to fishing sites is easy.

River angling effort is directed primarily to steelhead, coho, and chinook from midsummer to late October. Since 1969, various creel surveys have estimated or determined the angling effort, catch per unit effort (CPUE), gear fished, rate of release, and use of guide services (Pinsent 1970, Remington et al. 1974, Whately 1977, Lewynsky and Olmstead 1990, Tallman 1997). There are currently three licensed guides who operate on the river, and an additional guide who is inactive, with a total allocated quota of 393 angler days (Baxter 1997a). Seldom are quotas fully utilized, because water conditions, and thus fishing conditions, can deteriorate rapidly due to seasonal heavy rains in the Kispiox River watershed. Local anglers fish trout and char in the easily accessible lakes.

The Kispiox River is designated Class II waters from September 1 to October 31, and a steelhead stamp is mandatory. Tallman (1997) reported that in the fall of 1996, according to those anglers interviewed, all steelhead caught were released, 62% of the anglers were of foreign residence, fly fishing was the predominant method used (80%), and compliance with required regulations was fairly high (over 90%). The annual catch quota is one steelhead per year from the Skeena watershed, with no fishing in any stream from January 1 to June 15 (B.C. Fisheries 2005). No power boats are allowed and there is no fishing from drift boats or rafts. Steelhead angling is catch and release between July 1 and December 31.

Enhancement Activities

In 1977, Fisheries and Oceans Canada announced the Salmonid Enhancement Program (SEP) with the primary goal of doubling salmon production. Kispiox Hatchery was established as

Sockeye in lower Falls Creek

Table 2. Traditional fishing sites and fishing villages on the Kispiox River

Fishing Site Name	Site Location
Anspayaxw	Mouth of Kispiox River
Agwi'tin	Kispiox River, R & L banks
Xsi Ankalamsit	Kispiox River, R & L banks
Xsa Gailexan	
Xsa Angexlast	Kispiox River, R bank
Antkilakx	Kispiox River, L bank
Tsihl 'niit'in	Kispiox River, R bank
Xsa An Seegit	Kispiox River, L bank
Wiluuskeexwt	Kispiox River, R bank
Miinhlgwoogoot	Kispiox River, L bank
An'Uxwsdigehlxw	Kispiox River, R & L banks
Katgaidem	Kispiox River, L bank
Xsi Luukailgan	Kispiox River, R & L banks
Wiluuwak	Kispiox River, L bank
Nadak	Kispiox River, R bank
Sgansnat	Kispiox River, L bank
Luu'Andilgan	Kispiox River, L bank
Gitangwalk	Kispiox River, L bank
Lax Didax	Stephens Creek

one of the five major hatcheries in the Skeena watershed under the Community Economic Development Program. Under the auspices of SEP, habitat projects conducted throughout the watershed include incubation boxes, bioengineering investigations, biophysical studies, and habitat inventories.

Kispiox Hatchery was initiated as a pilot project operated by the Kispiox Band in 1977. Water quantity and quality problems were not resolved until 1983, when three wells were developed that supplied stable quality and constant water temperature. The hatchery was designed to increase the severely depressed Kispiox River chinook and coho stocks (DFO and MoE 1984). The hatchery continued to operate until 1995, when it was closed due to SEP program budget review cuts.

Re-opened in 1997 under the auspices of the GWA and with funding from a variety of sources, the hatchery allows for a flexible fish culture program. The total chinook and coho fry and smolts released from Kispiox Hatchery between 1984 and 2001 was 893,684 coho and 1,086,252 chinook. These chinook and coho

were prioritized to stock rebuilding efforts on the Nangeese River and Clifford and Skunsnat Creeks. Recent hatchery activities have focused on small-scale custom enhancement such as incubating, rearing, and outplanting sockeye fry for stock rebuilding initiatives.

Development Activities

The principal activities are logging, settlement, and linear development.

Forest Resource Development

The Kispiox River watershed is located within the Ministry of Forests Skeena Stikine Forest District. Upon completion of the railroad through the Skeena in 1914, forest development activity began with agricultural clearing by settlers, and by 1920 the pattern of land use and settlement was established. Small-scale lumbering led to small bush mills when the post-World War II demand for lumber skyrocketed. In the early 1950s, Columbia Cellulose was granted TFL #1, which initiated the centralization of license holding and milling capacity.

In 1958, 23 km of road were present on the Kispiox River, with logging operations concentrating on easily available, high-quality timber. In 1959, some 70 km of road bordered the river, while by 1966, approximately 90 km of road accessed the east side of the river (Taylor and Seredick 1968). Over the years and up to the present, industrial forest activities have waxed and waned, as the cut was concentrated on other watersheds such as the Kitseguecla and the Suskwa, and as the distance to timber off the highway increased.

The 1980s and the early 1990s saw the volume and rate of development expand dramatically, particularly across the northern, low-elevation portion of the watershed, from Murder Creek through to the Nangeese River. The early 1980s also saw completion of the Mitten Main connecting the Kispiox valley to Highway 37; this road facilitated development on the northwestern flank of the river and transport of logs to the saltwater Port of Stewart. The 1990s also saw logging and road development in Date and McCully Creeks.

In the Kispiox watershed, impacts are complex and result from the interactions

Anglers at the Kispiox-Skeena confluence

of naturally unstable soils and high-energy stream systems draining into low-gradient valley-bottom reaches that are incapable of transporting large amounts of sediment produced by poor logging practices. Completed and potential restoration activities under the auspices of the Watershed Restoration Program include road deactivation to prevent erosion and landslide potential, off-channel habitat restoration, riparian zone restoration to stabilize channels and diverse habitats, and stabilization of highly mobile stream channels and gravel bars often associated with logged alluvial fans. In the Kispiox watershed, fish access is the major overall impact from road crossings that were originally constructed as, or have become, barriers to fish migration.

A review of the Watershed Assessment Procedure results, uncompleted WRP works reported by Triton (2001), and the Kispiox LRMP equivalent clearcut area (ECA) strategy led to specific watershed-by-watershed

recommendations (Hudson 2002). The ECA strategy stated that no more than 22% of the forested land in a watershed will be in a clearcut hydrological condition (Ministry of Forests 2001b). Specific recommendations arising from the review were directed to the 17 subbasins in the watershed. Seven subbasins, including Brown Paint, Clifford, Corral, Cullon, Deep Canyon, Ironside, and Skunsnat, had either priority WRP works that were uncompleted or an ECA at or above 22%; no further conventional harvesting is recommended. A conservative rate of cut is recommended for the Date and McCully subbasins. The Nangeese, Hevenor, Lower Kispiox, and Sweetin subbasins were recommended for alternative silviculture systems due to high fisheries values, terrain stability issues, riparian concerns, or peak-flow risks.

Mineral Resource Development

Exploration and utilization of rocks, crystals, and minerals within the watershed have been

Onerka Lake

occurring since Gitxsan people settled into the watershed many thousands of years ago. Currently, mining development in the Kispiox watershed is limited to exploration of mining claims for coal and vein metallic mineralization located in the southern portion of the watershed close to Kispiox. The main type of mineral occurrence is in sedimentary rock types, with five coal showings, one marl showing, and one fireclay occurrence. There are three polymetallic veins containing silver, lead, zinc, and gold as well as three porphyry copper-molybdenum-gold showings. The watershed has a low mineral-potential rating.

Population and Settlement
The Kispiox valley has been home to Gitxsan people for thousands of years. Euro-Canadian settlers arrived following completion of the railroad in 1914, attracted by the agricultural possibilities. This population base remained relatively stable until the early 1970s, when rural living and hobby farming became a more popular lifestyle. Currently, approximately 650 people reside in Kispiox (SNDS 1998), and an additional 250 people reside on valley bottom lands north of the village and are mostly located adjacent or close to the Kispiox River.

Historically and up to the recent past, many Gitxsan people derived their income from the fishing and forestry sectors; however, severe job losses have curtailed this income. There are currently an estimated 18 – 23 relatively small ranches in the Kispiox, with approximately 460 breeding cows that graze on Crown land. Most residents derive their income from service sector employment in the Hazelton area. Land parcels are typically large (greater than 60 ha), with Agriculture Land Reserve restrictions regulating the majority of holdings. Population trends project growth for Kispiox and a stable rural resident community in the rest of the valley. Recreational and tourism-based incomes are projected to grow over the next decade.

Transportation and Utilities
Kispiox Trail is the major road passing for approximately 85 km up the east side of the Kispiox River from Kispiox. The road branches twice to accommodate two crossings of the Kispiox River and provides access to the west side of the valley. Both west-side roads, the Helen Lake Forest Service Road (FSR) and Mitten Main FSR converge 58 km upstream to provide access northwest out of the drainage and into the Nass watershed. The Kuldo FSR, located at 45.5 km on the Kispiox Trail, swings north providing access to the upper Skeena and Shedin drainages. There are numerous branch access roads in various states of deactivation and repair that accommodate forest development within the majority of tributary basins. Utility corridors consisting of transmission and phone lines parallel roads up the east side of the valley for approximately 40 km and, to a limited extent, on the west side of the river.

Babine
Watershed

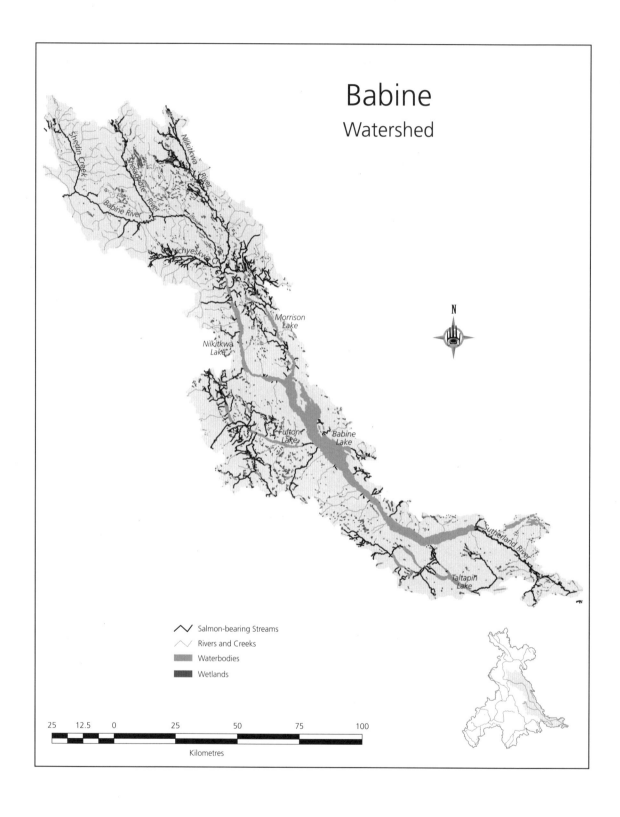

N

Shedin Creek

Nilkitkwa River

Shelagyote River

Babine River

Twain

Nichyeskwa C

Morrison Lake

Nilkitkwa Lake

Fulton Lake

Babine Lake

Sutherland River

Taltapin Lake

Salmon-bearing Streams
Rivers and Creeks
Waterbodies
Wetlands

25   12.5   0   25   50   75   100

Kilometres

15
Babine Watershed

Environmental Setting
Location
The Babine watershed is the largest tributary subbasin of the Skeena River. It is bounded on the east and southeast by the Nechako drainage, on the west predominantly by the Bulkley drainage, and to the north by smaller tributaries to the Skeena and Fraser drainages.

Hydrology
Babine Lake is the largest natural lake in B.C. and is drained by the Babine River, which flows 96 km into the Skeena River left bank about 65 km north of Hazelton. The Babine watershed is a sixth-order system with a catchment area of 10,477 km^2. Elevations range from 2,380 m in the Sicintine Range to 360 m at the Skeena River confluence with the Babine River at an elevation of 710 m. The Babine River is the largest tributary of the Skeena River, contributing approximately 15% of the Skeena River's mean annual flow (Levy and Hall 1985). The watershed is often separated into the upper and lower Babine.

The upper portion of the watershed, situated on the low-relief Nechako Plateau, has an area of approximately 6,584 km^2 and is composed of Babine Lake and its drainages. Its major tributary streams include the Morrison, Fulton, Pinkut, and Sutherland Rivers. The remainder of the Babine Lake watershed consists of relatively small creeks draining directly into Babine Lake. These creeks often tend to dry up in late summer and exhibit subsurface flow in parts of their lower reaches and alluvial fans.

The lower Babine watershed, with an area of 3,892 km^2, cuts northwesterly through the southern Skeena Mountains, draining portions of the Babine, Bait, Sicintine, and Atna Ranges and their major tributaries: the Nichyeskwa, Nilkitkwa, Shelagyote, and Shedin drainages. The surrounding glaciated mountains help maintain moderate summer streamflows. In comparison with the upper watershed, the lower watershed streams are relatively steep, with widely fluctuating flows. Originating from glaciers, these streams produce moderate amounts of natural sediment and are the primary contributors to the wide and rapid variation in water flows on the lower Babine mainstem.

Babine River peak discharges typically occur in May and June due to snowmelt, and then decrease until late September, when fall rains and early snowmelt increase streamflows until the end of October. Streamflows decrease in November and December, when precipitation falls as snow. Minimum discharges are recorded in January through April, prior to snowmelt.

Babine Village (Fort Babine) at the outlet of Babine Lake is the site of hydrometric station 08EC001, which has a 76-year record period. Another hydrology station is located close to the counting fence at the outlet of Nilkitkwa Lake (station 08EC013), which has a 33-year record (Environment Canada 2005). Several hydrometric stations are located on tributaries within the Babine drainage. The three stations located on the Fulton River have been discontinued; however, station 08EC004 on Pinkut Creek still exists.

The station at Babine Village shows annual average low flows of 7.53 m^3/s, while the monthly mean discharge record for June is 185 m^3/s. Peak discharge in tributaries is typically earlier, in May for the Morrison River and Fulton River flows. Approximately 65% of the surface inflow to Babine Lake enters south of Topley Landing, with 27.7% contributed from the Fulton drainage and 11% from Pinkut Creek, as measured at Babine Village (Stockner and Shortreed 1976).

The climate of the upper watershed is typical of the northwest Interior Plateau, with cool, moist summers and a relatively deep snow pack accumulating from October to May. Total precipitation is approximately 500 mm, of which rainfall accounts for 55% (Environment Canada no date). Lower Babine watershed climate is predominantly characterized by a shorter, cooler, and moister growing season than the upper watershed, and a longer, colder, and snowier winter. In the most western portion of the watershed, the Shedin drainage, the weather is moderated throughout the seasons due to coastal climatic influences.

Babine Lake is composed of numerous basins whose water bodies are discrete as indicated by various factors. Physical factors include seasonal thermal history; time of formation, depth, form, and stability of the thermocline; and time of formation and break up of its ice cover. Biological factors include composition and quantity of the pelagic zooplankton and discreteness of populations of its principal pelagic zooplankton-eating fish: sockeye (Johnson 1964). Babine Lake waters flow into Nilkitkwa Lake, which is also multi-basin in character; together they constitute a multi-basin system of at least 13 basins (Johnson 1964).

Stream Channels

The Babine River mainstem is 96 km in length, with a low gradient and no obstructions to anadromous fish passage over its entire length. Reach 1 on the lower mainstem has been moderately studied and documented in relation to the 1951 Babine Slide (White 1953 and 1964, Dyson 1955, Wild 1991, Moore 1993, Psutka and Rapp 1996, and Psutka 1996). In 1951, a rockslide occurred that dammed the river and obstructed the majority of returning salmon from reaching their spawning grounds in 1951 and 1952. Remedial efforts to maintain escapement were successful in late 1952 (Godfrey et al. 1954).

Reach 1, extending approximately 27 km from the Skeena River to 3 km upstream of the Babine Slide, carves a sinuous, confined channel with an average 0.5% gradient. For the most part, the river is incised 50–100 m below the valley floor, leaving local high slopes in bedrock and overburden that are susceptible to sliding (Psutka and Rapp 1996). Channel stability in reach 1 has a high hazard rating with a very

high-risk consequence. The major tributaries into reach 1, Shedin and Shenismike Creeks, are rated as moderately high-value fish habitat since they have high gradients with falls located a short distance above their mouths. Shenismike Creek is particularly steep and unstable with failing banks.

Reach 2 on the Babine River extends from downstream of Thomlinson Creek to 1 km upstream of the Shelagyote River. Reach 2 major tributaries include Thomlinson and Gail Creeks, as well as the Shelagyote River. Reach 3 on the mainstem extends from 1 km upstream of the Shelagyote River to Nichyeskwa Creek. Both reach 2 and 3 are characterized as a single-thread, irregularly sinuous channel pattern, incised into the bedrock, with glacial till and gravel deposits lying on the broad valley floor. These reaches possess an average gradient of 0.4%, with the channel and banks considered stable. There is a narrow and discontinuous floodplain, especially on the inside of channel bends. Hanawald Creek, Shahnagh Creek, and the Nilkitkwa River are the major tributaries. Reach 4 on the mainstem extends upstream to Nilkitkwa Lake and is a single-thread channel with a stable channel configuration. The upper Babine River, also known as "Rainbow Alley," is a 2.8 km constriction of Babine Lake caused by the Tsezakwa Creek fan (Wilford et al. 2000).

Watershed tributaries producing large volumes of sediment are mainly glacially fed and drain the mountainous, northern portion of the watershed. These include the Thomlinson, Shenismike, Gail, Nilkitkwa, Nichyeskwa, and Tsezakwa drainage streams. Weiland (1993) conducted and reported on a sediment point source study on the Nilkitkwa River and Nichyeskwa Creek, which provides a good understanding of the natural sediment regimes in the two drainages.

Water Quality

Water quality has been intensively investigated and studied in the Babine watershed, particularly in regards to Babine Lake, with limnological and fisheries studies dating back to the 1940s. The lake has a 461 km² surface area with a mean depth of 61 m and a water residence time of 18 years (Shortreed et al. 2001). The pH of Babine

Lake waters is slightly basic (mean=7.65), and the alkalinity range indicates moderate pH buffering capacity (Stockner and Shortreed 1976). The lake generally has a very low level of suspended solids.

Stockner and Shortreed (1976) reported Babine Lake as a dystrophic and oligotrophic lake, reflecting the water stain by input of organic matter and the low nutrient status, respectively. Most Babine Lake studies have been aimed towards understanding sockeye production, but there have been a number of studies of Babine Lake limnology because of the high importance to fisheries of this sockeye rearing lake as well as the presence of two open-pit copper mines that operated to the early and mid-1990s.

A detailed review of limnology and sockeye salmon ecology of Babine Lake was prepared by Levy and Hall (1987) as one component of Westwater's three-year study (1983–85) of log transportation impacts in the lake. In regards to the two copper mines, it was concluded that the low levels of copper in the lake, coupled with the considerable complexing capacity of the water, pose no acute threat to Babine sockeye (Davis and Shand 1978, cited in Remington 1996). Rescan (1992) also reviewed the historical environmental data of Babine Lake.

The Nilkitkwa River and Nichyeskwa Creek were the subject of water-quality monitoring studies of surface erosion, slope stability, and suspended sediment that were conducted between 1991 and 1997 (Weiland and Schwab 1991, Beaudry 1992, Weiland 1993, Maloney 1995, Weiland and Maloney 1997). Alteration of the natural sediment regime of the two streams due to forest management activities was a concern because of very high fisheries values. These studies inventoried and provided an understanding of water quality and the natural sediment regimes in the two drainages. Wilford et al. (2000) examined water quality, flow problems, and the geomorphology of Tsezakwa Creek fan following the spring flood that contributed to the mortality of over 200,000 coho fry at Babine Hatchery (Donas 2000).

Geography

An upper portion of the Babine watershed lies on the Nechako Plateau, and a lower section lies in the southern Skeena Mountains. The bedrock geology of the Nechako Plateau is comprised of flat or gently dipping Tertiary lava flows that cover older volcanic and sedimentary rocks of the Takla and Hazelton Groups and intrusive rocks of the Tertiary age. Glacial drift is widespread and a high percentage of bedrock is obscured (Holland 1976). The Nechako Plateau is generally below 1,500 m elevation. It is crossed by a series of north-northwest trending faults, which are the boundaries of down-dropped basins occupied in part by lakes and by a series of faults on the northern boundary, separating it from the uplifted Skeena Mountains.

The southern Skeena Mountains within the Babine watershed are comprised of the Atna, Babine, Bait, and Sicintine Ranges that abut the Nechako Plateau, presenting a striking visual appearance as they rise out of the plateau. Prominent northwesterly trending valleys, which are generally wide and drift-filled, divide these ranges.

The predominant biogeoclimatic zone, the Sub-boreal Spruce (SBS) zone, includes most of the lowland coniferous forests in the watershed, where subalpine fir and hybrid spruce are the major tree species. Subalpine fir tends to dominate older, high-elevation stands and moister sections of the zone. The Sutherland River, located in the southeast portion of the watershed, contains open aspen stands and grassland slopes with Douglas fir scattered throughout. Due to the pervasive natural and aboriginal fire history, as well as more recent logging activity, lodgepole pine and deciduous seral stands are extensive, particularly along stream terraces and on southern aspect slopes. Nonforested wetlands occur in the morainal landscape depressions, while dry grass/shrub meadows, though limited, are present on dry sites with favourable warm aspects.

In the upper Nilkitkwa, Shelagyote, and Shedin drainages, the SBS zone merges into the Engelmann Spruce–Subalpine Fir (ESSF) zone at upper elevations ranging from 900 to 1,300 m, depending on local topography and climatic conditions. The ESSF zone possesses a shorter, cooler, and moister growing season, with continuous forests passing into subalpine parkland at its highest elevations. Subalpine fir is the dominant tree species, with lesser amounts of lodgepole pine and white spruce hybrids in drier or fire-influenced areas. The low-elevation, southern section of the Shedin drainage is characterized by the transitional Interior Cedar Hemlock zone (ICH) zone, reflecting the close proximity of warmer, humid, maritime moisture. Major tree species in this zone are western hemlock, amabilis fir, and subalpine fir (Banner et al. 1993).

Fish Values

The Babine watershed has very high fish values and is a major producer of chinook, pink, sockeye, coho, and steelhead salmon, all of which are fished by aboriginal, commercial, and recreational fisheries. Rainbow and cutthroat trout, Dolly Varden, bull trout, and lake char, kokanee, lake and mountain whitefish, lamprey, burbot, sculpins, suckers, and shiners are also present in the system. The Babine watershed's rich salmon production is attributed to Babine Lake, which provides a moderating effect on water temperature, flows, and clarity. Babine Lake is one of the most intensively studied lakes in Canada. Babine Lake is the largest sockeye nursery lake in B.C. and currently produces about 90% of the sockeye returns to the Skeena River.

Chinook Salmon

Babine chinook are one of the important Skeena chinook populations. Relatively accurate escapement counts for chinook come from the weir located on the river below the outlet of the lake. The weir has operated from 1946 to the present, with the exception of 1948 and 1964. Nevertheless, there are a few minor difficulties arising from the weir escapement counts: firstly, the weir is located in the middle of the principal Babine chinook spawning grounds, and secondly, as a general rule of thumb, chinook salmon are not receptive to moving through weirs (Peacock 2002).

Babine chinook mean escapement between 1950 and 1959 was 8,040 fish. In 1960, the Babine subpopulation declined dramatically.

Escapements from 1960 to 1995 were between 500 and 2,500. Escapements have been improving since 1996, with an annual mean between then and 2005 of 3,220 chinook.

Adult chinook salmon begin their migration into the Skeena River usually in the third and fourth weeks of July, arriving at the counting weir throughout the month of August. There is uncertainty regarding the effects of the counting fence on chinook migrations. Chinook hold for extended periods below the fence, and it is not clear whether these fish are delayed because of the fence, or if they are pausing because the fence is a convenient normal holding area for adjacent spawning areas (Peacock et al. 1997). Spawning takes place principally from Nichyeskwa Creek upstream to the Nilkitkwa Lake outlet, typically peaking in mid-September and ending by mid-October (Shepherd 1975). Shepherd noted 90% of the spawning was above the fence to Nilkitkwa Lake, where spawning dunes are formed in years of high escapement.

Spawning also occurs occasionally on the Babine River downstream from Nichyeskwa Creek to the Nilkitkwa River (Shepherd 1975). In reviewing the period from 1934 of 1965, Smith and Lucop (1969) reported pockets of spawning for 12 km below the fence. Minor to moderate amounts of chinook spawners utilize the lower reach of Nichyeskwa and Boucher Creeks, as well as the Fulton River. In years of high returns, chinook will often spawn on the upper Babine River, locally known as Rainbow Alley. Rainbow Alley is characterized by excellent gravels with stable flows along the margin of the Tsezakwa Creek fan (Smith and Lucop 1969).

Shepherd (1975) reported that adult sampling shows that males (mean 629 mm) were significantly smaller than females (mean 742 mm). The run was dominated by four-year-olds (57%), three-year-olds (23%), and five-year-olds (13%). Ginetz (1976) noted that generally, Skeena chinook do not display consistent returns of a particular age class, except that the majority of returning adults are usually four- and five-year-olds.

Upon emergence, fry appear to migrate both upstream and downstream to rear in the Babine mainstem and tributaries, with the majority being "stream-type" migrants that have an extended freshwater phase of up to a year or more. Though Shepherd (1975) did not present density data, results of chinook fry trapping indicate that rearing habitats are utilized in the following order of importance: the upper Babine River (Rainbow Alley), the Babine River above the fence, Nilkitkwa Lake, and the Babine River below the fence.

Pink Salmon

The Babine run is one of the early Skeena pink salmon stocks and contributes approximately 4% of the total Skeena pink salmon escapement. Between 1950 and 1990, the mean even-year escapement into the Babine system has increased by a factor of 17. For the same period, odd-year pink adult escapements have increased by a factor of 13. In the 1990s, the increasing even-year trend leveled off somewhat. The slight expansion of distribution in years of high abundance has seen pinks spawning in the Babine River in the 1950s and moving upstream through Babine Lake to colonize Morrison, Pinkut, Nine Mile, Twain, and Pierre Creeks, as well as the Fulton River; however, 98% of both the even- and odd-year runs spawn in the Babine River. Spawning grounds other than the Babine mainstem appear to be utilized more frequently in years of high abundance.

Adult pink salmon usually pass through the counting fence on the Babine River between August 10 and September 10, with odd-year pinks typically entering and spawning 10 days earlier than even-year pinks (Smith and Lucop 1969). The principal spawning ground is from the forestry bridge (below the fence) upstream to Nilkitkwa Lake.

The high productivity of this section of river is mainly due to reliable water flows and low turbidity throughout spawning and incubation periods. Other minor spawning areas not mentioned above include the lower reach of Nichyeskwa and Boucher Creeks and selected pockets of the upper Babine River. Upon emergence from the spawning bed gravels, pink fry migrate downstream to the ocean.

Chum Salmon

The Babine River is close to the upstream limit of chum migration. Babine watershed chum salmon have never been abundant, with spawning observed in the Babine mainstem upstream and downstream of the counting weir. Average annual escapement from 1950 to 1959 was 28 chum salmon. These returning adult numbers have steadily decreased to an average escapement of 3 chum salmon in the 1990s.

Sockeye Salmon

The Babine watershed supports the largest sockeye salmon population in Canada. Babine sockeye studies and investigations began with the Fisheries Research Board of Canada in the 1940s. Extensive data have been gathered to date with references available in Levy and Hall (1985) and Wood et al. (1997). Tagging studies have identified three distinct runs (early, middle, and late timing) as subpopulations (Smith and Jordan 1973). Johnson's investigations (1956, 1958, and 1961) conclude that sockeye salmon production from Babine Lake was limited by the availability of suitable spawning habitat, and these results led directly to the Babine Lake Development Project (BLDP). This project, which developed artificial spawning channels, is located at Pinkut Creek and the Fulton River.

Sockeye salmon production from Babine Lake increased significantly because of the BLDP program. At least 90% of Skeena sockeye salmon now originate from the Babine-Nilkitkwa system compared with less than 80% prior to 1970 (McKinnell and Rutherford 1994). The relative increase in Babine Lake sockeye is due to an increase in abundance of enhanced sockeye and a decrease in most other (wild) Skeena sockeye subpopulations. The mixture of enhanced and wild stocks in the commercial mixed-

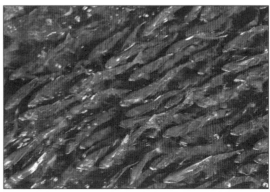

Sockeye at the Babine fence

stock fishing areas has generally depressed wild stocks, some to a greater extent than others – particularly non-Babine sockeye, coho, chinook, and steelhead stocks that migrate in the July and August peak of the enhanced runs. Management concerns regarding this situation are dual in nature: ensuring the conservation and continuity of non-Babine stocks, and maximizing the catch of enhanced Babine sockeye salmon (Wood 2001).

Accurate escapement counts for Babine sockeye salmon runs are derived from the counting weir located on the Babine River below the outlet of Nilkitkwa Lake. It has operated from 1946 to the present. Since 1966, spawning escapements to and fry emigration from the Fulton River and Pinkut Creek spawning channels have been counted through facilities maintained as part of the BLDP.

The annual average escapement at the Nilkitkwa counting fence in the 1950s was 449,000 sockeye, with depressed years in 1951 (141,415) and 1955 (71,352) due to the Babine Slide. Sockeye populations have steadily increased by a factor of three, with an average escapement during the 1990s of 1,425,613 sockeye. This total Babine sockeye population can be broken into enhanced and wild subpopulations. With the success of the spawning channels, average annual sockeye escapements to the Fulton River and Pinkut Creek have increased from 102,628 in the 1950s to 817,260 in the 1990s. Total wild (un-enhanced) subpopulations originate from 25 enumerated spawning areas in the watershed. The overall escapement of wild sockeye stocks in the Babine watershed has remained steady since the 1950s. From the 1950s through the 1990s, the average decadal escapement has been as follows: 258,448 in the 1950s; 308, 306 in the 1960s; 281,583 in the 1970s; 219,740 in the 1980s; and 301,423 in the 1990s (Spilsted cited in Wood et al. 1997).

Migration rates of Babine sockeye observed by Takagi and Smith (1973) indicate an average travel time of three weeks from the mouth of the Skeena to the counting fence, although differences in the median travel times were noted for the three runs; the early run took 24.5 days, the middle run 14–18 days, and the late

Babine River counting fence

run 21.3 days. The early run usually passes the counting fence from mid-July to mid-August and heads to the small streams located in the southern portion of Babine Lake. This is followed by a run of Morrison River, Tahlo Creek, Grizzly Creek, and Pinkut Creek sockeye, which typically passes the fence August 1 – 18. A large run of Fulton River sockeye passes the fence August 5 – 30. The last run appears at the fence after August 15 and customarily spawns in the upper and lower Babine River (Groot et al. cited in Levy and Hall 1985).

Smith and Jordan (1973) compared annual tagging results and noted remarkably precise adult migration timing from year to year. They suggested that because of differences in productivity and run timing for the various Skeena sockeye stocks, the commercial fishery could lead to over-exploitation and possibly deplete wild stocks.

Principal and minor spawning sites and mean escapements for wild, early Babine sockeye for the period of 1990 – 1999, are as follows: the Babine River below the fence (750); Babine Lake (5,000); Boucher Creek (43); Donalds Creek (53); Five Mile Creek (108); Fork Creek (N/I); Four Mile Creek (4,544); Nichyeskwa Creek (10); the Nilkitkwa River (155); Nine Mile Creek (1,486); Pendleton Creek (869); Pierre Creek (19,975); Shass Creek (4,683); Six Mile Creek (452); Sockeye Creek (2,279); the Sutherland River (900); Tachek Creek (2,036); Tsezakwa Creek (388); and Twain Creek (10,243) (DFO 2005).

Principal spawning sites for mid-Babine enhanced sockeye with annual mean escapement for the period of 1990 – 1999, are as follows: Fulton Channel #1 (15,416); Fulton Channel

#2 (116,287); Fulton above the weir (147,195); Fulton below the weir (143,500); Pinkut Channel #1 (78,451); Pinkut above the weir (27,348); Pinkut airlift (33,146); and Pinkut below the weir (125,315) (DFO 2005). Between 2000 and 2005, the escapement has been much the same (DFO 2005).

Principal and minor spawning sites and mean escapement for wild, middle-run Babine sockeye, 1990–1999, are as follows: Morrison Creek (8,990); Tahlo Creek (4,317); and upper Tahlo Creek (75). Principal spawning sites and mean escapement for wild, late Babine sockeye between 1990 and 1999 are as follows: the upper Babine River (Rainbow Alley) downstream to Smokehouse Island in Nilkitkwa Lake (201,000); and Babine River from the Nilkitkwa Lake outlet downstream to the counting fence (5,025) (DFO 2005).

Sockeye fry distribution and abundance patterns were investigated by Johnson (1956, 1958, and 1961) and Scarsbrook and McDonald between 1966 and 1977 (Scarsbrook and McDonald 1975, Scarsbrook et al. 1978). McDonald and Hume (1984) summarized fry abundance estimates. Fry feeding ecology, fry production characteristics, juvenile horizontal and vertical migration, parasitology, fry to smolt mortality, and smolt characteristics are comprehensively reviewed in Levy and Hall (1985) and Wood et al. (1997).

Despite its oligotrophic rating, the trophic status of Babine Lake based on the 1973 primary production measurements, which were the most complete, ranks Babine Lake as one of the most productive large lakes in B.C. (Stockner and Shortreed 1975). The distribution, abundance, and seasonal patterns of zooplankton occurrence strongly influence the growth and survival of juvenile sockeye during their nursery year (Foerster 1968). Limnological studies include intensive zooplankton sampling carried out by Johnson (1964) and Narver (1970), as well as McDonald's (1973) description of diel vertical migration patterns.

Rankin (1977) undertook a comprehensive analysis of the long-term changes in Babine zooplankton associated with the three- to four-fold increase in sockeye fry following construction of the spawning channels.

Comparisons were made between zooplankton collections obtained in the pre-enhancement years (1958–62) and the post-enhancement years (1973–74). The results indicated a decrease in zooplankton biomass, especially the large body-sized *Cyclops* (50%), *Diaptomus* (50%), and *Daphnia* (80%). Rankin (1977) attributed these changes to the selective removal of these species by the increased number of juvenile sockeye foraging in the pelagic zone of Babine Lake.

Shortreed and Morton (2000) stated that *Daphnia* biomass is low relative to other interior nursery lakes and comprises a much smaller proportion of juvenile sockeye diet. These data and escapements that are approaching the PR (Photosynthetic Rate) model optimum escapement suggest that fry production from natural spawning areas and the three spawning channels can fully utilize the lake's rearing capacity. The PR model uses a direct measure of lake productivity and is applicable to a wide range of lakes (Cox-Rogers et al. 2004). If fry recruitment to Babine Lake were to be further increased, lake fertilization would be required to maintain current fry growth and survival rates (Shortreed et al. 2001).

An average of one million sockeye smolts typically emigrate from Babine Lake every day over a 40 day period between May 5 and June 15 (MacDonald and Smith 1980). The pattern of migration is a bi-modal one, with a large peak of smolts emigrating out of the lake in mid-May, and a second one occurring at the end of June. Scale analysis of Babine Lake sockeye smolts (Dombroski 1952, 1954) indicate that the emigrating smolts are largely one-year lake fish, with less than 2% spending two years in the lake.

Coho Salmon

The Babine coho salmon aggregate constitutes approximately 3% of Skeena River coho escapement and is dominated by stocks that spawn in the upper Babine River. This coho stock is enumerated at the counting fence and used as an index of abundance in determining the status of Skeena coho (DFO 1999). Decadal mean escapement for the 1960s is 12,771; for the 1970s, 10,156; for the 1980s, 3,233; and from 1990 to 1998, 2,669 coho (DFO 2005).

Fence counts recorded between 2000 and 2005 averaged 2,195 annually, with ranges from 851 in 2000 to 4,636 in 2005.

The decline of the Babine coho stock from the early 1970s to 1998 averaged an annual reduction of 5%. The average age of a Babine coho at return is 3.3 years; consequently, for every generation, the stock size declined by an average of 16% (DFO 1999). The decline in coho stocks is attributed to a combination of Alaskan net fisheries, the northern B.C. troll fishery, the Skeena mixed-stock net fishery, and unknown but potentially important ocean survival factors. Tagging information available for Babine coho suggests that the stocks have a distinct ocean distribution off southeast Alaska. In 1998, with complete closure of the Skeena commercial fisheries, there was still an estimated exploitation rate of 60% on Babine coho, indicating an intense Alaskan net fishing impact (DFO 1999). Coho interception in Alaskan fisheries declined from 2001–2004 to 30–45%, probably aiding the rapid recovery of Babine coho stocks.

Coho pass through the counting fence throughout November, typically moving to the outlet of various tributaries to hold. Dependent on water flow conditions, coho will wait for the fall freshet before moving into the tributaries from late September to late November to spawn. In years of below-average flow, spawners will back off and utilize either the upper or lower Babine River or Morrison Creek, with minimal spawning loss (Finnegan 2002a). Principal spawning grounds in order of their production are the Babine River between the fence and Nilkitkwa Lake, the upper Babine River between Nilkitkwa and Babine Lakes, the Fulton River, and the Morrison River (Finnegan 2002a).

Coho fry emergence extends from April to July. Assessment of juvenile coho populations sampled throughout the watershed from 1994 to 1997 indicate that the majority of coho densities are under $0.30/m^2$ (Taylor 1995, 1996, 1997; Bustard 1997c). Coho smolt mostly after two winters in the Babine watershed and migrate downstream to the ocean with the spring high water to return as adults 14–17 months later.

Steelhead

The Babine watershed supports summer-run steelhead populations that enter the mouth of the Skeena in late June or early July and arrive in the Babine system beginning in August and continuing into autumn. Beacham et al. (2000) examined microsatellite markers that indicate Babine steelhead made up approximately 27% of the total Skeena steelhead adult returns. Babine steelhead are notable for their large body size and abundance, which makes them the basis of an internationally famous sport fishery from the late summer into the fall. Variation between different spawning stocks in the Babine watershed has been observed; for example, Shelagyote River steelhead exhibit shorter heads and varied spotting (Beere 2002). Radio-tagging studies reported by Beere (1991a, 1996) show steelhead overwintering throughout the mainstem.

Steelhead spawning occurs from March through May, coinciding with warming water and an increase in Babine River flows. The principal spawning ground is located between the counting fence and Nilkitkwa Lake. The largest concentration is 400 m below the lake outlet and adjacent to the Boucher Creek confluence. It is estimated that at least 40% of spawning occurs below the Babine fence, with spawning documented in Secret Creek, Hanawald Creek, and Shahnagh Creek, as well as the Shelagyote and Nilkitkwa Rivers (DeGisi 2000, Beere 2002). Results from radio tagging programs indicate that kelts generally migrate downstream promptly following spawning (Beere 1991a, 1996).

Steelhead fry emerge between mid-August and mid-September and are widespread throughout the middle section of the Babine River, in the smaller tributaries that offer suitable refuge. Juvenile steelhead freshwater residency varies from two to six years. Whately and Chudyk's (1979) study found 46% of the samples were 3.2s, 26.5% were 4.2s, and 10% were comprised of 3.3s. DeGisi (2000) using a much larger sample subset, found 60% were 3.2s, and 32% were 4.2s, though 10% return after one year and 25% return after three ocean years. It suggests that freshwater residency time is related to the location of juvenile steelhead

Fisheries
First Nations Traditional Use

First Nations traditional use and occupancy of the Babine watershed is extensive and well documented by oral history and early Euro-Canadian visitors. In general terms, the upper portion of the watershed, the Babine Lake drainage, was home to the Ned'u'ten, while the Gitxsan occupied the lower watershed. Although they differed linguistically, intercultural interactions were widespread resulting from the use of the same basic social structure, which had integral connections to the similar environment they inhabited. This shared social structure was composed of a matrilineal kinship society – exogamous clans divided into Houses, with crests, oral histories, and a land tenure system of territories, which were managed through a public forum process called the Feast. These separate aboriginal groups possessed distinctive characteristics and complexities that are important to note, but the social structure cut across major linguistic and cultural divisions (Rabnett 2000b).

The very abundant and predictable sockeye salmon stocks provided the Gitxsan and Ned'u'ten with the opportunity to harvest and preserve a large amount of high-quality food in a relatively short time of intensive effort. The two dominant sockeye runs (early and middle) were the major focus; they provided the majority of high-quality dried fish needed to sustain the Gitxsan and Ned'u'ten over the year and to produce a trade item. Following the passage of the bulk of the sockeye, coho appeared and were available well into the autumn, providing both fresh and dried fish. Rainbow trout, steelhead, lake trout, Dolly Varden char, lake char, and whitefish were also fished and processed at various localities.

For the Ned'u'ten, the salmon fishery at Nilkitkwa Lake formed the principal foundation of the traditional economy. Wud'at, also known as Tsa Tesli (where the lake ends), was the principal salmon season village on Babine Lake. It was located primarily on the right bank of the Babine River and is, for the most part, currently overlaid by the DFO's counting weir camp. Salmon fishing was conducted as a cooperative clan endeavour with the fish caught in weirs across Nilkitkwa Lake and the upper Babine River.

On the Babine River below Nilkitkwa Lake, the Tsayu or Beaver Clan operated a weir. Upstream from the Tsayu and close to the lake outlet, the Laksamasyu harvested fish from their weir. Further south, at the inlet to Nilkitkwa Lake and upstream of Smokehouse Island in the shallower water, the Gilserhu owned a weir that did not quite span the entire width of the river, as did the other three. The fourth weir, operated by the Laksamasyu, was positioned at the outlet of Babine Lake in the river section

Smokehouse and racks, Nilkitkwa Lake

near the present day hatchery site (Hackler 1958). Kobrinsky (1977) described "large weirs spanning the Fulton River near its confluence with Babine Lake, that served the village at that site."

Fisheries Officer Helgerson noted weir construction details downstream of Nilkitkwa Lake, when he reported on his trip to eradicate the weirs in 1906:

> The barricades were constructed of an immense quantity of material, and on scientific principles; I will endeavor to describe them. There were posts driven into the bed of the river, which is 200 feet wide, and from two to four feet deep, and running swiftly at the intervals of 6 or 8 feet.

> Then sloping braces well bedded into the bottom and then fastened to the top of the posts, then strong stringers all the way on top and bottom, in front of posts, then panels beautifully made of slats woven together with bark set in front of all, these were set firmly into the bottom, and reaching four feet above the water. This made a magnificent fence which not a single fish could get through.

> On the upper side of the dam were placed 12 big traps or fish bins. Opposite holes made in the panels for fish to enter the traps, prepared with slides to open and shut, and if the traps did not have a sufficient quantity of fish in them, when the women wanted more fish on the bank, the men would take their canoe poles, wade out in a line and strike the water, making a noise that could fill the traps in a moment, then shut down the slides, take a canoe on each side of bin, raise the false bottom, by some contrivance so as to elevate the fish, then load up canoes with gaff hooks. (Helgerson 1906).

In 1904 and 1905, a campaign was undertaken by the Department of Marine and Fisheries to remove the native fishing weirs and thus provide a larger harvestable surplus for canneries and guarantee the presence of native fishermen and plant workers (Newell 1993, Harris 2001). The legal action prohibiting weirs used by aboriginal fishermen and the sale of processed fish throughout the Skeena watershed, focused on the weirs in the Babine country. The dispute was somewhat settled with the Barricade Agreement of 1906; however, to this day, there are bitter feelings remaining with the Lake Babine Nation.

Since that time, the majority of Ned'u'ten food, social, and ceremonial fish needs have been procured with gillnets. Fort Babine Enterprises currently operates an Excess Salmon to Spawning Requirement (ESSR) fishery targeting sockeye jacks that are harvested from the Babine River at the counting fence. Beach seines and a small seine boat are used in the ESSR fishery near the Fulton River and Pinkut Creek spawning channels to harvest sockeye (Talon 2002).

For the Gitxsan, salmon represented the most important element of their diet. Kisgegas Canyon was the heartland of many adjacent Gitxsan villages and was likely the largest aboriginal settlement in the Skeena watershed (Rabnett 2001). Salmon fishing also occurred in locations on the Babine River, including Xsugwin Lik'l'insxw and Gwit Tsilaaswxt, which was located close to the Shenismike-Babine confluence; locations on the Babine mainstem, particularly at tributary mouths; and various locales on both the Shelagyote and Nilkitkwa Rivers.

In Kisgegas Canyon, where strong currents concentrated salmon along the steep banks, lashed wooden strip or woven basket traps were principally utilized to harvest large amounts of fish. Gear types suited to single fish harvest included specialized dip nets with a closable mouth (banna). The various traps and dip gear used depended on site location, fish quantities needed, the number of people available to fish the gear, and importantly, processing capacity (Morrell 1985). Ownership of fishing sites in the canyon is defined as one of the ancestral House prerogatives passed down through the generations. A weir crossed the Babine River below the canyon (Brown 1823), but little is known of its exact location and history; Muldon (2003) suggested this location is Anlagasimde'ek located downstream of Shedin Creek. Currently, salmon harvested for personal use and as part of the ESSR fishery are primarily dip-netted.

Recreational Fisheries

The sport fishery of the Babine watershed support large-scale recreational use from

Kisgegas smokehouses, 1915

residents and seasonal visitors. Angling is divided into two relatively distinct fisheries: a lake-based fishery and the Babine River fishery. The large, lake-based fishery, which is principally conducted from powerboats, typically runs from May until October and focuses on resident trout and char species. Bustard (1987) estimated anglers spent 21,000 and 15,000 angler days on Babine Lake in 1985 and 1986, respectively. Babine Lake supports one of the largest rainbow trout populations in the province with exceptional fly-fishing. Bustard (1989) noted that anecdotal information from long-time residents indicates that fish size and angling success rates have deteriorated in recent years. Char provide fine angling, and fishing for burbot, a fresh-water lingcod, is popular. Five fishing lodges, Granisle Marina, approximately 10 boat launches on the lake, and many angling guides, supply services and infrastructure to the Babine and Nilkitkwa Lakes' recreational fishery.

The Babine River fishery is conducted from the river shore, jet boats, and the occasional drift and inflatable boat. The river sport fishery is primarily directed to steelhead, though chinook may be harvested, and a sockeye fishery is usually held in years of abundance. This sports fishery, dominated by nonresident anglers, has steadily increased in popularity since the early 1970s, with the projected trend indicating further increase. Resident anglers account for 40% of the angling effort that takes place mostly in the reach upstream from the Nilkitkwa River in sight of the forestry bridge (MELP 1997).

Three angling guides run lodges and satellite camps on the Babine River with a fourth guide who has no lodge. Approximately 2,000 guided anglers (1995 – 1998) annually caught an average of 2,439 steelhead from 1990 to 1998, with less than 1% retention.

In relation to the high and growing level of angling activity on the river, the recent Management Direction Statement from B.C. Parks (2000) outlined three areas of concern regarding the sports fishery: impacts on fish populations, impacts on grizzly bears, and maintaining a quality recreation experience. Thirteen other provincial parks are spread throughout the watershed, seven of them with lakeshore access. Babine River Provincial Park (14,500 ha) and the recently created Sutherland Provincial Park (12,900 ha) are the largest

Anglers downstream of Babine River Bridge, 57 km

Enhancement Activities

Enhancement activities in the Babine watershed were initiated in 1907. The Babine Dominion Fish Hatchery was constructed and operated from 1907 to 1936 with a 10 million egg capacity. Morrison Creek sockeye generally supplied half the eggs needed; the remaining brood stock needs were acquired from the Babine River, Pierre Creek, and Fifteen Mile or Pinkut Creeks (Smith and Lucop 1969).

Between the mid-1930s and the 1950s, crews cleared streams on a regular basis to improve fish access, with emphasis applied to those streams with late summer–early autumn low water flows (DFO 1930–1960). The counting weir located on the Babine River was constructed in 1946 and has operated to the present (with the exception of 1948 and 1964). The fence was completely rebuilt in 1966–67 and replaced again in 1994.

Sockeye salmon studies in the late 1950s and early 1960s concluded that sockeye salmon production from Babine Lake was limited by the availability of suitable spawning habitat, and secondly, that the main basin of the lake was underutilized and could support additional sockeye fry (Ginetz 1977). These results led directly to the Babine Lake Development Project (BLDP). Construction of this approximately $10 million project, consisting of artificial spawning channels and dams to provide for water flow regulation, was located at Pinkut Creek and the Fulton River. The regulated water flows at Pinkut Creek and the Fulton River spawning channels provide stable conditions for the incubation of salmon eggs after they have been deposited in the gravel.

The first channel was completed on the Fulton River in 1965, the second on Pinkut Creek in 1968, and a third channel was completed on the Fulton River in 1971 (Ginetz 1977). Sedimentation problems in Pinkut Creek's enhanced channel led to a major rehabilitation program in 1976. Since 1973, an annual average of approximately 37,000 adult sockeye have been airlifted over the Pinkut Creek falls to allow utilization of a 6 km spawning area on upper Pinkut Creek.

High spawning sockeye densities in the enhancement channels have lead to health

protected areas; the other parks within the watershed are relatively small.

Current regulations applicable to Babine Lake and Nilkitkwa Lake are as follows: no fishing east of a line from Gullwing Creek to the south shore of Babine Lake; no fishing within a 400 m radius of the Pinkut Creek mouth; Rainbow Alley is open all year, though fly-fishing is allowed only from Babine Village Bridge to the counting fence, June 16 to September 30. Angling regulations specific to Babine River include a bait ban; no fishing between January 1 and June 15; fly fishing only from 80 m below the counting fence to the Nichyeskwa Creek confluence between June 16 and September 30. Class I waters apply from September 1 to October 31 with a mandatory steelhead stamp (B.C. Fisheries 2005).

problems over the last three decades. Three fish diseases have caused concern (Harrison 2002, Higgins, 2002). The viral disease, infectious hematopoietic necrosis (IHN), endemic to adult and juvenile sockeye, has been identified as an important cause of losses among sockeye alevins and fry at the Fulton River facility. As of this publication, no treatment methods have been developed to combat this disease, and fish culturists must rely solely on stock management procedures and manipulation of rearing conditions to minimize losses.

The protozoan parasite *Ichthyophthirius multifilis* (Ich) caused high pre-spawning mortality at Fulton River in 1994 and 1995 (Traxler et al. 1998) and continues to be a periodic problem. Although Ich is present in wild fish populations, this disease has a history of severe outbreaks in hatcheries and fish farms (Håstein and Lindstad 1991). The infestation at Fulton River seems to be related to high temperatures and crowding below the spawning channel (Higgins and Munby 2000, Higgins 2001). Changes in the spawning gravel material and spawner density may have contributed to control. Since 1997, *Loma salmonae* (Loma), a microsporidial parasite, has caused pre-spawning mortality of female sockeye. In 1997, 2000, and 2001, Loma and Ich were both present, and were the cause of pre-spawning mortality rates of 30% to 40% (Higgins 2002).

Fulton River Hatchery is small-scale and has been in operation discontinuously since the late 1970s. Babine River steelhead and chinook (>100,000) were hatched, reared, and then moved to the fence and net-penned to be imprinted, before being released at the fence for several years in the 1970s. In the late 1980s, the hatchery incubated and reared 30,000 Fulton River coho, with no apparent adult returns (Harrison 2002). Between 1999 and 2002 there have been efforts to increase coho abundance in the Morrison drainage. The hatchery held broodstock, incubated eggs, and reared fry to enable an annual average release of 65,000 fry into lower Tahlo Creek (Donas 2002)

Since 1983, Fort Babine Hatchery, a Community Economic Development Program (CEDP) project, has enhanced the Babine coho and chinook populations. Brood stock is obtained from adults held at the counting fence, then incubated and reared at the hatchery with an approximately 50/50 release of fry and smolts. Coho typically number 150,000 – 200,000, while 80,000 – 100,000 chinook are raised on an annual basis. In the Nilkitkwa River, 29,000 and 5,800 steelhead smolts in 1979 and 1980, respectively, were released close to the mouth (DFO 1991d).

Development Activities

Land-use and development activities in the Babine watershed are extensive, particularly in the lake drainage area. The principal land-use activity is logging, which is widespread throughout portions of the watershed with accessible timber. There are mineral resource, transportation, and settlement developments, particularly in the southern part of the drainage.

Forest Resource Development

The roots of forest development in the watershed started in the 1930s, with hand loggers falling timber into or close to Babine Lake and horse loggers hauling to the lake. Logging steadily increased following World War II, supplying

Spawning channel adjacent to the Fulton River

small local mills. Extensive selective logging operations focused on the accessible, low-relief, timbered ground south of Babine Lake. To a lesser extent, logging operations also occurred on the north side of the lake, with logs being boomed and towed across to sawmills on the south shore. Pendleton Bay was the hub of activity; log storage grounds and seven mills were situated 2 km north in Mill Bay, while another sawmill was 5 km north (BCFS 1956, Strimbold 2002). By 1955, logging camps and log dumps were prolific, with at least 21 operations in the midportion of the lake (BCFS 1956).

By 1960, Michell Bay, just north of Topley Landing, had a permanent sawmill. Milled wood was trucked out to the rail line at Topley, Burns Lake, and Decker Lake. Large-scale industrial logging occurred in the 1960s on the west side of the Babine Lake (Tsak to Babine River), as well as the north side of the lake, southeast of Hagen Arm, with the log booms towed down the lake to Pendleton Bay (Strimbold 2002). These 1960s cuts are characterized by their distinctive herringbone pattern.

Forest development activities in the 1970s took place mostly south and west of Babine Lake with the Pinkut River and Cross Creek drainages being heavily clearcut. The mid-Fulton and Guess Creek areas saw widespread clearcut activities, as did the west side of Nilkitkwa and upper Babine Lakes. The North Road, built in the early 1970s, linked Houston with Topley Landing and eased logging traffic. In addition, 1970s logging development adjacent to Hagen Arm progressed northward to the east side of Morrison Arm. These logs were dumped into Nose Bay, towed across the lake, de-watered in Michell Bay (close to the mouth of the Fulton River), and then trucked to Houston (Strimbold 2002).

In the early 1970s, the Babine Watershed Change Program was initiated in response to the increased logging and mining (Granisle Copper and Bell Copper) developments in the watershed. The purpose of the program was to collect and review baseline environmental data so that effects of environmental change on salmonids could be properly assessed (Smith 1973, 1975, 1976).

Forest development in the 1980s was widespread throughout the watershed, although concentrated logging of blowdown and beetle wood occurred in the upper Fulton River area. Logging was also intense in the Tanglechain–Smithers Landing area, the Taltapin-Henrietta area, the Fulton Lake south area, and, to a certain extent, the lower Nilkitkwa. In the mid to late 1980s, logging accelerated on the north side of the lake in response to mountain pine beetle outbreaks. This led to a three-year study by the Westwater Research Centre to determine the effects of log transportation on fish habitats and populations in Babine Lake (Levy and Hall 1985). Two barges operated by the forest industry currently provide transportation across the lake.

In the 1990s, logging continued with new roads and progressive clearcutting in the Morrison drainage and the north side of Babine Lake, the bulk of Nilkitkwa drainage up to the West Nilkitkwa River, and the Granisle area. In 1994, the Babine River Local Resource Use Plan was established for the Babine River downstream of the counting fence. This plan protected the river corridor from development activities and the area was subsequently confirmed as Babine River Corridor Park in 1999 (Ministry of Forests 2000a). The lower Babine River, downstream from Kisgegas, was bridged in 1997 to facilitate forest development north of the river. Logging development in the upper and lower Shedin watershed continued into 2001, until Skeena Cellulose Inc. financial difficulties caused operations to shut down.

The Watershed Restoration Program (WRP) in the Babine watershed has been applied since 1995. The overall gentle nature of the terrain, the coarse, well-drained soils, and the presence of large lake and wetland complexes appear to have mitigated the impacts of past forest development activities. One of the major impacts identified within the Babine watershed is impeded fish access—most often the result of road crossings that were originally constructed as, or have become, barriers to fish migration (Ministry of Forests 2001b). Restoration of fish access to these areas will re-establish access to many kilometres of productive upstream habitat. Because of the prevailing gentle terrain of the Nechako Plateau there are few open slope landslide failures or slumps related to logging. Logging and road-related landslides are present in the steeper terrain of the northern portion of the watershed.

Cumulative effects of logging-related sedimentation, if any, are presently unknown. Bustard (1986) documented high sediment loads due to road-related erosion and loss of small stream habitat in the upper Fulton drainage. Bustard also noted road-related erosion compounded by poorly drained soils and potential water temperature changes in the Morrison drainage.

Mineral Resource Development

Prospecting and mining were an important part of Euro-Canadian history in the Babine watershed. The watershed has received a substantial amount of primary and secondary exploration activity over the years. Past producers have been relatively small mining operations targeting gold- and silver-vein deposits at French Peak, Cronin Mountain, Thoen Mountain, Taltapin Lake, and Dome Mountain. Recent mining interest has centered more on large low-grade deposits worked by open pit methods, such as the Bell and Granisle deposits.

The Granisle Mine is located on McDonald Island in Babine Lake and operated between 1966 and 1982, when poor economics forced the closure. This open pit mine produced mainly copper with small amounts of gold. Granisle, an "instant town," was located on the west shore of the lake, with access via ferry between the town site and the mine. During winter, a bubbler system was used to maintain an ice-free ferry channel.

The Bell Copper Mine, located to the northwest of the Granisle Mine on Newman Peninsula, operated from 1970 to 1982 and was closed due to deteriorating market conditions. The mine reopened in 1985, with operations continuing until 1992, when available ore was exhausted. Production ore from this open pit mine was predominantly copper, with minor amounts of gold.

A number of studies were undertaken due to the presence of the two open-pit copper mines near the shore of Babine Lake and the risk to fish. Water quality investigations were reported by Stockner and Shortreed (1976), Hallam (1975), Hatfield (1989), Rescan (1992), and the Environmental Protection Service (Godin et al. 1985, 1992). No significant changes were observed in water quality. The mines' closure plans directed a relatively conservative approach to reclamation, ARD, and future monitoring. Environmental concerns regarding current toxicity and deleterious discharges are relatively minor (Stewart 2002). Lake Babine Nation community members continue to have concerns regarding the tailing ponds of the two decommissioned mines (Talon 2002).

The mineral development outlook in the Babine watershed is for continued primary and secondary development of mining prospects. Pacific Booker Minerals Inc. is proposing an open-pit mining and milling operation for the production of copper/gold concentrate from the Morrison deposit located less than 500 m northeast of Morrison Lake. It is anticipated that the Morrison deposit will be an open-pit mining operation with an ore production rate in the order of 25,000 tonnes per day or about 9 million tonnes per year.

Population and Settlement

Historically, Ned'u'ten people resided at various village sites and seasonal home places throughout the watershed, though particularly on the lakeshore. Current population centers continue to be based on the lake shoreline. The Granisle area has the largest population at 450, the majority of whom are retired. Topley Landing has approximately 50 residents, while Tachet has approximately 100 residents. Consumer services are limited, with these communities typically doing business in Houston or Burns Lake.

Fort Babine, located at the outlet of Babine Lake, has a population of approximately 140 people; however, in fishing season, the number rises to 300 people. Employment opportunities include a fishing lodge, fish hatchery, road maintenance, and village services. A chronic lack of employment, education, and training opportunities has drawn people off the reserve to larger centers such as Smithers and Burns Lake. Other full-time residents include recreational lodge service people and rural residents. In addition to the population of these year-round communities, there is a seasonal population of recreational cabin owners with approximately 110 properties. The summer season attracts many fishing enthusiasts and tourists.

Transportation and Utilities

The existing transportation network in the watershed reflects 70 years of steady improvement based on the First Nations trail infrastructure. Trails were initially widened for packhorses and later improved for wagons, then further improved for vehicular traffic. Overall, the development pattern has been spurred by a single motive: the extraction of forest products. Most of the watershed is currently roaded to support forest sector activities.

All major roads providing access into the Babine watershed originate from Highway 16. Other than the paved Granisle Highway, which accommodates traffic between Topley and Topley Landing–Granisle, the majority of roads are gravel and the responsibility of the Forest Service or forest sector. Forest resource roads continue to be developed throughout the watershed, and future trends point toward road maintenance responsibilities being off-loaded to industry or dropped altogether.

The Augier and Grizzly Forest Service Roads (FSR) as well as the Babine Fisheries Road access the southern shore of the lake at its eastern margin. Babine Forest Products facilitates log movement to its processing facility in Burns Lake with two barge crossings operating seasonally. The winter barge crosses from the Augier FSR north to the Fleming FSR. The summer barge crosses from the Grizzly FSR to the north shore of the lake. The North Road from Topley Landing to Houston is primarily used for one-way, southbound logging traffic. The barge crossing from Topley Landing to Nose Bay on the north shore accesses a large network of roads into the east-central Babine watershed and Fraser drainage.

The western portion of the watershed is accessed by three main roads: the Babine Lake Road, the Suskwa FSR, and the Salmon FSR, with the Kuldo FSR and Damsumlo FSR giving entry to the upper Shedin watershed. The Babine Lake Road, which leaves Highway 16 east of Smithers, provides cut-offs to the upper Fulton watershed (east and west), the upper Harold Price watershed, the Granisle Connector, and Smithers Landing. The road continues north as the Nilkitkwa FSR, accommodating access to Fort Babine, the Nichyeskwa drainage, and the Nilkitkwa drainage, with a western branch heading towards Shelagyote River. The Suskwa FSR, which leaves Highway 16 east of New Hazelton, furnishes access to a large area south of the Babine River. The Suskwa FSR also provides a cut-off to the Nilkitkwa Road with the recently constructed Nichyeskwa Connector. North from Hazelton, the Salmon FSR provides access to the Babine-Skeena confluence, Kisgegas Village, and northwestern drainages of the Babine River.

Utilities are limited within the watershed. Electricity is supplied from BC Hydro's provincial grid, with the main transmission line more or less paralleling Granisle Highway. This was constructed to service the two copper mines and communities close by. This line was extended northward to Smithers Landing and Fort Babine in 1979, and further extended eastward out of the watershed to provide service to Takla Lake communities. Telephone services are supplied to the Granisle area, and the prospect of service to Fort Babine is occasionally mentioned. Natural gas is not piped into the watershed. The outlook in regard to utilities is continuance of the status quo, with little future development foreseen.

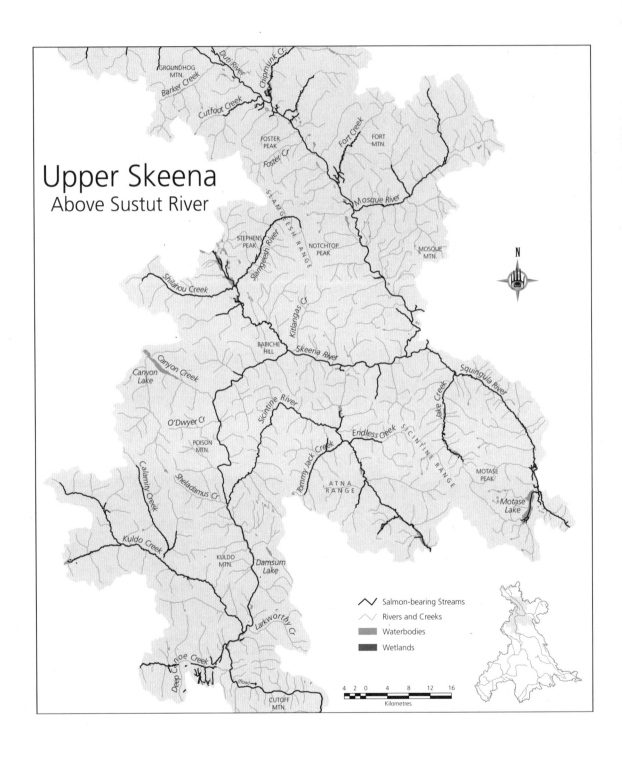

Upper Skeena
Above Sustut River

GROUNDHOG MTN.

Duti River

Chipmunk Cr.

Barker Creek

Cutfoot Creek

FOSTER PEAK

Fort Creek

FORT MTN.

Foster Cr.

Mosque River

S L A M G E E S H   R A N G E

STEPHENS PEAK

Slamgeesh River

NOTCHTOP PEAK

MOSQUE MTN.

Shilahou Creek

Kitlangas Cr.

Canyon Creek

BABICHE HILL

Skeena River

Squingula River

Canyon Lake

O'Dwyer Cr.

Scintine River

Jake Creek

POISON MTN.

Endless Creek

S C I N T I N E   R A N G E

Calamity Creek

Sheladamus Cr.

Tommy Jack Creek

A T N A   R A N G E

MOTASE PEAK

Kuldo Creek

Motase Lake

KULDO MTN.

Damsum Lake

Larkworthy Cr.

Deep Canoe Creek

(flow)

CUTOFF MTN.

Salmon-bearing Streams

Rivers and Creeks

Waterbodies

Wetlands

4 2 0 4 8 12 16

Kilometres

N

16
Upper Skeena River

Environmental Setting

The upper Skeena River is defined as the area encompassing the Skeena River and its tributaries from the Babine River upstream to Sustut River. This section includes the 124 km Skeena mainstem section and all tributaries.

Location

The upper Skeena River is located in north-central British Columbia. The area is bounded to the north by the Slamgeesh Range, to the west by the Nass River drainage, to the east by the Bear and Babine watersheds, and to the south mostly by the Kispiox watershed.

Hydrology

The upper Skeena River is mountainous with high relief. Elevations range from 366 m at the Skeena-Babine River confluence to 2,466 m at Shelagyote Peak in the Sicintine Range. The Skeena River, in its irregular route, bisects the upper Skeena area on a roughly north-south lineament to the Slamgeesh River and then heads eastward to the Sustut basin. Major streams within the watershed area include Kuldo Creek, the Sicintine River, Canyon Creek, the Slamgeesh River, and the Squingula River.

The hydrology of the upper Skeena River is dominated by snowmelt. The peak discharges of the major tributaries are typically in May and June due to spring snowmelt. Discharges are moderate as melting progressively rises in elevation as the season progresses. Discharges then decrease until September, when fall rains and early snowmelt increase streamflows through October. High streamflows and floods can occur due to rainstorms in the fall (October–November) or rain-on-snow events. These floods are rarely larger than the snowmelt floods, most likely in smaller streams and typically of short duration.

Streamflows decrease through November and December when precipitation falls as snow, with low discharges recorded January through March. Summer low flows are typically four to eight times greater than winter streamflows and are principally sustained by high-elevation snowmelt, while winter low flows are derived from groundwater, lakes, and unfrozen wetlands. The surrounding glaciated mountains help to maintain moderate summer streamflows. Originating from snowfields and glaciers, these streams contribute substantial amounts of fine sediment to the Skeena River.

Hydrometric station 08EB005 is located on the Skeena River just above the confluence with the Babine River. Monthly mean discharge (1970–1996) for March is 45.6 m³/s and 1,290 m³/s for June. This discharge can be compared to the monthly mean discharge for the Skeena River at Usk, station 08EF001, which is 160 m³/s in March and 2,830 m³/s in June (Environment Canada 2005).

The Skeena Mountains exert the major hydrological influences, and tributary streamflows have a moderately rapid response from water input due to the high gradients of the principal tributaries. Tributaries flowing into the Skeena River right bank include Deep Canoe Creek, Kuldo Creek, Sheladamus Creek, Poison Creek, O'Dwyer Creek, Canyon Creek, the Slamgeesh River, and Kitlangis Creek. Larkworthy Creek, the Sicintine River, and the Squingula River are the major Skeena River left-bank drainages. Most of the smaller Skeena River tributaries are short and steep.

The general climate of the watershed is transitional between the temperate, maritime, coastal climate and the colder, continental climate that characterizes the interior of the province. Summers are warm and fairly moist with a relatively long growing season. In the Skeena valley, snowpacks are moderate (up to

2 m) and the valley is prone to cold air ponding. The Kuldo, Canyon, and Slamgeesh valleys allow warm, humid, coastal weather systems to penetrate eastward into the Skeena watershed from the Nass basin, resulting in higher rainfall and heavy snowpacks in the mountains.

Stream Channels

Other than the Skeena mainstem, which is mostly entrenched in bedrock, stream channels of the major tributaries are erodible and generally occupy floodplains, with channel position more or less mobile. Sections of rivers with floodplains are generally flanked by terraces of older gravels, with the higher terraces always being older. Low terraces, a metre or more above the modern floodplain, may have formed in the past few thousand years and likely pick up additions of sediment during great floods (Gottesfeld and Gottesfeld 1990).

Skeena Mainstem

Channel change within the Skeena mainstem has been minimal since the end of the Little Ice Age (approximately 1850). A legacy of Pleistocene ice occupation is the possible derangement

Skeena upstream of Larkworthy Creek

of the Skeena River drainage from previously established lines. Holland (1976) described the zigzag course of the Skeena downstream of Kuldo Creek as cutting through the mountains in three different places that were evidently determined by the presence of ice in adjoining valleys. Geomorphology of the Skeena mainstem section was recently investigated by Fremier et al. (2004), who focused on the geomorphological and biological importance of tributary stream confluences between the Squingula and Babine Rivers.

Reach 7 of the Skeena River extends 23 km upstream of the Babine River to 4.8 km downstream of the Kuldo River. The channel has an average gradient of 0.1% and no obstructions to anadromous fish passage over its length (Resource Analysis Branch 1975–79). This reach of the Skeena River is characterized as a single-thread, confined, regular channel pattern, controlled by occasional bedrock outcrops or incised into the bedrock or reworked till. Boulders, cobbles, and bedrock dominate the substrate.

Reach 8 extends 6.5 km upstream to Larkworthy Creek. In this reach, the valley opens up and the Skeena River is irregularly meandering in a single, relatively wide (75–125 m) channel with an average gradient of 0.1%. The channel is bounded in the active floodplain that includes alluvial plains, low terraces, and the large Kuldo Creek alluvial fan, as well as various other fans where smaller tributary streams enter the Skeena River. Occasional islands and point bars characterize the channel.

Reach 9 extends 16 km upstream in a regular profile with an average gradient of 0.2%. The river is characterized as a single-thread, confined channel that is incised into mostly bedrock. For the most part, tributaries flowing into this reach are short and steep. Reach 10 of the Skeena River extends 6.8 km to above the Sicintine River and receives streamflow from Willowflat, Guish, and Sheladamus Creeks. The channel in this reach is typified by a regular profile and is inset into a narrow, rock-walled canyon. Reach 11 extends 71.5 km to the Sustut River confluence with an unknown gradient, but which is likely 0.1–0.15%. The regular channel

Upper Sicintine River, reach 5

profile is confined, and bedrock controls the sporadic meanders.

Kuldo Creek

Kuldo Creek flows eastward 46 km into the Skeena River right bank. It originates in Kuldo Lake, which is situated on the divide with the East Kispiox River. Major tributaries draining into the Kuldo Creek include Calamity Creek and several unnamed creeks draining glaciers in the southern Skeena Mountains.

Reach 1 of Kuldo Creek extends from the mouth at the Skeena River upstream 13.2 km with an average gradient of 0.9% in a regular, confined channel profile (Resource Analysis Branch 1975–79). Reach 2 extends upstream 18.2 km with an average gradient of 0.8% in a regular, moderately meandering alluvial channel, in a floodplain that ranges from 200 m to 600 m in width. Reach 3 extends 4.2 km in an irregularly meandering, confined channel with occasional cascades and an average gradient of 2.4%. Reach 4 extends 10.3 km upstream to Kuldo Lake with several impassable falls. The

regular channel profile is bounded and has an average gradient of 3%.

Sicintine River

The Sicintine River flows northeast, and then takes a right angle turn to flow to the southeast. It is 67.3 km in length and originates in upper and lower Sicintine Lake. Major tributaries flowing into the Sicintine River include Tommy Jack Creek and Endless Creek, which respectively provide drainage from the Atna and Sicintine Ranges.

Reach 1 is 2.1 km in length and encompasses the large glacio-fluvial fan at the mouth of the Sicintine River. The reach has slight meanders and a regular profile. Reach 2 extends 4.3 km in a confined, regular profile channel that is slightly meandering (Resource Analysis Branch 1975–79). The gradient is unknown; though there is a 1.2 km long series of rapids in the upper section. Reach 3 extends 25.2 km in length with a regular, sinuous, and straight channel profile. The channel occupies most of a narrow, steep-sided valley bottom that ranges from 80 to 600 m in width.

Motase Lake

Reach 4 begins 4.5 km downstream of Tommy Jack Creek and extends 10.8 km in length with a confined, sinuous channel. The reach receives streamflows from Tommy Jack and Endless Creeks, the two major tributaries. Reach 5 extends upstream 5.7 km in an unconfined, irregular meandering profile with zones of persistent woody debris accumulation. The floodplain ranges from 200 to 1,400 m in width with extensive areas of side channels and wetlands. Reach 6 is 5 km in length with a slightly sinuous, regular channel profile that is confined between steep valley walls and has several major bank failures.

Reach 7 extends 2.5 km in length with an irregularly sinuous pattern. Reach 8 extends 6.3 km to lower Sicintine Lake. The channel is unconfined in the relatively flat and broad valley bottom that contains extensive wetlands and fens. Reach 9 includes the very low gradient mainstem and lakes. Sixteen tributaries enter upper Sicintine Lake, and small side channels abound. The mainstem channel varies with deep pools and slight riffles, and the substrate is composed of fines, gravels, and organic debris.

Slamgeesh Mainstem and Tributaries

The Slamgeesh River flows 48 km in a southerly trending arc into the Skeena River right bank. The lower 23.4 km is in a wide valley that connects through to the Damdochax River, which is part of the Nass drainage. The upper part of the river is within the Slamgeesh Range. Where the Slamgeesh River enters the Slamgeesh Range, anadromous fish passage is blocked by a 10 m falls.

The lower Slamgeesh valley is relatively broad ranging, from 0.5 to 1.2 km wide. Reach 1 extends 19.5 km from the Skeena confluence to 300 m upstream of the Damshilgwit Creek confluence. The channel is characterized by irregular meanders and bounded by the valley walls or glacio-fluvial terraces. The floodplain varies from 0 to 850 m in width and has moderate amounts of side channels and wetlands, particularly in seasonal high water. Reach 2 extends 3.9 km upstream to the falls with a stepped and rock-bound profile.

The major tributaries to the Slamgeesh River are Shaslomal Creek and Damshilgwit Creek, the latter flowing out of Damshilgwit and Slamgeesh Lakes. From the outlet of Slamgeesh Lake, Damshilgwit Creek extends 450 m to the Slamgeesh River with a meandering channel. In the middle of this reach, Shilahou Creek joins Damshilgwit Creek, more than doubling the flow. Reach 2 is between Slamgeesh Lake and Damshilgwit Lake. It is 1.8 km long and occupies an irregular meandering channel. Beaver dams divert much of the flow through wide fens and swamps. Damshilgwit Creek continues as a small stream for about 3 km north of Damshilgwit Creek to the divide with Damdochax Creek. There is a history of shifting flows in the tributary that is at the very low gradient Nass-Skeena divide. At this time it flows north to the Nass, but probably contributed to the Skeena flow several hundred years ago.

Squingula River

The Squingula River is a left-bank tributary flowing into the Skeena River 4.8 km downstream of the Sustut confluence. The Squingula mainstem extends approximately 44.7 km to Motase Lake. Reach 1 extends 11.6 km in a single-thread, slightly sinuous pattern. The channel is entrenched in the lower third of the reach and confined in the upper portion with a 4 km section of rapids in the midportion. Jake Creek, the largest tributary that drains multiple glaciers, enters this reach. Reach 2 is 16.3 km in length with a regular, slightly sinuous channel pattern. The valley walls bound the lower half of reach 2 on a floodplain averaging 200 m in width, while moderately steep mountainsides confine the upper section.

Reach 3 extends 11.1 km with a sinuous profile and extensive wetlands and side channels. This reach contains major bank slumps, frequent logjams, and debris accumulation zones. Reach 4 extends 5.7 km upstream to the outlet of Motase Lake. The channel is mostly straight and unconfined across a broad floodplain and has a wide, low-gradient, large area of wetland with several side channels. Motase Lake is situated in a dramatic mountain bowl with multiple inlets draining numerous glaciers, as well as wetland complexes located at the northern and southern ends.

Water Quality

The upper Skeena River generally has good water quality; however, during floods or stream-bank failures, water is moderately turbid. Water originating in glaciated areas and areas of old, eroding glacio-lacustrine deposits, such as the Bear River valley, cause turbidity from late May through September. Industrial forestry development occurs downstream of the Sicintine River and adjacent to reach 1 of the Sustut River. In the early 1970s, there were significant impacts on water quality due to construction practices on the BC Rail extension.

Cumulative effects from the extensive logging operations in the Sustut area are unknown, but areas with sensitive soils and terrain are not presently being developed. Most of the Skeena headwaters remain in a largely pristine state with probably the best water quality in the watershed, though there are no long-term studies of water quality in this portion of the Skeena watershed. Temperature data for several localities are available in Finnegan (2002b) and in Hall and Gottesfeld (2003b) for the Slamgeesh Lake area.

Geography

The upper Skeena River area is a complex of mountains and valleys with a diversity of vegetation, climate, and scenery that varies with altitude. The area is composed of two broad physiographic landforms: the southern Skeena Mountains and a series of northwesterly trending trough-like valleys.

Mountain ranges of the Skeena Mountains include the range west of the Skeena River and south of the Slamgeesh drainage, as well as the Atna and Sicintine Ranges that lie, respectively, west and east of the Sicintine River. The headwaters of the Sicintine and Shelagyote Rivers are relatively broad and form a through valley as does the trough containing the Squingula and Nilkitkwa Rivers. The mountains of the Atna and Sicintine Ranges are bold and dramatic and possess the largest mountain glaciers in the Skeena watershed. Major valleys in the watershed area became established in the early or middle Tertiary period during uplift of the Skeena Mountains, but it was the recurrent glacial erosion and deposition of the Pleistocene era that modified the area into its present-day form (Clague 1984).

The ice that covered and flowed down the Skeena valley during the last glacial period forcefully eroded the mountain slopes and basins, leaving a legacy of glacial erosion and depositional features. The parallel rocky ridges of this section of the Skeena valley are drumlins formed beneath the glacial ice. They are especially conspicuous north of the Kuldo River on both sides of the Skeena River. Pleistocene ice has overridden most of the area a number of times and the rounded summits generally lying below 1,800 m are the result of ice moving across the mountain ridges. The peaks rising above the general summits have been strongly incised by montane glaciers in the past few thousand years. Many of these small glaciers are gone due to the warming climate of the past 150 years and are represented by cirques, fresh moraines, and ice-cored rock glaciers.

The upper Skeena River valley was significantly altered by ice moving down it, and subsequently by infills of glacial debris and alluvial deposits. Within this area, alluvial terraces of varying height mark standstills in the incision of the thick, late glacial deposits. These terraces are conspicuous in the Kuldo area, in patches further upstream along the Skeena, and, to a lesser extent, in the Slamgeesh and lower Squingula valleys.

Blankets and veneers of glacial till cover the main valley and mountain valleys, and extend up the valley sidewalls, though the surface expression conforms generally to the underlying bedrock surface. Thin soils, colluvium, and rock outcrops characterize the mountainsides sloping up from the Skeena River and tributaries. In the main Skeena valley between Deep Canoe Creek north to the Sicintine River, there are broad areas of gently to moderately sloping morainal blankets with organic deposits in depressions, often taking the form of bogs or wetlands.

The valley sides are typically irregular, with benches and ridges alternating with steeper sections. Snow avalanches and debris avalanches are conspicuous on steep slopes along the Sicintine River, the upper Kuldo River, and the Skeena above Canyon Creek; they often run out onto the valley floors.

Forests

The majority of the upper Skeena River area is covered with dense, coniferous forests. The mountainous high country is treeless and dominated by alpine vegetation, snowfields, and rock. Smaller but significant amounts of deciduous forests are interspersed in the Skeena River and major tributary valley bottoms, and in lower mountain slopes. Within the area, the principal forest types are represented by two Biogeoclimatic Ecosystem Classification (BEC) zones: the Interior Cedar Hemlock (ICH) zone, and the Engelmann Spruce–Subalpine Fir (ESSF) zone. Marginal amounts of the Sub-boreal Spruce (SBS) zone exist in the Sicintine and Squingula drainages and usually occupy floodplain and valley bottom sites.

The ICH zone characterizes the low-elevation coniferous and deciduous forests in the Skeena valley and tributary valleys. The ICH zone occupies valley floors and lower mountain slopes with a transitional coast-interior climate. These forests of coniferous and deciduous stands reflect the influx of warmer, humid, maritime moisture. Major tree species in this zone are western hemlock, spruce, pine, and subalpine fir.

Considerable deciduous and coniferous seral stands are present because of frequent aboriginal landscape burning over long periods of time. Seral coniferous stands are typically pine on the dry sites and spruce or subalpine fir (balsam) on the moister sites. Black cottonwood is generally found on the floodplain and tributary alluvial fans. Former aboriginal villages and campsites often have poplar stands and grassy vegetation.

The ICH zones merge into the ESSF zone at elevations ranging from 800 to 1,300 m, depending on local topography, aspect, and climatic conditions. This higher-elevation forest zone has a longer, colder, and snowier winter and a shorter, cooler, and moister growing season. Continuous forests pass into subalpine parkland at its highest elevations. Subalpine fir is the dominant tree species, with small amounts of lodgepole pine and white spruce hybrids in drier slope positions or fire-influenced areas. Black huckleberry, bunchberry, and five-leaved bramble are the dominant shrub layer species. Dry and wet grass meadows can be extensive within the area, particularly in the Damshilgwit, upper Sicintine, and upper Squingula areas. Nonforested wetlands occur on many slope positions and valley bottoms and consist of mostly sedges, along with a diversity of herbs. Avalanche tracks are common in many ESSF forests in the watershed, with vegetation ranging from shrubby slide alder to cow parsnip and hellebore.

Geology

The geology of the upper Skeena River area is composed of two main types of Mesozoic rocks: volcanic and sedimentary rocks of the Hazelton Group of mostly early to middle Jurassic age and sedimentary rocks of the middle to late Jurassic Bowser Lake Group. Cretaceous age granitic rocks compose minor amounts of the rock underlying the area and include rocks in the Sicintine Range and the Poison Mountain pluton (Evenchick et al. 2001).

Numerous faults and contacts prevail with a strong regional pattern oriented at 340°. The structure is dominated by block faulting, which controls the location of the major mountain valley systems, as well as the many rock suites and mineral deposits. Overall, metamorphism is light, aside from the contact effects near intrusive bodies.

Fish Values

Although it has several productive reaches, the upper Skeena area is a relatively moderate producer of salmon. Many tributaries to the

Skeena upstream of the Babine River are not known to have adult salmon presence; this is attributed to the generally steep gradient of streams tributary to the entrenched Skeena River. Habitat use, fish distribution, and fish populations are poorly known and documented for this area.

Fish values are very high on three tributary rivers: the Sicintine, Slamgeesh, and Squingula. Chinook, pink, sockeye, and coho salmon, and steelhead trout are present. Rainbow and cutthroat trout, Dolly Varden, bull trout, mountain whitefish, prickly and coast range sculpins, largescale and longnose suckers, threespine stickleback, and peamouth chub are also present in the area (DFO 1991d). Cutthroat trout, Dolly Varden, and bull trout are the three resident freshwater fish species of provincially designated conservation concern.

Fremier et al. (2004) reported that the number of salmonid species observed or collected in or near tributaries increased as they moved down the Skeena River from the Squingula confluence. Young of year salmonids, especially coho salmon, steelhead, and cutthroat trout, were the most common fish in tributaries and fans, while larger fish were found more often in the main river at the confluences. Juvenile salmonids seemed more abundant in the main river than in the tributaries (Fremier et al. 2004).

Slamgeesh watershed salmon smolts and adults are counted at the enumeration fence downstream of Slamgeesh Lake. The Sicintine and Squingula Rivers are often turbid, particularly in their upper sections, making enumeration difficult. In general, the most widely dispersed salmon species is coho, while Dolly Varden and rainbow trout are found in most fish-bearing waters.

Reconnaissance-level fish and fish habitat data are available for small Skeena River tributaries downstream of the Sicintine River (Lorenz 1998c, Lorenz 1998d). Generally, these small streams have waterfalls or high-gradient reaches occurring where the valley wall breaks the slope, thereby limiting fish passage. In addition, Lorenz (1998e) reported on fish and fish stream habitat for Larkworthy Creek and Deep Canoe Creek (Lorenz 1998f). Moderate-

sized tributaries including Kuldo, Sheladamus, Guish, and Willowflat Creeks received fish and fish habitat reconnaissance level surveys by Triton (1997).

Chinook Salmon

Chinook salmon use the Skeena mainstem as a migration route. Spawning is known to occur sporadically on tributary fans in this reach of the Skeena. Known chinook spawning occurs in the Sicintine, Slamgeesh, and Squingula watersheds. Historically, chinook were reported spawning in the Canyon Creek drainage (1929–34), according to B.C. 16 records. Deep Canoe Creek has a single record of 10 chinook spawners in 1990. Chinook juveniles (52–70 mm) have been sampled in the lower reaches of Larkworthy Creek (Lorenz 1998e), but they may be migrants from upriver streams. Lorenz (1998d) reported chinook juveniles utilizing numerous low-gradient, unnamed streams from Larkworthy Creek to the Sicintine River.

All of the upper Skeena chinook are stream type; that is, they rear for about one year in streams before migrating to the coast. Upper Skeena chinook return after one to five years at sea, though age 5_2 (= 1.3 in the European system) is the predominant (56%) age (Peacock et al 1997). Chinook with longer ocean residence times are often larger as adults.

Sicintine River
Sicintine River chinook escapement has not been estimated. Chinook have been observed at 63 km, and anecdotal information reports approximately 30 pairs of spawners in the low-gradient reach downstream of lower Sicintine Lake. Probable spawning locations exist throughout the mainstem, particularly in reaches 5, 6, and 7.

Slamgeesh River
Slamgeesh River chinook adult escapement records have been discontinuous since the mid-1930s, with a range since 1978 of 100–700 spawners. The varying abundance likely reflects the intensity and extent of enumerations. The principal spawning beds are located in the upper section of reach 1 and throughout reach 2 for 3.9 km to the impassable falls. Small numbers

Table 3. Slamgeesh escapements by species and year

Year	Sockeye	Chinook	Pink	Bull Trout	Coho
2006	330	1	0	113	960
2005	410	0	4	39	2,914
2004	485	0	0	58	1,854
2003	639	0	0	70	765
2002	881	0	3	53	1,154
2001	895	1	0	191	1,869
2000	946	22	10	22	849

of chinook (<100) spawn in the adjacent lower reaches of Damshilgwit Creek and Shilahou Creek. Chinook salmon have been observed in Slamgeesh Lake.

Squingula River
Squingula River chinook escapement has not been recorded; however, chinook have been observed at 37 km in reach 3, and spawning occurs in the upper portion of reach 4 below the lake outlet.

Pink Salmon

Pink salmon are relatively uncommon in the upper Skeena River area. They use the Skeena mainstem as a migration route for adults that primarily spawn in the Bear River. Pink salmon spawn in Deep Canoe Creek in the lower 0.9 km reach up to the 2.5 m falls, where 10 spawners were recorded in 1990. Canyon Creek had pink spawners recorded in 1929. Pink salmon also spawn in the Slamgeesh River below Slamgeesh Lake. Pink salmon recorded spawning in the Slamgeesh River range between 0 and 700. In the past few years there have been less than 20 spawners in even years.

Sockeye Salmon

Three distinct sockeye stocks spawn in the upper Skeena River area. Sockeye rearing lakes include Sicintine Lake, Motase Lake, and Damshilgwit and Slamgeesh Lakes in the Slamgeesh watershed. Furthermore there is evidence that sockeye spawned in or above Canyon Lake in the late 1920s and early 1930s. Reports by Fisheries Inspector A. R. MacDonnell in 1932 and 1934 described sockeye returns to Canyon Lake and referred to runs in 1928 and 1929. There are no known recent records for this stock.

Sicintine River
Little is known about the sockeye that spawn in or adjacent to Sicintine Lake. Smith and Lucop (1966) reported Sicintine sockeye spawning in ponds and the inlet narrows at the south end of the lake. There are also reports of Gitxsan traditional harvests of sockeye at the south end of Sicintine Lake. There are no known escapement records.

Slamgeesh River
Slamgeesh sockeye escapement records date back to 1909, but are intermittent till the mid-1930s. Discontinuous records from the mid-1960s to the present show a range between 0 and 4,800 spawners, with peak spawning in mid-September. Spawning is principally sustained in Damshilgwit Creek between the two shallow, productive lakes, Damshilgwit and Slamgeesh.

Damshilgwit Creek
Since 2000, the Gitxsan Watershed Authorities have conducted research on the system's salmon ecology, with a focus on sockeye and coho (Hall et al. 2002, Hall and Gottesfeld 2003, Hall 2004, and Hall and Gottesfeld 2005). Sockeye are present in both Slamgeesh and Damshilgwit Lakes. Kokanee or residual sockeye are also present in these lakes. Nearly all sockeye juveniles (97% to 100%) in Slamgeesh and Damshilgwit Lakes leave as smolts after one year in fresh water. They return after two or three years at sea, but some males (jacks) return after a single year at sea.

Mark and recapture estimates and weir counts for 2000–2003 showed escapement ranging between 435 and 1,000 sockeye. Sockeye smolts leave Slamgeesh Lake as the lake becomes ice free; 80% of the smolts are gone in the few days after the lake opens. Recorded smolt output from Slamgeesh, using the three years of data, ranged between 21,000 and 38,000. Smolts leaving the Slamgeesh are relatively large and have averaged 10 g (2001), 9.2 g (2002), and 8.6 g (2003). Table 3 shows sockeye and coho estimates based on weir counts adjusted through mark and recapture experiments. Other species escapements are weir counts alone.

Damshilgwet Creek connects Damshilgwet Lake, lower right, to Slamgeesh Lake

Squingula River
Motase Lake sockeye spawner escapement records began in 1970 and are fairly complete for the past 16 years. Escapement has ranged from 0 in some years to 3,000 (1973), but has generally been below 250 fish. Spawning has been observed at 1.5 km upstream of Motase Lake, though the principal spawning location is at feeder streams at the south end. Hancock et al (1983) noted spawning in the lake and in an eastern flowing tributary. Observations in the lake and the Squingula River are often difficult due to glacial coloring.

Coho Salmon
Coho juveniles and adults are relatively widely dispersed throughout accessible upper Skeena tributary areas. There are no known coho spawning grounds in the Skeena mainstem. It is likely that coho spawn in many, low-gradient tributaries flowing on the Skeena River valley floor. Coho migrate into the Skeena River between late July and the end of September, with the annual peak in late August, as recorded by the Tyee Test Fishery at the mouth of the Skeena River. In general, most upper Skeena tributary coho arrive in September or October. Coho are usually the last salmon to spawn in the fall, with spawning occurring from the end of September through December.

Skeena Tributaries
Coho have been observed at 63 km in reach 8 of the Sicintine River; preferred spawning and escapement records are not available. Coho adults most likely spawn in the tributary to the west and to the south of upper Sicintine Lake

(Sterritt 2003). The upper Sicintine River has been sampled in several years for juvenile coho presence and density (Taylor 1995, 1996, 1997; Wilson and Gottesfeld 2001, Wilson 2004). Juvenile coho numbers were low to moderate in all of these surveys that ranged from near 0 to $0.5/m^2$. It is not clear whether the habitat limits coho numbers or if this site has not yet recovered from the general crisis of low escapement in the late 1990s.

Larkworthy Creek sustains coho spawning in the lower reach, according to anecdotal Gitxsan reports. Currently, there are no known escapement counts. Coho have been observed in Kuldo Creek to 15 km; the abundance of preferred spawning locations is unknown. Juvenile coho (36–82 mm) were trapped in the lower reach of the left bank tributary south of Larkworthy Creek (Lorenz 1998c). Two coho spawner enumerations in Motase Lake recorded spawning in the feeder streams at the southern margin of the lake, with 200 in 1970 and 67 in 2001 (DFO 2005).

In contrast to most streams in the upper Skeena, coho numbers and densities are high in the Slamgeesh watershed, at least since 1999. The majority of coho spawn in the 1.8 km reach between Slamgeesh and Damshilgwit Lakes. In 1978, an estimate was made of 4,800 coho in this creek (DFO 2005). In 2000–2005, adult escapements to Damshilgwit Creek ranged

Damshilgwet Creek spawning beds

Slamgeesh River below Galaanhl Giist

between 750 and 1,900. In each of these years, between 23,000 and 46,000 smolts emigrated from Damshilgwit Creek.

Smolt emigration occurs over a prolonged period from late April to early July, but occurs mostly in May and June. About one half of the coho leave as smolts after one year of freshwater residence. The proportions of one-year-olds were 54% in 2003, 53% in 2002, and 31% in 2001 (Hall 2004). Coho smolts are relatively large, averaging about 20 grams. Smolt to adult survival was 3% for the 2001 smolt cohort and 3.3% for the 2002 cohort. Hall et al. (2002) and Hall and Gottesfeld (2003b) reported fry densities up to $15/m^2$ in Damshilgwit Creek in 2000–2002, although average densities are much lower.

Steelhead

The upper Skeena River area supports summer-run steelhead populations that enter the mouth of the Skeena from July through September, arriving in the area in September and continuing into autumn. Within the area, there are three known steelhead populations that spawn in the Sicintine, Slamgeesh, and Squingula watersheds. Steelhead spawning occurs in May and June, coinciding with warming water temperatures and an increase in streamflows. Steelhead fry emerge between mid-August and mid-September and are widespread throughout smaller tributaries that offer suitable refuge.

There are few good data recording steelhead escapements at individual streams. This is in large part because they spawn in spring at high water conditions, when counts are usually

not possible and they are typically spread out at many sites within a stream. In addition, little information is known concerning life-history characteristics, spawning, rearing, and overwintering habitats, as well as run timing, abundance trends, and minimum escapement requirements.

Skeena River Mainstem and Tributaries

The Skeena mainstem is used as a migration route for overwintering, an outmigration route, and a habitat for upper Skeena River and Sustut River steelhead. Some spawning is believed to occur on tributary alluvial fans. Fishery officer reports from the late 1920s and early 1930s documented heavy runs of steelhead arriving to spawn in May in Canyon Creek. Steelhead have been observed to Sicintine Lakes; however there is no information available on escapement, preferred spawning locales, or timing. According to Gitxsan traditional information, steelhead angling is productive in the Sicintine system. Little is known in regards to Squingula system steelhead other than anecdotal angling reports. Aquatic biophysical maps (Resource Analysis Branch 1975–79) show steelhead to Motase Lake. Anecdotal information for these two glacial systems suggests that spawning occurs in clear-water tributaries and/or groundwater-fed localities.

Steelhead presence is documented in the Slamgeesh system to Damshilgwit Lake (Resource Analysis Branch 1975–79). Gitxsan recount how steelhead would be gaffed at the outlet of Slamgeesh Lake in the spring. Steelhead now spawn in small numbers at this site and downstream several hundred metres (Damshilgwit Creek), though probably no more than 20 or 30 spawners are present each year.

Resident Freshwater Fish

In comparison to salmon, information is scant in regard to resident, nonanadromous or freshwater fish in both fluvial (river) and lacustrine (lake) habitats of the upper Skeena River area. Life history and habitat information is poorly known and undocumented. Dolly Varden, and, to a lesser extent, rainbow trout are the most widely dispersed species and are present in most fish bearing waters. Freshwater species and documented populations include rainbow trout *(Oncorhynchus mykiss)*, cutthroat trout *(Oncorhynchus clarki clarki)*, bull trout *(Salvelinus confluentis)*, Dolly Varden *(Salvelinus malma)*, mountain whitefish *(Prosopium williamsoni)*, northern pikeminnow *(Ptychocheilus oregonesis)*, longnose dace *(Rhinichthys cataractae)*, redside shiner *(Richardsonius balteatus)*, and prickly sculpin *(Cottus asper)*.

Rainbow trout have been documented in the following systems: Deep Canoe, Kuldo, Larkworthy, Guish, Willowflat, Sicintine, Tommy Jack, O'Dwyer, Canyon, Slamgeesh, Squingula, Jake, and numerous first-, second-, and third-order unnamed streams tributary to the Skeena mainstem.

Dolly Varden have been documented in many first-, second-, and third-order unnamed streams tributary to the Skeena mainstem, as well as the following tributaries: Larkworthy, Kuldo, Guish, Willowflat, Sicintine, Tommy Jack, O'Dwyer, Canyon, Slamgeesh, Squingula, and Jake. The range of Dolly Varden is obscured by past failure to distinguish bull trout from Dolly Varden.

Fremier (2004) reported that bull trout in particular exploit food resources at the Skeena mainstem-tributary confluence zones and that cutthroat are the most common fish in tributaries and fans. Bull trout are abundant in Slamgeesh Lake. There are probably several hundred breeding fish in this population. They rear for several years in small streams and then migrate to Slamgeesh Lake and the Slamgeesh River, where they are piscivorous – probably feeding mostly on small prickly sculpins but also on sockeye and coho smolts during the spring migration.

Fisheries
First Nations Traditional Use
Gitxsan salmon fisheries within the upper Skeena River area were concentrated on the Skeena mainstem and as terminal fisheries close to the outlets of the Slamgeesh, Sicintine, and Motase Lake. The majority of Gitxsan mainstem fishing sites were concentrated on four sections of the Skeena River: Gitangasx, Kuldo, Old Kuldo, and the section upstream of the Babine confluence.

Upper Sicintine Lake

Clearly defined alluvial terraces in the clearcut along the Skeena River north of Kuldo

Gitangasx is located approximately 12 km east of the Slamgeesh-Skeena confluence. Kuldo is located upstream of the Kuldo-Skeena confluence, and Old Kuldo is situated 10 km farther to the north on the mainstem left bank. These areas were ancient village sites and were mainly positioned to easily exploit the fish resource. These fishing sites took advantage of canyons where salmon could easily be gaffed, netted, and trapped. The number of Gitxsan fishing sites on the Skeena mainstem is unknown. Undoubtedly, there are many traditional sites not documented, but the four main fishing locales signify the importance of this upper Skeena mainstem section to Gitxsan fishing interests.

In the Slamgeesh drainage, Gitxsan traditionally resided at two ancient villages: Galaanhl Giist, located close to the Slamgeesh-Shilahou confluence, and Gitangwalk, situated 1 km downstream on the Slamgeesh mainstem. Gitxsan harvested the local salmon resources

utilizing a variety of sites and gear, focusing on the weirs and traps close to both village sites downstream of Slamgeesh Lake.

After the Department of Fisheries forced the abandonment of weir and trap fishing in 1906, Galaanhl Giist primarily supported a spear fishery, although trap fishing likely persisted in this remote northern area for some time. The salmon stocks spawning here formed the principal food resource that enabled people to make this their home. Prior to the 1950s when the Gitxsan were moved into centralized communities such as Gitanmaax and Kispiox, Galaanhl Giist functioned as a winter village. Cultural heritage resources include hundreds of cultural depressions, grave houses, trails, fishing gear, and lithic assemblages that are primarily composed of obsidian points, flakes, and microblades.

The mouth of Motase Lake, on both the north and south shores, is the location of the ancient Gitxsan fishing village site,

Xsimootixwit. This village was strategically located to take advantage of the fish resources and the fishing site attributes, which primarily supported a weir, and then after 1906, a spear fishery. There are no known historical references, though the village is located along the main north-south trending trail that passes along the east side of the Squingula and Nilkitkwa valleys, providing connectivity between Wud'at (Babine Lake) and the Sustut River–Bear River country. Known cultural heritage includes cultural depressions, trails, and a lithic assemblage composed of basalt flakes.

Anuux Du Guux Gan was an ancient village site located on the southeast shore of upper Sicintine Lake. The village took advantage of the fish resource and site characteristics that provided a fruitful fishery. Known cultural heritage includes cultural depressions, cabin sites, and trails from the Shedin, Squingula, Shelagyote, and lower Sicintine drainages.

Overall, only the Skeena mainstem fisheries continue into the present, with most food, social, and ceremonial fisheries conducted downstream of the Babine confluence and mainly targeting Babine sockeye.

Recreational Fisheries

There is limited, but growing, sport fishing within the area on the Skeena River mainstem. Angling is primarily directed towards chinook, coho, and steelhead. Bull trout, Dolly Varden, and a variety of other freshwater fish are fished to a lesser degree. Angling occurs seasonally on Kuldo and Canyon Lakes, incidental to guided hunting, with effort directed towards nonanadromous species. Steelhead regulations require mandatory release July 1 to December 31. General restrictions include no fishing in the upper Skeena area January 1 to June 15 (B.C. Fisheries 2005).

Development Activities

Historical post-contact development up to 1990 was limited to construction of the Dominion Telegraph Line in 1899, which provided telegraph service from Hazelton to the Yukon at the time of the Gold Rush, and continued up to the mid-1930s. The Telegraph Trail and cabins for the operators and linemen were in use

until the mid-1950s by Gitxsan and occasional trappers.

Since 1990, land-use and development activities in the upper Skeena River have been limited to the southwest portion of the area. The principal land-use activity is logging, which is concentrated in the Skeena valley bottom and lower-elevation tributary valleys. Mineral exploration is limited to the Atna and Sicintine Ranges and has been modest. Presently, linear development and settlements do not exist. Transportation has been the key to the recent modest growth in land-use and resource development.

Forest Resource Development

The southern portion of the area downstream of the Canyon Creek watershed is located within the Kispiox Forest District. To the north, the Fort St. James Forest District includes the Canyon, Slamgeesh, upper Skeena, and Squingula drainages. Within the Kispiox Forest District, general resource development is zoned for the valley bottoms, and special resource zoning is applied to higher elevations. The special resource zone has limited commercially viable timber opportunities and allows for scenic, wildlife, and recreation values (Ministry of Forests 1996). Logging in the upper Skeena River area within Kispiox Forest District has been moderate due to the poor sawlog quality, the high abundance of pulpwood, and the long distance to milling and conversion plants.

In the Fort St. James Forest District portion of the upper Skeena River, forest harvesting has not occurred. The BC Rail line operates sporadically as far as Minaret. Road development is controlled by the Northern Long Term Road Corridors Plan, a coordinated access management plan initiated by the Fort St. James Forest District (Ministry of Forests 2000b). The intent of the plan is to accommodate timber resource transportation routes and generally determine the direction of timber flow.

Mineral Resource Development

Euro-Canadian explorations for base and precious minerals have been ongoing since the turn of the 20th century, focusing mainly on high-grade silver and gold-rich veins, with

Squingula-Skeena confluence

most endeavors slowing down with the onset of the Depression. In the early 1960s, mineral exploration interest increasingly focused on porphyry copper-molybdenum occurrences.

Most known mineralization occurs in the eastern portion of the Sicintine Range, with the greatest concentration in the lower Squingula River area. This locale reflects alteration zones associated with either the Cretaceous Bulkley or Kastberg Intrusives. Fifteen mineral occurrences represent 12 showings, two prospects, and one developed prospect. The main types of mineral occurrences are polymetallic veins containing silver, lead, zinc, copper, and gold. Exploration in the early 1970s on the Red property, a developed prospect located east of the lower Squingula River, resulted in substantial copper reserve estimates and approximately 30 km of bladed trails in the lower Squingula River area.

Potential evaluations of metallic mineral for the watershed are broad based and nonspecific. The area has been delineated into mineral tracts on the basis of geology and distribution of known mineral occurrences. Most of the area has been rated as having moderate mineral potential.

Evenchick (2002) suggested that the geological framework data and initial petroleum resource assessments indicate significant undiscovered conventional petroleum potential in the Bowser and Sustut basins, part of which underlie this upper Skeena area. Presently, this potential is based on a generally sparse data set, and additional work is needed to evaluate the thermal and diagenetic history of potential systems.

Transportation

The upper Skeena River–Babine River to Sustut River area has vehicular access only to the low-elevation southwestern portion. This area is approached by the Kuldo Forest Service Road (FSR), which leaves the main Kispiox road at 41 km and continues northward on the west side of the Skeena River almost to Willowflat Creek at 58 km. The Damsumlo FSR crosses the Skeena River at 45 km and extends approximately 36 km northward through the upper Shedin drainage into Tommy Jack Pass. Access along the north side of the Skeena River from the Babine River extends approximately 15 km to Java Creek. The northern portion of the area is unroaded.

Fixed-wing float aircraft seasonally service the area. The only maintained fixed-winged airstrip is located at Minaret, north of the Sustut River. Future transportation trends in the Sustut drainage could potentially include the proposed Stewart Omineca Resource Road and the proposed Northern Long Term Road Corridors Plan.

In 2001, the District of Stewart, B.C., proposed the Stewart Omineca Resource Road (SORR) as a stimulus to northern regional economies, and commissioned a promotional and preliminary impact study (McElhanney et al. 2001). The total road length is 475 km; it would leave Highway 37 and use existing forestry roads that are in place to access the upper Shedin Creek–Tommy Jack Pass area. New construction would be needed between there and the Moosevale Creek–Sustut River confluence, except for a 12 km section of recently built forestry road (Birdflat Creek to the Bear River). The proposed SORR will undoubtedly have many impacts on the upper Skeena drainage and determine future development patterns.

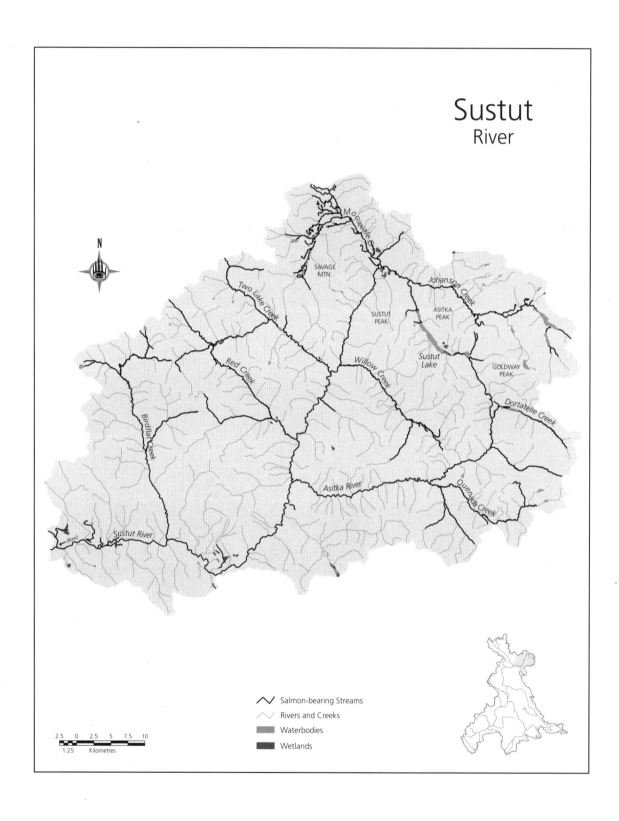

Sustut
River

Moosevale C

SAVAGE
MTN.

Johanson Creek

Two Lake Creek

SUSTUT
PEAK

ASITKA
PEAK

Red Creek

Willow Creek

Sustut
Lake

GOLDWAY
PEAK

Dortatelle Creek

Birdflat Creek

Asitka River

Quenada Creek

(flow) Sustut River

N

Salmon-bearing Streams
Rivers and Creeks
Waterbodies
Wetlands

2.5   0   2.5   5   7.5   10
1.25   Kilometres

17

Sustut River

Environmental Setting

The Sustut River is one of the major river systems in the high interior zone of the Skeena River watershed. The Sustut watershed includes the Sustut River and its tributaries other than the productive Bear River, which is discussed in the following chapter.

Location

The Sustut River drainage is located in north-central British Columbia within the northeastern headwaters of the Skeena watershed, approximately 200 km north of Smithers. The drainage is bounded to the north mostly by the Tatlatui and Swannell Ranges, to the west by the Skeena Mountains, to the east by the eastern Swannell Range, and to the south mostly by the Driftwood drainage, which is part of the Fraser watershed.

Hydrology

The Sustut River drainage is mountainous with high relief; elevations range from 580 m at the Skeena-Sustut confluence to 2,469 m at Sustut Peak in the Hogem Ranges. The Sustut River flows in an irregular route that bisects the Hogem Ranges in a roughly southwest direction, discharging into the left bank of the Skeena River. The mainstem is approximately 97 km in length from the outlet of Sustut Lake. The majority of the drainage is relatively high in elevation; the elevation at the Sustut River–Moosevale Creek confluence is approximately 1,160 m. Major tributary streams within the drainage basin include Birdflat Creek, the Bear River, the Asitka River, Red Creek, Two Lake Creek, Willow Creek, Moosevale Creek, and Johanson Creek. Most of the small first- and second-order tributaries flowing into the lower and middle reaches of Sustut River are short and steep.

Sustut River peak discharges occur in May and June due to spring snowmelt throughout the upper Skeena watershed. In most years, the peak spring freshets are the largest floods in the mainstem. Water levels in the main river channels and back channels fluctuate seasonally; typically, they are high from May to early July, drop for the summer months, rise to intermediate levels in the fall, and reach their annual low levels late in the winter season. There are no hydrometric stations located within the drainage.

The numerous ranges of the Skeena and Omineca Mountains exert the major hydrological influence within the drainage. Tributary streamflows have a moderately rapid response to water input due to the high relief and relatively steep channel gradients. This effect is dampened to a degree in the upper Sustut River, particularly in the Moosevale and Johanson valleys where extensive wetlands and/or lakes provide flood storage. Johanson Lake at 1,444 m and Sustut Lake at 1,301 m are the major lakes; smaller silty lakes include Darb Lake, upstream of Johanson Lake, and Spawning Lake, which is tributary to Johanson Creek. Asitka Lake at the head of the Asitka River is located 3 km east of Sustut Lake at approximately 1,440 m elevation and provides additional water storage capacity.

The general climate of the watershed is transitional between the slightly temperate, maritime, coastal climate and the dominant, colder, continental climate of the interior of the province. Summers are warm and fairly moist with a short growing season. Snowpacks are moderate in the lower Sustut valley (up to 2 m), which is prone to cold air ponding. There are no long-term weather records available.

Water Quality

The Sustut River drainage generally has good water quality; however, during flood events or stream bank failures, water is turbid in most

Bear-Sustut Confluence

of the major tributary flows. Industrial forestry development occurs adjacent to reach 1 of the Sustut River on both the north and south slopes. In the early 1970s, there were significant impacts on water quality due to the BC Rail extension construction practices and the lack of a bridge crossing the Sustut River. Chudyk (1974) noted that the Omineca Resources Access Road, which is located to the north and east of Johanson Creek and west of Moosevale Creek, yielded sediment when it was initially constructed, due to inadequate bridge and culvert provisions and seasonal high water levels. In general, the valleys of the Sustut watershed possess highly erodible soils with areas of glacio-lacustrine deposits that can yield high suspended sediment loads when disturbed.

Stream Channels

Other than the channels that are entrenched in bedrock, stream channels of the major tributaries are erodible and generally occupy active floodplains with the channel location changing on a relatively steady basis. The Sustut drainage is relatively pristine; consequently, channels have only experienced natural change. Many of the river and tributary creek sidewalls are steep, unstable, and composed of fine, silty, erodible soils.

Sustut Mainstem
Channel change within the Sustut mainstem has been minimal since the end of the Little Ice Age (approximately 1850) and is most evident in the lower Sustut floodplain. The overall gradient of the Sustut mainstem from the Skeena River to Moosevale Creek confluence is 0.7%; however, there are short sections with gradients up to 3.8%. From the Skeena River confluence upstream to the Bear River, the Sustut River is commonly referred to as the lower Sustut River. From the Bear River upstream to the canyon 500 m below Moosevale Creek, the river is

referred to as the mid-Sustut River. The upper Sustut River extends to Sustut Lake.

The Sustut River is classified as reach 1 for 18.5 km upstream of the Skeena River. The reach extends upstream to the small canyon approximately 600 m downstream of Birdflat Creek. The channel has an average gradient of 0.1% and no obstructions to anadromous fish passage over its length (Resource Analysis Branch 1975–79). This reach of the Sustut River is a wandering gravel bed river, mostly with a single, dominant channel. Low pine-covered terraces are abundant along this reach.

Reach 2 extends 24.3 km in length upstream to approximately 2.4 km upstream of Saiya Creek. In this reach, the channel is generally confined within the steep inner valley walls and is controlled by occasional bedrock outcrops or incised into the bedrock and till. The single channel is irregularly sinuous, rapids are common, and the gradient averages 0.7%. Reach 2 receives the flow from the Bear watershed and from many small first- and second-order tributary streams that are, for the most part, short and steep.

Reach 3 extends 10.5 km in length upstream to the mouth of Red Creek and receives the discharge from the Asitka River. The 0.5% gradient single-thread channel is confined by the valley walls, and a series of constrictions and rocks are found in the canyon below and above the Asitka River. Reach 4 extends 29.5 km upstream to within 3.5 km of the Moosevale Creek confluence with an average gradient of 0.9%, and it receives the discharge of Willow and Two Lake Creeks. The reach is irregularly sinuous, entrenched, and confined by the valley walls except in the vicinity of the two major tributaries. Steelhead Canyon is about 8 km above Red Creek. Reach 5 extends 4.5 km in length with a steeper gradient of about 1.3%, especially at the series of cascades below the Moosevale confluence.

The Sustut and Moosevale valleys are both relatively wide with a complex of wetlands and eskers. Reach 6 is 3.9 km in length and extends upstream to the confluence of Johanson Creek. The reach is characterized by irregular meanders and a low-gradient channel across a floodplain, which averages 500 m in width. Reach 7 is

5.4 km in length, with an average gradient of 1% and extends upstream to Mud Lake, which is a shallow and wide section of the mainstem. A further 0.8 km of low-gradient channel surrounded by extensive wetland and seepage areas approaches Sustut Lake.

Sustut Lake is situated at an elevation of 1,301 m and receives the discharges of the upper Sustut River and Transparent Creek, as well as numerous, very small streamlets and seepage zones. It is characterized by depths up to 9 m in the southern portion, an area of deeper water in the central part, and a maximum depth of 18.6 m with extensive shallows at the northern end.

Johanson Creek

Johanson Creek receives discharge from two main tributaries: Solo Creek, which drains Spawning Lake, and Darb Creek, which drains Darb Lake. Reach 1 of Johanson Creek extends 7.6 km to the first major right-bank tributary. A meandering channel that is inset within the 150–300 m wide floodplain characterizes this reach. Reach 2 extends 9.6 km to within 2 km of the mouth of Solo Creek, which drains the somewhat silty Spawning Lake. The channel is sinuous and is bounded by a narrow floodplain that ranges in width from 80 to 500 m. Much of this reach passes through a series of subalpine meadows

Reach 3 extends 5.4 km with a confined, slightly sinuous channel. Reach 4, which extends 3.6 km upstream to the outlet of Johanson Lake, is unconfined and meandering. Johanson Lake is in the alpine fringe at 1,444 m. It is 3.8 km in length, averages 0.5 km in width, and has several small islands. There is a lack of shallows, and the shores in most areas drop off sharply to moderately deep water. The main tributary into the lake is Darb Creek, which drains the usually silty Darb Lake.

Asitka River

The Asitka River receives streamflow from two major tributaries: Quenada and Dortatelle Creeks. Quenada Creek is approximately 25 km in length and drains Whistler Basin as well as Carruthers Pass. Reach 1 of the Asitka River extends from the Sustut River upstream 14 km and is generally confined by the valley walls, resulting in little or no floodplain. Reach 2 extends 7.6 km upstream to the mouth of Quenada Creek. The channel is sinuous and bounded by a floodplain that ranges in width from 120 to 380 m; it is bordered by steep valley walls. Reach 3 extends 10.2 km in a slightly sinuous, confined channel with a series of small falls in the lower section.

Reach 4 extends 8.4 km upstream and receives the discharge of Dortatelle Creek, which is approximately 13 km in length. The main channel is slightly sinuous and passes across a floodplain that averages 300 m in width. The Asitka River in this reach is partially blocked by beaver dams as well as several valley sidewall slump zones. Reach 6 is 3 km in length with a stepped profile and confined channel. Reach 7 is 3.8 km in length, extending to the outlet of Asitka Lake. Beaver dams often obstruct this reach. Asitka Lake is located at 1,300 m elevation, 3 km east of Sustut Lake. The shoreline is irregularly round (0.75 km radius) and broken by many small, shallow bays. The lake has a maximum depth of 7.9 m.

Geography

The Sustut watershed is a complex group of mountains and valleys with a diversity of vegetation, climate, and scenery that vary with altitude. The general elevations of both valley bottoms and peaks rise from the Sustut-Skeena confluence in the southwest corner of the drainage to a maximum in the upper Sustut valley and Sustut Peak area. Sustut Peak, Asitka Peak, and Goldway Peak frame the upper Sustut and Asitka valleys and are the highest mountains in the drainage.

Tsaytut Spur, the Connelly Range, and the Sikanni Range form northwest trending ridges or mountain masses, whereas to the northeast the mountains are rectilinear or irregular in outline. The complex drainage pattern forms an interlocking network of the deep and commonly broad U-shaped valleys. The rugged mountains are characterized by knife-edged ridges, spires, abrupt crumbling slopes, and nearly vertical cliffs.

The Sustut River and its tributaries divide the Swannell Ranges and further separate them from the Skeena Mountains to the west, and the McConnell Range and the Osilinka Ranges to the east. The lower Sustut valley (Skeena to Bear River) is relatively broad, with low to moderately steep mountain slopes. In addition, there are broad areas of gently to moderately sloping morainal blankets with organic deposits in depressions, often taking the form of bogs or wetlands.

The mid-Sustut valley possesses moderate-to-steep valley walls, and tributary valleys are generally narrow and steep. The upper Sustut valley and most of the upper tributary valleys open into large alpine grasslands and shrublands, meadows, swamps, and wetlands.

Ice flowed across the drainage from the north (Tatlatui Range) during the last glacial period, forcefully eroding the mountain slopes and basins and leaving a legacy of glacial erosion and depositional features (Lord 1948). Pleistocene ice has overridden most of the drainage a number of times, and the rounded summits generally lying below 1,800 m are the result of ice moving across the mountain ridges. Glaciers moving along the major valleys caused oversteepening of mountain slopes and contributed to the present-day U-shaped valley profiles. Above 1,500 m, alpine glacial activity of the last few thousand years significantly modified the high ridges.

Tributary valleys throughout the drainage were significantly altered by ice moving down them, and subsequently by glacial debris and alluvial deposits. Kame terraces formed between the ice and the valley wall and may have undergone modification by slumping subsequent to the ice melting. These terraces are the discontinuous terraces seen on both sides of the lower Sustut valley adjacent to or approaching the tributaries, as well as along some of the major tributaries, particularly the Asitka River, Two Lake Creek, and Willow Creek. Glacial and post-glacial deposits cover the lower, gentle-to-moderate slopes and infill the broader sections of valley floors. Two major eskers more than 4 km in length lie south of Moosevale and Johanson Creeks. These eskers are sinuous and steep-sided, with narrow crested ridges that generally consist of cross-bedded sand and gravels. They were laid down in glacial meltwater either at the retreating edge of the ice sheet or in subglacial tunnel channels.

Blankets and veneers of glacial till cover the main valley and mountain valleys, and extend up the valley sidewalls, though the surface expression generally conforms to the underlying bedrock surface. Thin soils, colluvium, and rock outcrops characterize the mountainsides sloping up from the Sustut River and tributaries. Bedrock is exposed along deeply incised streams and on steep-sided hillslopes.

Forests

Extensive alpine areas, snowfields, and rock dominate the higher-elevation portions of the Sustut watershed. The higher valleys are covered by willow and birch scrub tundra. The less extensive lower-elevation part of the basin is covered with dense coniferous forests. There are smaller but significant amounts of deciduous forests in the Sustut and major tributary valley bottoms and lower mountain slopes. Within the forested part of the drainage, two Biogeoclimatic Ecosystem Classification (BEC) zones are represented: the Sub-boreal Spruce (SBS) zone and the Engelmann Spruce – Subalpine Fir (ESSF) zone.

The lower-elevation biogeoclimatic zone, the SBS zone, covers the lowland in the Sustut valley below the Bear River and a narrowing wedge of low elevation along the Sustut River as far as Two Lake Creek. Subalpine fir and hybrid spruce are the major tree species. Subalpine fir tends to dominate older, high-elevation stands and moister sections of the zone. Nonforested wetlands occur in the morainal landscape depressions, while dry grass/shrub meadows, though limited, are present on dry sites with favourable, warm aspects.

Due to the pervasive natural and aboriginal fire history, as well as more recent logging activity, lodgepole pine and deciduous seral stands are extensive, particularly along stream terraces and on southern aspect slopes. Lodgepole pine is prevalent on the mountain slopes and lower terraces on the north and south sides of the lower Sustut River, upstream approximately to the Bear River. Seral coniferous stands are typically pine on the dry ground and spruce or subalpine fir (balsam) on the moister

sites. Black cottonwood is generally found on the floodplain and tributary alluvial fans.

The SBS zone merges into the ESSF zone at upper elevations ranging from 800 to 1,000 m, dependent on local topography, aspect, and climatic conditions. In the lower Sustut, the SBS-ESSF merge occurs at 800 m on the north slope, and at 1,000 m on the south slope. The ESSF higher-elevation forest zone has a longer, colder and snowier winter and a shorter, cooler and moister growing season. The zone is continuous forest at its lower and middle elevations; it then passes into subalpine parkland at higher elevations. Subalpine fir is the dominant tree species, with small amounts of lodgepole pine and white spruce hybrids in drier slope positions or fire-influenced areas.

Black huckleberry, bunchberry, and five-leaved bramble are the dominant shrub layer species. Mosaics of tree islands interspersed with dry or wet grass meadows or shrublands can be extensive and common within the drainage, particularly in the Moosevale, upper Sustut, Johanson, and upper Asitka areas. Nonforested wetlands occur on many slope positions and valley bottoms and consist of mostly sedges, along with a diversity of herbs. Avalanche tracks are common in many ESSF forests in the drainage, with vegetation dominated by shrubby slide alder, sedges, cow parsnip, and false hellebore.

Geology

The geology of the Sustut watershed is extremely complex. This watershed lies at the eastern edge of the Stikine terrane and crosses into the Quesnel terrane. In relation to the older physiographic usage, these are the Intermontane Belt and the Omineca Crystalline Belt. The younger Stikine terrane rock units have been described in the sections on the geology of the mid and upper Skeena regions. The Sustut River transects through all of the different rock groups in the Stikine terrane. The lower part of the Sustut is in sedimentary rocks of the Bowser Lake Group. The Sustut River then crosses the southern end of the Sustut basin at the Bear River. The Sustut Group are nonmarine sediments of Cretaceous and early Tertiary ages, which are better exposed on the Stikine Plateau to the north.

From the Bear River upstream, the Sustut River passes deeper and deeper onto the rock sequence of the Stikine terrane. From the Bear River to Two Lake Creek, Hazelton Group volcanic rocks of Jurassic age form the mountains. The oldest part of the Stikine terrain is found from Two Lake Creek to the Sustut Lake–Moosevale valley depression (Monger 1976). These units are poorly exposed in the areas to the west. The Takla Formation is formed from Triassic rocks, which are marine mudstones, ash deposits and lava flows underlain by Permian Asitka Group Marine Lava flows, and volcanic pyroclastic (ash) deposits.

The broad valleys occupied by Asitka Lake, Sustut Lake, and Moosevale Creek are major fault zones related to the Pinchi fault, which separates the Stikine terrane from the Quesnel terrane on the east (Monger 1976, Gabrielse et al. 1991). In the Mesozoic era, there were probably long distances of horizontal displacement along this fault zone with the western portion moving northward. The Omineca Mountains in this portion of the Quesnel terrane are formed by granites of the Hogem batholith intruded into a series of Triassic or older marine volcanic rocks.

Numerous faults and pervasive shearing prevail with a strong regional pattern at 340°. The geologic structure is dominated by complex block faulting, which controls the location of the major mountain valley systems as well as the many rock suites and mineral deposits (Richards 1975). Overall, metamorphism is light, aside from the contact effects near intrusive bodies.

Fish Values

The Sustut watershed has high fish values, particularly in the upper portion of the system. The drainage supports moderate but significant populations of sockeye and coho salmon, as well as significant numbers of chinook and steelhead trout. Rainbow and lake trout, Dolly Varden, bull trout, mountain whitefish, burbot, longnose dace, prickly sculpins, and peamouth chub are also present in the drainage system (FISS 2003). In general, the most widely dispersed salmon species is coho, while Dolly Varden/bull trout

and rainbow trout are located in most fish-bearing waters. There are two resident freshwater fish species of conservation concern: Dolly Varden and bull trout.

Since 1992, one or two adult fences have been operated seasonally in the upper Sustut system. Presently, the mainstem fence is located 700 m upstream of the Moosevale Creek–Sustut River confluence and provides the most inclusive count (Holtby et al. 1999). The DFO and the Fisheries Branch of the Ministry of Water, Land and Air Protection (WLAP) have cooperatively conducted annual escapement programs for chinook, sockeye, and summer-run steelhead, along with interspersed juvenile sampling.

Chinook, coho, and steelhead rearing in the high-elevation (>1,000 m) upper Sustut drainage experience a short growing season and cold water temperatures. Consequently, the rearing period of these fish is generally a year or more longer than in warmer parts of the Skeena watershed (Williamson 1998). Beyond a basic understanding of the migration timing and spawning, little is known about the biology of these high-elevation populations.

Chinook Salmon

Sustut watershed chinook escapement was carried out discontinuously from 1978 to the early 1990s, when considerable effort was put into adult enumerations at the counting fence located approximately 500 m upstream of Moosevale Creek. Since 1994, the aggregate of upper Sustut chinook spawners recorded at the Sustut mainstem lower fence has ranged from 570 to 1,639 with an annual average of 993 chinook. Escapement numbers are unavailable for 1996 and 1997. In most years the chinook count through the fence peaks in mid-August. The Sustut fence is usually put into service at the first of August and thus misses the earliest (July) chinook.

Chinook salmon migrate through the length of the Sustut mainstem. In general, adult chinook spawn in approximately 10 km of habitat that extends from 0.5 km downstream of the Moosevale Creek–Sustut River confluence to 1 km upstream of the Johanson Creek–Sustut River confluence and in the lower portion of Johanson Creek. The counting weir is within the

Asitka River partly dammed by beaver activity

chinook spawning area and may be affecting the pattern of habitat use. Scattered chinook spawning may also occur in Moosevale Creek (Atagi 1998).

The principal chinook juvenile abundance and habitat use studies in the upper Sustut were conducted by Williams et al. (1985), Shirvell and Anderson (1990b, 1991a, 1991b), Bustard (1994b), and Williamson (1998), using a variety of locations and techniques. Based on their inclined plane trap sampling, Williams et al. (1985) concluded that the upper Sustut River was not important overwintering habitat for juvenile chinook salmon. Shirvell and Anderson (1990b and 1991b) censused chinook juveniles with a nighttime drift diving technique they found more representative of actual abundance. Juvenile chinook salmon abundance was 1 fish/2.3 m of stream in 1990, while in 1991, they sampled 1 fish/1 m of stream close to the Johanson and Moosevale Creek–Sustut River confluences.

Williamson (1998) utilized a 1.5 m rotary trap and gee traps to determine relative abundance and movement of juvenile salmonids. Juvenile chinook density estimates for Moosevale Creek were $0.47/m^2$; for the Sustut mainstem they ranged between 0.48 and $2.78/m^2$. Williamson found considerable downstream migration of chinook fry in August and September. He suggested that the high juvenile chinook densities in the Sustut mainstem sites might be related to the substantial amounts of large and small woody debris present in those site reaches.

Pink Salmon

Pink salmon pass through the lower Sustut River to spawn in the Bear River. Pink salmon are apparently not present in all years of either the odd- or even-year lineages.

Sockeye Salmon

Three or four distinct sockeye stocks spawn in the upper Sustut River drainage. Sockeye rearing lakes include Asitka Lake, Sustut Lake, and the Johanson watershed lakes: Spawning and Johanson. Significant differences between protein allele frequencies of Sustut Lake and Johanson Lake sockeye were not detected by Rutherford et al. (1999), perhaps due to limited sample size. Microsatellite DNA separation is generally more sensitive than protein electrophoresis, and might show distinct stock status.

Bustard (1994b) reported that the majority (91.5%) of upper Sustut system sockeye smolts are age-1 fish, with the remainder (8.5%) age-2. The average fork length of the sockeye smolts was 76.9 mm for age-1, and 101.9 mm for age-2. Rutherford et al. (1999) reported higher abundance of two-year fresh water resident sockeye in Johanson and Sustut Lakes. In 1993, catch estimates for sockeye smolts using a 2.4 m rotary screw trap located upstream of the Moosevale-Sustut confluence suggest that the peak of sockeye smolt downstream movement occurred during the period from May 24 to June 9 (Bustard 1994b).

Escapements have been recorded since the Skeena River Investigations in the 1940s. Brett (1952) reported that the average escapement to the Sustut Lakes in 1946 and 1947 was approximately 5,000 sockeye. Aggregate sockeye stock counts are discontinuous from 1950 to the early 1990s, and population trends are difficult to distinguish. Since 1992, counts have been made each year at the Sustut River counting fence. The 14-year annual average sockeye run size is 1,895 as shown in Table 4.

Asitka Lake

The fence counts shown in Table 4 do not include Asitka Lake spawners, since the Asitka River branches from the Sustut below the counting weir. The annual escapement from visual estimates (20 years recorded since 1950) averages approximately 300 sockeye, with a range of 1 to 700. Spawning locations noted by Hancock et al. (1983) are the three largest bays within the lake to the south, southwest, and to the west. The DFO (1991d) and Smith and Lucop (1966) also noted spawning grounds in the upper section of river below the lake outlet.

Sustut Lake

Sustut sockeye escapement records date back to the late 1940s and then are intermittent till the early 1960s. From the early 1960s to early 1990s, the annual average escapement was 606 sockeye. Escapement data from the lower fence counts during the 1990s show the annual average to be at least double that number of spawners. The peak of the sockeye run passes through the fence in late August. Spawning is principally sustained in Sustut Lake in patchy sections along the southern shoreline. Foskett (1948) reported shallow water spawning along the eastern edge and at the southern end where Seepage Creek enters the lake through the gravel, as well as waters beneath 4.5 m in this region. Hancock et al. (1983) showed patchy spawning locations generally throughout both the east and west sides of the lake, except for the mid portion.

Johanson Lake

Johanson Lake sockeye escapement records are fairly consistent since the late 1950s. Spawners have ranged from 2 to 800, ranging most years from 250 to 300. Since 1990, a fence has been operated discontinuously on Johanson Creek, but these data are not available. Spawning locations have been reported both in the lake and

View across Sustut Lake

immediately downstream of the outlet (Foskett 1948, Hancock et al. 1983, DFO 1991d).

Coho Salmon

Coho juveniles and adults are relatively widely dispersed throughout the accessible upper Sustut tributaries. There are no known coho spawning grounds in the lower and middle Sustut mainstem, but it is likely that coho spawn in the many accessible low-gradient tributaries situated on the Sustut River valley floor and in back channels of the lower Sustut. Known spawning locations include the upper Asitka system, the upper Sustut mainstem, Moosevale Creek, and Johanson Creek.

Adult coho migrate into the lower Skeena River between late July and the end of September, with the annual peak in late August, as recorded by the Tyee Test Fishery. In general, most upper Skeena tributary coho arrive in September and early October in a protracted run that peaks in mid-September (Holtby et al. 1999). Coho are usually the last salmon to spawn in the fall, spawning from the end of September through December. Enumeration of coho spawners in streams is notoriously difficult, and true escapement numbers are commonly underestimated.

Asitka Creek sustains known coho spawning in the upper reach, possibly in the section above Asitka Lake, and most likely in mid reaches in the years that beaver dams obstruct fish passage. According to historic anecdotal Gitxsan and prospector reports, there are often large numbers of salmon spawning 15 km below Asitka Lake downstream of beaver dams. The DFO (2005) reported two escapements records of 33 and 20 coho spawners, but spawner populations are essentially unknown.

Moosevale Creek has two coho escapement records of 20 and 50 fish. Coho adults have been observed to 17 km. A known spawning ground is located 3.5 km upstream of the abandoned airfield on an unnamed tributary (DFO 1991d).

Two Johanson Creek escapement records show 10 and 18 coho in the 1990s, with the coho spawning in unknown locations in the mainstem and in Solo Creek. The Sustut mainstem sustains known coho spawning from downstream of the Moosevale confluence to

Table 4. Sustut River fence counts 1992–2005

Year	Sockeye	Coho	Steelhead	Chinook
1992	2,590	30	487	100
1993	2,169	18	476	199
1994	3,737	118	598	956
1995	523	24	658	
1996	3368	33	515	
1997	965	5	701	
1998	2,777	64	1,252	570
1999	221	30	896	609
2000	476	12	377	1,020
2001	1,258	9	769	1,639
2002	674	64	812	988
2003	4,992	119	1,104	1,106
2004	1,604	25	1,042	483
2005	1,175	88	271	383

Sustut Lake. Escapement records show five enumerations starting in 1960, with a range from 5 to 300 fish (DFO 2005).

Aggregate escapement records from the Sustut mainstem fence underestimate coho abundance because the weir is removed before the coho migration is over. The 14 years of data in Table 4 show the annual average is 45 coho with no apparent trend. Taking into account the 37 km of stream habitat and at least 20 km of lake margin above the fence, the carrying capacity of the system ranges from 1,000 to 1,500 fish with 9–13 females/km (Holtby et al. 1999). Holtby et al. (1999) described the fence aggregate escapement as less than 10% of the carrying capacity, or at a level that is consistent with other areas of the depressed upper Skeena.

Steelhead

The upper Sustut River drainage supports a summer-run steelhead population that enters the mouth of the Skeena early in comparison with other Skeena steelhead arrivals. The upper Sustut population is used as an index stock to monitor run strength of early migrating steelhead in the upper Skeena River (Parkin and Morton 1996). The upper Sustut steelhead are of particular concern to provincial fishery managers due to this early timing that coincides with the large mixed-stock fishery that harvests Skeena runs of sockeye and pink salmon. Improved access

Sustut River counting weir

to the Sustut River via the BC Rail line and the Omineca Mining Access Road has also raised concerns since the early 1970s, which led to an angling ban upstream of the Bear-Sustut confluence.

Steelhead populations, run timing, and spawning locations in the Sustut watershed are comparatively well known in relation to other Skeena subbasins. The counting fence upstream of the Moosevale-Sustut confluence is primarily intended to enumerate these steelhead. The counting fence also provides data on the size and sex distribution of steelhead, and the effect of the stream stage and temperature on migration (Parkin and Morton 1996). Diewert (2000) examined the number of gillnet-marked fish and the regression relationship between the Sustut steelhead index and the Tyee Test Fishery index.

Estimates of steelhead escapement are aggregates for the Sustut system above the counting fence. Escapement estimates have been continuous since 1992, with an annual average of 685 adult steelhead, with a range of 377 to 1,252 fish.

Steelhead arrive in the upper Sustut from early August through mid-October. They overwinter predominantly in Sustut and Johanson Lakes, particularly at their outlets (Chudyk 1972b, Spence et al. 1990, Bustard 1992b, Lough 1993). Lough (1993) reported that one radio tagged steelhead overwintered in the Sustut mainstem downstream of the Moosevale-Sustut confluence.

Steelhead spawn in May and June, coinciding with warming water temperatures and increasing streamflows. Bustard (1994b) indicated that spawning in the upper Sustut

River upstream of the Johanson-Sustut confluence likely peaked in the week of May 21, and was completed with most fish off the redds and holding in deep water by June 15.

The two principal spawning locations are the Sustut River below Sustut Lake and the mainstem of Johanson Creek (Bustard 1994a). Bustard noted that there is limited spawning in the mainstem below the Johanson-Sustut confluence, and that more spawning apparently occurs in upper Sustut River than in Johanson Creek. No other steelhead spawning locations are known on the Sustut mainstem. Bustard (1994a) suggested steelhead may spawn in side channels below the Bear River confluence and that suitable habitat exists on the north side of the Sustut River. However, northside tributaries may receive limited spawning activity due to fish passage concerns through BC Rail culverts and drainage structures. Gitxsan anecdotal reports state that steelhead spawn in Darb Creek and are present in Darb Lake during February and March.

Developing eggs remain in the gravel for 2.5 to 3 months through spring freshet, with steelhead fry emerging throughout August. Downstream migration appears to be widespread, to sites along the mainstem and to smaller tributaries that offer suitable refuge. Bustard (1992b) noted relatively low fry densities adjacent to steelhead spawning locations, though fry densities were high in the mainstem near Moosevale Creek.

Timing and abundance of the downstream migration of steelhead smolts have been studied by Williams et al. (1985) utilizing an inclined plane trap, as well as by Bustard (1994b), and by Dubeau and Johannes (1996), who used a rotary screw trap. The large numbers of parr and fry moving downstream in the upper Sustut mainstem likely indicate that preferred downstream habitats are important for growth and survival in this high-elevation nursery area. Scales collected over the last 20 years show that upper Sustut steelhead spend, on average, three years in freshwater prior to moving to the ocean (Baxter 1997b). Tautz et al. (1992) estimated the mean smolt age for upper Sustut steelhead to be 4.5 years, or about one year more than the direct scale estimates.

Bustard (1992a, 1993b, 1994b) sampled steelhead fry and parr densities in the upper mainstem and tributaries. Johanson Creek had high densities in 1993, but most tributaries downstream to Two Lake Creek had low densities.

Resident Freshwater Fish

With the exception of salmon and steelhead, information is scant in regard to fish in both fluvial and lacustrine habitats of the Sustut River drainage. Some of the drainage is poorly known and may contain populations of special interest or status that are presently unknown or undocumented. Ecological and life-history information that permits good conservation planning is simply not available.

Freshwater species and documented populations inhabiting the Sustut drainage include rainbow trout *(Oncorhynchus mykiss)*, bull trout *(Salvelinus confluentus)*, lake trout *(Salvelinus namaycush)*, Dolly Varden *(Salvelinus malma)*, mountain whitefish *(Prosopium williamsoni)*, longnose dace *(Rhinichthys cataractae)*, burbot *(Lota lota)*, redside shiner *(Richardsonius balteatus)*, peamouth chub *(Mylocheilus caurinus)*, and prickly sculpin *(Cottus asper)* (FISS 2003). Dolly Varden and, to a lesser extent, rainbow trout are the most widely dispersed species and are present in most fish-bearing waters.

Fisheries
First Nations Traditional Use

Gitxsan salmon fisheries within the Sustut River drainage were concentrated at two main locations on the Sustut mainstem: Gaps Ganeex and Wilna Guuk. Dispersed fishery camps targeting anadromous and freshwater fish were located close to the outlets of Johanson and Sustut Lakes, at Asitka Lake and River, and at the mainstem canyon section downstream from the Moosevale Creek–Sustut River confluence.

Gaps Ganeex is located approximately 1.75 km downstream of the Birdflat Creek–Sustut River confluence in the small canyon on the lower Sustut mainstem. This fishing and river-crossing site supported the ancient village of Anx Wil Djabas, which was located on the north bank of the Sustut River, west of Birdflat Creek. The canyon was crossed by a cantilevered suspension bridge that connected major Gitxsan travel trails on the north and south sides of Sustut River.

The ancient bridge crossing was recorded by O'Dwyer (1901) and discussed by Buckham (1950) in his Indian engineering research. Anx Wil Djabas was unfortunately destroyed by Keene Industries when they bladed the site for a camp during construction of the BC Rail line (Helmer and Mitchell 1972). The fishing site is currently designated as B.C. Archaeological Site HbSs-01, while the pine cambium foraging site with approximately 1,200 Culturally Modified Trees (CMTs) on the north bank of the Sustut River is designated B.C. Archaeological Site HbSs-04 (Norcan 2000).

Steelhead Canyon on the Sustut River is located about 6.5 km downstream of the Two Lake Creek–Sustut River confluence. This was an important salmon and steelhead fishing site. When families finished drying their groundhog meat in the late summer and fall, they fished at Steelhead Canyon and then dried fish for the winter. They also fished there while trapping in the winter (Abraham 1995). Abraham indicated that the mouth of Two Lake Creek is also a salmon fishing net site.

Gitmusundat, an ancient Gitxsan sockeye and steelhead fishing site, is located on the slough at the mouth of Willow Creek on the Sustut River. Abraham (1995) considered this site to be one of the best fishing spots on the river. The confluence of Johanson Creek and the Sustut River is an important fishing place, with

Bridge at Gap Ganeex, 1899

two traditional camps mostly utilized in the fall to fish steelhead.

Subsequent to the Department of Fisheries forcing the abandonment of weir and trap fishing in 1906, the fishery was primarily conducted with spears, and then later with gillnets. The salmon stocks spawning in the upper Sustut formed the principal food resource that enabled people to make this their home.

Dispersed traditional fishing sites in the upper Sustut drainage included the outlets of Johanson and Sustut Lakes, Asitka Lake and River, and the mainstem canyon section downstream of the Moosevale Creek–Sustut River confluence. Oral histories recount fishing at Johanson and Sustut Lakes. Foskett (1948) reported in the course of fieldwork on the upper Sustut sockeye nursery lakes that a family of 10 Indians took chiefly coho and steelhead with both spears and gillnets in Johanson Creek just below Johanson Lake.

Gitxsan maintained a camp on the west bank at the outlet of Sustut Lake. This area has recently been the site of a mineral exploration camp. Foskett (1948) noted that Indians gillnetted steelhead during the winter at this site. He also noted that sockeye and coho were fished with spears by Indians in Asitka Lake and downstream in the Asitka River when beaver dams slowed fish passage. Foskett stated that shifting traplines and a decrease in sockeye runs to Asitka Lake largely account for the abandonment of fishing in this area.

Recreational Fisheries

The high fisheries value in the lower Sustut, in combination with the high natural resource values – wildlife, remoteness, water quality, and beauty of the landscapes – results in uncrowded fishing conditions and exceptional fishing. There is no fishing permitted on the Sustut River or its tributaries above the BC Rail bridge at the mouth of the Bear River. Due to the remoteness and the necessity to access the lower Sustut by air, the majority of anglers are guided clients. Currently two angling guide lodges operate on the lower Sustut River: the Suskeena Lodge and the Steelhead Valhalla Lodge. Recent catch per unit effort has been below one fish per angler day. The majority of anglers are non-Canadians, and

this trend is expected to continue. The Sustut River is designated Class I waters September 1 to October 31, and a steelhead stamp is mandatory during this period. Regulations require mandatory release July 1 to December 31, and general restrictions include no fishing in the upper Skeena drainage January 1 to June 15 (B.C. Fisheries 2005).

Development Activities

Land-use and development activities in the Sustut River drainage are limited to the lower and upper portions of the drainage. The principal land-use activity is logging, which is concentrated in the lower Sustut. Mineral exploration is focused in the Hogem Ranges swest of Sustut Lake. Linear development and settlement do not exist. Transportation has been the key to growth in recent land-use and resource development.

Forest Resource Development

The Sustut River drainage is located within the Fort St. James Forest District. Within the Fort St. James Land and Resource Management Plan (LRMP) area, general resource development is zoned for the valley bottoms. Special resource zoning to allow for scenic, wildlife, and recreation values is applied to higher elevations where commercially viable timber opportunities are limited (Ministry of Forests 1999). The Fort St. James LRMP essentially zones the Sustut River drainage as multi-value, which is essentially a pro-timber development policy.

Large-scale forest exploitation did not occur until the early 1990s, when the Sustut area forests were optioned off to various licensees to cut 900,00m³/year on a 20-year nonreplaceable license condition. This forestry development was conceived to mitigate the effects of overcutting in the Prince George Forest District. Since the early 1990s, forest development has high-graded the economically valuable pine stands on the gently sloping, north side of the lower Sustut River, with timber being shipped on the BC Rail line to Fort St. James and Prince George. The Sustut mainstem downstream of Birdflat Creek was recently bridged, and highgrading activities on the south side commenced.

The BC Rail line that served logging as far north as Minaret is being kept open at least for the short term. Track is being removed from areas between Minaret and Mosque Creek. The Northern Long Term Road Corridors Plan is a coordinated access management plan initiated by Fort St. James Forest District (Ministry of Forests 2000b). The intent of the plan is to accommodate timber resource transportation routes and generally determine the direction of timber flow. Other than Saiya Creek and the lower Asitka River drainage, roads are planned from the lower Sustut to the upper Sustut, as well as access into all tributaries with commercially viable timber.

Mineral Resource Development

Exploration and utilization of rocks, crystals, and minerals within the Sustut drainage have been ongoing since the Gitxsan established themselves many thousands of years ago. The Sustut drainage is unique in that it provided Gitxsan people with native copper, probably from the area of the current Sustut Copper prospect.

Euro-Canadian explorations for base and precious minerals have been ongoing since the turn of the 20th century and were focused mainly on high-grade silver and gold veins. Mining exploration slowed down with the onset of the Depression. In the early 1960s, mineral exploration focused increasingly on porphyry copper-molybdenum deposits.

There are 52 mineral showings in the Sustut drainage, including two developed prospects: Sustut Copper and Sustut Coal. The greatest densities of mineral occurrences are located in the upper Sustut, upstream of Red Creek and the Asitka River. These occurrences are largely silver, gold, and copper polymetallic vein-type deposits. Sustut Copper, west of Sustut Lake, owned by Doublestar Resources is a volcanic massive sulfide-type deposit that has drill-indicated reserves greater than 54.4 million tons at 1.25% copper (Doublestar 2001). The deposit geology is described by Harper (1977) and Wilton and Sinclair (1988).

Evaluation of metallic mineral potential for the watershed by the B.C. Ministry of Energy, Mines and Petroleum Resources is extremely

Rapids on lower Sustut River

generalized. The area has been delineated into mineral tracts on the basis of geology and distribution of known mineral occurrences. Mineral assessments of the upper Sustut drainage show high metallic mineral values, though there is a narrow area of land extending from Moosevale Creek southward to Sustut Lake, where metallic mineral values are classified as low (Ministry of Energy, Mines and Petroleum Resources 2002).

Evenchick et al. (2002) suggested that the geological framework and initial petroleum resource assessment indicate significant undiscovered conventional petroleum potential in the Bowser and Sustut basins, part of which underlie the Sustut drainage. Presently, this potential is based on a generally sparse data set, and additional work is needed to evaluate the thermal and diagenetic history of potential systems.

Transportation

The Sustut drainage has limited transportation infrastructure consisting of the BC Rail Dease Lake extension and the Omineca Mining Access Road (OMAR). In 1970, construction of BC Rail's

Road bridge crossing the Sustut River, downstream of Birdflat Creek

Dease Lake extension reached the watershed. The railway construction resulted in severe environmental impacts to fish, wildlife, and water quality, due to irregularities in design and construction practices. The construction also damaged the nonrenewable Gitxsan cultural heritage resources on the north bank of the Sustut River. The marginal drainage structures on tributary creek crossings have caused past and present fish passage problems.

In 1977, BC Rail and the province abandoned construction of the line indefinitely. In the early 1990s the southern portion of the rail line was put into limited service for log transport from the Minaret and Birdflat Creeks area. BC Rail presently hauls about 200,000 m³/year of timber to Fort St. James and Prince George. BC Rail's contract with the timber companies requires a minimum freight volume and extended until 2007. The rail line is shut down for spring break-up, from the third week of March to the third week of June. An upgrade

to allow year-round service is estimated to cost approximately $22 million. In 2007, most of the easily accessible timber had been harvested and all forestry activities were moved southward to concentrate on the pine beetle salvage operations on Nechako Plateau.

Vehicle access to the upper Sustut drainage gradually developed as the Omineca Mining Access Road (OMAR) was extended and used over the years. The first rough road approaching the area was the Omineca Road from Fort St. James to Germanson Landing. The road was extended to Aiken Lake in 1959–1962. In 1970–71, Falconbridge extended the road to Moose Valley in the Findlay drainage. In 1986, Cheni Mines extended the road to Sturdee Valley and Lawyers Pass. The southern section of the road, south of Moose Valley, was opened to the public in 1986.

The Kemess-South gold project is located 37 km north of the Moosevale Creek–Sustut River confluence. The project is an open pit gold-copper mine near lower Kemess Creek, a tributary of Attichika Creek, which flows into the east shore of Thutade Lake. In 1993, El Condor Resources proposed to build the 60 km Sloane Connector, which was to connect the OMAR with BC Rail's Dease Lake extension.

In 2001, the District of Stewart, B.C., proposed the Stewart Omineca Resource Road (SORR) to provide road access to the Kemess Mine and divert the flow of ore to Stewart. Little actual development took place in the ensuing five years, and the future of the proposal is in doubt.

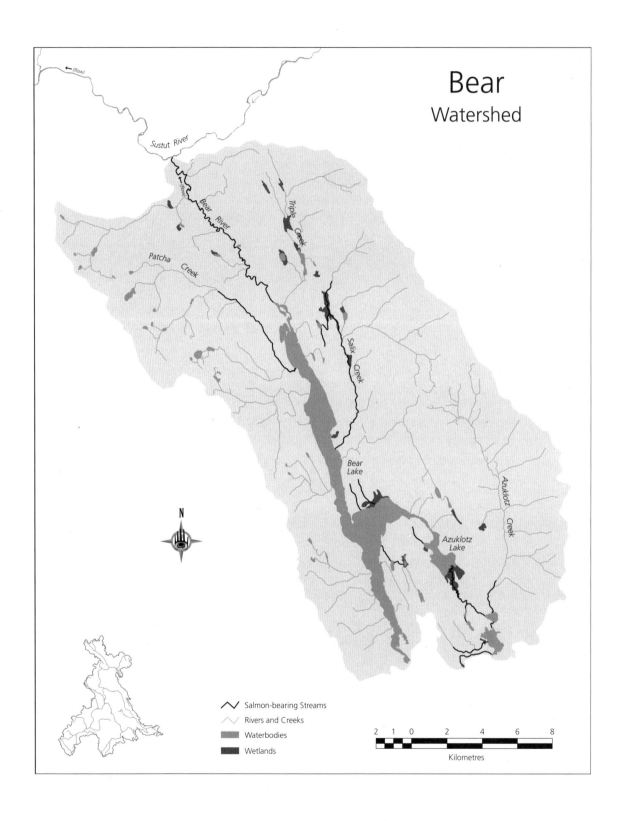

Bear
Watershed

Sustut River

(flow)

(flow)

Bear River

Triple Creek

Patcha Creek

Salix Creek

Bear Lake

Azuklotz Lake

Azuklotz Creek

N

Salmon-bearing Streams
Rivers and Creeks
Waterbodies
Wetlands

2 1 0   2   4   6   8
Kilometres

18

Bear Watershed

Environmental Setting

Location

The Bear River watershed is located in the northeastern headwaters of the Skeena River in central British Columbia. It is bounded to the south by the Driftwood River basin, to the north by the Sustut River drainage, to the east by the Connelly Range, and to the west by the southeastern spur of the Skeena Mountains.

Hydrology

The Bear watershed is a fifth-order system with a catchment area of approximately 452 km². Elevations range from Peteyaz Peak at 2,241 m to 690 m at the Bear-Sustut confluence. The Bear River drains 10 km northwesterly into the Sustut River, which then discharges westerly into the upper Skeena River. Tributaries flowing into the Bear River are minor, with the exception of Patcha, Salix, and Azuklotz Creeks. Patcha Creek, the largest, is the only one that carries glacial silt.

Maximum freshet discharges occur in mid-June corresponding with high-elevation snowmelt; flows then decrease until September, when fall rains and early snowmelt increase streamflows through October. Summer low flows are typically four to eight times greater than winter streamflows and are principally sustained by high-elevation snowmelt draining from the Skeena Mountains. Streamflows decrease through November and December, when precipitation falls as snow. Stable winter low flows are derived from groundwater, lakes, and unfrozen wetlands. There is no history of hydrometric stations in the watershed. The watershed has a sub-boreal climate with relatively low precipitation, warm summers, and long, cold winters with a heavy snowpack.

The dominant hydrological feature in the watershed is Bear Lake, which regulates river flows and levels. The lake lies at 779 m elevation, is 19 km in length, and averages 1 km in width, although it is 4.5 km in width at Tsaytut Bay. Two deep basins at either end of the lake are separated by a shallow section. Limnological data collected in 1978 and 1995 indicate the lake has a mean depth of 14.8 m, an average summer thermocline depth of 7.5 m, a mean surface temperature of 14.6° C, an 8.8 m euphotic zone depth, and a substantial, cool hypolimnion. The average photosynthetic rates and average macrozooplankton biomass in Bear Lake are among the highest in the Skeena system (Shortreed et al. 2001).

Stream Channels

The Bear River is 10 km in length and exhibits a single-thread channel with an irregular, sinuous channel pattern. It is generally low gradient; however, the lower reach and the upper 1.6 km of the river are steeper in gradient (Shepherd 1975). Pink salmon access upstream is restricted by one obstruction near the top of the river. Williams et al. (1985) delineated the Bear River into two reaches. The lower section of reach 1 is steeper than the upper section and is channelized, with a boulder substrate and little cover other than overhanging streamside vegetation. The upper section of reach 1 is characterized by slow, meandering, long, wide runs punctuated by relatively deep pools, with substrate composed primarily of gravel and some cobbles. In addition to numerous beaver ponds, side channels, oxbows, and pools, stream complexity is also provided by logjams, undercut banks, and abundant streamside vegetation (Williams et al. 1985). Reach 2 has less than 1% gradient and is channelized, with predominantly boulder substrate at its upper and lower ends. The middle 1.5 km of reach 2 consists of wide, slow runs with gravel substrate and minor amounts of log debris.

View downstream from upper reach 1 of the Bear River

Geography

The Bear watershed is comprised of two mountain masses split by a trench-like fault valley. The valley is located along a major crustal break that separates the Stikine terrane on the west from the Quesnel terrane on the east. This fault zone extends for several hundred kilometres southward past Takla Lake and Stuart Lake. The mountains to the west, part of the Skeena Mountains, are composed of the middle Jurassic Hazelton Group volcanic and sedimentary rocks. The Connelly Range to the east is comprised of upper Cretaceous Sustut Group sedimentary rocks, presumably overlying Asitka Group volcanic rocks at depth. The Thumb, a dramatic tower located at the north end of the Connelly Range, is formed by a Tertiary volcanic neck. Dispersed small stocks of Tertiary age Kastberg granitic intrusions are evident surrounding Bear Lake (Lord 1948) and are responsible for the mineralization on Tsaytut Spur to the west.

The fluvial and surficial geomorphology is strongly influenced by its recent glacial history. Ice moving southward eroded the Bear Lake valley and scoured all slopes. Thick blankets of glacial till cover the main valley and mountain valleys and extend up the valley sidewalls, though the surface expression conforms generally to the underlying bedrock surface, with bedrock exposure along deeply incised streams and on steep-sided hillslopes (Holland 1976). During deglaciation, streamflow was to the south into the Takla Lake basin, the Stuart River, and then the Nechako River.

The predominant biogeoclimatic zone, Sub-boreal Spruce (SBS), covers most of the lowland coniferous forests in the watershed. Subalpine fir, lodgepole pine, and hybrid spruce are the major tree species; subalpine fir tends to dominate older, high-elevation stands and moister sections of the zone. Due to the pervasive natural and aboriginal fire history, lodgepole pine and deciduous seral stands are extensive, particularly along stream terraces and on southern aspect slopes. Nonforested wetlands occur in the morainal landscape depressions, while dry grass/shrub meadows, though limited, are present on dry sites with favourable warm aspects (Meidinger and Pojar 1991).

The SBS zone merges into the Engelmann Spruce – Subalpine Fir (ESSF) zone at mid elevations ranging from 900 to 1,300 m, depending on the local topography and microclimate. A shorter, cooler, and moister growing season is found in the ESSF zone with continuous forests passing into subalpine parkland at higher elevations. Subalpine fir is the dominant tree, with lesser amounts of lodgepole pine and white spruce hybrids in drier or fire-influenced areas (Meidinger and Pojar 1991).

Fish Values

The Bear River watershed is a relatively small, but biologically rich river system that has high-value fish habitat. Fish species utilizing this habitat include sockeye, kokanee, coho, pink, and chinook salmon, steelhead, rainbow trout, Dolly Varden, bull trout, lake char, burbot, and lake and mountain whitefish. Hatlevik (1999) reported a very large bull trout population in the Bear River.

Chinook Salmon

Bear River chinook salmon are one of the largest chinook populations in the Skeena system. It is estimated that as much as 85% of the Sustut chinook stock spawn in the Bear River. The Bear River has exceptionally high chinook spawning densities, which are probably the maximum achievable by this species (Shervill and Anderson 1990a). Four field studies have documented chinook in the Bear River: the Skeena River Salmon Investigation conducted between 1944 and 1948 (Foskett 1948), a counting fence operation in 1972 (unknown reference), juvenile salmonid studies in 1984 (Williams et al 1985), and Shirvell and Anderson's (1990a) chinook salmon and habitat study.

Chinook escapements appear to have been high in the 1950s (average 18,750) and considerably lower since, except for a recovery in the 1990s (average 11,300). More recent escapements (2000 to 2005) have averaged 5,820. In the 1950s, the Bear River chinook escapement was the largest component in the Skeena watershed. At present, the Bear River is third or fourth in chinook abundance behind the Kitsumkalum and Morice and perhaps the Kispiox Rivers.

Chinook salmon enter the Bear River throughout August with peak spawning in early September. Concentrated spawning occurs from 2–3 km downstream of the lake outlet adjacent to the airstrip, which has excellent gravel and many dunes resulting from redd construction (Shepherd 1976, Williams et al. 1985). The remainder of chinook spawning takes place from this locality downstream to 2 km above the Sustut confluence. Shepherd (1975) reported that the Bear River chinook spawners had smaller body size than fish from other similar age Skeena chinook runs, and that male fish were significantly smaller than females.

Peak emergence of chinook fry occurs during mid-June. Williams et al. (1985) trapped migrating fry with an inclined plane trap and found that most fry migrate downstream after July upon reaching a threshold size of 50 mm. Shervill and Anderson (1990a) sampled one juvenile chinook per 1.4 m of stream in Bear

View eastward to the Thumb

Bear River falls 3.5 km downstream of the lake, 1946

River and suspected that the river is not an important chinook overwintering area. Their sampling methods utilized stream walk visual observations, daytime drift diving, electrofishing, night-time drift diving, and minnow trapping.

Pink Salmon

Bear River pink salmon account for a few percent of the Skeena system total escapement. Odd-year escapement is the dominant cohort with an annual mean escapement of 46,600 adult pink salmon in a range of 200 to 500,000. Even-year pink annual escapements have ranged from 700 to many years of no returns. From 2000 to 2005, there was a single count in 2001 of 200 pinks.

Pink salmon generally enter the Bear River in mid-August with peak spawning typically occurring in late August to early September. The principal spawning grounds are scattered in the lower and middle river, and pink salmon seldom seem to pass the falls 3.5 km below the Bear Lake outlet. Emergence from gravels is coincident with ice break-up. Peak emergence levels occur in early May, followed by seaward migration (Williams et al. 1985).

Chum Salmon

Adult chum and fry have been reported in the Bear River, although at very low densities (Williams et al. 1985, Shirvell and Anderson 1990a).

Sockeye Salmon

Sockeye abundance in the Bear system was historically greater than shown by the last 50 years of escapement data. To facilitate estimation of the sockeye salmon run to Bear Lake, a fence was constructed across the head of Bear River during 1947 and 1948. Direct counts, combined with recaptures from a fence-tagging program, demonstrated that a high proportion of the returning sockeye salmon spawned in the lake (Fisheries Research Board 1947, 1948). Brett (1952) estimated the number of Bear Lake spawning sockeye at 42,000, with only Azuklotz Creek supporting a >1,000 fish stream spawning population. Bear Lake sockeye spawners decreased greatly in the 1950s and have not recovered. Since 1950, Azuklotz Creek has had variable escapement with no clear trends till the mid-1980s, when escapement increased to above pre-1950 levels. At least some of the Azuklotz fry rearing takes place in Azuklotz Lake, which is separated from Bear Lake by a low-gradient stream channel a few hundred metres long. Annual average escapement from 2000 to 2005 was 2,012 sockeye with a range between 486 and 3,630, representing a decreasing trend from the 1990s.

Sockeye adults typically return to Bear Lake in mid to late August, with peak spawning in mid-September. In the past, Bear Lake spawners used various beach and deep-water grounds scattered along the western lakeshore. Upper and lower Azuklotz Creek are now the principal spawning grounds, and Salix Creek supports minor numbers of spawners when flow conditions are high enough to permit entrance.

Fry emergence is followed by a one-year lake residency (Rutherford et al. 1999). The 1995 sampling program found sockeye fry stomachs were 60% full and contained mostly *Daphnia* and *Heterocope*, both large and preferred food items. Although fry densities were relatively low with a mean of 132 fry/ha, mean late summer weight was 3.9 g, which is above average for sockeye nursery lake fry in the Skeena system (Shortreed et al. 1998).

Coho Salmon

Knowledge about coho escapement numbers and spawning distribution is limited. Coho arrive in the Bear River generally throughout September and head to their spawning grounds. Scattered spawning occurs in the Bear River and in tributaries feeding Bear Lake, which include the lower reaches of Salix Creek, the unnamed creek across the lake from Salix Creek, Azuklotz Creek, and the unnamed tributary flowing into lower Azuklotz Creek (Finnegan 2002). There were serious conservation concerns with Bear system coho in the late 1990s, due to very few coho adult spawners. The aggregate escapements have steadily increased; since 2000 the average annual counts show 2,194 from a range of 851 to 4,636 returning adults. Williams (1985) reported that peak fry emergence was in mid-April, with another larger peak in mid-June and coho smolts migrating down the Bear River in May.

Steelhead

The Bear watershed supports one of the large summer-run steelhead populations in the Skeena watershed. The Bear River is well known for its large and abundant steelhead. Once steelhead that are bound for the Bear River enter the lower Skeena River, it is estimated that it takes about one month to reach the Bear River (Lough 1980, 1981). Bear River steelhead and lower Sustut River steelhead are grouped into a distinct subpopulation due to run timing and unique life-histories (Baxter 1997b). Bear Lake is a known overwintering area (Chudyk 1972a), and it has been suggested that steelhead also overwinter in Sapolio Lake (Bustard 1993c) as well as in the Sustut River mainstem at the Bear confluence (Spence et al. 1990).

Bustard (1993c) estimated steelhead started spawning in the Bear River in mid-May, peaked in late May, and ended in early June. Chudyk (1972a) reported that 3,000 adult steelhead were spawning in the spring salmon spawning "ridge" on the Bear River. Turnbull reported 700 steelhead spawners in late May 1989 in the mid reach where chinook spawn (B.C. MoE). Kelts are thought to migrate seaward immediately following spawning (Beere 2002).

Bustard (1993c) estimated that in the Bear River, fry emergence was July 24 to August 5, with a peak around July 30. Williams et al. (1985) found that steelhead fry moved downstream into the lower Sustut River; they suggested that fry production in the Bear River was critical for seeding Sustut mainstem habitat. Steelhead smolt migration peaked at the beginning of May, before the onset of the freshet (Williams et al. 1985).

Fisheries
First Nations Traditional Use

Traditionally, First Nation peoples from Wil Dahl' Ax (Fort Connelly) occupied and used the Bear watershed. It is generally thought that pre-contact Bear Lake, by the nature of its location, was peripheral to Gitxsan, Sekanni, and Carrier territories, and may have been used, to some extent, by all of these nations at different times. Economically, the Gitxsan claimed exclusive trading privileges with the Carrier, Babine, and Sekanni, and ascended inland as far as Bear Lake to trade (Jenness 1937, 1943).

The strong chinook run was the basis of a vigorous fishery as evidenced by the 10 recorded fishing and processing sites (Rabnett et al. 2001). The abundant and predictable sockeye salmon stocks provided the aboriginal fishery at Bear Lake with opportunity to harvest and preserve a large amount of high-quality food in a relatively short time of intensive effort. The sockeye served to sustain the Gitxsan throughout the year and produce a trade item. Following the passage of the bulk of the sockeye, coho were available well into the autumn, providing both fresh and dried fish. Rainbow trout, steelhead, lake trout, and Dolly Varden char were also fished in the lake.

Although various fisheries occurred at the Bear River and Bear Lake, a large fishing effort was expended at the weir immediately below the Bear Lake village, also known as Wil Dahl' Ax. Posts were pounded into the river bottom and overlaid with panels secured on the upstream side. These often supported a walkway across

Bear Lake Indians pulling strings of chinook upstream, 1946

the top, enabling access to barrel-type traps. These traps were fitted with a movable panel through which fish could be dipped, gaffed out, or released, depending on whether the species was desired. Patrick (2001) noted that one trap in the fence provided enough fish for everybody. The only other known weir location was at the shallows between Azuklotz and Bear Lakes; this fence also used only one trap (Patrick 2001). Wilna Guuk is located on the south bank at the Bear River–Sustut River confluence and lies primarily on the west side of the Bear River. This ancient fishing village exploited sockeye, chinook, coho, and steelhead from both the Bear and Sustut Rivers. Bear River Indian Reserve No. 3 was laid out to retain the ancient village site.

Recreational Fisheries

The recreational fishery in the watershed is limited by access, and a no fishing regulation applies to the Bear River. Access for unguided and guided sport anglers is usually by aircraft or helicopter. A fishing lodge located near the northwest corner of the lake, provides clients with game fish opportunities on Bear Lake and day trips to other nearby popular lakes, such as Babine Lake.

Development Activities

Development activities are proposed forest harvesting and expansion of the existing transportation infrastructure. Mineral occurrences do exist, although no properties have been developed. The Bear Lake village is occupied year round in most years.

Forest Resource Development

From 2003 to 2006, logging activities were concentrated on the 6 km long valley terrace, west of and upstream of the Sustut-Bear confluence. These activities resulted in the majority of pine forests being cut and roaded.

Transportation and Utilities

In 1970, construction of BC Rail's Dease Lake extension reached the watershed. The railway construction resulted in severe environmental impacts to fish, wildlife, and water quality due to irregularities in design and construction practices. Portions of the roadbed were constructed relatively close to sections of the Bear River, where plastic glacial-lake deposits of stiff clay and silt underlie sand and gravel terraces. Consequently, earth slides and rotational failures into the river caused massive sediment deposition. The marginal drainage structures on tributary creek crossings, as well as deranged subsurface flows and seepage, caused more failures and sediment problems over a number of years. In 1977, BC Rail and the province abandoned construction of the line indefinitely. In the early 1990s the railroad was completed almost to the mouth of the Sustut River. In the past few years, the railroad has been used to transport logs out of the Minaret and Birdflat Creek area.

Upper Skeena
Headwaters

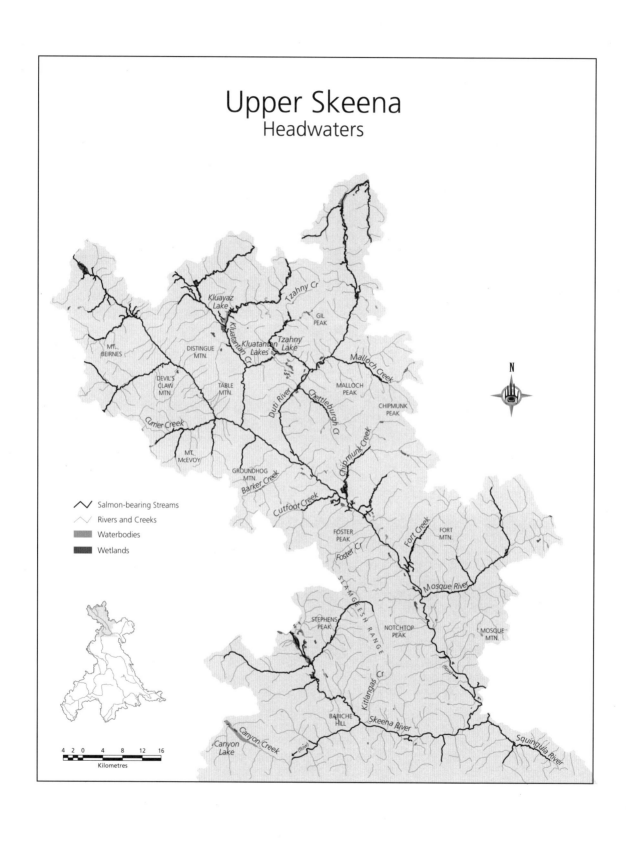

Salmon-bearing Streams
Rivers and Creeks
Waterbodies
Wetlands

Kilometres
4 2 0   4   8   12   16

N

Kluayaz Lake
Tzahny Cr
GIL PEAK
MT. BEIRNES
DISTINGUE MTN.
Kluatantan Cr
Kluatantan Lakes
Tzahny Lake
Malloch Creek
DEVIL'S CLAW MTN.
TABLE MTN.
Duti River
Chettleburgh Cr
MALLOCH PEAK
CHIPMUNK PEAK
Currier Creek
MT. McEVOY
GROUNDHOG MTN.
Barker Creek
Chipmunk Creek
Cutfoot Creek
FOSTER PEAK
Foster Cr
Fort Creek
FORT MTN.
Mosque River
STEPHENS PEAK
STAMGEESH RANGE
NOTCHTOP PEAK
MOSQUE MTN.
Kitlangas Cr
(flow)
BABICHE HILL
Skeena River
Canyon Creek
Canyon Lake
(flow)
Squingula River

19

Upper Skeena Headwaters

Environmental Setting

The upper Skeena headwaters area is defined as encompassing the Skeena River and its tributaries from the Sustut River confluence to the upper headwaters. The upper Skeena River is one of the major river systems in the high interior zone of the Skeena River watershed.

Location

The upper Skeena headwaters area is located in north-central British Columbia approximately 200 km north of Smithers. The Sustut-Skeena confluence is 466 km upstream of the mouth of the Skeena River. The drainage is bounded to the north by the Spatsizi Plateau, to the south and west by the Slamgeesh and Groundhog Ranges, and to the east by the Tatlatui Range. The upper Skeena drainage forms the northwestern headwaters of the Skeena watershed.

Hydrology

The upper Skeena headwaters area is mostly mountainous and relatively high in elevation, ranging from 580 m at the Skeena-Sustut River confluence to 2,375 m at Melanistic Peak in the Tatlatui Range. The Skeena River flows southeast bisecting the Skeena Mountains. The Skeena River mainstem in this portion is approximately 133 km long. The broad and flat divide between the Spatsizi River (tributary to the Stikine River) and the Skeena River is within a single valley at 1,372 m elevation.

Major north-slope tributary streams within the drainage basin include the Mosque River, Chipmunk Creek, the Duti River, and the Kluatantan River, which flow into the Skeena left bank. Major right-bank tributaries include Foster, Cutfoot, Barker, Currier, and Beirnes Creeks. Most of the small first- and second-order tributaries flowing into the Skeena River are short and steep.

Skeena River discharges usually peak in late May or June due to spring snowmelt. In almost all years, the peak spring freshets are the largest floods in the mainstem. Water levels in the main river channels and back channels fluctuate seasonally; typically they are high from late May to early July, drop for the summer months, rise to intermediate levels in the fall, and reach their annual low levels late in the winter season. There are no hydrometric stations located within the upper Skeena drainage. The surrounding glaciated mountains help to maintain moderate summer streamflows. Streams originating from glaciers, particularly in the Kluatantan and Duti watersheds, produce moderate amounts of natural sediment that contribute to the downstream wash load.

The numerous ranges of the Skeena Mountains located within the watershed exert major hydrological influences, with tributary streamflows having a moderately high response from water input due to the high relief and steep channel gradients of the tributaries. This effect is slightly dampened upstream of the Kluatantan River, where extensive wetlands provide water storage and valley bottom gradients are low. Water storage in lakes is relatively small, with the only notable lakes being North and South Duti Lakes, Kluayaz Lake, Kluatantan Lakes, and Tzahny Lake.

The general climate of the watershed is on the dry end of the transition between the moist, coastal climate and the colder, continental climate that characterizes the interior of the province, particularly at elevation. Summers are warm and fairly moist with a short growing season. Snowpacks are moderately dry and vary in the valley bottom from 1.5–2.5 m in depth (Wilson and Marsh 1975). Slight relief and the U-shaped valley bottom profile lead to prevalent cold air ponding. There are no long-term weather records available.

Stream Channels

Other than the channels that are entrenched in bedrock, streams of the major tributaries have alluvial channels and generally occupy active floodplains. Other than the construction activities of BC Rail's Dease Lake extension, the upper Skeena drainage has not received development. Channel bedload movement from natural sources is common.

The BC Rail construction practices caused a number of stream bank failures and sediment contributions to the mainstem. Gates and Reid (1985) undertook a reconnaissance-level investigation to determine the extent of natural repair over the preceding decade. They noted continuing chronic impacts related to the rail grade. Within the drainage, the main BC Rail construction problem was apparently the undersizing of culverts and drainage structures resulting in most culverts failing and general rail grade erosion. Several deep-fill culverts with massive fills experienced failures, contributing short-term sediment and creating chronic siltation problems. Magdanz (1975) noted unnecessary erosion on the many access roads caused by poor ditching techniques and the marginal drainage structure installations.

Skeena Mainstem

The overall gradient of the upper Skeena mainstem from the Sustut River confluence to the headwaters is 0.5%; however, there are limited sections with gradients up to approximately 3.5%. Reach 12 is the section of the Skeena River from the Sustut-Skeena confluence to 65.5 km upstream. The channel

Kluayaz Lake

has an average gradient of 0.5% and no obstructions to anadromous fish passage over its length during moderate or high flows (Resource Analysis Branch 1975–79).

A partially eroded rock sill across the Skeena mainstem just north of Chipmunk Creek may be a partial barrier to fish at low water. This reach of the Skeena River is characterized as a single-thread, slightly sinuous channel, bounded by wide gravel terraces or by the valley walls. Occasional islands and point bars characterize the channel with cobbles and gravels dominating the substrate. The floodplain varies in width from 0 to 0.5 km, though broad areas up to 1.5 km in width are evident at the mouth of Chipmunk Creek and other tributary mouths.

Reach 13 extends 16 km in length upstream to the Kluatantan River. In this reach, the valley is narrow, with the channel generally confined within the valley walls and controlled by occasional bedrock outcrops or incised into the bedrock or reworked till. The single channel is irregularly sinuous, with occasional rapids and an unknown gradient. Reach 13 receives the flow from many small first-, second-, and third-order tributary streams that are short and steep or possess steep gradients at the main valley slope break.

Reach 14 extends 29.3 km in length upstream to the mouth of Ethel Creek, and the gradient is 0.1–0.5%. The channel is characterized as a wandering gravel-bed river that is confined by the valley walls or high bench terraces. The floodplain varies in width from 0 to 500 m, and the gradient is unknown.

Reach 15 extends 22 km upstream to the headwaters and the Spatsizi River divide. The reach is irregularly sinuous, wandering across a floodplain that ranges from 0.1 to 1.5 km in width and contains extensive side channels and wetlands. The channel is noticeably smaller in width and depth upstream of Porky Creek.

Mosque River

The Mosque River extends 42.4 km upstream into the alpine, and is divided into five reaches. Reach 1 extends 7 km, is stepped in profile with many rapids, and is mostly entrenched between distinct glacial terraces. A canyon at 3.8 km extends approximately 0.5 km, and fish passage

Lower Mosque valley

is unknown. The BC Rail bridge (Mile 140) crossing this reach near the mouth has persistent debris accumulations that pose a hazard to abutments at high-water flows. Reach 2 extends 11.3 km with a floodplain that is 0–0.5 km wide and constricted by the valley walls. Numerous bank and valley wall failures are scattered along the reach, as are debris accumulation zones and persistent beaver dam complexes.

Reach 3 extends 3.9 km and is characterized by a stepped profile that is entrenched in the valley walls. The 300 m long canyon at 75.9 km lies downstream of the only section of floodplain. Reach 4 extends 4.1 km with a smooth gradient and a floodplain that averages 120 m in width and is controlled by steep mountain slopes. Reach 5 extends into the alpine through extensive, high-elevation wetlands that receive avalanche debris and sediment.

Duti River

The Duti River mainstem extends 63 km to drain a large mountainous area and the headwater lakes: North and South Duti Lakes. The three major tributaries include Chettleburgh Creek, 15 km in length; Tzahny Creek, 18 km in length that drains Tzahny Lake; and Malloch Creek, 20 km in length.

Reach 1 of Duti River extends 14.5 km in a single-thread, smooth profile channel with a relatively moderate gradient that is bordered by gravel terraces. Reach 2 extends 8.5 km to the mouth of Tzahny Creek; the nearly straight channel lies in a floodplain that ranges from 0 to 600 m in width and contains substantial wetlands. Reach 2 receives the discharges

from Chettleburgh and Tzahny Creeks. Reach 3 extends 6.8 km upstream of the mouth of Malloch Creek with a series of falls that are probably impassable to all fish species. The reach has a floodplain that is on average 1 km wide with sections controlled by the valley walls and ridges. The irregular stream course winds through extensive wetlands.

Kluatantan River

The Kluatantan River extends 61.6 km upstream to drain a large mountainous area and approximately a dozen small ice fields in its headwaters. Historically, the Kluatantan was known as the eastern fork of the Skeena (Malloch 1912). The major tributaries include Tantan Creek, which drains the upper and lower Kluatantan Lakes; Kluayaz Creek, which drains Kluayaz Lake; and Campbell Johnston Creek.

Reach 1 of the Kluatantan River extends 11.25 km upstream to the mouth of Lonesome Creek (1.75 km downstream of Tantan Creek). It has a slightly sinuous, single-thread channel with a stepped profile and is entrenched within a narrow valley. A relatively long (3 km) canyon area culminates downstream with a 0.4 km series of chutes approximately 1 km upstream of the Skeena confluence. Tributary streams to the east and west are mostly short and steep.

Reach 2 extends 9.1 km in a single-thread channel with a smooth profile and a floodplain that ranges from 0 to 500 m wide and is constrained by the valley side slopes. Reach 2 receives the discharge from the Tantan Creek drainage, which includes the Kluatantan Lakes. Reach 3 extends 7.6 km with a smooth channel profile. The valley bottom, which extends eastward into the Kluayaz drainage, is broad and flat with substantial wetlands. Reach 3 receives the discharge from the Kluayaz Creek drainage, which includes Kluayaz Lake.

Reach 4 extends 8.7 km with a gently sloping profile. The valley walls limit channel movement across a floodplain that averages 1 km wide. This reach has extensive wetlands and side channels amid open fens, black spruce bogs, and shrubby willows. Reach 5 extends 3.8 km past a series of rock falls that are most likely impassable to anadromous fish. The reach is characterized by a smooth profile; the stream

Skeena River above Currier Creek

has multiple channels across a floodplain that averages 800 m in width and is composed of an extensive wetland complex.

Tantan Creek System
Tantan Creek extends 8.5 km into the subalpine and includes the upper and lower Kluatantan Lakes. The system usually has clear water. The drainage of a third nearby lake, Beaverlodge Lake, is into Tzahny Lake and then the Duti River. Reach 1 of Tantan Creek extends 1.5 km to the outlet of the lower Kluatantan Lake; it has a slightly sinuous pattern with the channel inset in a floodplain that averages 150 m in width. The reach is susceptible to beaver dam activity, which, at low flows, can block fish passage. The reach possesses a wide variety of riffles, pools, and abundant cover (Bustard 1975).

The lower Kluatantan Lake comprises reach 2. The shoreline is irregular, and the lake is 1.5 km in length with an average width of 250 m; it is 120 m at the west end and widens to 350 m at the east end. Mid-Tantan Creek is represented by reach 3, which extends 1.75 km to the outlet of the upper Kluatantan Lake. Reach 3 exhibits extensive wetlands, side channels, and sections of channel widening. The upper Kluatantan Lake represents reach 4, which is 750 m long and averages 400 m in width with an irregular shoreline. Reach 5 extends 0.9 km, with the channel moving through a valley-bottom wetland complex of swamps and ponds.

Kluayaz Creek System
Kluayaz Creek flows southwesterly 31.6 km from the alpine, where it drains several small ice fields resulting in frequent glacial coloring throughout the mainstem and Kluayaz Lake. Reach 1

extends 600 m upstream from the Kluatantan River to Kluayaz Lake through a wetland and back channel complex. The reach is low gradient and unconfined with good protective cover. Collingwood (1974) noted that the Kluayaz Lake outlet stays open virtually all year.

Reach 2 is represented by Kluayaz Lake, a high-elevation (1,007 m) lake approximately 2.5 km in length; on average it is 750 m wide at the southwestern end and 270 m wide at the northeastern end. Cold and mostly silt-laden streams feed the lake. Reach 3 extends 12.4 km with a smooth longitudinal profile and is unconfined, meandering across a floodplain with side channels, ponds, and wetlands. Reach 4 extends 7.9 km and is confined by the valley walls and plateau areas. A falls located 1 km above the reach break may be an obstruction to fish passage.

Water Quality
The upper Skeena headwaters generally have good water quality; however, during flood events or stream bank failures, water quality can be compromised in most of the tributary streams. In the early 1970s, there were significant impacts on water quality due to construction of BC Rail's Dease Lake extension. When the line was constructed, there was liberal stream channel sediment movement, due to inadequate drainage structure provisions and poor construction practices. In general, the Skeena headwaters area possesses highly erodible soils of glacial origin.

The majority of the soil is glacial till composed of a mixture of clay, silt, sand, gravel, and boulders. Pockets of glacial lacustrine deposits, located at many of the tributary stream mouths, can yield high sediment loads when disturbed. An example of this is the lower third of the Mosque River, which has cut through the till and flows through steep, deep banks of clay. Most of the Skeena headwaters remain in a largely pristine state with probably the best water quality in the watershed.

Geography
The upper Skeena headwaters area is a varied assemblage of mountains and valleys producing a diversity of climate, landforms, and vegetation. The Skeena River bisects the Skeena Mountains,

composed of the Slamgeesh and Groundhog Ranges to the west and the Tatlatui Range to the east. The majority of the high peaks rise above 1,800 m elevation; the predominant vegetation cover is alpine. The general elevations of the valley bottoms rise from the Sustut-Skeena confluence in the southeast corner of the drainage to a maximum at the upper Skeena-Spatsizi divide. The elevation difference between the valley bottoms and surrounding ranges is generally 800 – 1,000 m.

The mountain ranges generally form northwest trending ridges or mountain masses that parallel the Skeena mainstem. Both broad and narrow tributary valleys intersect the ranges, forming a complex and sometimes parallel drainage pattern, as exhibited by the Kluatantan and Tzahny drainage systems. The mainstem and tributary valleys are typically U-shaped, rising rather abruptly with slopes to 50%, which frequently increase with elevation. The valleys

generally have thick fills of glacial sediments in contrast to the mountains, which are rock with a veneer of till and colluvium. Upstream of the Duti River, the valleys are broad; while downstream, the streams follow narrow and steep valleys.

The Skeena valley is relatively broad, with low to moderately steep mountain slopes at lower elevations, often with broad areas of gently to moderately sloping morainal blankets and organic deposits in depressions that take the form of bogs or wetlands. In general, tributary streams possess higher gradients in their lower reaches and many meander through marshy flats and wetlands to head in cirque-like amphitheatres. The upper Skeena and Kluatantan valleys open into large alpine shrublands, meadows, and wetlands.

During the last glacial period, ice flowed down from the north across the drainage, forcefully eroding the mountain slopes

Glacier adjacent to Alma Peak

and basins and leaving a legacy of glacial erosion and depositional features (Lord 1948). Pleistocene ice has overridden most of the drainage a number of times, and the rounded summits generally lying below 1,800 m are the result of ice moving across the mountain ridges. Glaciers moving along the major valleys caused oversteepening of mountain slopes and contributed to the present-day U-shaped valley profiles. Above 1,500 m, alpine glacial activity of the Holocene era significantly modified the high country, resulting in the knife-edged ridges, spires, abrupt crumbling slopes, and nearly vertical cliffs that characterize the rugged mountains. Tarns and hanging valleys are evident, with the mountains showing the erosional effects of valley and alpine glaciation.

Forests

The upper Skeena headwaters area is mountainous and dominated by alpine vegetation, rock, and snowfields. In the valleys, a dense coniferous forest of subalpine fir (balsam) is present. There are a significant number of deciduous forest stands in the Skeena and tributary valleys that originated from fire or other disturbance. The widespread Biogeoclimatic Ecosystem Classification (BEC) zone is the Engelmann Spruce – Subalpine Fir (ESSF) zone. The Sub-boreal Spruce (SBS) zone is present in the lower elevation portions of the valleys. There are small amounts of the northern vegetation types, including Boreal White and Black Spruce (BWBS) and Spruce-Willow-Birch (SWB), in the upper Duti and Kluatantan valleys.

The SBS zone covers most of the lowland coniferous forests. Subalpine fir and hybrid spruce are the major tree species, though subalpine fir tends to dominate most stands in the drainage. Nonforested wetlands occur in the morainal landscape depressions, while dry grass/shrub meadows are present to a limited degree on dry sites with favourable warm aspects.

There are considerable deciduous and coniferous seral stands present due to frequent aboriginal landscape burning over long periods. Seral coniferous stands are typically pine and spruce in the Skeena valley bottom south of the Mosque River and dominantly subalpine fir upstream from there. Aspen is the prevalent tree

that has regenerated in many of the burnt stands downstream of the Duti River, particularly in the Mosque Burn. Black cottonwood is generally found on the floodplain and tributary alluvial fans.

The SBS zone merges into the ESSF zone at upper elevations ranging from 800 – 1,000 m, depending on local topography, aspect, and climatic conditions. In the Skeena valley, the SBS-ESSF transition occurs at approximately 890 m, which on the mainstem is 12 km upstream of the Kluatantan mouth. The only tributary valleys that have an appreciable amount of SBS are the Mosque River, and Fort, Alma, and Chipmunk Creeks.

The ESSF forest zone has a longer, colder, and snowier winter and a shorter, cooler and moister growing season. The forest is continuous at its lower and middle elevations and becomes more open as it passes into subalpine parkland at higher elevations. Subalpine fir is the dominant tree species, with small amounts of lodgepole pine and white spruce hybrids in drier slope positions or fire-influenced areas.

Mosaics of tree islands interspersed with dry or wet grass meadows or shrublands can be extensive and common within the drainage. Nonforested wetlands occur on many slope positions and in the valley bottoms, where vegetation consists mostly of sedges and a diversity of herbs. Avalanche tracks are common in many ESSF forests in the drainage, with vegetation ranging from shrubby slide alder to cow parsnip and hellebore.

The BWBS zone, a lowland to the montane zone, is present in the lower half of the Kluatantan and Duti drainages. The climate is northern continental, with frequent outbreaks of artic air masses, short growing seasons, and long, very cold winters. BWBS forests are characterized predominantly by white spruce with a large component of aspen and lodgepole pine and poor tree growth (Banner et al. 1993). Fires are frequent in the BWBS, maintaining a variety of forest age classes and successional stages. Dry grasslands and scrub vegetation are common on south and southwest facing slopes in the Kluatantan and Duti valleys.

The BWBS passes with increasing elevation into the Spruce-Willow-Birch (SWB) zone at

Kluakaz-Skeena confluence

approximately 1,000 m. The climate of the SWB is an interior subalpine type with mean annual precipitation of 760 mm, 60% falling as snow (Banner et al. 1993). Winters are long and cold, and summers are brief and cool with frequent cloud cover. Moist Pacific air often causes sudden, frequently violent local storms during the summer. Lower elevations of the SWB are forested mainly by white spruce and subalpine fir, with varying amounts of lodgepole pine, aspen, and black spruce on valley bottoms and lower slopes. Deciduous scrub birch and willow shrubs, mainly 1 – 4 m tall, dominate on higher-elevation sites.

Geology

Sedimentary rocks of the middle to late Jurassic Bowser Lake Group underlie nearly all (98%) of the upper Skeena drainage. The Bowser Lake Group, formed approximately 157 – 136 ma in the middle to late Jurassic Period, is a series of marine and nonmarine sedimentary rocks formed by massive and rapid erosion of land to the east. This deposit is made up of mudstones, sandstones, and conglomerate rock. During the course of massive deposition, the environment changed from offshore to nearshore, then to lowland fluvial sites.

These rocks were folded, with the dominant fold trend to the northwest, and thrust faulted, showing extremely complex forms. In parts of the Groundhog Range and the Skeena valley, broad open folds predominate (Holland 1976). A strong pattern of north to northwest block faulting and occasional cross faults break up the mountains and form the wide valley systems.

Fish Values

To the extent which it is known, the upper Skeena headwaters has high fish values. Sockeye, chinook, and coho salmon, steelhead trout, rainbow trout, Dolly Varden, bull trout, mountain whitefish, and longnose sucker are present in the drainage system (FISS 2003).

View northwest across Skeena headwaters

In general, the most widely dispersed salmon species is coho, while Dolly Varden/bull trout and rainbow trout are located in many fish-bearing waters.

All fish populations and habitats and enumerations are poorly known and recorded. Beyond a basic understanding of some presence, migration timing, and spawning locations, little is known about these high-elevation populations. Discrete anadromous stocks are potentially vulnerable to risks from disturbance and overharvest.

Chinook Salmon

Early August usually marks the peak of chinook arrivals in the Skeena headwaters. Most chinook in the upper Skeena headwaters spawn in August. Chinook salmon are present in the Skeena mainstem, which is used for a migration route upstream to Porky Creek, where they have been observed at the mouth (Sekerak et al. 1984). Anecdotal information indicates that chinook are farther upstream than Porky Creek and spawn in the mainstem near Kluakaz Creek. Chinook are present in the Kluatantan River at Tantan Creek and below Kluayaz Lake; however, spawner numbers, timing, productivity, and preferred habitats are unknown.

Sockeye Salmon

Two sockeye stocks spawn in the Kluatantan drainage; though it is unknown to what extent they are genetically differentiated. The sockeye rearing lakes are the upper and lower Kluatantan Lakes and Kluayaz Lake. Hancock et al. (1983) noted sockeye spawning grounds on the southern shore of the lower Kluatantan Lake and in Kluayaz Lake along both longitudinal shorelines. Collingwood (1974) observed sockeye salmon arriving in mid-September in Kluayaz Lake in 1973 and 1974. The only escapement record is 1970, which showed 50 adults in the lower Kluatantan Lake and 600 adults in Kluayaz Creek and Lake. The scant information available is insufficient to facilitate conservation planning and provide management objectives.

Coho Salmon

Coho migrate into the lower Skeena River between late July and the end of September, with the annual peak in late August as recorded by the Tyee Test Fishery. In general, most upper Skeena tributary coho arrive upriver in September. Coho juveniles and adults may be relatively widely dispersed throughout the accessible upper Skeena tributary drainages. There are no known coho spawning grounds in the Skeena mainstem. It is likely that coho may spawn in the many accessible, low-gradient tributaries situated on the Skeena River valley floor.

Known locations with juvenile coho presence include Chipmunk Creek (at 4 km), the Duti River (at 6 km), the Kluatantan River (in Kluayaz Lake), Kluatantan Lake, Currier Creek (at 3 km), Beirnes Creek (at 2.5 km), and Otsi Creek (at 6 km), with most of these records showing presence in the lowest stream reach. Escapement records indicated 300 adults in Kluayaz Creek in 1970 (DFO 1991d). For 2006, records indicate 560 in Kluayaz Lake, 230 in Kluatantan Lake, and 180 in Chipmunk Creek (Finnegan 2007). Limited juvenile coho sampling has occurred upstream of Kluayaz Lake and in the lower Kluatantan mainstem close to the Skeena confluence.

Steelhead

Steelhead presence in the upper Skeena drainage has been noted in the Mosque River at 0.6 km and in Kluayaz Lake in the Kluatantan system (DFO 1991d). The upper Skeena River drainage supports a summer-run steelhead population that likely enters the mouth of the Skeena early in comparison with other Skeena steelhead populations. There are no data from Floy tag recaptures in the TAGS database, so an estimation of timing through Area 4 cannot be directly made (Baxter 1997c). The TAGS database was established in 1970 to record all steelhead tagging, recoveries, and DNA samples. Information from angling records and the lone radio-tagged steelhead tracked to the Kluatantan system suggest that these steelhead move

through the commercial fishery during the peak of commercial fishery effort.

There are no steelhead escapement estimates; though Whately (1975) suggested a total spawning population of 150 steelhead subsequent to reviewing angling guide records for Kluayaz Lake. Overwintering occurs in Kluayaz Lake (Lough 1979) and possibly in the Kluatantan Lakes, the Kluatantan River canyon, and the Skeena River mainstem. Steelhead spawn in May and June, coinciding with warming water temperatures and an increase in streamflows. Spawning locations within the drainage are unknown.

In 1984, Tredger (1986) sampled four sites in the lower Kluatantan watershed for juvenile steelhead; results showed that fry densities were highest downstream of the Tantan Creek confluence, with generally low densities in the lower Kluatantan mainstem and downstream of Kluayaz Lake. Parr densities were generally low at all sampled sites, though marginally higher downstream of Kluayaz Lake.

Resident Freshwater Fish

In comparison to salmon, information is scant in regard to resident freshwater fish in both river and lake habitats of the upper Skeena River drainage. Much of the drainage is poorly known and may contain populations of special interest or status that are presently unknown or undocumented. Freshwater species documented include rainbow trout *(Oncorhynchus mykiss),* bull trout *(Salvelinus confluentus),* Dolly Varden *(Salvelinus malma),* and mountain whitefish *(Prosopium williamsoni)* (FISS 2003).

Fisheries
Gitxsan Fishery

Gitxsan salmon fisheries within the upper Skeena River drainage were concentrated at six known locations on the Skeena mainstem. Fishing villages/stations were located at the mouth of Alma and Currier Creeks and at the mouth of the fourth left-bank creek upstream of the Mosque River. Fishing sites were also located at the mouth of the Kluatantan, Mosque, and Duti Rivers. Dispersed fishery camps that mainly targeted coho and steelhead were located upstream on Chipmunk Creek, Tantan Creek,

Kluayaz Creek, and several unnamed creeks. These known village and fishery sites were positioned mainly to exploit efficient capture sites.

Recreational Fisheries

Due to the remoteness and the necessity to access the upper Skeena by air, the majority of anglers are guided clients. Currently, there is one guide licensed to operate in the Kluatantan drainage with an allocation of 55 guided angler days. The angling days are used between August 1 and October 16 and based out of Kluayaz Lake, where the main guiding lodge is located.

The Kluatantan River is designated Class II waters all year, and a steelhead stamp is mandatory from September 1 to December 31; a bait ban applies during this time as well. Regulations require mandatory steelhead release July 1 to December 31 (B.C. Fisheries 2005).

Development Activities

Land-use development in the upper Skeena headwaters drainage is limited to a linear strip east of the Skeena River used as a rail right-of-way. Mineral exploration has been focused northwest of the Kluatantan River in the Groundhog Range. There is no settlement in the area.

The Northern Long Tern Road Corridor Plan proposes a mainline road up the east side of the Skeena to the Mosque River and up the west side to Cutfoot Creek. The rationale for these roads is to access mature and merchantable timber that contributes to the Prince George TSA cut.

Forest Resource Development

The upper Skeena headwaters area is located within the Fort St. James Forest District. Forest development began in the adjacent lower Sustut drainage in the early 1990s. Road and cutblock development is currently located at the lower Sustut–upper Skeena drainage divide. Current forest development plans (FDPs) show the mainline heading to within 5 km of the Mosque River with 21 cutblocks positioned along the road. The future trend regarding forest activities is unknown and complicated by beetle activity in the southern portion of the Forest District.

Mineral Resource Development

Euro-Canadian explorations for base and precious minerals within the upper Skeena drainage have been ongoing since the turn of the 20th century. The most noteworthy mineral deposit is the Klappan-Groundhog coalfield, an oblong (roughly 30 by 80 kilometres) area extending southeast from the headwaters of the Klappan and Little Klappan Rivers to Groundhog Mountain. The Groundhog coals are considered high-grade anthracite. Early exploration efforts were intense before the onset of World War I. By 1913, approximately 1,813 km^2 were covered in over 460 claims that are part of the Groundhog deposit (Malloch 1913).

In 1948, Buckham and Latour (1950) conducted a mapping and data collection program in the Groundhog coalfield wherein they documented 192 coal showings. Interest in the coal potential of the area was rekindled in 1966 and continued into the mid-1980s. Assessment of the Groundhog coalfield potential by a consortium of companies in 1970, who conducted mapping, sampling, and diamond drilling programs, estimated upwards of 3.6 billion tonnes of speculative coal reserves in the southeastern Groundhog coalfield. The coal reserves in the Klappan, Groundhog, Jackson Flats, and Sustut deposits are estimated at more than 10 billion tonnes (Ryan 2000). Gulf's measured economic reserves at Klappan are 64 million tonnes with total reserves of 630 million tonnes (Gulf Canada Inc.1986).

The Bowser and Sustut basins potentially represent the largest petroleum exploration target area within the Intermontane region. This area has received renewed interest in the last few years as a result of new thermal maturation data indicating that large portions of these basins are within the oil and gas window (Evenchick et al. 2002). Much of this area, particularly the Bowser basin, was previously considered to be overmature with respect to oil and in the upper end of the gas window (Hannigan et al. 1995). These new data suggest the potential for hydrocarbon resources beyond those described in the report by Hannigan et al. (1995). Further examination of catalogued surface and subsurface samples and core samples recognized more oil staining (Osadetz et al. 2004; Evenchick et al. 2004). These occurrences also confirmed the new thermal model generated for these areas.

In 2003, the province released a call for proposals for drilling rights in the Klappan and upper Skeena area totaling more than one million ha, which were subsequently awarded to Shell Canada. In 2004, Shell drilled three exploratory test wells in the Klappan and Spatsizi drainages and conducted seismic surveys. In 2005 and 2006, Tahltan elders blockaded Shell from drilling four proposed test wells in the Skeena drainage. "Our land is our kitchen," elder James Dennis told the Shell team. "When you bring your poison onto our land you are poisoning our kitchen." Coalbed methane development in the U.S. and Alberta has caused methane to migrate into groundwater and then contaminate streams – an unacceptable risk in the Skeena headwaters.

Transportation

The upper Skeena headwaters area has a limited transportation infrastructure consisting of BC Rail's Dease Lake extension that was constructed from 1970 to 1972. In 1977, BC Rail and the province abandoned construction of the line indefinitely.

The railway construction in the early 1970s resulted in severe environmental impacts to fish, wildlife, and water quality due to irregularities in design and construction practices. The construction also eliminated Gitxsan cultural heritage resources on the east bank of the Skeena River. The poorly designed drainage structures on tributary creek crossings, as well as deranged subsurface flows and seepage, continue to cause fish passage problems.

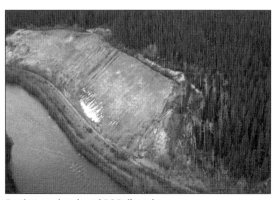

Erosion on abandoned BC Rail grade

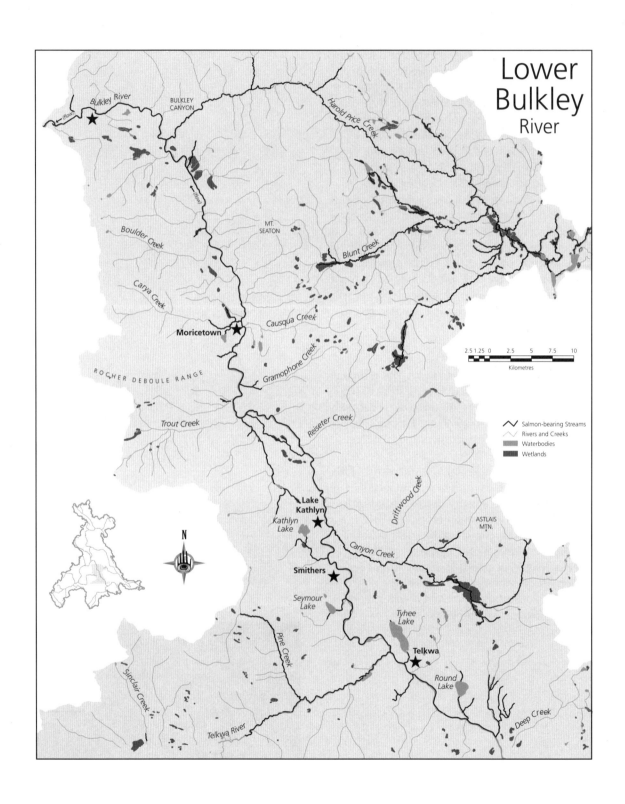

Lower
Bulkley
River

Bulkley River

BULKLEY CANYON

Harold Price Creek

Boulder Creek

MT. SEATON

Blunt Creek

Carya Creek

Causqua Creek

Moricetown

ROCHER DEBOULE RANGE

Gramophone Creek

Reiseter Creek

Trout Creek

2.5 1.25 0 2.5   5   7.5   10
Kilometres

∧ Salmon-bearing Streams
∧ Rivers and Creeks
▨ Waterbodies
▨ Wetlands

Driftwood Creek

Lake Kathlyn

Kathlyn Lake

ASTLAIS MTN.

Canyon Creek

N

Smithers

Seymour Lake

Tyhee Lake

Pine Creek

Telkwa

Sinclair Creek

Round Lake

Deep Creek

Telkwa River

20

Lower Bulkley River

Environmental Setting

The lower Bulkley River includes the Bulkley River and its tributaries from the Bulkley-Skeena confluence upstream to the Telkwa River. The Bulkley River is a major tributary to the Skeena River and flows into its left bank at Hazelton, 285 km upstream of the mouth.

Location

The lower Bulkley River is located in west-central British Columbia, and extends from Hazelton 99 km southeast to Telkwa. It is bounded to the north and east mostly by the Babine drainage and to the west by the Zymoetz and Kitseguecla drainages. To the south, the upper Bulkley watershed bounds the area.

Hydrology

The lower Bulkley drainage is mountainous with high relief. Elevations range from 2,504 m at Brian Boru Peak in the Rocher Déboulé Range and 2,362 m at Netalzul Mountain in the Babine Range to 252 m at the Bulkley-Skeena confluence. Moricetown Canyon lies at approximately 366 m, while the Bulkley-Telkwa confluence lies at 500 m elevation. Most tributary streams are relatively short. The only relatively major subbasin is the Suskwa, which cuts through and drains much of the Babine Range.

The one-in-ten-year, seven-day average, low-flow estimates for the Bulkley River are 13.7 m^3/s at Quick, which is upstream of Smithers, and 15 m^3/s at Smithers (Nijman 1986). The maximum daily discharge for the Bulkley River at Quick is 957 m^3/s. Monthly mean discharges range from a low in March of 28.8 m^3/s to a high in June of 366 m^3/s (WSC gauging station 08EE004, 75-year record, Environment Canada 2005). Overall, the hydrology is dominated by snowmelt. Mountains in the Hudson Bay Range, the Rocher Déboulé

Range, and the Babine Range exert major hydrological influences. Tributary streamflows have a moderately high response from water input due to the high gradients and the lack of water storage in lakes and wetlands of the major tributaries.

Peak discharges from the Bulkley River and the major tributaries typically occur in May and June due to snowmelt, and then decrease until late September, when fall rains and early snowmelt increase streamflows until the end of October. Streamflows decline in late November and December, when precipitation falls as snow; minimum discharges are recorded in January through March, prior to snowmelt.

The Bulkley valley lies in the rain shadow of the Coast and Hazelton Mountains. The relatively low amount of precipitation causes the discharges of low- and medium-elevation tributaries to drop off sharply following the spring snowmelt. There are seven hydrometric stations located in the lower Bulkley drainage that record flows on the mainstem and on some high- and low-elevation tributaries.

The coastal-interior climate transition is reflected in the distribution of the major ecological zones. The wide valleys of the Skeena, Bulkley, and Suskwa Rivers allow warm, humid coastal weather systems to penetrate, thereby losing their warmth and moisture gradually. The climate is predominantly characterized by a mild, moist growing season, and light-to-moderate snowpacks. Depending on elevation, snowpacks typically last from six to eight months, and average annual precipitation varies from 500 to 1,200 mm, as there is a climate gradient up the mountain slopes. Precipitation is greatest in the fall and early winter and then generally uniform throughout the year. Summer convective storms are common but rarely deliver more than 20 mm of rainfall in a day. Weather is recorded by stations in the Hazeltons, Smithers,

Station Creek culvert passing under Highway 16 with 1.5 m outfall drop that does not allow fish passage

Telkwa, and Suskwa valley. At Smithers Airport, located in the valley bottom, mean annual precipitation is 522 mm, with 331 mm of rainfall and 191 mm water equivalent of snowfall.

Stream Channels

Most of the stream channels are lightly to moderately incised into the hillslopes, terraces, and valley bottoms. Valley bottom deposits are largely the result of glacial sediment supply that filled in the wider portions of the valley bottom. In the past 10,000 years, rivers downcut through most of the glacial fills creating the river benches that favour settlement. The floodplain occupies only part of the valley floor. In some areas, the river impinges upon and erodes the thick, older glacial deposits. Where this happens along the lower Bulkley River, the resulting bluffs contribute much of the fine sediment transported by the river.

The tributaries flowing into the Bulkley River from the mouth upstream to Moricetown are generally short, high-energy, and steep-gradient streams that level out only in the last kilometre or less before entering the Bulkley mainstem. Sharp relief, with well-defined drainages, marks most of these tributaries. Most streams possess stable channel conditions throughout much of their length; however, active fans often characterize the lower reaches. The Suskwa River is a large subwatershed with significant tributaries of its own.

Moderate size streams flowing into the Bulkley River right bank, starting at the downstream end, include Nine Mile Creek, the

Suskwa River, Corduroy Creek, Luno Creek, Sharpe Creek, Kwun Creek, Casqua Creek, Gramophone Creek, Reiseter Creek, Driftwood Creek, and Canyon Creek. Moderate size left-bank Bulkley River tributaries include Mudflat Creek, Porphyry Creek, Boulder (East) Creek, Corya Creek, John Brown Creek, Trout Creek, and Toboggan Creek.

The majority of the lower reaches of Bulkley River tributary stream channels have been impacted to an unknown degree from land-use activities – principally transportation, agriculture, and urban developments (Mitchell 1997). Agricultural activities have frequently caused the loss of riparian areas, which contributes and is linked to increased runoff, reduced streambank stability, and streambank failures. Urban development has led to channelization and loss of riparian areas. The major linear developments occurring in the watershed are CN Rail and Highway 16, and, to a lesser extent, BC Hydro transmission lines.

Bulkley River

From the Skeena confluence upstream to the Telkwa River, the Bulkley mainstem is a single-thread, irregularly sinuous channel with an average gradient of 0.43%. Bedrock outcrops and sills control the channel gradient and location. Sediment production is relatively high due to bank undercutting and surface erosion of fine-textured materials. Bank failures are, for the most part, due to natural processes and events.

Reach 1 extends from the Skeena confluence 47 km upstream to immediately above Moricetown Canyon. From the Skeena confluence upstream to the Suskwa confluence, the Bulkley River flows through a deeply incised canyon that is approximately 18 km in length. The lower portion is known locally as Hagwilget Canyon and the upper portion as the Bulkley Canyon. The majority of the canyon cuts through bedrock, though the riverbanks are occasionally formed of bedded layers of gravels and sand. The canyons are characterized by swift water and occasional falls, rocks, and rapids.

Hagwilget Canyon, below Hagwilget Village, was altered in 1959 by the Department of Fisheries blasting in an unsuccessful attempt to reverse the abrupt decline of Bulkley River sockeye populations (Harding 1969). At the upstream end of this reach, Moricetown Canyon consists of a bedrock constriction and a short series of cascades that drop approximately 6 m over a length of 50 m. Department of Fisheries engineers first blasted a fish pass out at the falls in 1929. More elaborate fishways were constructed prior to the 1951 fishing season.

Reach 2 extends 52 km to the mouth of the Telkwa River. The channel is generally confined upstream as far as Canyon Creek and then occupies a narrow floodplain ranging from 200 to 500 m in width. The floodplain is mostly composed of gravel and sand and is bordered by the valley side slopes or river terraces, the latter developed by agriculture and residential interests.

Suskwa River

The Suskwa River cuts southwesterly through the Babine Range. Major tributaries into the Suskwa River include Natlan Creek, Thirty-One Mile Creek, and Thirty-Three Mile Creek flowing into the right bank, while Skilokis Creek and Harold Price Creek flow into the left bank. The Suskwa River mainstem is approximately 38 km in length with the 19 km upstream of the Harold Price confluence usually referred to as the upper Suskwa. Generally, the stream channel is incised into the valley bottom, which is either bedrock or glacial deposits that pose no obstructions to fish passage.

View to the west (downstream) of Hagwilget Canyon

From the mouth upstream, reach 1 is characterized as being occasionally confined, with an active floodplain and back channels and an average gradient of 1%. It extends from the mouth of the river upstream about 3.5 km. The irregularly sinuous wandering channel has occasional islands and frequent point and mid-channel bars. Reach 2 runs upstream to Fifteen Mile Creek and is primarily a canyon entrenched into bedrock. Reach 2 has major bank or valley-wall slump zones and an average gradient of 1%, with no floodplain. Reach 3 passes from Fifteen Mile Creek to Natlan Creek, and has an average gradient of 1%. This reach presents an irregularly sinuous channel that is frequently confined by valley bottom rock outcrops, discontinuous floodplain, and sporadic alluvial terraces.

Reach 4, with an average gradient of 0.6%, is upstream of Natlan Creek and composed of the 1 km long canyon that is confined by the valley walls. Reach 5 has a wandering gravel-bed river configuration and is generally sinuous and largely unconfined, except occasionally by valley bottom benches. This reach is relatively active, having changed channel position since 1975 (Gottesfeld 1995). Reach 5 has an average gradient of 0.6%, is almost 7 km in length, and is bordered upstream by the Harold Price confluence. Reach 6 extends 1.6 km upstream and is occasionally confined between high gravelly terraces. The channel is straight and braided, with extensive gravel and boulder deposition. Reaches 7 to 12 in the upper Suskwa are considerably steeper with gradients ranging from 2.5% to 3.3%. The channel is generally confined or entrenched upstream to Thirty-Three Mile Creek. Upstream of Thirty-Three Mile Creek, a floodplain is developed as the valley opens up into the relatively broad Suskwa Pass.

Harold Price Creek

The Harold Price drainage is the portion of the Suskwa watershed east of the Babine Range on the Nechako Plateau. The relatively high elevation of the watershed maintains a considerable snow pack, which contributes 60% of the overall Suskwa watershed streamflow. Blunt Creek, which flows to the east from where its valley cuts a low pass back through the

Babine Range, is the major tributary to Harold Price Creek. Reach 1 of Harold Price Creek extends 5.8 km in an irregularly sinuous pattern with occasional islands and frequent point and mid-channel bars at an average gradient of 1%. The wandering channel is intermittently confined and unconfined; the active floodplain averages 100 m in width and contains active side and back channels.

Reach 2 extends 4.8 km in an irregularly sinuous pattern, with an average gradient of 0.8% and occasional gravel bars. The floodplain is discontinuous where present and is frequently confined by the valley walls. Reach 3 extends 6.2 km upstream to the Harold Price falls. The reach is characterized by an irregular, sinuous, confined channel with an average gradient of 1.1%. The Harold Price falls, which are entrenched into bedrock and frequently back up a large deposit of gravel and woody debris, is reach 4. The falls were a three step series (5 m, 2 m, 2 m) until 1977, when the Salmonid Enhancement Program funded blasting of the falls and the massive logjam immediately upstream. This blasting effectively flushed the upper Harold Price, although downstream reaches appear to have reached a new equilibrium (Weiland 1995).

Reach 5 extends 2.4 km to the mouth of Maish Creek in an irregularly sinuous pattern and with an average gradient of 1%. Glacio-fluvial terraces and a discontinuous narrow floodplain generally confine the channel. Reach 6 extends 8.3 km upstream to the mouth of Blunt Creek in an irregularly sinuous pattern and with an average gradient of 0.5%. The channel is alternately confined between

Moricetown Canyon

low glacio-fluvial and fluvial terraces and unconfined with a few active back channels and sloughs that are a result of naturally occurring avulsions. Reach 7 extends 9.3 km to the mouth of Torkelson Creek with irregular meanders through a series of swamps and broad wetlands.

Natlan Creek
Natlan Creek is a large southward flowing tributary to the Suskwa River. Reach 1 extends upstream to the Denison Creek confluence. This reach presents a sinuous channel pattern that is occasionally confined or entrenched, with an average gradient of 1.7% and a discontinuous floodplain. Above reach 1, Denison Creek, Natlan Creek, and Iltzul Creek are all confined or entrenched by valley walls, with moderate gradients that range from 3% to 10%. The channels are dominated by riffles with occasional pools and runs.

Since the late 1970s, the Natlan Creek channel has been undergoing modification in its upper 7 km and also from 2 km above Denison Creek downstream to the mouth (Gottesfeld 1995). There is evidence of moderate widening along with a moderate increase in coarse sediment within the channel. This channel widening is coincident with a significant increase in road-related landslide activity. The stream bank failures involve floodplain deposits and bluffs of terraced fluvio-glacial material overlying thick till deposits.

Along the north side of Iltzul Creek, the existing quiescent slumps have potential for large-scale mass wasting. The slumps are in thick till, as well as in an area of drainage concentration. If these slumps reactivate, there would be serious long-term consequences for downstream anadromous fish habitat (Gottesfeld 1995).

Water Quality
The monitoring station most relevant to the lower Bulkley is at Quick, which is upstream of Telkwa. It shows the water as soft, with a near neutral pH and moderate colouration due to wetlands in the drainage; water quality is relatively good (Wilkes and Lloyd 1990). Upstream influences include mining, linear development, urbanization, agriculture, and

Harold Price Creek clearcut to both banks

logging; these could potentially change or modify water quality of the lower Bulkley River.

Six urban settlements all discharge wastewaters into the Bulkley River: Telkwa, Houston, Smithers, Moricetown, New Hazelton, and Hagwilget. Two municipal solid waste landfills and two wood waste landfills may also affect Bulkley River water quality. There are seven water withdrawal licenses from Telkwa downstream to Canyon Creek on the Bulkley mainstem.

Jumbo Creek, which drains a portion of the Denison Creek uplands, is an example of modified water quality in the drainage. Forest development activities in five cutblocks in the mid-1980s modified the water quality and flows in Jumbo Creek, which drains an area of 650 ha. The total logged area represents 167 ha or 25.7%. This is well above the limit of 18% clearcut harvesting that begins to affect streamflow according to Bosch and Hewlett (1982).

The impacts of logging have affected snowmelt and rain-on-snow generated peak flows, summer low flows, and sediment generation. These in turn have accelerated erosion, caused channel changes, and produced aggradation (Rabnett 1997b). These effects were complicated by a logging-related avulsion originating from Xsa Anax Siin (unnamed on government maps) into Jumbo Creek, causing a debris torrent. Other effects include lowered water quality for domestic use and negative effects on juvenile coho, chinook, and steelhead rearing habitat.

Water quality of the lower Bulkley watershed is generally good, as shown by data from site 0400204 located near Smithers; however, during flood events or streambank failures, water quality is compromised. Overall, describing water quality in the lower Bulkley drainage is complex due to the generally intensive land use in the mainstem valley bottom and wide-ranging forest development activities at middle to high elevations. Cumulative effects are unknown. How water quality is affected in relation to land-use activities, particularly forest development is not well understood. Major concerns include runoff changes in logged and roaded areas, temperature changes in streams, sedimentation, and effects to the physical stream structure.

Grassy Mountain slide, 2001

Geography

Located within the watershed are a diverse assemblage of basins and ranges and a portion of the Nechako Plateau. This produces a diversity of vegetation, climate, and landscape. The lower Bulkley drainage is composed of three broad physiographic landforms characterized by isolated mountains that are separated by the prominent Bulkley, Harold Price, and Suskwa valleys.

The lower Bulkley drainage is mostly mountainous with high relief. The Bulkley River valley forms the physiographic boundary between the southern Skeena Mountains to the north and east and the Hazelton Mountains to the south and west. The Babine Range lies within the southern Skeena Mountains. The Hazelton Mountains are composed of the Bulkley Range, which in turn includes the smaller Rocher Déboulé Range and the Hudson Bay Range.

The Bulkley valley and the upper Harold Price are located on the northwestern margin of the relatively low-relief Nechako Plateau, where it fingers into the mountains. The Nechako Plateau, which is generally thought of as being below 1,500 m in elevation, passes by transition into the Skeena Mountains, approximately between Smithers and Moricetown, though the boundary is a gradual one. To the south and east, the topography consists of broad and rolling landforms; while to the north and west, the valley is relatively constricted from Moricetown to the Skeena confluence.

The Babine Range to the northeast and the Hudson Bay and Rocher Déboulé Ranges to the southeast present striking views as they rise out of the Bulkley valley. These mountains, along with Mount Thoen and Netalzul Mountain within the Suskwa drainage, possess substantial pocket glaciers mostly on their northeastward-facing cirque basins. The ice that covered and flowed down the Bulkley and Suskwa valleys during the last glacial period forcefully glaciated the mountain slopes and basins, leaving a legacy of glacial erosion and depositional features.

Pleistocene ice overrode most of the drainage a number of times. Glaciers moving along the major valleys caused oversteepening of mountain slopes and contributed to the present-day valley profiles, while the rounded summits are the result of ice moving across the mountain ridges. Above 1,500 m, alpine glacial activity significantly modified the high country. At the peak of the last glaciation, ice in the Bulkley valley flowed coastward through a number of low passes.

Glacial and post-glacial deposits cover the lower gentle-to-moderate slopes and infill the broader sections of valley floor. Deposits formed before the last glaciation are found at depth in parts of the Bulkley valley near Smithers. Thick blankets of glacial till from the Ice Age cover the main valley and mountain valleys and extend up the valley sidewalls, though bedrock is exposed along deeply incised streams and on steep-sided hillslopes. A morainal veneer is often found on many mid-slope positions, while upper areas are dominated by thin soils, rock, and colluvium.

The Nechako Plateau is an area of low relief, with large expanses of rolling country where glacial drift is widespread and the majority of bedrock is obscured. During the Fraser Glaciation, the plateau experienced complex patterns of glacier movement and sedimentation associated with multiple phases of ice flow (Stumpf 2001). During deglaciation, remnant ice disrupted drainage patterns, ponded proglacial lakes, and created eskers and meltwater channels. In many of the wider portions of the lower Bulkley valley, river terraces formed after deglaciation. They are generally composed of several metres of gravel and occasional pockets of organic terrain overlying thick deposits of till and other glacial sediment.

Forests

Other than the mountainous alpine country, the majority of the lower Bulkley drainage is covered with dense, coniferous forests. Smaller but significant patches of deciduous forest occur over the low-elevation valley bottoms and lower mountain slopes along the Bulkley valley and on the north side of the Suskwa River. Within the watershed, principal forest types are represented by three Biogeoclimatic Ecosystem Classification (BEC) zones: the Interior Cedar Hemlock (ICH) zone, the Sub-boreal Spruce (SBS), and the Engelmann Spruce–Subalpine Fir (ESSF) zone.

The ICH zone occupies valley floors and mountain slopes with a transitional coast-interior climate. The ICH zone characterizes the low-elevation coniferous and deciduous forests

as far east as Moricetown in the Bulkley valley and in the Suskwa drainage as far east as Harold Price Creek. Major tree species in this zone are western hemlock, spruce, pine, and subalpine fir with local stands of red cedar. Considerable deciduous and coniferous seral stands are present because of frequent aboriginal landscape burning over long periods of time. The SBS zone characterizes the lowland coniferous forests of the lower Bulkley watershed. Subalpine fir and hybrid spruce are the major tree species; subalpine fir stands tend to dominate older, high-elevation stands and moister sections of the zone. Due to a relatively intense natural and aboriginal fire history, lodgepole pine seral stands are extensive, particularly on stream terraces and south aspect slopes.

The ICH and SBS zones merge into the ESSF zone at higher elevations ranging from 900 to 1,300 m, depending on local topography and microclimate. Subalpine fir is dominant, with lesser amounts of lodgepole pine and white spruce hybrids in drier or fire-influenced areas.

Geology

The lower Bulkley watershed is underlain by rocks of the Stikine terrane almost entirely of Mesozoic age. The middle Jurassic rocks were deposited in a former island arc system. Volcanic activity produced great quantities of volcanic flow rocks, such as basalt and andesite, as well as ash and mudflows. In the late Jurassic age, these island arcs were pushed into and accreted with the North American continent. The Bowser Lake Group is a series of marine and nonmarine sedimentary rocks that were formed during and after this accretion. Massive and rapid erosion caused the deposition of volcanic sand and mud. The Bulkley valley area was located along the southern edge of the Bowser basin, and shoreline sediments prevail (Gottesfeld 1985).

In the early Cretaceous age, tectonic uplift caused the area to be pushed up, initiating mountain building. The primarily volcanic Skeena Group formed between 120 and 100 ma. This group consisted of a series of lava flows interspersed with sediments in a shallow sea and coastal plain river environment that included large swampy areas where peat and plant life accumulated to form coal beds.

In the late Cretaceous age, from 100 to 65 ma, the area was uplifted with volcanic and plutonic activity creating the Bulkley and Babine Intrusives, which are chiefly composed of granitic and porphyritic rocks. These intrusions into the sedimentary and volcanic rock sequence occurred at relatively shallow depths. Alteration zones around the intrusions host metallic mineral deposits. These alteration zones are in close proximity and relationship to the three major faults that occur throughout the watershed.

Fish Values

Fisheries values are high within the lower Bulkley drainage. Coho, pink, sockeye, and chinook salmon, as well as steelhead characterize anadromous fish presence. Freshwater resident fish presence includes cutthroat trout, Dolly Varden, rainbow trout, mountain whitefish, bull trout, prickly sculpins, longnose dace, lamprey, and red-sided shiner (FISS 2005). In general, the most widely dispersed salmon species is coho, while Dolly Varden, rainbow trout, and mountain whitefish are widely distributed nonanadromous species.

Chinook Salmon

The lower Bulkley River is the migration route for adult chinook passing through to the upper Bulkley and Morice systems and to tributaries in the lower Bulkley drainage that sustain chinook spawning. There is no known chinook spawning in the lower Bulkley mainstem. Significant amounts of chinook are harvested at Moricetown Canyon in the aboriginal fishery. Bulkley River tributaries supporting chinook include the Suskwa system and John Brown Creek, where they have been observed at 0.4 km; however, spawner abundance and spawning locations are unknown.

Adult chinook salmon usually begin their migration into the Skeena River in the third and fourth weeks of July, arriving at the Suskwa River throughout the month of August, with the usual spawning peak from mid to late August. Suskwa River chinook spawners use the upper half of reach 5, which is the wandering gravel bed downstream of the Harold Price confluence, and the lower 3 km of Harold Price Creek.

These two spawning grounds dominate chinook production. Hancock et al. (1983) noted minor chinook spawning grounds on Natlan Creek upstream to Denison Creek.

Suskwa River chinook escapement estimates have been recorded discontinuously since 1960, and since that time, escapement has been generally low. In 1960, Suskwa chinook population was estimated at 400. Escapements from 1961 to the present range from 10 to 250 (DFO 2005). Suskwa chinook spawner numbers have not recovered in the past two decades as many other Skeena chinook stocks have. The present level of chinook escapement does not suggest a stable population level or long-term survival of this population.

Pink Salmon

Pink salmon are exclusively two years old at spawning time, meaning that odd- and even-year stocks are genetically separate. Pink salmon arrive in the lower Bulkley drainage in mid to late August and small populations spawn in seven known tributaries.

Lower Bulkley River

Station Creek supports pink salmon spawning from the mouth upstream to a falls at 0.5 km. Wide fluctuations in escapement have ranged from 3,500 in 1955 to 25 in many years. Spawning has been observed at 0.2 km in Boulder (East) Creek with unknown abundance. Casqua Creek supports pink spawners in the lower 0.2 km with five escapement records since 1950 that range from 0 to 1,000. Casqua Creek is subject to seasonal water fluctuations, which

Beach seining below Moricetown Canyon

in the past have been exacerbated by industrial mining and logging activities.

John Brown Creek has no documented counts, but pink salmon were observed by Triton (1998c). Pink salmon have been observed on the fan at the mouth of Trout Creek, where spawning may or may not occur. Toboggan Creek supports spawners to 8 km, but most spawning occurs on the fan at the mouth. Since 1950, the escapement has varied from 0 to 20,000, though in most years, it ranges from 300 to 1,000 pinks. Kathlyn Creek supports a small population of pink salmon that apparently spawn 1.2 km upstream to the Chicken Lake Creek confluence. Escapement has ranged from 0 to 2,500 in five enumerations since 1950.

Suskwa Watershed

Adult pink salmon usually migrate upstream on the Suskwa River, arriving August 15 with the peak of spawning in mid-September. The principal spawning ground is the lower Suskwa mainstem in reaches 1 and 2, particularly from the mouth upstream for 2.8 km. Other minor spawning areas include reach 3 up to the canyon cascade east of Natlan Creek. Jantz et al. (1989) noted pink salmon presence in Harold Price Creek, though this may only occur in years of high abundance. Spawner escapement has been recorded discontinuously since 1967. Spawners have ranged from 0 to 5,000, though in most years 100–500 pink salmon return.

Coho Salmon

Coho juveniles and adults are relatively widely dispersed throughout the lower Bulkley drainage. Coho migrate into the Skeena River between late July and the end of September as recorded by the Tyee Test Fishery, with the annual peak of the migration in late August. Daily counts of coho through the Moricetown Canyon fishways during the late 1950s to the mid-1960s and more recently show coho arriving in late July to early August, with peak abundance during mid-August.

Coho typically move to the outlet of various tributaries or pools to hold. Depending on water flow conditions, coho will wait for fall floods before moving into the tributaries from late September to late November to spawn. Coho are usually the last salmon to spawn in the fall, from the end of September through December. Coho life histories, particularly during the freshwater period prior to smolting, are not well known in the Bulkley watershed.

Lower Bulkley River

Overall, Bulkley system coho seriously declined in the 1970s and fluctuated at low levels of abundance until 1998 when they began a steady increase to the present. The extent of the decline and recovery in the lower Bulkley drainage is difficult to evaluate, as only Kathlyn and Toboggan Creeks have a near complete escapement data set from 1950. At this time, most of the coho production (74%) comes from Toboggan Creek, which has been augmented by the Toboggan Creek Hatchery since 1988.

Toboggan Creek is a somewhat singular subdrainage in the Bulkley watershed, in that it has a hatchery, operates a wild smolt trap, and enumerates returning adults. The hatchery, which has the capacity to rear 155,000 coho and chinook smolts on a yearly basis, augments chinook and coho stocks in various Bulkley subbasins. An adult counting fence, operating since 1988, provides accurate enumeration of coho adults. Beginning in 1995, hatchery smolts have been marked with CWT and adipose fin clips prior to release, and wild smolts have been enumerated with fyke and rotary screw traps (SKR 1999). This coho index program lends itself to studies in freshwater survival, age distribution at smoltification, migration timing, and recruitments of coho juveniles.

Coho spawning enumerations in Toboggan Creek have been ongoing since 1950, with fence counts since 1988. Since 1950, coho adult escapement has gradually increased with annual averages of 2,711. Spawning is scattered throughout the mainstem up to Toboggan Lake and into the tributaries that include Owen, Elliot, and Glacier Gulch Creeks.

Station Creek supports coho spawning from the mouth upstream 1.6 km to the Highway 16 crossing, which blocks fish passage since the culvert is a migration and velocity barrier (Mitchell 1998). Escapement ranges between 0 and 300 coho, with an annual average (from a discontinuous enumeration record) of less than

250 fish. In 1993, coho were transplanted from Toboggan Creek to Station Creek, and, since 1998, a counting fence facilitated enumeration and allowed easier transportation of a proportion of the returning adults above the impassable culvert (Houlden and Donas 2002b). There are anecdotal reports of coho spawning in Corduroy Creek, Luno Creek, and the lower reach of the unnamed creek west of Corduroy Creek. Trout Creek sustains coho spawning 1.6 km upstream to the falls, with a range between 0 and 300 spawners, although there have been no enumerations since 1970.

Reiseter Creek sustains coho, with irregular escapement records ranging between 0 and 400 spawners. Spawning is scattered throughout the lower 2.5 km with seasonal water fluctuations that can present difficulties to coho entering the creek (Hancock et al. 1983). Driftwood Creek supports coho spawners with escapement ranging between 50 and 300, though most years average 150 according to discontinuous enumerations. Hancock et al. (1983) noted that spawning locations extended upstream approximately 25 km in Driftwood Creek.

Canyon Creek supports coho spawning with escapements that range from 0 to 400 coho, though most enumerations are from the 1950s and 1960s. Hancock et al. (1983) reported two principal spawning locations, between 1.5 and 2.5 km and between 6.5 and 7.5 km on the mainstem. Canyon Creek is subject to seasonal water fluctuations and beaver activity. In the late 1980s, Canyon Creek was stocked with coho fry from Toboggan brood stock.

Kathlyn Creek has a near continuous escapement record since 1950, which ranges between 10 and 500 coho. Since 1986, coho fry and smolts from Toboggan Creek wild broodstock have been transplanted into the Kathlyn Creek system, the majority of which were released into Kathlyn and Club Creeks (Toboggan Creek Hatchery 1999, DFO 2002). Preferred spawning locations are in Simpson, Kathlyn, and Chicken Creeks downstream of the lake and in Kathlyn and Club Creeks upstream of Lake Kathlyn in pockets of good gravel.

Suskwa Watershed

Discontinuous records show aggregate coho escapements for the Suskwa subbasin that range between 25 and 2,500, with no recorded escapements since 1992. Coho spawning is principally located in the upper half of the wandering gravel bed (reach 5) in the Suskwa mainstem. Dispersed spawning occurs upstream to 10 km on Natlan Creek and possibly in the Suskwa mainstem 1.5 km upstream of the Harold Price confluence, though channel morphology in the latter reach has greatly changed over the last decade due to a combination of natural and logging-related flood events.

The primary spawning location in Harold Price Creek is the lower 2.5 km stretch of good gravel. Coho also spawn to a limited degree in unknown locations of the upper Harold Price, most likely in Blunt Creek and/or its tributaries.

In 1999, the Suskwa Coho Synoptic Survey was established to provide information on coho salmon abundance and distribution in the Suskwa River watershed (McCarthy 2000). This short-term program established and assessed 21 sample sites focused on side channels and tributaries along 20 km of the Suskwa River and Harold Price Creek. The site on lower Jumbo Creek had juvenile coho densities over 1 fish/m²; the densities at all other sites had less than 0.5 fish/m². Sites sampled in Blunt Creek yielded coho juveniles in very low numbers.

Steelhead

Lower Bulkley River steelhead are significant in a provincial and regional context—as a food source and cultural symbol to the local First Nations and for their high angling popularity. The lower Bulkley drainage supports a summer-run steelhead population that enters the mouth of the Skeena in late June or July, arriving in the Bulkley system beginning in August and continuing into autumn.

Beere (1991b) reported that of the 23 summer-run steelhead radio-tagged in the lower Bulkley mainstem downstream of Moricetown Canyon, only five migrated upstream through the canyon. At least 12 of the 23 migrated downstream to the Suskwa-Bulkley confluence or entered the Suskwa River to overwinter. This is consistent with the findings of Lough

(1980), who found that four of the six steelhead radio-tagged at the Suskwa-Bulkley confluence overwintered in the Bulkley mainstem as far as 12 km upstream.

Overall, the Tyee Test Fishery provides the best estimates of steelhead escapement to the Skeena watershed. Total summer-run escapement estimates based on the Tyee index data are available starting in 1956. Overall, Skeena steelhead declined from about 1985 to 1992. Changes to steelhead populations in tributary watersheds in the Skeena are hard to identify. The most useful source of data is the Steelhead Harvest Analysis (Ministry of Water, Lands and Air Protection 1991). In general, the pattern of catch reported for the Bulkley River shows an increase in fishing effort and catch over the past 35 years, with the pattern of total catch, including released fish, resembling the Tyee Test Fishery estimates for summer-run steelhead. Mitchell (2001a) and SKR (2001) reported Bulkley-Morice steelhead population estimates based on the Wet'suwet'en Fisheries tagging program conducted downstream of Moricetown Canyon.

There are few good data that records steelhead escapements at individual streams. This is in large part because they spawn in spring at high water conditions when counts are usually not possible, and they are typically spread out at many sites within a stream.

Bulkley Watershed

Steelhead adults have been observed at the mouth of Station Creek, though actual spawning locations are unknown (DFO 1991c). The lower 2 km of Gramophone Creek sustains steelhead spawners. Trout Creek has steelhead presence upstream to the falls, and the reach above the falls was stocked with steelhead fry and parr intermittently since the mid-1980s (DFO 2002). The Kathlyn Creek system has steelhead presence in Kathlyn Lake, and spawning has been observed upstream of the Chicken Creek confluence.

Toboggan Creek supports steelhead spawning to 1.5 km above Toboggan Lake. Between 1995 and 2001, Toboggan Creek Hatchery operated a counting fence to enumerate adults. Spawning is primarily

scattered on the mainstem upstream of the counting fence, which is located at 2.5 km from the mouth.

Suskwa Watershed

Suskwa steelhead are summer-run fish that enter the Suskwa system in late August or early September. Suskwa steelhead are noted for their large body size, deepness through the body, and in the past, their abundance. This made them the basis of an international sport fishery that continues into the present, though on a smaller scale (Chudyk 1978).

The 1979 steelhead radio telemetry project noted that all steelhead wintered at or below the canyon just upstream of Natlan Creek or relatively close to the Suskwa-Bulkley confluence (Lough 1980). Steelhead spawning occurs the following March through May, coinciding with warming water temperatures and an increase in Suskwa River flows. Spawning is concentrated primarily in the lower 3 km of reach 1, in reach 3 particularly adjacent to the mouth of Jumbo Creek, and in the lower 5 km of Harold Price Creek. Sporadic spawning occurs in reach 3 downstream of Natlan Creek and in reach 6 upstream of Harold Price Creek. Results from the radio-tagging program indicated spawning is dispersed in side and main channels on the mainstem throughout the system below the Harold Price falls (Lough 1980).

In 1982, the average age of 27 steelhead spawners sampled was 4.2 years, while the repeat spawners averaged 8.6% for 1977, 1979, and 1982 (Schultze 1983). Steelhead fry emerge between mid-August and mid-September and widely disperse throughout the system, including the smaller tributaries that offer suitable refuge.

In the mid-1970s, lack of angler success and conservation concerns in regard to Suskwa steelhead initiated a revitalization program under the auspices of the Salmonid Enhancement Program (SEP). Through the SEP, the B.C. Fish and Wildlife Branch (BCFW) developed a three-point program designed to increase the number of returning steelhead and thereby improve angler success (Chudyk 1978). The three-point program included an assessment of potential stocks and habitat within the watershed, the removal of the Harold Price

Hagwilget Canyon fishing platforms, c. 1920

falls as a fish barrier, and the colonization of steelhead in the upper Harold Price. Presently, adult steelhead do not appear to pass above the falls.

Resident Freshwater Fish

In comparison to salmon, information is scant in regard to resident freshwater fish in both the river and lake habitats of the lower Bulkley drainage. Freshwater species and documented populations inhabiting the lower Bulkley drainage include rainbow trout *(Oncorhynchus mykiss)*, cutthroat trout *(Oncorhynchus clarki clarki)*, bull trout *(Salvelinus confluentis)*, Dolly Varden char *(Salvelinus malma)*, river lamprey *(Lampetra ayresi)*, Pacific lamprey *(Lampetra tridentata)*, mountain whitefish *(Prosopium williamsoni)*, longnose dace *(Rhinichthys cataractae)*, redside shiner *(Richardsonius balteatus)*, and prickly sculpin *(Cottus asper)*.

Fisheries
Gitxsan and Wet'suwet'en Fisheries

The salmon stocks passing through and spawning in the lower Bulkley formed the principal food resource that enabled Gitxsan and Wet'suwet'en people to make the area their home. The Gitxsan and Wet'suwet'en salmon fishery at Hagwilget Canyon was likely one of the largest aboriginal fisheries on the Skeena system, along with the large fisheries at Kisgegas and Wud'at on the Babine River. Relative to its history, the Hagwilget Canyon fishery currently functions on a small scale. In the past, the Moricetown Canyon fishery fulfilled the food, societal, and ceremonial (FSC) needs of the Wet'suwet'en; however, recent sockeye escapements in the Morice and upper Bulkley systems have been so low as to preclude sockeye fishing.

Gitxsan and Wet'suwet'en traditional salmon fisheries within the lower Bulkley drainage were concentrated at four principal locations on the Bulkley mainstem: Hagwilget Canyon downstream of Hagwilget Village, Hagwilget Canyon at the village and upstream, the Bulkley mainstem upstream of the Suskwa confluence, and Moricetown Canyon.

The salmon fishery in Hagwilget Canyon downstream of Hagwilget Village was fished by Gitxsan, who utilized more than 20 sites on both the north and south banks (GWA 2003a). Gitxsan fished nine sites on the right bank in Hagwilget Canyon opposite Hagwilget Village, and the Wet'suwet'en fished 12 sites on the left bank (Gitksan and Wet'suwet'en Chiefs 1907). The majority of these canyon sites were not fished after the winter of 1958–59, when the Department of Fisheries blasted the rocks that helped to concentrate fish close to the canyon walls.

An Naa Wildam Ax, also known as Dizkle, was a large village located on the Bulkley mainstem, 1.5 km upstream of the Suskwa confluence. This primarily Wet'suwet'en village fished eight nearby sites, as well as operating a weir. Further upstream in Moricetown Canyon and below the canyon, Wet'suwet'en fished 22 known trap and gaff sites. Presently, the majority of FSC fishing at Moricetown Canyon is with dip nets.

According to current knowledge, dispersed fisheries operating on the Bulkley mainstem included nine camps between Boulder Creek and Moricetown Canyon and 11 camps upstream of the canyon to the Telkwa River confluence (Wet'suwet'en Fisheries 2003). These dispersed fisheries that mainly targeted coho and steelhead were often positioned at tributary mouths to easily exploit the fish resource. Dispersed fisheries away from the Bulkley mainstem included the fisheries at the outlets of Toboggan (Tabeegh Tay) and lower Reiseter Lakes (Rabnett et al. 2001).

Suskwa Watershed

Gitxsan primarily fished the Suskwa mainstem. Wet'suwet'en fished the upper and lower Harold Price. The largest fishery was located at An Djam

Lan, also known as C'ee Ng'heen, which was adjacent to the Harold Price–Suskwa confluence. This fishery seasonally targeted chinook and coho utilizing a weir, and steelhead using nets under the ice. Downstream at the Natlan Creek confluence, and at Anx Milit, steelhead were fished in the early spring prior to ice break-up (GWA 2003a).

Dispersed seasonal fisheries in Harold Price Creek included steelhead fisheries in the canyon located at approximately 9.5 km and downstream of the falls at 16.7 km. Sites that seasonally exploited nonanadromous fish included Ax Kegh Nee Tay, also known as Laydlii, which was located at the Blunt–Harold Price confluence, and Loots Wenii Cguunget Hozaay, located at the outlet of Touhy Lake,

Part of a hlamgan trap in Hagwilget Canyon, c. 1920

Moricetown Canyon fishery, 1946

Recreational Fisheries

The Bulkley and Suskwa Rivers support large-scale seasonal recreational angling by local and area residents and nonresident visitors. The high-value Bulkley River fishery is conducted from the river shore or by power and float craft. The Suskwa River sports fishery is almost entirely shore-based.

The river fishery is primarily steelhead angling, though chinook may also be harvested. Pink and coho openings are held in years of abundance. The majority of steelhead angling effort occurs over approximately a 10-week period from early September through mid-November. Peak activity is commensurate with the period most anglers feel provides the best combination of steelhead availability – good weather and clear water – usually from mid-September to late October. Water clarity of the Bulkley River is usually good relative to other nearby steelhead rivers, but the Telkwa River, which provides the majority of turbidity from heavy rainfall or warm weather, can produce poor angling conditions.

The use of boats – both power and drift boats – by local anglers, guides, and nonresidents is a major feature of the steelhead fishery. Boat launches are located close to the Skeena confluence, adjacent to the Suskwa Forest Service Road (FSR) crossing of the Bulkley River, on the left bank downstream of Moricetown Canyon, at Trout Creek, and at Lunen Road immediately east of Smithers. Most of the fishing activity occurs in the most navigable water, from Trout Creek upstream to the Telkwa River (Morten and Parkin 1998). At the Bulkley-Skeena confluence, Morten and Parkin (1998) and Hall

and Gottesfeld (2001) reported that the majority of anglers fishing for coho are local residents.

Most angling on the Suskwa River occurs on the lower 4 km section downstream of the canyon. Construction of the Suskwa FSR bridge across the Bulkley River in 1967 provided improved access and greatly increased steelhead fishing opportunities. Presently, the status of Suskwa steelhead is unknown. Angling for steelhead typically occurs from mid-September to the end of October, though the timing is dependent on river clarity.

Seven angling guides are licensed to operate on the Bulkley River with a set quota of guided rod days that total 1,504 for the entire river. Although annual angling regulations specific to the Bulkley River may change, they generally include a bait ban and no fishing from January 1 to June 15. Class II waters apply from September 1 to October 31, with a mandatory steelhead stamp. Angling regulations specific to the Suskwa River include a bait ban and Class II waters year round; a steelhead stamp is also mandatory September 1 – October 31 (B.C. Fisheries 2005).

Enhancement Activities

Enhancement activities in the lower Bulkley watershed consist of the two major fish passage projects in Hagwilget and Moricetown Canyons and the Toboggan Creek Hatchery program. The present Moricetown Canyon fishways were constructed in 1951. Concerns regarding the low abundance of Morice-Nanika sockeye in 1955 led to the decision to remove obstructions in Hagwilget Canyon. The canyon rocks that were blasted in 1959 effectively eliminated the aboriginal fishery in the canyon and facilitated pink salmon colonization in the upper Bulkley and Morice subbasins.

Toboggan Creek Hatchery, which has the capacity to rear 155,000 coho and chinook fry and smolts on a yearly basis, augments chinook and coho stocks in various subbasins of the watershed. Most fry and smolt raised at the hatchery are released in the upper Bulkley drainage. The species summaries include details of juvenile releases. Minor enhancement programs include the incubating and rearing of salmon through community and school programs.

The Salmonid Enhancement Program altered fish passage at the Harold Price falls in attempt to introduce coho and steelhead to the upper Harold Price. Modification of the one-step falls to a three-step falls allowing fish passage was successful; coho adults have been observed upstream of the falls. From 1979 to 1988, 257,000 marked steelhead fry were released above the falls, but returns were dismal. Up to the present time, adult steelhead have not been observed upstream of the falls.

Development Activities

Land use in the lower Bulkley drainage is relatively intensive compared with other Skeena watershed subbasins. Over the last 40 years, overall land-use and development activities in the drainage have greatly increased. The principal land-use activity is logging, which is generally concentrated in the Suskwa and Harold Price drainages, and to a somewhat lesser extent in the Bulkley valley. Urban development is concentrated in the Hazeltons, Moricetown, and the Smithers-Telkwa areas. Farms and rural residential settlement areas are scattered along Highway 16 east to Moricetown, and then concentrated along both sides of the Bulkley River eastward. Industrial land use occurs to a limited degree close to the urban areas.

Areas protected from development for their ecological, cultural heritage, and/or recreation values are scattered throughout the drainage. Small protected areas include Ross Lake (357 ha), Boulder Creek (48 ha), Netalzul Meadows and Falls (339 ha), Burnt Cabin Bog (646 ha), and Call Lake (62 ha). The Babine Mountains and Driftwood Provincial Parks (approximately 39,000 ha) comprise the larger protected areas that are relatively high profile due to their diverse recreational activities, good road access, and proximity to Smithers.

The early part of the 20th century saw extensive mineral exploration in the Babine and Rocher Déboulé Ranges; presently there is little mineral exploration or development in the drainage. Linear development along the southern bank of the Bulkley River accommodates the CN Rail line, Highway 16, and the utilities infrastructure of BC Hydro and Telus. Transportation development has been the key to land-use and resource development. Coal bed methane extraction has recently been proposed for the Telkwa coalfield.

Forest Resource Development

The lower Bulkley drainage is located within the Skeena Stikine Forest District of the Ministry of Forests. Currently, the Ministry of Forests relies on land-use management objectives defined in the Kispiox and Bulkley Land and Resources Management Plans (LRMPs).

From the late 1960s to the present, the area and volume logged annually increased. The post-World War II demand for lumber contributed to the gradual logging of low and mid-elevation stands in the Bulkley valley. In recent decades the trend toward centralization of forest license holding and milling capacity has greatly increased the intensity of logging. Since the early 1970s, road building and logging activities within the Suskwa drainage provided timber to mill operations primarily located in the Hazeltons. In the late 1970s, forest development greatly increased in the upper Harold Price area, with timber flowing to sawmills in the Smithers area.

These activities have largely led to the disappearance of the viable commercial forests in the Suskwa and Harold Price subbasins as well as the low and mid-elevation slopes in the Bulkley valley. The majority of forest development has been focused on the easily accessible, commercially viable timber with the result that most forestry development options for the near and mid-term future have been foreclosed. Overall, current forestry activities are limited to salvage of beetle killed stands and nontimber forest products.

Between 1995 and 2001, the Watershed Restoration Program (WRP) was involved in assessing logging-related disturbances in relation to fish, fish habitat, and upslope sediment-producing areas within the Suskwa watershed. The effects of the 30-year history of disturbance were assessed; however implementation of watershed restoration activities has been limited. A large contribution of the WRP has been to increase awareness in the forest sector of the best management practices regarding water quality, fish, and fish habitat. Watershed health has benefited from the program principally through road deactivation activities. However, road and logging-related landslides and earth flows continue to sporadically contribute sediment to streams and riparian areas.

Mineral Resource Development

Euro-Canadian explorations for base and precious minerals started late in the 19th century. Early efforts focused mainly on high-grade silver and gold-rich veins, with most endeavours slowing down with the onset of the Depression. Silver-lead-zinc ore was discovered on Hudson Bay Mountain in 1905, and in 1908 similar ore was discovered on Nine Mile Mountain at Hazelton. Between 1910 and 1930, development was carried out from time to time on 75 groups of claims in these areas (Kindle 1940). Explorations in the mid-Babine Range led to claims on Mount Thoen and Netalzul Mountain that were worked from 1916 to 1930. Claims in the southern Babine Range were active by 1916 and worked discontinuously until 1974, when the Cronin Mine closed.

There is known mineralization in the Babine, Hudson Bay, and Rocher Déboulé Ranges, reflecting alteration zones associated with the Bulkley and Babine Intrusives. The main types of mineral occurrences are polymetallic veins containing silver, lead, zinc, copper, gold, and tungsten, and a few large porphyry deposits. The majority of occurrences are located on Hagwilget Peak, Hudson Bay Mountain, and the southern Babine Mountains, followed by smaller amounts on Mount Thoen and Netalzul Mountain. Most of the watershed has been rated as having moderate-to-high mineral potential due to the rocks of the Bulkley Intrusive.

Population and Settlement

The Bulkley and Suskwa valleys have been home to Gitxsan and Wet'suwet'en people for thousands of years, and land and resource utilization within the drainage was extensive and complex. Euro-Canadian settlers mostly arrived following completion of the railroad in 1914, or later, attracted by agricultural, mining, and various work opportunities. This population base remained relatively stable with low growth until the early 1970s when logging opportunities increased greatly, followed by the expansion of the government sector in the late 1970s.

The current settlement pattern reflects the past, wherein the valley bottoms contained the easiest travel routes and the most productive agricultural land, as well as providing the best cultural and economic opportunities. Currently, approximately 13,860 people reside within the drainage, in the Hazeltons, Moricetown, Smithers, Telkwa, and surrounding rural areas (RDKS 2002, BC Stats 2005).

Historically and up to the recent past, many Hazelton area people derived their income from the fishing and forestry sectors; however, severe job losses in these sectors has curtailed this income. Presently, most residents derive their income from service sector employment in the Hazelton area. Old and New Hazelton serve as the health, education, and service centers for residents in the northern portion of the drainage. Population trends project growth for aboriginal communities and a stable rural resident community. Presently, housing vacancies exist in New Hazelton and South Hazelton due to the downturn in forestry activities. Recreation and tourism-based incomes are projected to grow over the next decade.

In the southern section of the drainage, Smithers (pop. 5,612) and Telkwa (pop. 1,200) are the main settlement areas. Smithers serves as the health, government, education, recreation, and service center for the regional area extending to Moricetown and Fort Babine. Forestry is clearly the most dominant economic activity in terms of employment and community income, followed by agriculture. Tourism, retail trade, government services, and transportation help to keep the economy relatively buoyant and stable. The agricultural land base, of which approximately 50,000 ha are used as farmland, is mostly suited to dairy farming and ranching, with a focus on cow-calf production and forage crops. A significant amount of Crown rangeland supplements private grazing land.

Population growth trends are projected to be stable, though closure of the major lumber mill or further downsizing of regional government services could have a moderate impact on the population. Agricultural interests and rural residents help to give stability to the economic base.

Transportation and Utilities

The geography of the watershed is sufficiently adverse that both transportation routes and settlement patterns must conform to its restrictive mountain and valley characteristics. The existing transportation network in the watershed is based on the original Gitxsan and Wet'suwet'en trail network and reflects 120 years of steady development. Trails were initially widened for packhorses, and in some cases later improved for wagons, then further improved for vehicular traffic. The railroad line was completed in 1914, considerably increasing the population and economic activity. Highway 16 is the major road providing access into the lower Bulkley drainage. Forest resource road development throughout the watershed has tapered off and future trends point toward road maintenance responsibilities being off-loaded to industry or dropped altogether. Linear development within the drainage parallels the Bulkley River, the major river trunk system. Highway 16, CN Rail, BC Hydro, and Telus utilize a corridor that is located on the west and south side of the river, taking advantage of the geography and settlement patterns.

Channelization and installation of culverts blocking or hindering fish passage are seen at numerous locations along the rail line and highway. An example of a perched culvert can be found at the Highway 16 crossing of Porphyry Creek. Both Waterfall and Station Creeks near New Hazelton have had a history of habitat alterations related to railroad and highway construction, and more recently to changes in New Hazelton's water supply and sewage disposal systems (Remington 1996).

CN Rail serves the region and connects rail shipments to the rest of the country and to saltwater at Prince Rupert and Kitimat. The route is presently used for the movement of grain to Prince Rupert, for manufactured goods moving eastward, and for locally produced wood products moving in both directions. In the mid-1980s, portions of the track were upgraded to a standard sufficient to permit the carrying of coal by unit train from northeast British Columbia. The sophisticated communications and movement control system was also recently upgraded. With these modifications, the rail line presently has the capacity to handle many more trains than it does.

The Suskwa drainage has a well-established road network that was constructed to facilitate forest development activities. It connects to Highway 16 near the river mouth. The Suskwa Forest Service Road (FSR) leaves Highway 16 east of New Hazelton and crosses out of the watershed at Natlan Pass near the 38 km mark. This provides linkage to the Babine watershed and the Nichyeskwa Connector, which provides connection to the upper Babine River. From Highway 16 between Smithers and Telkwa, the Babine Lake Road through McKendrick Pass provides access to the upper Harold Price drainage via the upper Fulton FSR.

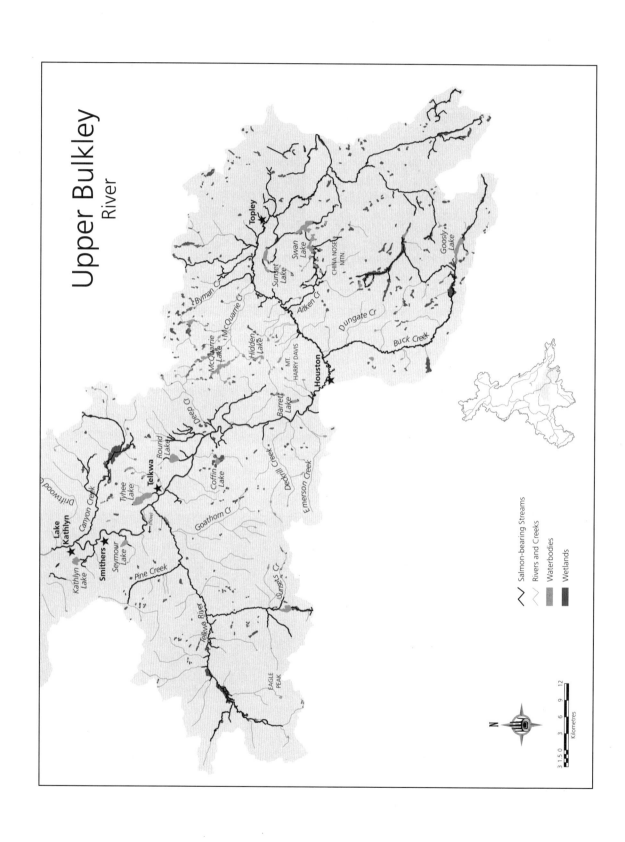

Upper Bulkley
River

Topley

Swan
Lake

Sunset
Lake

CHINA NOSE
MTN.

Goosly
Lake

Byman Cr

Aitken Cr

Dungate Cr

McQuarrie Cr

McQuarrie
Lake

Hidden
Lake

MT.
HARRY DAVIS

Houston

Buck Creek

Deep Cr

Barrett
Lake

Driftwood Cr

Round
Lake

Telkwa

Coffin
Lake

Dewill Creek

Emerson Creek

Canyon Creek

Tyhee
Lake

Goathorn Cr

Lake
Kathlyn

Smithers

Seymour
Lake

(flow)

Pine Creek

Kathlyn
Lake

Sunset Cr

Telkwa River

EAGLE
PEAK

Salmon-bearing Streams

Rivers and Creeks

Waterbodies

Wetlands

N

3 1.5 0   3   6   9   12
Kilometres

21

Upper Bulkley River

Environmental Setting

The upper Bulkley River includes the Bulkley River and its tributaries upstream of the Bulkley-Telkwa confluence, and the Telkwa watershed. This section does not include the Morice watershed, which is discussed in the next chapter. The Bulkley River is a major tributary to the Skeena River and flows into its left bank at Hazelton, 285 km upstream of its mouth.

Location

The upper Bulkley River is located in west-central British Columbia, and extends from the Telkwa-Bulkley confluence generally southeastward and upstream for 123 km to Bulkley Lake. The Telkwa-Bulkley confluence is approximately 99 km above the Bulkley-Skeena confluence, giving a total Bulkley River mainstem length of 222 km to Bulkley Lake. The drainage is bounded to the north by the Babine drainage, to the south and east by the Nechako drainage, and to the west by the Zymoetz and Morice drainages. Highway 16 and the CN Rail parallel the upper Bulkley River and pass through the towns of Topley and Houston.

Hydrology

The upper Bulkley drainage is mostly subdued, rolling country that lies on the Nechako Plateau, which is part of the Interior Plateau system. The Telkwa and Howson Ranges to the west afford the only areas of high relief. Elevations range from 2,523 m in the Telkwa Range and 2,796 in the Howson Range to 500 m at the Bulkley-Telkwa confluence. The Morice-Bulkley confluence lies at 620 m, while Bulkley Lake lies at 776 m elevation, with the majority of upper Bulkley drainage land lying between 600 and 1,500 m elevation.

The Telkwa River is a major streamflow contributor to the Bulkley River. In the Telkwa watershed, coastal climatic conditions and the comparatively high elevations influence overall precipitation with more than 1,500 mm total annual precipitation. Winter snowfalls form heavy snowpacks and significant glaciers in the mountains. Coastal storms often move into the upper Telkwa watershed and produce runoff that can contribute most of the storm flows of the Bulkley River. Beaudry and Schwab (1990) estimated that more than 50% of the peak streamflow volume is generated in the upper third of the watershed. Baseline streamflow assessments conducted in the Telkwa watershed in relation to the Telkwa coal deposit have been reported by Klohn Leonoff (1985), MacLaren (1985), and Agra (1998).

In contrast, the primarily continental climate of the Nechako Plateau is characterized by greater extremes of temperature and annual precipitation of 500–650 mm, with less than half of it falling as snow. In the upper Bulkley watershed, snow depths range from 0.5 m to 1 m at elevations below 900 m, with snow cover usually extending from mid-November to mid-April. Precipitation is greatest in the fall and early winter, and then tapers off to a low in April and May, before sharply increasing through the summer and into the fall. Summer convective storms are common but rarely deliver more than 10 mm of rainfall in a day.

Weather stations have operated or presently operate in Smithers, Quick, Houston, and Burns Lake. At Smithers Airport, located in the valley bottom, mean annual precipitation is 522 mm, with 331 mm of rainfall and 191 mm water equivalent of snowfall. Equity Silver Mine, located at 1,300 m elevation and 35 km southeast of Houston, receives an average annual precipitation of 710 mm, which consists of 300 mm rain and 410 mm of snow accounting for 70% of the precipitation.

Overall, the hydrology is dominated by snowmelt. The precipitation pattern usually results in a moderate storm-flow discharge distribution, and, other than in the Telkwa watershed, accounts for the lack of storm-flow events in the annual peak flow series. Mountains along the western margin of the watershed in the Hudson Bay, Telkwa, and Howson Ranges exert a major hydrological influence on the Telkwa River, capturing coastal rainstorms that can lead to floods.

Peak discharges from the Bulkley River and its major tributary, the Telkwa River, typically occur in May and June due to snowmelt and then steadily decrease until late September, when fall rains and early snowmelt increase streamflows until early November. Streamflows decrease in late November and December, when precipitation falls as snow, with minimum discharges recorded in January through March, prior to snowmelt.

Many of the smaller, low-elevation tributaries to the Bulkley River peak earlier in the spring melt than the mainstem. The relatively low amount of precipitation causes the discharges of low- and medium-elevation tributaries to drop off sharply following spring freshet. There are 14 hydrometric stations located in the upper Bulkley drainage recording flows on the mainstem, as well as high- and low-elevation tributary streams. The hydrometric station at Quick (08EE004), which is located approximately 13 km upstream of Telkwa, has one of the longest periods of record in the Skeena basin.

One-in-ten-year, seven-day average, low-flow estimates for the Bulkley River are 0.1 m³/s near Houston, 13.7 m³/s at Quick, and 15 m³/s at Smithers (Nijman 1986). The maximum daily discharge for the Bulkley River at Quick is 957 m³/s. On average, the Morice River flow accounts for 84% of the Bulkley streamflow at its confluence with the Bulkley and over 95% of flows at certain times.

Other than the Telkwa and Morice Rivers and Buck and Maxan Creeks, the majority of tributaries into the Bulkley River are less than 30 km long and are characterized by summer/early autumn low flows. These low flows are typically similar to winter streamflows, and both are principally derived from groundwater, lakes, and unfrozen wetlands. In some years during low flow conditions, McQuarrie, Maxan, Richfield, and Byman Creeks are partially dewatered and impassable to fish. An overall trend in declining discharge volumes has been noted for the upper Bulkley River (Remington 1996). The Bulkley River at Quick has a trend toward decreased September streamflows since 1930, which is discussed in the beginning of this book.

McBean et al. (1992) suggested that climate change might potentially reduce late summer streamflow in the B.C. Interior Plateau region. If this is occurring, it exacerbates the historical low streamflow situation in the Bulkley River and its tributaries upstream of the Morice River. Remington and Donas (2001) noted that it is difficult to separate hydrological variables due to climate change from land-use activities that also affect the hydrological regime in the upper Bulkley River.

Sufficient water flows and levels are required to allow upstream migration of salmon spawners over beaver dams, shallow riffles, small cascades, and waterfalls. Generally, streams with low flows and low water levels are also subject to extremes in water temperature, especially when riparian vegetation has been removed. During the warm summer months, and frequently into the fall, low flows may lead to increased stream temperatures, which may stress or kill salmonids, reducing reproductive success. Conversely, low flows and low water levels may lead to freezing in the winter, which reduces available habitat and potentially kills juveniles (Tamblyn and Donas 2001).

Stream Channels
The upper Bulkley River heads in Bulkley Lake and flows generally northwestward for 57 km to the confluence with the much larger Morice River. Right-bank tributaries from Bulkley Lake downstream include Ailport, Watson, Cesford, Richfield, Robert Hatch, Johnny David, Byman, McQuarrie, and Barren Creeks. Left-bank tributaries flowing into the Bulkley downstream to the Morice River include Maxan, Crow, Aitken, McKilligan, and Buck Creeks. The two major tributaries upstream of the Morice River are Maxan Creek and Buck Creek, both draining relatively large subbasins. The Bulkley

River then flows northwestward 45 km to the Telkwa River. Summary and detailed descriptions of reach habitat characteristics for various subbasins are available in Tredger (1982), Bustard (1985), Agra (1996), MacKay et al. (1998), and Tamblyn and Jessop (2000).

The uppermost part of the Bulkley valley has thick accumulations of deglaciation gravels, some of which formed in streams flowing eastward to the Fraser watershed. The apparent drainage divide at that time, approximately 10,000 years ago, was west of Topley. Valley bottom deposits formed by glacial sediments have been partially excavated by post-glacial streams, creating the floodplain and river benches that favor settlement and agriculture. Generally, most of the stream channels are lightly to moderately incised into the hillslopes, terraces, and valley bottoms.

Upper Bulkley River
The Bulkley River upstream of Telkwa has some of the most intense land use in the Skeena Basin on both public and private land, predominantly in the form of agriculture, linear development, and forest development activities. The 123 km mainstem to Bulkley Lake is generally low gradient throughout.

Reach 3 of the Bulkley mainstem extends 45 km from the Telkwa River upstream to the Morice River. Reach 3 is a single-thread, irregularly sinuous channel with an average gradient of 1%. The reach is characterized by gentle flows with intermittent rapids and occasional islands and bars. The river is lightly to moderately incised in the valley bottom with a discontinuous floodplain. Right-bank tributaries include Vallee, Thompson, Deep, Robin, McDowell, and Tyhee Creeks. Left-bank tributaries include Emerson, Dockrill, and Helps Creeks, and the Telkwa River. Emerson and Dockrill Creeks drain the east slope of the Telkwa Range and are cold-water creeks. Tamblyn and Jessop (2000) reported on the comprehensive fish habitat, riparian, and channel assessment on the 14 main tributaries that drain into reach 3.

Reach 4 extends from the Morice River confluence (locally known as the Forks) upstream 9.1 km to the mouth of Buck Creek at an average gradient of 0.5%. Reach 4 is characterized as an irregularly sinuous gravel-bed channel with occasional island and bar formations (MacKay et al. 1998). Buck Creek is the only significant tributary, on average supplying 19% of the upper Bulkley River flows. MacKay et al. (1998) noted that suspected cumulative impacts to reach 4 include loss of riparian areas, confinement of the river channel between the valley wall and the railway/highway corridor, increased sediment delivery, and increased peak flows from Buck Creek and the upstream reaches of the Bulkley River.

Reach 5 extends upstream 19.8 km in length to North Bulkley. The channel is typified by regular to tortuous meanders. The channel has an average gradient of 0.1% with primarily point and lateral bars and frequent oxbows (MacKay et al. 1998). The floodplain varies from 0.5 to 1 km in width with lateral movement of channels often constrained by the railway and highway fills. Channel structure is composed of avulsions, cutoffs, and side and back channels, as well as relatively large logjams. Reach 5 tributaries include McKilligan, Barren, and Aitken Creeks.

Reach 6 extends 7.9 km in length to the mouth of Byman Creek with an average gradient of 0.4%, and receives tributary flows and sediment input from McQuarrie and Byman Creeks. The channel, which is constrained between the valley walls and the highway and railway subgrades, slightly meanders in a floodplain ranging from 0.2 to 0.9 km in width. MacKay et al. (1998) suggested that the periodic removal by CN Rail of the large logjams removes key elements of water and sediment storage and energy dissipation, decreases channel complexity, and directly contributes to the loss of fish habitat as the channel becomes more incised and disconnected from the floodplain.

Reach 7 extends 29.2 km to the mouth of Bulkley Lake and is described as an unconfined, regular-to-tortuous meandering stream with oxbows and a very low gradient channel averaging 0.2%. The floodplain varies from 0 to 1.1 km in width, and channel lateral movement is often constrained by the extensive channelization associated with the CN Rail line, which has nine river crossings. From Forestdale upstream to Bulkley Lake, the substrate is composed mainly of gravel, while downstream the channel bed consists mostly of sand and fine sediments. Intact riparian areas are interspersed between heavily developed farmland. Reach 7 receives the streamflows from Crow, Ailport, Cesford, Richfield, Robert Hatch, and Johnny David Creeks.

Within reach 7, the Bulkley falls comprise a narrow rock sill that crosses the Bulkley River at a right angle. It is located upstream of Watson Creek and 11.3 km downstream of the Bulkley Lake outlet and has often been discussed as a potential salmon migration barrier. Dyson (1949) and Stokes (1956) surveyed the falls and the upper Bulkley system and concluded that the Bulkley falls pose a partial obstruction to migrating fish. Chinook, which migrate into the system early in the summer when water levels are high, are able to ascend the falls in substantial numbers. The falls are almost completely impassable to salmon during low water flows. Coho have not been observed above the falls since 1972, which appears to be a function of streamflow levels at the falls and the presence of beaver dams on the system (Pendray 1990).

Telkwa River

The Telkwa River flows generally eastward 70 km from the Coast Mountains to enter the Bulkley River at Telkwa. The river receives streamflow from 127 tributaries and is low gradient for most of its length. Other than Howson Creek, the tributaries flowing into the Telkwa River are mostly short, moderately steep-gradient streams that level out only in the last kilometre or less before entering the Telkwa mainstem. In general, most of these tributaries are of moderate relief with well-defined drainages. Channel stability conditions vary throughout, with unstable stream banks commonly contributing moderate to large amounts of sediment (Beaudry et al. 1991). Beaudry et al. (1991) noted that the large sediment loads carried by the Telkwa River during spring and fall runoff are dominated by natural erosion sources in all major tributaries to the river.

The sediment loads are principally caused by large active slump earth flows, gully erosion in alpine and subalpine areas, and streambank undercutting. The red soils derived

Upper Bulkley falls

by weathering of the Telkwa Formation volcanic ash deposits give a distinctive colouration to floods from the Telkwa River that is recognizable at the Bulkley-Skeena confluence (Hazelton).

Pine Creek is the most important chronic suspended sediment source due to active earth flow movements in the lower canyon (Beaudry et al. 1991). Saimoto (1996a), working under the auspices of the Telkwa Watershed Restoration Program, reported that the Goathorn-Tenas subbasin, followed by the Cumming and the Hubert-Coffin subbasins, were the most heavily impacted by forestry-related activities and also had the highest potential for restoration.

Reach 1 of the Telkwa River extends with a low gradient (<1%) 7.6 km upstream to the mouth of Goathorn Creek, and is a complex wandering gravel bed reach with multiple channels. Goathorn Creek is a major sediment source to the Telkwa River, which is evidenced by frequent channel shifts and widening, as well as dynamic side-channel development (Bustard and Limnotek Research and Development Inc. 1998).

Reach 2 of the Telkwa River is a single-thread channel extending 16.6 km to the mouth of Howson Creek. The channel is characterized by moderate, irregular meanders across the floodplain that is limited by the valley walls (Resource Analysis Branch 1975–79). Saimoto (1996a) noted many sections of unstable banks along the mainstem and the major tributaries, Pine and Cumming Creeks.

Reach 3 of the Telkwa River extends upstream 10 km to slightly above Sinclair Creek. The reach is typified by a single-thread channel with irregular meanders and is unconfined across a floodplain ranging from 0 to 1 km wide that contains extensive wetlands in the midsection. Reach 3 tributaries include Jonas, Arnett, Winfield, and Sinclair Creeks.

Reach 4 extends 14.5 km upstream in a slight to tortuous meandering pattern with an average gradient of 0.2%. The channel is unconfined across the floodplain, which averages 1.2 km in width and is, for the most part, a large interconnected wetland complex. Tributaries that contribute streamflows include Tsai, Milk, and Elliott Creeks, all of which drain glacial areas. Reaches 5–9 extend approximately 21 km upstream into the alpine.

Water Quality

Tamblyn and Donas (2001) indicated that the top water quality concerns in the Bulkley watershed are water temperature, suspended sediment loads, nutrient levels, attached algae (periphyton), and pathogens. Anthropogenic factors that have modified water quality include mining, linear, urban/rural, agriculture, and forestry developments.

Upper Bulkley River

At least in comparison with other Skeena watershed subbasins, water quality in the upper Bulkley drainage has been relatively well studied. A wide range of concerns and issues related to the nature and extent of developments are reported in Remington (1996) and Remington and Donas (2001). Nijman (1986) set provisional water quality objectives and made monitoring recommendations for the Bulkley River. Designated water uses included drinking water, aquatic life, wildlife, recreation, livestock, irrigation, and industrial use. In reach 4 between Houston and the Morice River, the use of Bulkley River water for drinking was excluded, presumably due to discharge from the Houston sewage treatment plant.

The Bulkley River at Quick shows the water as soft (mean hardness 26.5 mg/L), with a near neutral pH (mean=7.3), and moderate coloration due to wetlands in the drainage (Wilkes and Lloyd 1990). Water sampling of Bulkley River water at Telkwa in 2000 and 2001 was conducted by Dayton and Knight (2002), who recommended full treatment to alleviate water quality hazards to residents. In October 2001, Remington (2002) sampled Bulkley River at Telkwa and reported that fecal coliforms and *E. coli* concentrations were 1.6 and 1.6 CFU/100 mL (90th percentile) respectively. No other physical or chemical parameters exceeded drinking water guidelines.

The impacts of agriculture on water quality in the upper Bulkley drainage principally stem from land clearing and various aspects of livestock operations, such as manure handling, feed yard runoff, and cattle access to riparian zones and streams for watering. Gaherty et al. (1996) noted that winter feedlots may contaminate water bodies due to manure runoff into nearby streams; high nutrient loads may result as documented by Remington and Donas (1999, 2001). However, Gaherty et al. (1996) assessed the contribution of grazing lands to water quality concerns to be inconsequential in the upper Bulkley due to low cattle densities on rangelands and limited winter use.

Licensed water withdrawals in the Bulkley drainage above Houston for irrigation, domestic, stock watering, and water delivery is 0.527 m^3/s (Brocklehurst 1998). This is approximately 2.4 times the average 7-day, 10-year low flow of 0.216 m^3/s derived from the 1980–1993 Water Survey of Canada April–September data (Remington 1996). Remington and Donas (2001) noted that water allocation is potentially double the average supply that is currently available during summertime low streamflows.

Several other water quality studies have been conducted on the Bulkley River and its tributaries. Bustard (1996a) conducted dissolved oxygen and water temperature measurements during the winter at 22 locations. Remington (1997, 1998) reported on water quality and accumulated periphyton surveys in 1996 and 1997 in the Bulkley River and tributaries. Portman and Schley (2001) and Portman (2002) conducted water quality and fish sampling with the main objective of providing insight into the potential of impacts of agricultural spring runoff. Remington (2002) conducted drinking water source quality monitoring in the Bulkley River basin. Dayton and Knight (2002) conducted a water quality monitoring program

Bulkley Lake outlet

for the village of Telkwa. Finnegan (2002b) summarized water temperature data collected between 1994 and 1999 from four sites in the upper Bulkley drainage.

Overwintering studies in the upper Bulkley River show that winter water temperatures generally remain below 20° C. This may result in poor growth conditions for rearing fish and may possibly compound other factors, such as low discharge and oxygen depletion, which make winter survival difficult (Donas and Saimoto 1999).

The Rivers Ecosystem project was initiated by MELP to develop a toolbox of impact assessment methods that could be used to measure changes in the aquatic ecosystem. In 1997, Dykens and Rysavy (1998) conducted water quality inventories for Buck and Klo Creeks that assessed water chemistry, substrate composition, chlorophyll *a*, and benthic invertebrates. This work was continued in 1998 and then conducted by Agra (1999b). In 1999, Buck Creek was dropped, and Ailport and Klo Creeks were assessed (Agra 2000b). Rysavy (2000a) and Bennett and Ohland (2002) assessed and calibrated a multimetric benthic invertebrate index of biological integrity for the upper Bulkley drainage.

In addition, water quantity and quality sampling of Maxan and Foxy Creeks were conducted from 1972 to 1977 as part of the Maxan Lake multi-land use study, and was reported on by Abelson (1977) and McNeil (1983). Agra (2000a) re-installed staff gauges at historic Water Survey of Canada (WSC) stations on Richfield, Maxan, and Buck Creeks, as well

as a new station on Buck Creek to gather data to generate discharge hydrographs. Agra (1999a) conducted streambed substrate composition studies and sampled interstitial dissolved oxygen in various upper Bulkley locations, with control sites in Maxan and Buck Creeks.

Several studies have assessed and documented the effects and impacts of land use on tributaries of the Bulkley River. Mitchell (1997) assessed riparian and instream impacts to streams from Boulder Creek to Maxan Creek in the Bulkley headwaters. Tamblyn and Jessop (2000) conducted detailed fish habitat, riparian, and channel assessments on 10 Bulkley River tributaries from Telkwa to the Morice River. MacKay et al. (1998) conducted fish and fish habitat assessment work from the Morice River upstream to Richfield Creek.

Agra (1996) conducted fish and fish assessment work in the Maxan watershed, reviewing impacts from logging. Croxall and Wilson (2001) summarized and prioritized potential future restoration efforts in the upper Bulkley watershed east of Richfield Creek. These studies were conducted under the auspices of the provincial Watershed Restoration Program to document impacts to riparian, channel, and fish habitat, and provide restoration prescriptions.

Between 1972 and 1982, copper sulphide concentrates from Noranda's Bell and Granisle Mines were stored at two transfer stations located near Topley, quite close to the Bulkley River and apparently contaminated large volumes of the soils. Morin and Hutt (2000a, 2000b) conducted aquatic studies and concluded that water quality and aquatic life were receiving insignificant impacts.

In 1981, acidic drainage, commonly called acid rock drainage (ARD), was discovered at Equity Silver Mine. Interactions of the ARD and water quality are thoroughly reviewed by Remington (1996). Bustard conducted fish population studies in Foxy and Buck Creeks from the mid-1980s to 2002 (Bustard 1984b, Bustard 1993d, Bustard 1987–2002). Equity Silver Mine recently constructed a high-density sludge treatment plant to neutralize the ARD, which is projected to be deleterious for the long-term (the next 200 years or more).

Remington and Donas (2001) summarized the upper Bulkley drainage as a "high risk" watershed because of the easily erodible soils and the P-rich soils in some subbasins that contribute unusually high total and soluble phosphorous concentrations. In addition, they noted that relatively small increases in nitrogen concentrations from rural residences and agriculture might have a big effect on algae growth. This is seen in thick accumulations of benthic algae at some monitored sites of the mainstem near Houston. Remington and Donas (2001) described these sites as being accompanied by a shift in community composition toward mesotrophic or eutrophic taxa.

Telkwa River

Suspended sediment levels in the Telkwa River can be high during and following spring snowmelt. The generally milky color of the Telkwa River during the summer and early fall is caused by high-elevation glacier melt. Beaudry et al. (1991) investigated sediment sources within the Telkwa system and concluded that no direct link could be made to logging activities and that natural sediment sources dominated the sediment profile.

Water quality studies in the Telkwa drainage were conducted by Wilkes and Lloyd (1990) to gather baseline data for the province. The Telkwa River has soft water (mean=38.8 mg/L), a mean pH of 7.5, mean alkalinity of 42 mg $CaCO_3$/L, and low nutrient levels. Water quality data collection and monitoring in relation to the proposed Telkwa coal mining developments were undertaken by MacLaren Plansearch (1985) and Klohn Leonoff (1985); these reports were summarized by Manalta Coal (1997).

In addition, Bustard (1985) and Bustard and Limnotek (1998) reported on aquatic inventory resources studies, which focus on periphyton and benthic invertebrates in the lower Telkwa River drainage. Water quality data collected for the Telkwa Coal project indicate that surface and groundwater attributes are highly variable and naturally high in iron, manganese, and dissolved aluminum. Remington (1996) provided a succinct discussion of Telkwa Coal baseline water quality.

Upper Bulkley Lakes

The majority of Bulkley tributaries have lakes in their headwaters. Saimoto (1993) surveyed 10 of these lakes with the focus on fish species composition and recruitment factors. These lakes included Sunset, Gilmore, Swans, Lars, Old Man, McBrierie, Elwin, Watson, Day, and Bulkley Lakes. Hatfield (1998) conducted secondary lake inventories and compiled general, limnological, and aquatic flora data.

Bulkley and Maxan Lakes are the two headwater lakes located in shallow valleys characteristic of the area. Maxan Lake is approximately 8 km^2 with a maximum depth of 15.2 m and drains into Bulkley Lake via Maxan Creek. Bulkley Lake is approximately 4 km^2 in area with a maximum depth of 14 m and a mean depth of 7.2 m (Tredger and Caw 1974). Broman, Old Woman, and Conrad Lakes are connected by wetlands and ephemeral streams to Bulkley Lake (Agra 1996). The majority of streams in this upper portion of the drainage support extensive beaver activity.

Remington and Donas (2001) summarized the water quality analysis for Foxy and Maxan Creeks and the Bulkley River downstream of Bulkley Lake. They classified the trophic status of Bulkley Lake as eutrophic, with the total phosphorus concentration during spring overturn of 34 µg/L. The Water Quality Guidelines for total phosphorus is 10 µg/L during spring overturn of lake waters (Remington and Donas 2001). The lake waters support dense phytoplankton, causing reduced transparency and anoxic hypolimnion waters during periods of thermal stratification due to the high level of organic productivity.

Mid-Bulkley Lakes

Boyd et al. (1985) prepared water quality assessment and objectives for Tyhee and Round Lakes that focus on designated water uses including drinking, aquatic life, recreation, irrigation, and industrial use. Concerns with eutrophication due to nutrient contributions from agricultural and residential development prompted the establishment of the Tyhee Lake Management Plan (Rysavy and Sharpe 1995) and the Round Lake Management Plan (Kokelj 2004).

Geography

The upper Bulkley drainage is divided between two physiographic areas: the Telkwa and Howson Ranges in the west and the Nechako Plateau in the north and east. The prominent Bulkley River valley dissects the drainage in a general southeast-northwest line cutting through the Nechako Plateau and low, isolated mountains.

The upper Bulkley drainage upstream of Houston lies within the Nechako Plateau, which is characterized by rolling topography with elevations between 550 and 1,500 m. A blanket of glacial drift is widespread on the surface, and, other than along ridges and knobs, the majority of bedrock is obscured. Runka (1974) described the soils of the upper Bulkley valley as extensive fluvial and glacio-lacustrine floodplain deposits of fine-grained sands and silts that are particularly sensitive to soil disturbance and highly erodible.

During the Fraser Glaciation, the plateau experienced complex patterns of ice movement associated with multiple phases of ice flow moving southeastward from the higher mountain ice centers of the Hazelton and Skeena Mountains (Stumpf 2001). During deglaciation, remnant ice disrupted drainage patterns, ponded proglacial lakes, and created eskers and meltwater channels.

The Telkwa and Howson Ranges west of the Bulkley valley between Houston and Smithers are the prominent mountain features providing

View across the Bulkley River to the Telkwa River

streamflow into the upper Bulkley drainage. These ranges, particularly the Howson Range, possess substantial glaciers originating mostly on their northeastward-facing cirque basins. These ranges were forcefully glaciated during the Fraser Glaciation causing oversteepening of the mountain slopes and contributing to the present-day U-shaped valley profiles. At the peak of the Fraser Glaciation, ice moved up the Telkwa valley and through the mountains to the coast. The ice flow divide between westerly flowing ice and easterly flowing ice was over the Babine Mountains east of the Bulkley valley.

Most of the Telkwa River valley is filled with thick glacial deposits. Glacial till extends up the valley sidewalls, though the surface expression conforms generally to the underlying bedrock surface, with bedrock exposure along deeply incised streams and on steep-sided hillslopes. Thin soils, colluvium, and rock outcrops characterize the upper mountainsides.

Forests

The majority of land in the upper Bulkley drainage is covered with dense, coniferous forests, except the mountainous high country in the Telkwa and Howson Ranges. Smaller, but significant amounts of deciduous forests occur over the valley bottoms and lower slopes of the Bulkley and tributary valleys.

Within the watershed, principal forest types are represented by the two principal Biogeoclimatic Ecosystem Classification (BEC) zones: the Sub-boreal Spruce (SBS) and the Engelmann Spruce–Subalpine Fir (ESSF) zones. There is a small amount of the Coastal Western Hemlock (CWH) zone located in the upper Telkwa valley (Banner et al. 1993).

The SBS zone characterizes the lowland coniferous forests in the Telkwa and Bulkley valleys, as well as most of the Nechako Plateau. Subalpine fir and hybrid spruce are the major tree species; subalpine fir stands tend to dominate older, high-elevation stands and moister sections of the zone. The CWH zone is located in the upper Telkwa valley to 1,000 m elevation, in the valley bottoms and mid-mountain slopes; amabilis fir, alpine fir, and western hemlock are the most common species. The CWH and SBS zones merge into the ESSF zone at higher elevations ranging from 900 to 1,300 m, depending on local topography and climatic conditions. The ESSF zone possesses a shorter, cooler, and moister growing season, with continuous forests passing into subalpine parkland at its highest elevations. The main ecological attributes signifying this submaritime, high-elevation zone are a short, wet, cool growing season with considerable amounts of snow and comparatively low ecosystem productivity.

Geology

The geology of the upper Bulkley drainage is composed of three main assemblages of rocks: the Hazelton Group, the Skeena and Kasalka Groups, and the Ootsa and Endako Groups. The mainly Jurassic age Hazelton Group consists of mostly volcanic rocks formed in an island arc environment. These rocks are exposed along the western margin of the watershed and in the cores of uplifted mountain masses. The Telkwa Formation and Nilkitkwa Formation are the thickest units in the Hazelton Group. They are composed of marine and terrestrial sediments and thick volcanic ash deposits that were formed 157–136 million years ago (Gottesfeld 1985).

The island arc rocks are overlaid by younger volcanic rocks, such as the Skeena and Kasalka Groups formed 136–100 ma in a series of lava flows. These volcanic rocks are interspersed with Bowser Lake sedimentary strata to the north. The Bowser Lake sediments were formed in near-shore marine and coastal plains that included large swampy areas where peat and plant life accumulated to form coal beds.

From 100 to 65 ma, uplifting, volcanic, and plutonic activity created the Bulkley Intrusives that are chiefly composed of porphyritic rocks with minor amounts of granitic rocks. These intrusions into the sedimentary and volcanic rocks occurred at depth, and provided high heat levels to produce alteration zones of local mineral deposits including metallic ore bodies. Overall, metamorphism is light, aside from the contact effects near intrusive bodies.

A major period of uplift occurred when the Coast Mountains formed to the west. Subaerial volcanic rocks mainly composed of the Ootsa and Endako Groups erupted, producing the high Interior Plateau rocks. Over much of the Nechako Plateau, these flat-lying or gently dipping Tertiary lava flows cover the older volcanic strata and sedimentary rocks (Holland 1976). The plateau was disrupted by a series of major faults, which down-dropped blocks of rocks, causing the formation of major valleys such as those of the Bulkley River and Babine Lake. The dominant block faulting structure controls the location of the major mountain valley systems, as well as the many rock suites and mineral deposits. Numerous faults and contacts prevail with a strong regional pattern at 340°.

Fish Values

Fisheries values are very high within the upper Bulkley drainage. Coho, sockeye, pink, and chinook salmon, as well as steelhead and Pacific lamprey, are characteristic anadromous fish (FISS 2005). In general, the most widely dispersed salmon species is coho, while Dolly Varden, rainbow trout, and mountain whitefish are present in most fish-bearing waters.

Based on available historical records, it appears that low water and obstructions to fish passage in the upper Bulkley River above the Morice have been an ongoing issue. Typical reports from fishery officers (Elliot and Gelley) and memos from the district engineer (Dyson) and the district biologist (Stokes) between 1949 and 1955 indicate that the abundant coho, chinook, and sockeye salmon that used to frequent the upper Bulkley River above Richfield Creek were declining considerably. Subsequently, and continuing into the 1970s, the many logjams and beaver dams above and below the falls were cleared in the winter by CN Rail crews, various DFO programs, and local ranchers.

Chinook Salmon

The upper Bulkley River is an important migration route for two chinook stocks: the spring run that passes through to the upper Bulkley above the Bulkley falls and a summer-run to the Morice River and the Bulkley River above and below the Morice confluence. Run timing appears to be split at the Moricetown Canyon fishways at about July 30 (Peacock et al. 1997). The upper Bulkley early run is genetically distinct from the larger and later run. The status of the early Bulkley run is unknown.

Estimates of upper Bulkley River summer chinook escapements have been recorded continuously since 1945. Escapement was comparatively low from the mid-1960s through 1988; since then, there has been a substantial recovery. There were record high escapements in 2000 and 2001 of 2,560 and 5,600, respectively. Counts since then showed 1,100 in 2002, 1,280 in 2003, and 620 in 2005, with no survey in 2004. Chinook spawn in the mainstem, Buck Creek, Byman Creek, Richfield Creek, Maxan Creek, and Foxy Creek, with the latter four creeks being subject to seasonal fluctuations in water levels and flows.

Buck Creek supports a small chinook population ranging from 12 to 100 spawners recorded since 1970 on a discontinuous basis. Spawning is scattered throughout the mainstem as far upstream as the falls at the top end of the second canyon (reach 8, ~36 km). The series of cascades in reach 3 at 7.3 km is impassable in some years due to water conditions. Byman Creek has historical references to chinook spawning, and juveniles have been recorded in reach 1 up to the highway crossing (DFO 1991e). Current escapement status is unknown.

Richfield Creek historically supported moderate numbers of chinook spawners, ranging from 0 to 100 in the lowest reach near the Bulkley confluence (Hancock et al. 1983). There is no recorded escapement since 1964, and current escapement status is unknown. Maxan Creek and its major tributary, Foxy Creek, have both supported chinook spawners historically (Dyson 1949, Stokes 1956). There is one escapement record since 1950: 50 chinook in 1988. The preferred spawning location in Maxan Creek appears to be the boulder-gravel patches between the outlet of Maxan Lake and Foxy Creek confluence. In recent years, Maxan Creek has been subject to beaver activity, seasonal low flows, and drying.

Between 1987 and 2002, considerable quantities of chinook smolts, and, to a lesser extent, fry, were outplanted into the upper Bulkley mainstem – principally between McQuarrie and Richfield Creeks (O'Neill 2003). The upper Bulkley enhanced chinook stock serves as a coded wire tag indicator stock (Peacock et al. 1997).

Pink Salmon

The movement of pink salmon in the upper Bulkley drainage is primarily through the mainstem upstream into the Morice system. Spawning has been documented in the Telkwa River and Buck Creek systems and in the upper mainstem to the Bulkley falls, though escapement data are scarce.

Bulkley River

Pink salmon spawning on the Bulkley mainstem between the Telkwa and Morice Rivers are confined to five small discrete locations (DFO 1991e). The upper river has historical references to scattered pink salmon spawning up to the Bulkley falls (Hancock et al. 1983, DFO 1930–1960). MacKay et al. (1998) noted that pink spawners are found to just above Buck Creek in the available riffles. Spawning was observed in the lower kilometre of Deep Creek in 1999 (Tamblyn and Jessop 2000).

Telkwa River

In the Telkwa system, pink salmon spawn in the mainstem above and below the Goathorn Creek confluence. Bustard (1984c) noted that gravel areas interspersed along the mainstem downstream of Goathorn Creek, particularly in active side channels, are utilized to a certain extent by spawners. Bustard also noted that specific spawning sites would probably change from year to year given the frequency of channel changes and the range of flow conditions. In years of high escapement to the Bulkley River, the lower Telkwa River is most likely heavily utilized by pink salmon spawners.

Pine Creek supports pink spawning in the lower 2 km. Pink salmon spawn in the lower 2.6 km of Howson Creek (DFO 1991e). Bustard (1984a, 1998) reported spawning in lower Tenas Creek and 58 spawners (1983) in the lower 0.6 km of Goathorn Creek. Bustard (1984) suggested a crude estimate of between 500 and 1,000 pink salmon for the lower Telkwa River in 1983.

Buck Creek

Pink salmon have been recorded spawning in the lower kilometre of Buck Creek and are suspected to be present up to the cascade in reach 3 at 7.3 km (MacKay et al. 1998). The only escapement record is from 1963 with 100 individuals.

Sockeye Salmon

Sockeye salmon used to spawn in Maxan Creek and most likely spawned in Bulkley and Maxan Lakes. Recorded escapements ranged between 50 and 600 until 1978. The stock or stocks then appear to have collapsed, and records in the 1980s show few or no fish returning. In 2001, several sockeye were spotted at the coho counting weir in Houston that may have been heading upstream to Bulkley Lake. Maxan Creek does not have sufficient flow to allow sockeye passage in some summers. This was reportedly the case in 2001, a relatively wet year (Joseph 2001b). High water temperatures could also cause access problems. Bulkley Lake and Maxan Lake sockeye are at high risk of extirpation. Recent observations reported by Finnegan (2006) indicate sockeye spawning in the mainstem downstream of McQuarrie Creek.

Coho Salmon

Juvenile and adult coho salmon are the most widely dispersed salmon species in the upper Bulkley drainage. Coho behavior and the variability in their life histories, particularly in the freshwater period prior to smolting, are not well known in the upper Bulkley watershed.

Coho fry emergence extends from April to July with an estimated 15–27% average egg-to-fry survival rate. Saimoto and Jessop (1997) suggested that based on the relatively early spawning time and suspected times of emergence, coho eggs and alevins are in the gravel for periods of six to seven months in the upper Bulkley drainage. Juveniles are widely distributed in accessible, slow stream waters and in various side and back channels. Many of the small tributaries flowing into the Bulkley River serve as auxiliary juvenile coho habitat as downstream migrants move into these streams.

Upper Bulkley River

From 1949 to 1970, coho spawner escapement was recorded in 13 out of 21 years in the upper Bulkley mainstem. The dominant limiting factor appeared to be the low water levels. Historical escapement estimates for the upper Bulkley coho aggregate, including Maxan and Buck, ranged as high as 7,500 in the 1950s, though the annual average was 2,850 coho for the 1950s and 1960s. These visual escapement estimates are almost certainly underestimates of real abundance. No adult coho have been recorded in Maxan Creek since 1972, and juvenile sampling efforts from 1987 to 1990 did not record coho presence (Pendray 1990).

The upper Bulkley coho aggregate is made up of populations spawning and rearing in the mainstem channels and in Buck, Aitken, McQuarrie, Byman, Richfield, Ailport, and Maxan Creeks. Overall, the upper Bulkley subbasin coho aggregate showed a serious decline from the mid-1960s to 1998, with an apparent increase beginning in 1998. Holtby et al. (1999) conservatively estimated the wild coho escapements to the upper Bulkley, and evaluated a decrease in returns of 11% per year from 1970 to 1998. Escapements increased between 1998 and 2005, and annual returns averaged 1,358 coho with a range of 317 to 2,508.

During the past few decades, the distribution of adult and juvenile coho has been limited to the downstream portion of the Bulkley River channel below Richfield Creek (Saimoto and Saimoto 2001). This is most likely due to low flows in late summer/fall and to a lesser extent, winter streamflows. Pendray (1990) noted that in years of relatively high summer streamflows, upper Bulkley tributaries appeared to be heavily utilized by juvenile coho, with rearing densities much higher than in the mainstem. Pendray (1990) reported that the best coho juvenile densities found in the mainstem were at the riprap sites, which provided artificial cover.

Since 1989, an annual average of 30,000 coho fry and smolts have been outplanted in the upper Bulkley mainstem (McQuarrie to Richfield Creeks) from upper Bulkley stock raised at the Toboggan Hatchery (O'Neill 2003). Holtby et al. (1999) noted that it would be interesting to know if the synchrony of enhancement, which began with the 1989 smolt release, and the rapid decline in wild abundance thereafter was just a coincidence, and if so, what was the probable cause of the decline.

A counting weir on the upper Bulkley River located at Houston has been operated annually since 1989, except for 1991. The primary function of the fence operation has been to capture brood-stock for hatchery production. Holtby et al. (1999) reported that the total escapement in 1998 was 317, of which 139 coho were the progeny of wild spawners – a number that was slightly greater than the brood year escapement.

The proportion of hatchery coho in the escapement has been an issue of concern. In most years since enhanced coho began returning, over 60% of the escapement has consisted of the hatchery stock. Donas (2001a) reported that between 1997 and 2001, the average proportion of hatchery coho counted at the fence was 71%. Another point of concern has been that the coho pool below the fence and are reluctant to pass upstream through the fence. This has necessitated seining operations to move fish above the weir (Ewasiuk 1998, Glass 1999, Glass 2000, Donas 2001a). It is uncertain if the coho falling back downstream spawn elsewhere or regroup for later upstream movement.

Studies concerning the assessment of overwintering habitat and distribution of juvenile coho in the upper Bulkley drainage (above the Morice River) were conducted by Saimoto and Jessop (1997) and Donas and Saimoto (1999, 2001). Saimoto and Jessop reported on fish presence and densities at 15 sample sites and found no juvenile coho above the McQuarrie Creek confluence. Overall coho densities in the mainstem were relatively low.

Telkwa Watershed
Adult coho destined for the Telkwa drainage generally ascend the Bulkley River in October. Spawning is observed from October through December and usually peaks in mid to late November (Manalta 1997). Coho escapement estimates have ranged from 100 in many years to 9,450 in 2001. The annual escapement between 2000 and 2005 has been 5,612 from a range of 1,088 in 2000 to 14,840 in 2005 (DFO 2005).

Bustard (1983b, 1985) reported on detailed coho spawning surveys conducted in 1982 and 1984 that focused on the lower Telkwa River area, particularly Goathorn, Tenas, and Pine Creeks. No juvenile or adult coho were found in the lower Telkwa River area, probably due to the severe icing conditions prevalent during the winter in the lower river. Bustard (1985) reported that surveys found the majority of coho spawning in the upper mainstem reaches (reaches 2 and 3) of the Telkwa River, as well as in lower Elliot Creek. Much of the mainstem spawning occurred from above Jonas Creek to 5 km above Milk Creek (30 – 47 km), with the heaviest use from above Jonas Creek to Sinclair Creek (30 – 34 km). The upper reaches of the Telkwa River were also preferred by juveniles. This may be due to less severe icing conditions, the abundance of groundwater inflows, smaller substrate material, and the generally excellent rearing areas available (Bustard and Limnotek 1998).

Two off-channel ponds developed to create juvenile coho habitat are located on the Telkwa River floodplain at 10 km and 11 km on the Telkwa Forest Service Road (Bustard 1996b, Bustard 2000). The ponds were constructed in 1993 by the DFO as a pilot juvenile enhancement project, with additional improvements to increase water flows and deal with beaver activities (Bustard 1997a). Success has been measured with mark and recapture, immigration studies, and emigration studies, which show large increases in pre-smolt production reflecting strong fry and yearling recruitment into the ponds (Bustard 2000).

Buck Creek
Buck Creek is thought to be one of the most potentially productive salmonid streams in the upper Bulkley system (MacKay et al. 1998). Buck Creek supports spawning and rearing coho with escapements that ranged from 75 to 600 up to the late 1970s. Since then, there has been one record: 50 coho in 1982. During the 1960s and 1970s, the annual average escapement was approximately 275 coho. The DFO (1991e) reported that at certain water levels the reach 3

Maxan Creek downstream of Maxan Lake

cascades cause fish passage difficulties. Hancock et al. (1983) noted coho spawning areas in the lower reach: at approximately 3.5 km, above and below the Dungate Creek confluence on Buck Creek, and in the lower reach downstream of the falls on Dungate Creek.

Buck Creek enhanced coho stocks were outplanted in 1999, 2000, 2001, and 2002 with a mix of fed fry and smolts. These coho, along with juvenile chinook and rainbow trout/steelhead, were then sampled to assess emigration potential in 1999 and 2000, utilizing a rotary screw trap positioned off the first bridge on Buck Creek (SKR 2000, Donas 2001b).

The Buck Creek release pond was constructed in 1999 to improve performance of coho fry and smolt releases into Buck Creek (Donas 2001b). The intent was to assist in increasing the survival rate by improving the release technique. SKR (2000) reported that released coho vacated the release pond within a few days following their release. The release pond was destroyed in the Buck Creek 2002 spring flood (Tamblyn 2003).

Steelhead

Steelhead of the upper Bulkley River drainage are highly valued. The upper Bulkley drainage supports a summer-run steelhead population that enters the mouth of the Skeena in late June to early August, arriving in the Bulkley system beginning in August and continuing into autumn (Lough 1981). In general, there is scant information concerning discrete stocks, life history, and instream movements.

The pattern of catch reported for the Bulkley River shows an increase in fishing effort and catch over the past 35 years, with the pattern of total catch, including released fish, resembling the Tyee Test Fishery estimates for summer-run steelhead (Anonymous 1998). Recently, Mitchell (2001a) and SKR (2001) reported on Bulkley-Morice steelhead population estimates based on the Wet'suwet'en Fisheries tagging program conducted downstream of Moricetown Canyon. Mitchell (2001a), using the Petersen Method, estimated a population of 22,630 steelhead upstream of Moricetown Canyon.

Upper Bulkley River
The Bulkley mainstem is used as the migration corridor for the large Morice watershed steelhead population. The sport steelhead fishery in the Bulkley River mainstem downstream of Morice River is known worldwide for providing high-quality fishing and good fishable water conditions. Unfortunately, information regarding steelhead adult escapement and distribution, overwintering habitat, and juvenile densities is scant.

In the Bulkley River upstream of the Morice confluence, steelhead spawners have been, or are present, in the mainstem, in Buck, McQuarrie, Byman, Richfield, and Ailport Creeks, and possibly in Johnny David and Robert Hatch Creeks (Tredger 1982, DFO 1991e, Mitchell 1997). Tredger (1982) conducted a reconnaissance-level assessment in the Bulkley upstream of the Morice that focused on outlining the standing crop of steelhead juveniles and estimated carrying capacity. Tredger expressed difficulty in getting any confident estimates of steelhead juvenile populations due to problems differentiating steelhead from resident rainbow populations, particularly near headwater lakes. Tredger made rough estimates of basin-wide smolt outputs and adult escapements based on the standing crops of fry, which in turn were based on the output of carrying capacity from minnow trapping data; his data suggest 92,100 fry, 4,100–11,800 smolts, and between 155 and 1,260 adults.

Steelhead spawn on the Bulkley mainstem between the Telkwa River and the Morice River near Hubert (DFO 1991e). Three years of sampling by Bustard and Limnotek (1998) of steelhead juveniles in Hubert Creek indicates that the abundance and distribution are highly variable from year to year due to habitat conditions and presumably the number of fry recruiting upstream from the Bulkley River.

Telkwa Watershed
Telkwa steelhead are summer-run fish that enter the Telkwa system in late August or early September (Spence 1989). Most steelhead apparently overwinter in the Bulkley River downstream of the Telkwa River, though some steelhead have been observed overwintering in the Telkwa River (Read 1982).

The Bulkley River is one of the most heavily fished steelhead rivers in the Skeena system, and the section in the vicinity and downstream of the Telkwa River confluence is one of the most heavily fished reaches (O'Neill and Whately 1984). Although the Telkwa River receives limited steelhead angling, it is likely that Telkwa steelhead holding in the Bulkley mainstem form an important component of the steelhead fishery on the Bulkley River (Bustard and Limnotek 1998).

Movement into steelhead spawning areas is variable in time (March to May) and also dependent on temperature and water conditions. Estimates of steelhead abundance for the Telkwa system have not been conducted; though Bustard (1985) suggested steelhead spawning populations in the order of 52 in Goathorn Creek, 107 in Tenas Creek, and 347 in the lower Telkwa River. These estimates were established by projecting assumed survival rates from parr to returning adults.

Bustard and Limnotek (1998) summarized the fish and habitat assessments from the early 1980s and 1997, noting that lower Goathorn Creek, Tenas Creek, and the lower Telkwa River are important steelhead streams. In addition, their studies suggest that Tenas and lower Goathorn Creeks are the most productive steelhead tributaries in the Telkwa watershed. The data also suggest that lower Goathorn and Tenas Creeks, along with the Telkwa River mainstem and side channels, are very important steelhead rearing areas with average steelhead fry densities ranging from 20 to 50 fry/100m^2.

Resident Freshwater Fish
In comparison to salmon, information is scant in regard to resident freshwater fish in both river and lake habitats of the upper Bulkley drainage. Freshwater resident fish presence is represented by cutthroat trout, lake trout, Dolly Varden, rainbow trout, mountain whitefish, bull trout, kokanee, burbot, prickly sculpins, longnose dace, largescale sucker, white sucker, longnose sucker, river lamprey, Pacific lamprey, lake chub, peamouth chub, northern pikeminnow, prickly sculpin, and red-sided shiner (FISS 2002, WLAP 2002).

Fish inventories in the Telkwa system focused on or including freshwater resident fish have been reported by Triton (1998d), SKR (1998), Bustard (1985), and Bustard and Limnotek (1998). In the Bulkley subbasin, Tamblyn and Jessop (2000) sampled streams from the Telkwa River upstream to the Morice confluence. Tredger (1982) noted freshwater fish presence in the course of his juvenile steelhead carrying capacity sampling. Saimoto (1993) and Hatfield (1998) surveyed many of the upper Bulkley drainage lakes and sampled for freshwater resident fish. Juvenile fish population studies that included resident fish in Foxy and Buck Creeks were reported by Bustard (1984b, 1987–2002, and 1993d) as part of the Equity ARD environmental monitoring.

Fisheries
Wet'suwet'en Fisheries
The salmon stocks and anadromous lamprey passing through and spawning in the upper Bulkley formed the principal food resource that enabled Wet'suwet'en people to make the area their home. Both major aboriginal fisheries, at Hagwilget and at Moricetown Canyons, were negatively affected by alterations to the canyons to speed fish passage. Nevertheless, in the recent past, the Moricetown Canyon fishery fulfilled the food, societal, and ceremonial (FSC) needs of the Wet'suwet'en; however, over the last several years sockeye escapements to the Morice system have been so low as to preclude sockeye fishing.

Table 5. Selected Wet'suwet'en traditional fisheries locations and species

Site Location	Traditional Site Name	Fish Species
Maxan Lake outlet	Tasdleegh	SK, CO, CH
Maxan–Foxy Creeks confluence	Tsaslachque	SK, CO, CH, trout
Bulkley Lake outlet	Nehl' dzee tez diee	SK, CO, CH.
Gilmore Lake		Trout
Sunset Lake	Alk'at	Trout
Elwin Lake	Deetts'eneegh	Trout
Fishpan Lake	Laytate Ceek	CO
Emerson Creek–Bulkley confluence	Decen Neeniinaa	SK, CO, PK, CH, ST
McQuarrie Lake outlet	Deeltsik	CO, ST, trout, char
Buck Creek, below falls		CO, ST, CH
Bob Creek–Buck confluence	Dzenk'et Hoz'aay	CO, ST, CH
Goosly Lake outlet	Neelhdzii Teezdlii Ceek	CO, CH, KO, RB
Klo Lake	Tsee zuulceek ben	Trout
Barrett Lake outlet	C'eli t'oots Ta'eet	CO
Round Lake outlet	Coostl'aat ben	CT, RB
Round Creek–Bulkley confluence		Trout
Coffin Creek–Bulkley confluence		Trout
Tyhee Lake outlet	Kyo kyut tezdlii	Trout and suckers
Bulkley River, downstream of Quick	Ses biit kwe	SK, CO, CH, PK
Bulkley River, upstream of Morice River at Hwy. 16 crossing	Needz Kwe	SK, CO, PK, CH, ST
Howson Creek–Telkwa confluence	Neetay	CO, ST, PK
Telkwa River at Jonas Flats	Tasdleegh	CO
Elliot Creek–Telkwa confluence	Sggwelii dziikw taceek	CO

Upper Bulkley River

Wet'suwet'en salmon fisheries within the upper Bulkley drainage were concentrated at seven principal areas that contained multiple fishing sites on the Bulkley mainstem. These sites include the Telkwa River confluence, a site 3 km northwest of Barrett, the Morice River confluence, the mouth of Buck Creek, the Bulkley falls, the outlet of Bulkley Lake, and the outlet of Maxan Lake (Gitksan and Wet'suwet'en Chiefs 1987).

Many other secondary fishery sites were operated on the Bulkley mainstem, usually close to tributary streams. Dispersed fisheries were operated that focused particularly on coho and resident freshwater species. Many of these dispersed sites are only recognizable in the present by cache pits or other cultural heritage evidence. Selected Wet'suwet'en fishery locations and types of fish harvested are shown in Table 5 (Gitksan and Wet'suwet'en Chiefs 1987, Mills

and Overstall 1996, Rabnett et al. 2001, Office of Wet'suwet'en 2001, Wet'suwet'en Fisheries 2002).

Telkwa Watershed

Wet'suwet'en fished various sites on the Telkwa mainstem including Neetay at the Howson Creek confluence, Sggwelii Dziikw Taceek at the Elliot Creek confluence, Tasdleegh at the Jonas Creek confluence, and a site at the Goathorn Creek–Telkwa confluence. Dispersed traditional fisheries were operated in lower Goathorn Creek, Pine Creek, Howson Creek, and Mooseskin Johnny Lake.

Recreational Fisheries

The Bulkley River supports large-scale, seasonal recreational use. The majority of recreational angling on the Bulkley mainstem is divided into two relatively distinct fisheries of equal proportions: the nonresident angling fishery and the area resident fishery (Morten and Parkin

1998, Morten 1999). Anglers predominantly favour the stretch from the Telkwa River to the Morice confluence on the mainstem. This reach generally possesses stable water conditions in terms of flows and clarity, except in the heaviest of rainstorms.

Pedestrian and vehicle access by anglers is generally achieved by crossing private land in most cases and Crown land in some instances. The Bulkley mainstem (from Telkwa to Morice) is navigable by powerboats and drift boats for its length, and access is available at regular intervals. The use of both power and drift boats by local anglers, guides, and nonresidents are a major feature of the steelhead fishery. Morten and Parken (1998) reported that the use of powerboats as compared to drift boats is approximately 2:1; their survey data also confirm that the primary method of access to the river is by boat rather than by foot.

The river sport fishery is primarily directed first to chinook until early August, and then to steelhead, with pink salmon openings held in years of abundance. Chinook angling primarily occurs around the vicinity of the Bulkley-Morice confluence, though angling effort is scattered throughout. Total chinook effort is comparatively small in relation to steelhead effort, and participants are mostly B.C. residents. Tallman (1996) and Morten and Parken (1998) conducted creel surveys that focused on the Bulkley mainstem chinook and pink salmon sports fisheries and reported on effort, catch, and catch rates.

The majority of steelhead angling occurs over approximately a 10-week period from early September through mid-November. Peak activity is commensurate with the period that most anglers feel provides the best combination of steelhead availability, good weather, and clear water – usually from mid-September to mid-October. Angler activity for steelhead is intense, and a public consensus planning process was established in late 1996 to deal with angler crowding and provide for long-term angling management on the Bulkley River. The Angling Use Plan for the Bulkley River (Anonymous 1998) was signed off in 1998; however, the senior manager of the Ministry of Environment,

Lands and Parks refused to implement it, and many plan participants have sour feelings.

Seven angling guides are licensed to operate on the Bulkley River with a set quota of guided rod days that total 1,504 for the entire river. Although annual angling regulations specific to the Bulkley River may change, they generally include a bait ban and no fishing January 1 – June 15. There is no fishing upstream of the Bulkley-Morice confluence. There is no angling from boats between the Morice River and the CN Rail bridge at Barrett from August 15 to December 31. The Bulkley mainstem is Class II waters from September 1 to October 31, with a mandatory steelhead stamp and steelhead release between July 1 and December 31 (B.C. Fisheries 2005).

Enhancement Activities

Enhancement activities in the upper Bulkley watershed consist primarily of juvenile chinook and coho outplants as well as the construction and maintenance of the Telkwa River coho rearing ponds and the Buck Creek release pond. These projects have been discussed in the Telkwa and Buck Creek coho summaries. The enhancement projects that had large impacts to fish passage and abundance in the upper Bulkley are the rock removal in Hagwilget Canyon and construction of the fishways at Moricetown Canyon.

Development Activities

Land use in the upper Bulkley drainage is relatively intensive in comparison with other Skeena watershed subbasins. Over the last 40 years, land-use and development activities in the drainage have increased significantly. The current principal land-use activity is logging, which is generally concentrated in the Telkwa valley and low to mid elevations in the Bulkley valley. There is extensive agricultural development in the valley bottom, which has the highest agriculture capability in the Skeena watershed (Remington 1996). Urban development is concentrated in the Telkwa and Houston areas. Farms and rural residential settlement areas are scattered along Highway 16 east to Forestdale and generally parallel both sides of the Bulkley River. Industrial land use is limited and close to the urban areas.

The early part of the 20th century saw extensive mineral exploration in the Telkwa Range; presently there is little mineral exploration or development in the drainage. Coal bed methane exploration has recently been proposed for the Telkwa coalfield. Linear development along the Bulkley River accommodates the CN Rail line, Highway 16, and the BC Hydro and Telus utilities infrastructure.

Wet'suwet'en Land Use

Wet'suwet'en land and resource use has occurred for approximately 10,000 years (Allbright 1987, Trusler et al. 2002). Wet'suwet'en forest utilization within the upper Bulkley drainage was extensive and complex. Cultural use features can be seen as complex series of ancestral and historical threads that form the cultural landscape fabric. In both the pre-contact and post-contact periods, trails formed the travel and communication network of the region. Trails facilitated travel, trade, social interaction, and access to spiritual and ceremonial sites, home places, and resource gathering locales (Suskwa Research 2002).

Wet'suwet'en used fire as a tool to shape their environments and improve opportunities to harvest plant and animal resources. Evidence presented in a number of ecological studies (Williams et al. 2000; Haeussler 1987; Pojar 2002, Trusler 2000) shows that the landscape burning activities of the local First Nation established and contributed to the maintenance of extensive seral landscapes.

Forest Resource Development

Logging in the upper Bulkley drainage is an important land-use and economic driver. The upper Bulkley drainage is located within the Ministry of Forests, Nadina and Skeena Stikine Forest Districts. Currently, the Ministry of Forests provides land-use management zoning, objectives, and strategies through the Lakes, the Morice, and the Bulkley Land and Resources Management Plans (LRMPs).

The community of Houston developed as a rail tie-cutting center in the early 1900s. The post-World War II demand for lumber

contributed to the gradual logging of low and mid-elevation stands in the Bulkley valley. The first planer and gang mills were brought to the area in the 1940s, and by 1958, there were 84 small sawmills operating in the area (Ministry of Sustainable Resource Management 2002). In the late 1960s, Bulkley Valley Industries (BVI) was formed to produce an integrated forest products complex including lumber, plywood, stud, and pulp mills, though this plan fell through.

Since then, the trend to centralization and the monopolistic pattern of forest license holding and milling capacity have laid the foundation for the current situation. The BVI mill was sold to Northwood and is currently owned by Canadian Forest Products. The mill is currently the largest sawmill in the world. In 1978, Houston Forest Products opened a second sawmill. From the late 1960s to the present, the area and volume logged has increased from small-scale operations to a large-scale operation with an industrial road network and intensive clearcutting within most portions of the drainage (Hols 1999, Office of the Wet'suwet'en 2001).

The majority of forest development has been directed to the easily accessible, commercially viable timber. Major licensees log approximately 90% of the annual allowable cut, with the remainder cut by the small business and woodlot programs (Ministry of Sustainable Resource Management 2002). Presently, most forest harvesting and road building operations focus on timber affected by the mountain pine beetle (MPB).

The strength and stability of the forest sector on the Nechako Plateau is supported by two principal factors: over 50% of forest stands is composed of pine, which allows for production of high-quality timber; and the geography of the plateau lends itself to relatively easy access to forest lands. In addition, timber supply is relatively plentiful in comparison to other Skeena subbasins.

From 1995 to 2001, the Watershed Restoration Program (WRP) was involved in assessing and repairing logging-related disturbances to fish, fish habitat, and upslope sediment-producing areas. Within the Telkwa watershed, Saimoto (1996a) and SKR (1998) reported on fisheries, fish habitat, and riparian zone assessments, while Silvicon (2001) reported on site works completed. Completed projects included bank stabilization, cut slope and landslide rehabilitation, and bridge abutment repairs.

In the Bulkley valley, the WRP sponsored the mid-Bulkley overview assessment of fish and fish habitat conducted by MacKay et al. (1998) for the area from the Morice River upstream to Richfield Creek. Agra (1996) conducted fish and fish assessment work in the Maxan watershed, with Croxall and Wilson (2001) summarizing and updating restoration efforts in the upper Bulkley watershed.

Mineral Resource Development

Euro-Canadian exploration for base and precious minerals has been active since the early years of the 20th century. Early efforts were focused mainly on high-grade silver, copper, and gold-rich veins, as well as placer gold, with most endeavors slowing down with the onset of World War I, the Depression, and World War II. Silver-lead-zinc ore was discovered close to Goosly Lake, and similar ore was discovered on Grouse Mountain around 1912 (Hols 1999). Exploration and mineral development in the Telkwa Range centered on copper-silver prospects west of Mooseskin Johnny Lake and copper-silver-gold prospects in the headwaters of Goathorn Creek.

The Telkwa coal deposit was located early in the century and exploited on a sporadic basis from 1918 until the mid-1970s – operating as the McNeil Coal Mine until 1930, then as the Bulkley Valley Collieries. Manalta Coal Ltd. currently holds the coal reserves in the lower Telkwa River area. Proposed mining plans include conventional open-pit methods with diesel powered mining equipment. The outstanding potential risk is the generation of acid rock drainage (ARD). Whether and how it might be mitigated is not clear. Aquatic baseline resource studies relevant to the Telkwa coal deposit are summarized by Bustard and Limnotek (1998).

The early 1960s saw a resurgence of exploration focused on porphyry, copper-zinc-silver style deposits. In 1961, copper and zinc values were found on the ridge east of Goosly Lake. After two decades of extensive proving of ore bodies, Equity Silver Mine operated from early 1980 to early 1994, producing primarily silver with additional gold and copper. The mine used open-pit, surface mining methods, with a limited underground program. An on-site mill produced concentrate from the excavated ore.

Features remaining from the mine development include two open pits, one backfilled pit, a contiguous series of waste rock dumps, the mill site, and a tailings impoundment. Acid rock drainage was evident in 1981 and large-scale mitigation efforts to neutralize the acidity were put in place. Since that time, extensive environmental assessments have been conducted at the mine site and the receiving environments drained by Foxy and Buck Creeks. Except for flooded surfaces, access routes, and the facilities for drainage collection and treatment, all disturbed surfaces have been capped with a compacted soil cover and revegetated.

Oil and Gas Development

In late 2003, the B.C. Ministry of Energy, Mines and Petroleum Resources issued a call for proposals to develop coal bed gas resources near Telkwa. In early 2004, it announced that Outrider Energy would have the rights to explore and develop coal bed methane near Telkwa. If the Telkwa project were to go ahead, it will be the first in the province. A recent survey shows a broad cross section of local First Nations and residents oppose the development because it threatens the economy, lifestyle, and land. Given the potentially serious impacts of coal bed gas development, the proposed development undermines and threatens the integrity of the lower Bulkley and Telkwa drainages.

Population and Settlement

The upper Bulkley valleys have been home to Wet'suwet'en people for thousands of years. Euro-Canadian settlers arrived in numbers following completion of the railroad in 1914, attracted by agricultural, mining, tie hacking, and other work opportunities. This population base remained relatively stable with low increments of growth until the early 1970s, when the forest sector expanded, followed by the government sector in the late 1970s, and the opening of Equity Silver Mine in 1980.

The current settlement pattern reflects the past in that the valley bottoms contain the easiest travel routes, the most productive land, and the best economic opportunities. Currently, approximately 5,750 people reside from Telkwa east to Topley and in the surrounding rural areas within the drainage (RDBN 2002, BC Stats 2005). Generally, many people from the main settlement areas, Telkwa and Houston, derive their income from the logging sector. Both these towns are relatively small, resource-based communities, and the opening and closure of a major industry adjacent to them could have a significant impact upon the population and community stability.

Houston is a key supply and service center, with seven schools, a satellite community college campus, a municipal airport, and First Nations, provincial, and federal social services. In recent years, Houston has promoted itself as the "Steelhead Capital of B.C.," and there is a growing tourism market in fishing and other types of outdoor recreation. Most of the rural population between Telkwa to Topley is located alongside or close to Highway 16, in low-density rural housing with an income base derived from agriculture.

Smithers in the west and Burns Lake to the east serve as the government social and resource service centers for the upper Bulkley area. The anticipated future growth rate is projected at a modest 1% per annum by the Bulkley Nechako Regional District and BC Stats (RDBN 2002, BC Stats 2005). The Houston/Topley/Granisle Rural Official Community Plan has planned for and designated residential and commercial use areas that are capable of providing on-site water and sewage disposals systems that will likely provide future growth opportunities.

In the Bulkley valley, a large proportion of the bottomland is devoted to agriculture, which consists of forage and livestock production. The majority of agricultural land is included in the Agricultural Land Reserve (ALR), which was established as a provincial reserve on potential and existing farmland. A sizeable amount of Crown rangeland supplements private land for grazing. Overall, logging is the largest land user and major contributor to the economic base; it contributes over 50% in terms of employment and community income. Two major lumber mills located in Houston directly employ approximately 600 people.

Transportation and Utilities

Highway 16 is the major road providing access into the upper Bulkley drainage. Over the last two decades, Highway 16 has assumed an increasingly important role for residential traffic, recreational vehicles, and commercial trucking operations. This reflects the trend towards the centralization of goods and services to regional focal points, the increased use of Highway 37N, and the growing use of recreational vehicles.

Granisle Highway heads north from Highway 16 at Topley, providing access to the west shore of Babine Lake. The east side of the lake is accessed via the private barge at Mitchell Bay, which is operated by the forest industry. The North Road heads north from Houston to Mitchell Bay and is primarily utilized by forest industry traffic. Other significant secondary roads are the Buck Creek Road that accesses the upper Buck watershed and the Telkwa Forest Service Road that accesses the Telkwa valley as far west as Milk Creek. Forest resource road development throughout the upper Bulkley drainage is extensive, and most subdrainages are developed to various degrees.

Linear development within the drainage parallels the Bulkley River. Highway 16, CN Rail, BC Hydro, and Telus utilize the corridor, taking advantage of the geography and settlement patterns. The natural gas pipeline stays south of the Bulkley River, crosses the Telkwa River, and then goes through Telkwa Pass into the Zymoetz watershed.

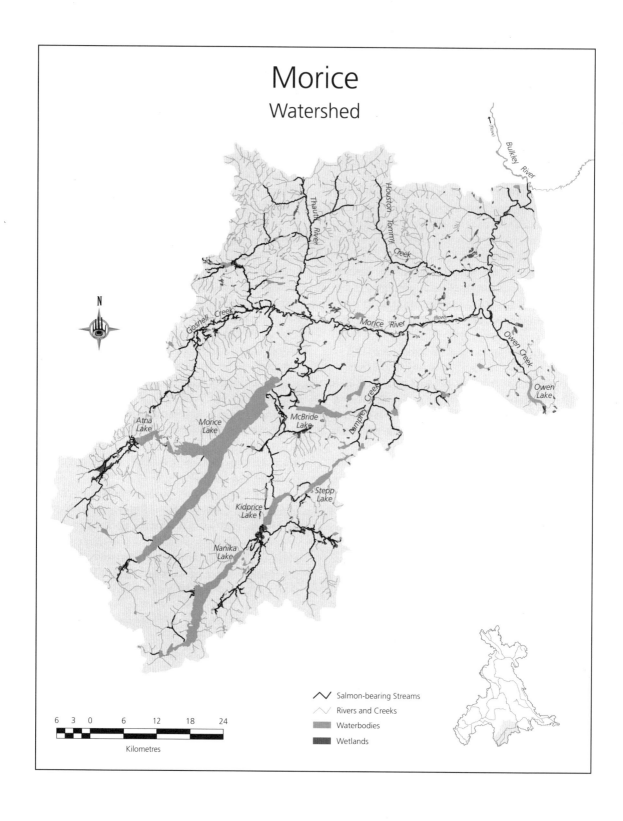

Morice
Watershed

N

Bulkley River

(flow)

Thautil River

Houston Tommy Creek

Gosnell Creek

Morice River   (flow)

Owen Creek

Owen Lake

McBride Lake

Lamprey Creek

Atna Lake

Morice Lake

Stepp Lake

Kidprice Lake

Nanika Lake

	Salmon-bearing Streams
	Rivers and Creeks
	Waterbodies
	Wetlands

6 3 0   6   12   18   24

Kilometres

22

Morice Watershed

Environmental Setting

Location

The Morice watershed is located in west-central British Columbia south of Houston. The watershed is bounded to the west by the Telkwa River and Burnie River drainages and to the east and south mainly by Nechako River tributaries. To the north, the watershed is bounded by the Bulkley River drainage.

Hydrology

The Morice watershed is part of the Bulkley River drainage basin, which is fed by streams originating in both the Interior Plateau and the glacier fields of the Coast Mountains. From the outlet of Morice Lake, the Morice River flows northeastward 80 km to join the Bulkley River near Houston. The Bulkley River flows 150 km northwestward to enter the Skeena River at Hazelton. Although the Morice is the larger tributary at the fork of the Bulkley River near Houston, the Bulkley River name is used for the tributary that flows eastward along the travel route to the interior.

The Morice River is a sixth-order stream that drains a catchment area of 4,349 km^2 and comprises the southwestern portion of the Bulkley River watershed. Elevations range from approximately 2,740 m at the western border to 560 m at the Bulkley confluence. Morice Lake (762 m) is the largest lake in the system and is the origin of the Morice River. Major tributaries include the Atna River, Nanika River, Thautil River, Lamprey Creek, Owen Creek, and Houston Tommy Creek.

Annual discharge peaks at 250 to 550 m^3/s during the early summer snowmelt season and after the occasional fall frontal storm; however, much of the flow is buffered by storage in Morice Lake. Late winter low-flow conditions have discharges of 15–25 m^3/s (Gottesfeld and Gottesfeld 1990). Hydrometric stations are located on the Nanika River at the outlet of Kidprice Lake (station 08ED001), on the Morice River at the outlet of Morice Lake (station 08ED002), and just upstream of the Bulkley River confluence (station 08ED003, discontinued).

The contribution of high-elevation snowmelt and ice melt runoff is important in maintaining adequate summer water levels in the mainstem and side channels of the Morice and Nanika Rivers. Rainstorms in the fall and decreasing evapotranspiration yield moderate flows. The Morice River contributes, on average, more than 90% of the flows to the Bulkley River at their confluence, and up to 99% of flows at certain times (Nijman 1986). There is a steep precipitation gradient from west to east, as well as from the high alpine to the valley bottom in the drainage. Annual total precipitation ranges from 2,250 mm in the Coast Mountains to under 500 mm along the lower Morice River.

Three large headwater lakes — Morice, Nanika, and Kidprice — provide most of the lake storage in the Morice and Bulkley systems. Morice Lake lies in a deep trench between the Morice Range to the west and Tahtsa Range to the east. Nanika and Kidprice Lakes occupy another trench east of the Tahtsa Range. Morice Lake is surrounded by glaciated mountains that drop steeply into the lake from an elevation of 1,200–1,500 m. The lake has a surface area of 96 km^2 and drains a basin area of 1,872 km^2. Morice Lake is deep, with an average depth of 69 m, and relatively cold, with an average summer seasonal surface temperature of 10.2° C (Shortreed et al. 1998). The two main lake tributaries are the lower Nanika River and the Atna River. The Nanika River contributes about 50% of the total water inflow into Morice Lake. Other major lake-headed tributaries within the Morice system are Owen Creek, Lamprey Creek, and McBride Creek; these are buffered to some extent from extreme floods by lake storage.

The Morice River, downstream of Morice Lake, has several large tributaries including the Thautil River and Houston Tommy Creek, which drain mountainous areas. Two lake-headed tributaries, Owen Creek and Lamprey Creeks, drain Nadina Mountain and southern plateau areas. Tributary descriptions can be found in Morris and Eccles (1975), Carswell (1979), Wet'suwet'en Treaty Office (1996), Nortec (1998), and Bustard and Schell (2002).

Stream Channels

The Morice River mainstem is 80 km in length with a very low gradient (<0.2%) and no obstructions to anadromous fish passage over its entire length. The river channel and floodplain dynamics were the subject of several studies in the early 1990s, which describe the flood plain history and elucidate the patterns and processes of channel change (Gottesfeld and Gottesfeld 1990, Weiland and Schwab 1992). Reach 1 is situated between the outlet of Morice Lake and the Thautil River and is a single-thread channel with a stable channel configuration. The substrate is mainly cobble with some gravels, deep pools, rock outcrops, and steep banks.

Reach 2 extends from the Thautil River downstream to the Fenton Creek confluence. This reach is characterized as a wandering gravel-bed river with one to several channels, frequent channel changes, gravel bars, forested islands, eroding banks, logjams, and a network of seasonally flooded channel remnants over the floodplain (Weiland and Schwab 1992). The bedload of reach 2 is coarse (over 97% is coarser than 2 mm), consisting mostly of gravel and cobbles. Cobble lithologies show that the Thautil River provides as much as 98% of the reach 2 bedload (Gottesfeld and Gottesfeld 1990). Reach 3 of the Morice River, which extends from Fenton Creek to the Bulkley River confluence, is a single-thread channel that maintains a relatively stable channel configuration.

Nanika River is 23 km in length to the Nanika falls and is commonly divided into four reach zones. Reach 1 is approximately 4 km in length, multi-channeled with islands, and has a floodplain often exceeding 400 m in width. Reach 2 is confined with a deep, straight channel. Reach 3 is a multi-channeled reach with numerous islands, gravel bars, and side channels set into a floodplain sometimes wider than 400 metres. Reach 4 is a confined reach with deep, fast flows and with the Nanika falls (11 m) at its head.

Of the many tributaries feeding into the Morice mainstem, only the Thautil River is a large producer of bedload and washload sufficient to create turbid conditions. Mountainous valleys throughout the watershed have glacio-fluvial fans that were mostly constructed during deglaciation. These fans are now more or less active with building or down-cutting zones in or adjacent to the stream channel. Fan stability is dependent on two key factors: the influences of the delivery of water, and the supply and delivery of sediments to the fan. Forest development activities influence snow accumulation, snowmelt, and water movement, which in turn influence erosion potential and sediment movement. Given the natural tendency of streams in British Columbia to increase their geomorphic response to disturbance as sediment progresses downstream (Lisle et al. 2002), it is likely that the sediment supply and movement have increased throughout the watershed.

Water Quality

Morice River water is soft; the pH is near neutral, while mean alkalinity, a measure of pH buffering capacity, is low. Morice River water is typically very clear, although TSS readings can be high during freshets. Nutrient levels overall are extremely low — in many cases less than the detection limits (Remington 1996). Morice Lake shows high dissolved oxygen content (90 – 100%) and cool water temperatures. Results of Wilkes and Lloyd's (1990) five-year water quality sampling program were summarized by Remington (1996), who concluded that the Morice River water quality is excellent.

Primary and secondary productivity in Morice Lake is limited by low nutrient inputs. The shallow north end of the lake is consistently warmer than the south end, except in winter. The Nanika River is the only tributary to contribute measurable phosphorous into Morice Lake. In

Atna Lake

the northern portion of Morice Lake, the water chemistry, the greater phytoplankton production, and the zooplankton are substantially influenced by the supply of nutrients received from the Nanika River (Cleugh and Lawley 1979). The carcasses of spawned-out salmon are the nutrient source in the Nanika River. The low intrinsic productivity of Morice Lake is compounded by the low numbers of sockeye returning to spawn and depositing their marine-derived nutrients in the Nanika system.

Water quality issues in the watershed have been minor and focus on logging and mining land use. Bustard (1986) assessed and reported on stream protection practices in the Morice Forest District. Sediment from roads, due to both construction and inadequate maintenance, was cited as the main impact from logging activities on streams in the Morice watershed. Saimoto (1994) assessed four tributaries of the Morice River and found that about half the sites examined had been impacted by roads or cutblocks in some manner.

Geography

The Hazelton Mountains within the Morice watershed are comprised of a complex group of small ranges: the Telkwa Range, the Morice Range, and the northern portion of the Tahtsa Range. Relief is relatively high in these ranges, with rugged peaks partially covered in glacial ice. The mountainous portions of the watershed are underlain by Mesozoic sedimentary and volcanic rocks, intruded by isolated stocks and small batholiths of granitic rock from the Cretaceous age (Holland 1976). The Coast Mountains (Kitimat Ranges) on the western edge of the Morice watershed are underlain by granitoid rocks of the Coast Plutonic Complex. The Nechako Plateau extends into the northern and eastern portions of the watershed, with elevations mainly below 1,500 m. Over much of the Nechako Plateau, Tertiary lava flows cover the older volcanic and sedimentary rocks of the Takla and Hazelton Groups and intrusive rocks of the Tertiary age.

The predominant biogeoclimatic zone, Sub-boreal Spruce (SBS), covers most of the lowland coniferous forests in the watershed. Subalpine fir and hybrid spruce are the major tree species;

subalpine fir stands tend to dominate older, high-elevation stands and moister sections of the zone. Due to a relatively intense natural and aboriginal fire history, lodgepole pine seral stands are extensive, particularly on stream terraces and south aspect slopes. Small areas of grassland and shrub-steppe are found on warm, dry sites scattered along the Morice River, the Owen valley, and occasionally in other major tributaries. The SBS zone merges into the Engelmann Spruce – Subalpine Fir (ESSF) zone at higher elevations ranging from 900 to 1,300 m, depending on local topography and climatic conditions.

The ESSF zone possesses a shorter, cooler, and moister growing season, with continuous forests passing into subalpine parkland at its highest elevations. Subalpine fir is dominant, with lesser amounts of lodgepole pine and white spruce hybrids in drier or fire-influenced areas. The Coastal Western Hemlock (CWH) zone characterizes the low-elevation sites along the southern sections of Morice Lake and the Atna drainage, reflecting the close proximity of maritime moisture from the coastal Kildala and Kemano drainages. Major tree species are western hemlock, amabilis fir, and subalpine fir (Banner et al. 1993).

Fish Values

The Morice watershed has high fisheries values and is a major producer of chinook, pink, sockeye, and coho salmon, as well as steelhead trout, which are fished by the aboriginal, commercial, and recreational fisheries. The Morice watershed has excellent spawning and rearing habitat for the five salmon species, which is extremely valuable. Bustard and Schell (2002) provided detailed descriptions of Morice fish populations' status and key habitats by species and life stage.

Recent Nanika-Morice sockeye stock returns have declined and are a cause for aboriginal and conservation concerns. This is likely due to a combination of results, including its run timing overlapping with the mixed-stock fishery, poor ocean survival, and poor lake productivity. In the past, the watershed was one of the largest Skeena coho producers, but there has been an overall decline in abundance since the 1970s.

Chinook Salmon

Morice River chinook salmon are the most important single salmon stock in the watershed, contributing approximately 30% of the total Skeena system chinook escapements in the 1990s. In the recent past, this stock has constituted as much as 40% of the total Skeena River chinook escapement (DFO 1984). In the late 1950s, an estimated escapement of 15,000 Morice River chinook spawners was recorded. From 1960 through the mid-1980s, an average of 5,500 spawners returned, after which chinook spawner escapement increased, reflecting substock rebuilding following the closure of directed chinook net fisheries. Between the mid-1980s and 2001, Morice River chinook spawners increased to the historic levels of the late 1950s (~15,000) returns. From 2002 to 2005, average annual escapement decreased to 7,325 from a range between 4,800 and 10,000 chinook.

Adult chinook salmon begin their migration into the Morice River system about mid-July and spawn from August to October; peak spawning was observed by Shepherd (1979) to be mid-September, with die-off by mid-October. Spawning principally occurs in the upper 2 km of the Morice River downstream of the lake outlet. Most of the riverbed at this site is characterized by a series of large gravel dunes oriented perpendicularly to the direction of flow (DFO 1984). These dunes are constructed by chinook during redd excavation. Scattered minor spawning also occurs downstream to Lamprey Creek and downstream of the falls in the Nanika River.

Morice chinook mostly spend less than one year in freshwater and return mainly as four- or five-year-olds (85% in 1973 and 1974). In comparison with other Skeena chinook stocks, the Morice River produces more six-year-olds than other systems in the Skeena (12% average versus 3% average) and fewer two- and three-year-olds (3% versus 17%) (Shepherd 1979).

Chinook fry migrate or are displaced downstream upon emergence between mid-April and early July, though peak emergence typically occurs from late May to early June. Downstream movement of the one-year-old smolts occurs between mid-April and mid-August, though it appears to peak in early June. Survey results from Smith and Berezay (1983) indicate that chinook fry overwinter throughout most of the Morice River mainstem. However, the reach between the Thautil River and Owen Creek, with abundant side channels and log debris, is considered the most productive rearing area.

Pink Salmon

The Morice pink salmon run is significant among the smaller pink producing systems in the Skeena. The odd-year pink run to the Morice River has been expanding since construction of the Moricetown Canyon fishway in 1951 and was augmented with the removal of key rocks by blasting at Hagwilget Canyon in 1959. Pink salmon were first seen in the lower Morice River in 1953 and had reached Owen Creek by 1961 and Gosnell Creek by 1975 (Shepherd 1979). By the mid-1980s, this steady expansion of range saw pink spawners colonizing the Nanika River spawning grounds.

Adult pink salmon usually migrate upstream into the Morice system in late August to early September. Pink spawning is reported to take place through September (DFO 1991b), with over 90% of the escapement spawning in reach 2 side channels. Small numbers of spawners have also been observed in Gosnell Creek, the Nanika River, and in the mainstem downstream of the lake. Winter observations of pink redds in heavily utilized side channels indicate that dewatering of redds, and probable losses of eggs and alevins with reduced flows, occurs more often at these sites than in the deeper main channel spawning areas. Upon emergence from gravels, pink fry migrate directly to the ocean, returning to spawn as two-year-old fish.

Chum Salmon

The Department of Fisheries of Canada (1964) reported that a small number of chum utilize the lower Morice River, but very little is known regarding their distribution. Kussat and Peterson (1972) noted that the chum escapement had never been enumerated, but observations indicate that the population numbers only a few hundred fish. Shepherd (1979) noted that he did not observe chum salmon in the Morice system. At the Moricetown Canyon, no chum were observed from 1992 to 1995, and only three

Morice River and Lake

were observed in 2001, which were possibly strays from the Kispiox system.

Sockeye Salmon

The Morice-Nanika sockeye stock is the largest sockeye run in the Bulkley basin. The Morice sockeye stock is composed of two subcomponents: Nanika River spawners and Morice Lake and Atna Lake beach spawners. Bustard and Schell (2002) suggested that Morice Lake beach spawning sockeye might comprise a significant component of the Morice sockeye run during some years. This is now backed up by the Moricetown Canyon mark–recapture program that shows 35% of the total sockeye spawn in locations other than Nanika. Many of these are thought to be Morice Lake and Atna Lake beach spawners (Finnegan 2006).

Historically, sockeye returning to the Morice watershed numbered on the order of 50,000 to 70,000 fish and comprised as much as 10% of the total Skeena River escapement as shown in Figure 20 (Brett 1952). The population collapsed in 1954, and in the following 20-year period (1955–1975), an annual average of 4,000 sockeye returned to the watershed (DFO 1984). Average annual returns in the 1980s were 2,500 fish, while the annual average returns in the 1990s were 21,500 fish. This robust increase in the 1990s fell off in 2000. Returns to the

Nanika appear to be decreasing; since 2000, escapements have ranged between 3,000 and 10,000 with an annual mean of slightly more than 5,000 sockeye.

Since the mid-1950s, Morice-Nanika sockeye abundance has mostly fluctuated at levels below historical escapements, with low fry densities in relation to Morice Lake juvenile sockeye rearing capacity. Constraints to sockeye production stem from the high exploitation rates in the Alaskan, Canadian, and First Nation fisheries and low production from the ultra-oligotrophic Morice Lake. The Morice Lake sockeye stock's spawning and rearing habitat is in its natural condition; it has not been impacted by development activities.

Following emergence, sockeye fry emigrate from spawning beds into Morice Lake from late May to late July, usually coincident with peak annual flows (Shepherd 1979). Morice Lake serves as the freshwater rearing lake for sockeye spawned in the Nanika River, Morice Lake, and possibly an unknown amount from Atna Lake. Morice Lake juvenile sockeye studies were conducted primarily in the 1960s, 1970s, and early 1980s and reported on by Palmer (1986b), Crouter and Palmer (1965), Shepherd (1979) and Envirocon (1984a, 1984b) respectively. Shortreed et al. (1998, 2001) and Shortreed and Hume (2004) reported on more recent sockeye juvenile sampling conducted in 1993 and 2002.

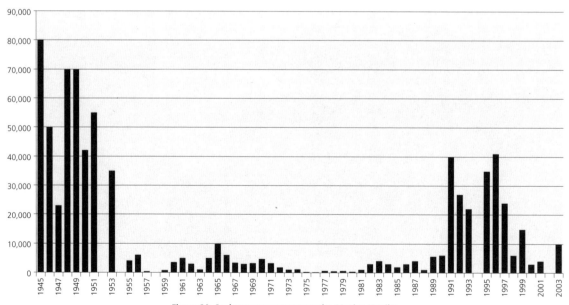

Figure 20. Sockeye escapement on the Morice-Nanika

Lake rearing habitat capacity and fry production relationships are presented in Cox-Rogers et al. (2004). In Morice Lake, the understanding of juvenile sockeye rearing and smolt production dynamics, such as age and growth, distribution and abundance, movement timing, and predation, is still evolving.

Due to the low of nutrient input into Morice Lake, phytoplankton and zooplankton biomass levels are low, resulting in very slow growth rates for sockeye fry (Costella et al. 1982). In contrast with other Skeena sockeye stocks, which spend one year in freshwater, over 85% of Nanika River sockeye spend two years in Morice Lake, and 90% return as four- (2.2) and five- (2.3) year-olds (Shepherd 1979). Age-0 fall fry are the smallest in any sockeye nursery lake in B.C.; the large percentage of two-year-old smolts in Morice Lake is also indicative of its low productivity (Shortreed et al. 1998). Sockeye smolts migrate out of Morice Lake from late April to August with a peak migration in May (Shepherd 1979, Smith and Berezay 1983).

The Wet'suwet'en, whose territory includes the Bulkley basin, have fished Morice-Nanika sockeye stocks at Hagwilget and Moricetown Canyons and at numerous terminal sites for at least six thousand years. Morice-Nanika sockeye are critically important for food, social, and ceremonial (FSC) needs, and stock restoration is a high priority to the Wet'suwet'en as it is the last significant anadromous sockeye salmon population remaining on their traditional territory.

Concerns regarding Morice-Nanika sockeye abundance have been raised since the mid-1950s, primarily by the Wet'suwet'en and the Department of Fisheries and Oceans (DFO), though their approaches to fisheries and stock rebuilding efforts have differed. In 2005, Wet'suwet'en Fisheries and the Department of Fisheries and Oceans initiated the Morice-Nanika Sockeye Recovery Plan (MNSRP). The MNSRP process provides a framework for aboriginal, government, industry, and public groups to work together towards stock recovery. The key elements of the process to date have been determining current conditions reported by Rabnett (2005), evaluating enhancement options (Rabnett 2006a), and developing a communications strategy.

Coho Salmon

The relative contribution of coho from the Morice River system to overall Skeena coho escapement is approximately 4%. In reviewing the escapement data, a declining trend from the 1950s to the present is apparent in Morice system coho populations (DFO 2005). The decline is in absolute numbers as well as relative to the overall Skeena escapement. The highest 10-year period of abundance in escapement numbers, the 1950s, shows an annual average escapement of 10,700 fish. In the 1970s, the average annual escapement was approximately 4,300 fish, with the annual escapement diminishing to 518 fish in the 1980s. It remained low in the 1990s with an average annual escapement of 672 fish. Since 1999, the aggregate coho escapement has steadily increased through 2005, except for Gosnell coho, which have remained relatively depressed.

Coho enter the Morice system in mid-August through mid-September, generally holding in the mainstem and in Morice Lake, and then, depending on water flow conditions, move with fall freshets into the tributaries to spawn. In years of below-average streamflows, most coho spawners (85%) have been observed in the prime spawning grounds downstream of the lake outlet, with scattered spawning along reach 2 side channels (Envirocon 1980). In these low-flow years, often the only tributary streams with adequate flow for coho access and spawning are Gosnell Creek, the Thautil River, and Houston Tommy Creek. In years with higher flows, other tributaries used for spawning include Owen Creek, McBride Creek, and the Nanika River. Documented spawning areas occur in all tributary streams of the Morice River (Shepherd 1979); however, this is likely to depend on adequate adult escapement and fall freshets coinciding with the late October and November spawning period.

Coho fry emergence extends from April to July. Juveniles are widely distributed throughout the Morice mainstem, as well as in most of the tributaries and lakes in the system during years of suitable recruitment. Rearing in these streams

Atna Creek

occurs for one to two years. Habitat preferences are well defined and include side channels, side pools, ponds, and sloughs with instream cover providing an important key habitat component (Shepherd 1979, Envirocon 1980). Overwintering coho prefer side channels, which makes them susceptible to reduced winter flows and cold temperatures that may result in dewatering and freezing of their winter habitat. This is a major constraint for coho smolt production in the Morice River, as significant mortalities have been documented (Bustard 1983a).

Steelhead

The Bulkley-Morice likely accounts for 30% to 40% of the total escapement of steelhead in recent years based on population estimates for the Bulkley River, genetic markers, and data from the Tyee Test Fishery (Beacham et al. 2000, Mitchell 2001). The significant summer-run of the Morice system moves into the river in mid-August and continues into the autumn (Whately et al. 1978). Overwintering appears to occur throughout the mainstem, particularly downstream of Gosnell Creek, with evidence that steelhead also utilize Morice Lake (Lough 1981, Envirocon 1984b). With the exception of Gosnell

Creek, tributaries do not support overwintering steelhead due to insufficient discharge (Envirocon 1980, Tetreau 1999).

Steelhead spawning coincides with an increase in Morice River snowmelt flows, typically from late May to early June. Results from Envirocon (1980) sampling surveys indicate widespread spawning distribution through the mainstem and tributaries. According to DFO stream survey maps, critical spawning habitat occurs in the upper Morice River and scattered downstream pockets to the Thautil confluence, as well as the lower reach of Gosnell Creek (DFO 1991b). Key spawning tributaries are Shea Creek, Owen Creek, the upper Thautil River, and upper Lamprey Creek (Bustard and Schell 2002). Repeat spawners among Morice River steelhead comprise 6.6% of the total returns, with females outnumbering male repeat spawners by a ratio of 2:1 (Whately et al. 1978).

Steelhead fry emergence in the Morice mainstem occurs primarily between mid-August and mid-September, while emergence in some tributaries may occur as early as late July due to earlier spawning and warm water temperatures. Tredger (1981 – 87), Bustard (1992 and 1993), and Beere (1993) described juvenile steelhead fry and parr distribution, densities, and size estimates from a network of index sites. Most Morice steelhead remain in freshwater for three (24%) or four (70%) winters prior to smolting, which is a longer freshwater residency time than in the six other summer-run steelhead rivers studied in the Skeena system (Whately 1978). Rearing occurs throughout the mainstem and tributaries, though the Thautil River and Owen, Lamprey, and Gosnell Creeks account for most of the steelhead fry (85%) and parr (75%) sample catch (Envirocon 1984b).

Resident Freshwater Fish

Freshwater species and documented populations inhabiting the Morice drainage include rainbow trout, bull trout, lake trout, Dolly Varden, mountain whitefish, Pacific lamprey, longnose dace, burbot, redside shiner, lake chub, longnose and largescale sucker, and prickly sculpin (FISS 2003). Resident freshwater fish information for the watershed is relatively abundant in relation to many Skeena subbasins. This is mainly due to the extensive surveys associated with the Kemano II project that was proposed in the 1970s.

Resident rainbow trout are associated with a number of the larger lakes in the Morice system, particularly Morice, Owen, and the Nanika-Kidprice Lakes system upstream of the Nanika falls (Bustard and Schell 2002). Tredger (1981) noted that rainbow trout populations are present in upper Lamprey Creek and Lamprey, Phipps, and Bill Nye Lakes.

Bustard and Schell (2002) noted that Dolly Varden are the most widely distributed fish species in the Morice watershed and dominate the catches in many of the smaller tributaries. Dolly Varden are also present in many of the larger lakes throughout the system such as Owen, Lamprey, Shea, and the Nanika-Kidprice Lakes system. Dolly Varden tend to exhibit relatively low densities in most Morice stream systems. Length frequency information collected by Bustard in the Gosnell (1999) and in the Thautil (1997b) shows that less than 2% of the total samples exceeded 150 mm in length. Those studies indicate that most spawning occurs from mid-September into October.

Kokanee are reported in Morice Lake (FISS 2002) and in Shea Lake (DeGisi and Schell 1997). Lake trout are present in Morice, Owen, Atna, and McBride Lakes. Mountain whitefish are widespread throughout the drainage, with presence noted in Morice, Atna, Shea, Owen, and Lamprey Lakes. Schell (2002) noted that whitefish have been found to be the most abundant fish species in the Morice River. Bustard and Schell (2002) reported that the highest concentrations of juvenile and adult whitefish appear to be in the Morice mainstem and use of the tributaries appears to be minor. Pygmy whitefish are reported in Owen and Morice Lakes, and lake whitefish are present in McBride and Morice Lakes. Pacific lamprey are abundant and widely distributed throughout the drainage with concentrated spawning occurring in Lamprey Creek, Owen Creek, and the Morice River (Envirocon 1984b). Studies separating Dolly Varden and bull trout in the Morice watershed were principally conducted by Bustard (1997b, 1999) in the Thautil River and in Gosnell Creek, respectively. Bahr (2002) conducted radio-telemetry studies in the drainage to determine habitat areas utilized by bull trout. Ninety-three adult bull trout were captured and implanted with radio transmitters, which allowed identification of staging, spawning, and overwintering areas. These fish demonstrated a preference for cold water by spawning in colder tributaries and migrating in the fall to spawning areas with suitable water temperature and discharge.

Fisheries
First Nations Traditional Use

First Nations traditional occupation and use of the Morice watershed is extensive. Morice River watershed territory is held by three Wet'suwet'en clans: Gilseyhyu, Laksamshu, and Gitumden. Seven House territories from these clans overlay the Morice drainage. Wet'suwet'en traditional salmon fisheries within the Morice drainage are not well documented; however, Naziel (1997) recorded approximately 20 site locations. These included concentrated sites on the Morice River, many of which were adjacent to tributary streams, as well as dispersed sites located on tributary lakes and streams.

Morice-Nanika River sockeye are a large part of the aboriginal food fishery. The current decline of Nanika River sockeye, due to a combination of factors such as mixed-stock fishery interception as well as ocean and lake productivity issues, has deeply impacted the Wet'suwet'en First Nation. For the last several years, the Native Brotherhood of B.C., in association with the United Fisherman and Allied Workers Union, North Coast gillnet groups, and fish processing companies, have supplied the Wet'suwet'en with 8,000 sockeye (PFRCC 2001). In turn, the Wet'suwet'en have not directed a food fishery on the Morice-Nanika sockeye stocks. With this cooperation, reduced harvest rates on these stocks may be addressed at the terminal fishery (river) level in a way that is more difficult to achieve in the mixed stock fishery.

Recreational Fisheries

The Morice River watershed with its many tributaries and lakes provides valuable sport fishing opportunities to resident and nonresident anglers. Provincially, the Morice River is one of the significant streams for chinook and steelhead angling enthusiasts. Coho, chinook, and steelhead are seasonally fished. The sports fishery directed towards steelhead is intensive and includes angling effort on the Bulkley River for Morice system steelhead. Further angler pressure is applied to Morice steelhead when other popular steelhead fishing rivers such as the Kispiox and Zymoetz become turbid with fall storm floods.

Rainbow trout, lake trout, and Dolly Varden are also actively angled, particularly in Morice

The Nanika falls

and other headwater lakes. Aspects of the Morice steelhead sport fishery have been noted by Fennelly (1963), Pinsent and Chudyk (1973), and Lewis (2000). Creel surveys were conducted by Remington et al. (1974), Whately et al. (1978), and Envirocon (1980).

The Morice River is designated as Class II waters September 1 to October 31 and requires both a Classified Water license and a steelhead stamp. Fly fishing only is allowed between Gosnell and Lamprey creeks from September 1 to 31. No fishing zones are in place between Morice Lake and Gosnell Creek from January 1 to September 30 and from Gosnell to Lamprey creeks between January 1 and August 31. Throughout Morice River there is no angling from boats allowed from August 15 to December 31 and a bait ban year-round (B.C. Fisheries 2005).

Enhancement Activities

The major enhancement projects affecting the Morice system were improving fish passage in Hagwilget and Moricetown Canyons. In 1951, the present Moricetown Canyon fishway was constructed. Concerns regarding the low number of returning Nanika sockeye in 1955 led to the decision to remove perceived obstructions in Hagwilget Canyon. The canyon rocks were blasted in 1959, effectively eliminating the aboriginal fishery in the canyon. There was no apparent increase of sockeye populations following this action. However, intended or not, these actions facilitated the colonization of the Morice drainage with pink salmon.

As the Nanika sockeye population continued to decline with escapements of less than 1,000 fish for the three consecutive years of 1957 to 1959, a pilot hatchery was constructed on the lower reach of Nanika River. In operation from 1960 to 1965, the hatchery was not successful, most likely due to the use of transplant stock from Pinkut Creek in the Babine system. The selected stock was unsuitable with emergence three to four weeks early, differences in Pinkut adult life history (largely 1.2 and 1.3 age fish), and Pinkut fry being small and thin compared to Nanika sockeye fry (Shepherd 1979). The hatchery was closed in 1966 pending evaluation of returns.

In 1969, Owen Creek was studied as a potential site for the construction of artificial spawning facilities for steelhead (Pinsent 1969). Owen Creek was deemed not suitable due to water supply problems at low flows; however, recommendations included correction of the Morice River Road crossing near the mouth of Owen Creek, which continued to create a fish passage problem. This access problem was corrected with replacement of the culvert at the mouth of Owen Creek by a bridge in the late 1990s. Morice Lake was aerial fertilized in 1981 and 1985, and although the project was not comprehensively followed up, preliminary results indicated that phytoplankton increased 35%, while zooplankton levels went up 60% (Shortreed 2002). Fish responses to this fertilization are unknown.

Studies to assess steelhead enhancement opportunities in the Morice watershed were undertaken from 1980 to 1983 (Tredger 1981, 1981–86). The stocking of the Morice River system began in 1983 with 70,000 summer steelhead fry released above barriers (locations unknown) and in underutilized areas. In 1984, this program continued with an additional 62,000 fry released (DFO and MoE 1984). In 1985, 3,900 steelhead fry were released, while 21,000 were released in 1986 (FISS 2001). The majority of fry releases were to the Morice mainstem, Gosnell Creek, and Houston Tommy Creek. A post-stocking cost-benefit assessment suggested the program be terminated.

In 1983, 850 chinook smolts from Morice brood stock were released into the Morice River; this is the only known chinook enhancement project (DFO 2002). Coho enhancement included two off-channel habitat developments close to Owen Creek that were largely unsuccessful as well as coho fry and smolt outplants. These outplants in 1999 and 2000 totaled 36,000 releases into Owen Lake. Various enhancement projects were proposed by Alcan to mitigate and compensate impacts from the potential Kemano Completion Project; however, the hydroelectric project was not implemented due to fisheries, economic, and social concerns (DFO 1984, Envirocon 1984b).

Development Activities

Principal land and resource use in the Morice watershed result from logging, mining, and transportation and utilities development. The Morice watershed is located in the Nadina Forest District with land and resource use management direction established in the Morice LRMP.

Forest Resource Development

The foundations of the permanent forest industry were laid by tie hackers, who cut small amounts of lodgepole pine for railroad ties at the lower end of the Morice watershed from 1925 through the Depression years. Following World War II, there was a great demand for lumber; however, most of the logging was centered in the Bulkley valley and in the Buck Creek area (Hols 1999). The Morice River Road from Houston to Morice Lake was constructed from 1954 to 1958. Logging started shortly thereafter, with sawmills located at Collins and McBride Lakes and the lumber trucked to Planer Row in Houston. Three small operators cut timber off the road on Morice Mountain, and a small mill operated at Owen Flats.

By the mid-1960s, Bulkley Valley Pulp and Timber had bought up the small mills and attached timber quotas; then they built a new mill in 1970, the predecessor of the current Canadian Forest Products (Canfor) mill. In 1963, the Bulkley Valley Pulp and Timber Company obtained rights to 104,000 km^2 of timber and proposed large-scale log drives on the Morice River. The Department of Fisheries was quick to reject the idea citing serious problems on the Kitsumkalum River (Department of Fisheries Canada 1964). Prior to 1968, logging in the Morice was mainly conducted off trails, creating herringbone patterns. By 1968, clearcut logging was dominant, utilizing the easily accessible timber stands located adjacent to the Morice River Road. Years of planning by Weldwood and Eurocan resulted in the opening of Houston Forest Products in 1978.

In 1983, the Swiss Fire burnt 18,000 ha on both sides of the Morice River, precipitating three years of intensive salvage operations. Logging operations in the 1980s were widespread, with rigorous development in the Morice North area (Chisholm Lake). The 1990s saw the initial development of the Morice West area with continued widespread operations throughout the watershed. In the recent past, development has been targeting various types of beetle infestations, with timber extraction from many, small infested areas. Recent timber harvesting and current proposed forest development plans (FDPs) in the Morice West area have focused on the Gosnell, Thautil East, and Shea drainages.

The Morice Watershed Restoration Program was initiated in 1994 with SKR conducting assessment work on Cedric, Lamprey, Fenton, and Owen Creeks (Saimoto 1994). An overview assessment was conducted by the Wet'suwet'en Treaty Office (1996), which was followed by detailed assessments in 1997 and reported on by Nortec (1998) and by the British Columbia Conservation Fund (BCCF 1999) for the Nanika and Lamprey subbasins. These assessments of logging-related damage to fish habitat led to a small number of site works to alleviate fish passage problems at Fenton Creek and an unnamed creek at 28 km on the Morice West Road, as well as for some minor riparian rehabilitation (Ministry of Forests 2001b).

Mineral Resource Development

Mineral exploration activities have been extensive in the Morice watershed; however, economic considerations and mining circumstances have resulted in only one producing mine: the Silver Queen property. Substantial amounts of money and effort have been spent on developing this silver, lead, and zinc deposit located east of Owen Lake, but a sustainable mining operation has yet to be developed there.

Recent mineral exploration activity consisting of a drill program and geophysical surveys has centered at the Lucky Ship molybdenum property. It is located on the ridge between Nanika River and Morice Lake with access off the Cutthroat FSR. Molybdenum mineralization is associated with a rhyolite plug that cuts through Hazelton Group volcanic, pyroclastic, and sedimentary rocks. The Lucky Ship was discovered by prospectors in 1956 and was explored intermittently by Amax, and later Canamax, from the late 1950s to the early 1980s.

This reactivated development is being spurred by the current high price of molybdenum.

Population and Settlement

Historically, Wet'suwet'en people resided at various village sites and home places throughout the watershed, but in the early 1950s, lifestyles changed and many Wet'suwet'en moved into the Bulkley valley. Less than 20 people currently reside in the Morice watershed; few seasonal workers "camp out," with the majority of workers traveling into and out of Houston.

Located 4 km east of the Morice-Bulkley confluence, Houston is home to approximately 4,000 people (Statistics Canada 1996). This community is heavily dependent on the forest resource, with minor amounts of agriculture and mining contributing to the economy (Horn 2001). Population growth trends are projected to be stable; however, closure or downsizing of major industrial or public sector operations would have a significant impact upon the population. The forest sector, which is based on two mills located in Houston, directly employs approximately 600 people and indirectly employs approximately 800 people (Regional District of Bulkley-Nechako 1998).

Transportation and Utilities

A network of roads developed by the forest sector lies in most valley bottoms and extends onto the majority of plateaus. These gravel roads branch from the main road, the Morice-Owen Road, which connects to Highway 16 west of Houston. Access through the majority of the Morice River and tributary valleys is relatively easy due to the timber harvesting road network.

Between 1954 and 1958, the Forest Service began the half-million dollar construction of the Morice River Road from Houston to Owen Lake, which included the construction of the road to the Collins Lake – McBride Lake area. The most remote portion of the road, built in 1958 to Morice Lake, was not originally part of the plan, but was built for fire crew access when the 10,000 ha Clore Fire broke out (Smith 2002). The road to Morice Lake was then used for access to a hatchery on the Nanika River.

The building of the Morice River Road and the many subsequent smaller access roads opened a new timber supply area, with many small bush mills appearing throughout the area. The Morice-Nanika, Lamprey, Nado, Morice West, and Cedric branch roads were constructed in 1961.

Since 1996, Huckleberry Mine, located 120 km southwest of Houston, has used the Morice-Owen Road daily to haul ore to Stewart. A BC Hydro transmission line serving the mine was constructed in 1997 and, for the most part, closely follows the road right-of-way. Improvements and upgrades on the Morice-Owen Forest Service Road (FSR) occurred in 1978 and 1997, with the latter accommodating the greatly increased traffic to the Huckleberry Mine site. In 1990, the Morice River West FSR was extended, crossing the Morice River and Gosnell Creek.

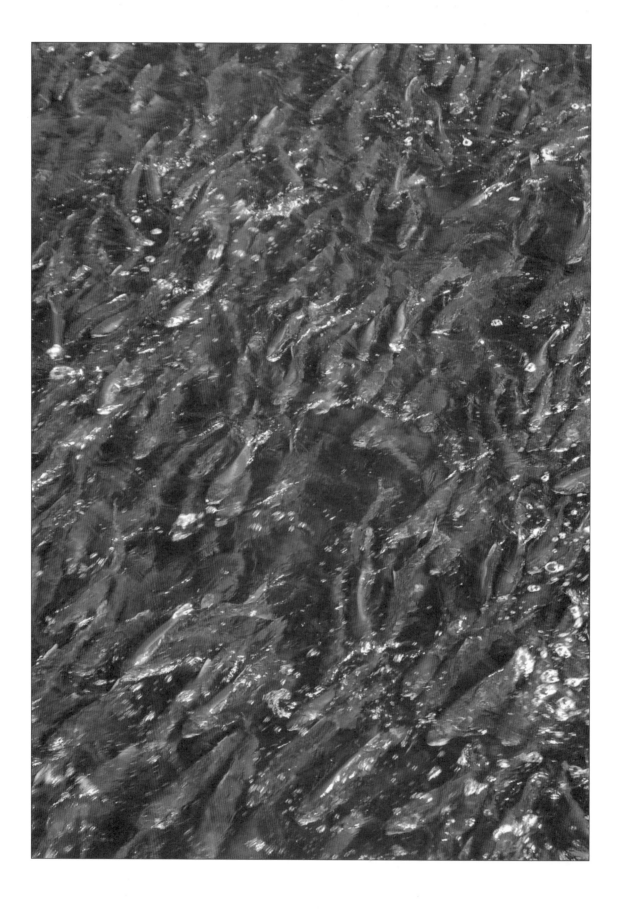

Conclusion

The Skeena watershed is the second largest watershed in B.C. — both in area and salmon production. Presently, the Skeena supports the freshwater life stages of six Pacific salmon species: chum, chinook, sockeye, pink, coho, and steelhead. These salmon provide ecosystem links that connect the 600 km long Skeena River with a large portion of the northeast Pacific Ocean. Another 22 species contribute to the freshwater fish community, inhabiting stream and lake habitats within the Skeena watershed.

The Skeena is blessed with a multitude of lake-fed tributaries with a wide variety of climatic, biological, and hydrological conditions. In relation to other North American salmon rivers, the Skeena is unique for its presence and abundance of salmon and for its overall low habitat disturbance. For people in the Skeena, salmon are an iconic species that is intertwined with their sense of identity.

Salmon evolved significantly during the Pleistocene ice ages and are well adapted to occupy new territory as ice sheets recede. The first people to reach the Skeena probably found salmon widespread. The postglacial spread of salmon was apparently so fast that, to some extent, modern distributions reflect brief connections between watersheds that functioned early in the postglacial period about 10,000 to 11,000 years ago (McPhail and Carveth 1993b).

Salmon ecosystems evolved under the pressures of salmon-based First Nations economies. The annual catch in British Columbia, assuming a large aboriginal population, may have been of the same order of magnitude as recent commercial catches (Hewes 1973, Haggan et al. 2006). The Coast Tsimshian, the Canyon Tsimshian, Gitxsan, Wet'suwet'en, and Ned'u'ten people typically harvested their salmon close by their spawning grounds with weirs, traps, and dip net and spear gear. This harvest, on a stock-by-stock basis, allowed for conservation, optimal utilization, and adaptation to variable abundance and changing natural conditions. Despite a system emphasizing careful use and conservation (Trosper 2002), this level of intensive harvest would have affected wild salmon populations and selected against unproductive stocks in ways much different from the system that was to follow.

In the 1870s and 1880s, the right to market fish on the coast was transferred from native fishermen to the new industrial fisheries and canneries (Harris 2001), a fishery management regime that continues to the present. The large-scale intensive coastal fisheries resulting from this industrial expansion imposed a new mixed-stock fishing pattern. The high harvest volumes that ensued in the following decades diminished the runs of sockeye and then chinook and steelhead. This intensive fishery probably led to the extirpation of many salmon stocks that could not withstand the increasingly high harvest rates. Improved fisheries management in the past 50 years, following long-term catch declines, has apparently stabilized most salmon stocks. But many of these stocks are stable at diminished levels of abundance.

Mixed-stock fishing continues to be a key factor in the suppression of many salmon sub-populations and thus reduces aggregate abundance. If the mixed-stock fisheries are replaced in part with terminal or semi-terminal fisheries, many smaller suppressed stocks will rebound. This conservation objective is the principal argument for moving part of the coastal fishery upriver, a change that appears likely to proceed.

Skeena salmon stock productivities vary in complex predictable and unpredictable ways and are thus differentially impacted by fishing and development pressures. In general, the farthest inland stocks experience losses by predation along their longer fluvial migratory paths and have lower productivity, perhaps because of cold upland habitats, which are associated with slower growth and longer fresh water residence.

Salmon stocks vary from year to year in their freshwater productivity. Particularly important in this arena are highly variable rates of egg and alevin survival, which are often related to fall and winter floods, exceptionally cold winter weather, and fluctuating snow depths. Snow accumulation, snowmelt timing, and summer weather conditions affect water storage, discharge, and thus productivity in streams and lakes. In the last decade, the importance of changing marine survival rates has been acknowledged by most biologists (Beamish et al. 1999, Mueter et al. 2002); it is likely that the decade scale oscillations in North Pacific Ocean circulation patterns as well as changes in more local coastal circulation are behind much of the pattern of variation in run sizes among Skeena salmon.

Habitat degradation, especially from logging of valley bottomlands, has had impacts on the stream-rearing salmon, coho, chinook, and steelhead. But the specific impacts are often difficult to quantify considering the near lack of baseline information. Now that the easily accessible, lowland forest stands have been nearly exhausted and there have been several decades for recovery, stream fish values seem to be improving. Compared to other parts of the northern hemisphere, Skeena salmon stocks are in relatively good shape; however, there are known and likely a few unknown problems. Overall, a few stocks are known to have been extirpated in the past 50 years.

Skeena sockeye are the backbone of the commercial coastal fisheries and are the most important fish for Skeena First Nations. Two sockeye stocks have been extirpated, and, other than for the enhanced Babine stocks, abundance generally fluctuates at low levels. Most of the 27 non-Babine stocks are depressed, but conservation concerns have focused on a few stocks. Fishing pressure has particularly affected the Kitwanga Lake sockeye, now at about 10% of their former level, which return at the same time as the Fulton River sockeye, the largest Babine Lake stock. Degraded habitats are implicated in the decline of the Lakelse Lake and the upper Bulkley sockeye stocks. Problems of the Morice-Nanika stocks are the most difficult to define and correct as the large Morice Lake has inherent productivity limits that are exacerbated by poor ocean survival and past fishing practices.

Skeena River chinook are the largest group of chinook salmon in northern B.C. Skeena chinook are large stream salmon and have the most complex life histories. Approximately 75% of the Skeena chinook aggregate spawn in the Kitsumkalum, Morice, and Bear River systems. Abundance has increased since 1985 when the Pacific Salmon Treaty reduced coastal fishing, and now typically ranges from 40,000 to 80,000 chinook annually. Kitsumkalum River chinook are the exploitation rate indicator stock for the B.C. North Coast; this stock includes a mix of hatchery (CWT) and natural production, with tagged smolt survival rates that range from 0.3% to 2.5%. Despite this overall improvement, some of the many small, poorly known populations are a cause of concern. Upriver stocks, especially Bear River, have greatly declined.

Chum are the least abundant of the Skeena salmon stocks, and since 1990 average escapement has been decreasing. Skeena chum returns are characterized by significant annual and interannual variability. The decline in abundance is basin wide, suggesting that much of the problem is in the marine realm and is likely related in part to highly variable survival rates in their early marine life. Chum are very vulnerable to gill nets, and their decline might be considered a part of the mixed-stock conundrum. The most important spawning areas are in the Ecstall River and in the multi-channeled mainstem reach downstream of Terrace. The counting weir on the Kitwanga River has greatly contributed to our understanding of upriver chum abundance.

Coho salmon are the most widely dispersed salmon throughout the watershed. Declining abundance from the 1950s culminated with the closure of commercial fisheries in 1998; since then, there has been a strong rebound with relatively high off-shore marine survival. The rapid recolonization is encouraging for conservation efforts. Likely causes of the collapse of coho stocks were the high exploitation rates in the Alaskan and Canadian fisheries combined with habitat damage and a decline in ocean survival rates.

Pink salmon is the most abundant salmon species in the Skeena. The majority of pink salmon production occurs in a few large populations such as Lakelse, Kitwanga, Kispiox, Babine, Morice, and the lower Skeena downstream of Terrace. They are exclusively two years of age when they return to spawn, with neither even or odd-year runs having a well-developed dominance. Key Skeena pink salmon characteristics include their wide spatial distribution, exceptional variability in annual stock recruitment, and the large amounts of marine nutrients delivered by adults.

Steelhead in the Skeena watershed consist of two runs: the summer-run that returns in July and August and comprises 80% of the steelhead aggregate, and the winter-run that returns October to May and mostly spawns in the tributaries downstream of Terrace. Steelhead abundance is poorly known and quantitative data are few. The only reliable escapement information is based on estimates generated from the Tyee Test Fishery. There are few good data on individual stream escapement. This shortage of abundance information does not facilitate comprehensive management activities by the B.C. government, which has jurisdiction over steelhead.

Threats to the well-being of Skeena salmon come from environmental changes and resource development initiatives. Environmental damage is now widespread in the southern portion of the watershed; however, for the most part, it is not serious enough to interfere with normal ecosystem functions. Mining proposals for large projects have become more numerous and more conspicuous in recent years. Currently, the largest development threats are proposed coal bed methane exploration and production in the upper Skeena headwaters, which would greatly disturb the pristine chinook, steelhead, and coho habitat, and proposed metal mining adjacent to Babine Lake, the most productive sockeye rearing lake in Canada.

Given the negative effects on the genetic fitness of wild salmon from mixing with hatchery stocks, hatchery enhancement efforts intended to restore wild salmon abundance has become a topic of increased interest and concern among salmon biologists in the past few years (Naish et al. 2007, Hiroshi et al. 2008). In the Skeena, there are no production scale hatcheries at present, but there are and have been several community-based hatcheries. The effects of hatchery production from the small-scale hatcheries are probably minor; the effects of hatcheries in the 1920s through the 1960s are unknown. It seems that we should take care to avoid a situation where the survival of salmon is dependent on artificial means, with the attendant problems of domestication.

Proposed development of the Skeena estuary is worrying. A large expansion of the Prince Rupert port facilities is expected; this will impact important pink salmon juvenile habitat south of Ridley Island. More serious has been the threat of net-pen salmon farms in the Skeena estuary in critical juvenile habitat for pink, chum, and sockeye salmon. The 2002 proposal for salmon farms in the southern part of the Skeena estuary has been deferred and may well not go ahead, but other proposals may turn up. Petroleum exploration and well development in Hecate Strait could seriously affect productivity and juvenile salmon survival. The concern over these development proposals shows the necessity for a large-scale, watershed-wide planning process for the Skeena.

Mitigating the impacts of social and economic activities on Skeena salmon abundance and diversity and their relationships to habitats and ecosystems is challenging. The task is to maintain Skeena salmon diversity and biodiversity while conserving productive and critical habitat, as well as functioning marine, freshwater, and terrestrial ecosystems. The Wild Salmon Policy—which was proposed by DFO in 2005 and is being implemented as this book goes to press—is a step forward. It represents an institutional change for the Department of Fisheries and Oceans from a government organization designed to support the commercial fishing industry to an organization that manages the marine environment and is responsible for salmon conservation. A successful transition would be welcome as a needed step to meet the present and future challenges of salmon management in British Columbia. The increasing

role of First Nations in fisheries and habitat
management, driven by a decade of successful
court challenges, is a hopeful political change
that should have positive effects on retaining
salmon population diversity and conserving
critical habitat.

References

Abelson, D. 1976. Lakelse Lake water quality study. Technical report. Water Resources Service, Pollution Control Branch. Victoria, B.C.

Abelson, D. 1977. Maxan Creek water quality study 1972–1975. Water Resources Service, Ministry of Environment. Victoria, B.C.

Abraham, P. 1995. Interview with J. Dewhirst *in* Dewhirst, J. 1995. An aboriginal sustenance impact assessment of the Kemess South gold-copper project: A status report. Prepared for El Condor Resources. Archeo Tech Associates. Victoria, B.C.

Acer Resource Consulting Ltd. and Gitsegukla Band Council. 2000. WRP level 1 fish and fish habitat assessment for Kitsuns Creek, Kitseguecla River watershed. Prepared for the Ministry of Environment, Lands and Parks.

Agra Earth and Environmental Ltd. 1996. Level 1 fish population and riverine habitat assessment for Maxan watershed. Prepared for Yin Waghunlee Habitat Enhancement Corporation. Burns Lake, B.C.

Agra Earth and Environmental Ltd. 1998. Telkwa Coal Mine surface water monitoring program 1997. Prepared for Manalta Coal Ltd. Calgary, Alberta.

Agra Earth and Environmental Ltd. 1999a. Substrate composition mensuration including McNeil coring, freeze coring and interstitial dissolved oxygen measurement of three spawning redds on the upper Bulkley River and Maxan Creek in October and November 1998. Report prepared for CFDC Nadina.

Agra Earth and Environmental Ltd. 1999b. Guidebook of impact assessment of river ecosystems of the Bulkley, Morice and Lakes Districts of the Skeena Region of the Ministry of the Environment, Lands and Parks. Smithers, B.C.

Agra Earth and Environmental Ltd. 2000a. Reinstallation of historic Water Survey of Canada staff gauges, upper Bulkley River. Smithers, B.C.

Agra Earth and Environmental Ltd. 2000b. 1999 aquatic impact assessment of river ecosystems in the Bulkley, Morice and Lakes Districts, Skeena Region of the Ministry of the Environment, Lands and Parks. Smithers, B.C

Alexander, R.F. and K.F. English. 1996. Distribution, timing and numbers of early-run salmon returning to the Kitsumkalum River watershed in 1995. Prepared by LGL Ltd. for DFO, Skeena Green Plan. Prince Rupert, B.C. Draft. 82 pp.

Alfred, H. 1999. Trout-Corya Landscape Unit field trip notes.

Allbright, S.L. 1987. Archaeological evidence of Gitksan and Wet'suwet'en history. Opinion evidence for *Delgamuukw et al v. the Queen.* Unpublished report on file, Gitxsan Treaty Office library. Hazelton, B.C.

Anonymous. 1964. Skeena River sockeye and pink salmon. In Inventory of the natural resources of British Columbia. 15th B.C. Natural Resources Conference.

Anonymous. 1977. Sailing directions. British Columbia Coast (North Portion). Vol. 2. 7th edition. Department of Fisheries and Environment. Institute of Ocean Sciences. Sidney, B.C.

Anonymous. 1998. Angling Use Plan, Bulkley River. Draft, June 1998.

Anonymous. 2001a. Freshwater fishing regulations synopsis. B.C. Government.

Anonymous. 2001b. Skeena Bulkley Region Resource Management Plan, 2002–2006.

Anonymous. 2002. Freshwater fishing regulations synopsis. B.C. Government.

Argue, A.W., C.D. Shepard, M.P. Shepard, and J.S. Argue. 1986. A compilation of historic catches by the British Columbia commercial salmon fishery, 1876 to 1985. Internal DFO report.

Aro, K.V. and J. McDonald. 1968. Times of passage of Skeena River sockeye and pink salmon through the commercial fishing area. Fish. Res. Board Can. MS Rep. 984. 169 pp.

Atagi, D. 1998. Moosevale Creek juvenile fish sampling (fall 1998). Ministry of Environment, Lands and Parks. Skeena Region stream files. Smithers, B.C.

AES. 2005. Twenty year summary of selected northwest weather stations. Atmospheric Environment Service. Environment Canada.

Bahr, M. 2002. Examination of bull trout *(Salvelinus confluentus)* in the Morice River watershed. Biology Program, UNBC, Prince George, B.C. Prepared for Canadian Forest Products, Houston Forest Products, and B.C. Ministry of Water, Land and Air Protection. Smithers, B.C. Unpublished manuscript.

Bams, R.A. and A.S. Coburn. 1962. Experimental hatchery operations. *In* F.C. Withler, ed. Studies in salmon propagation. Annual Report of the Biological Station. Nanaimo, B.C. 1961–1962: D1.

Banner, A., W. McKenzie, S. Haeussler, S. Thomson, J. Pojar, and R. Trowbridge. 1993. A field guide to site identification and interpretation for the Prince Rupert

Forest Region. Land Management Handbook No. 26. Ministry of Forests. Victoria, B.C.

Barbeau, C.M. 1917. Growth and federation of the Tsimshian Phratries. Proceeding of the 19th International Congress of Americanists. Washington, D.C. 402–408.

Baxter, J.S. 1997a. Kispiox River steelhead: Summary of current data and status review, 1997. B.C. Environment. Skeena Fisheries Report SK-100. MELP, Skeena Region. Smithers, B.C.

Baxter, J.S. 1997b. Upper Sustut, lower Sustut and Bear River steelhead: Summary of current data and status review, 1997. Skeena Fisheries Report SK-98. MELP, Skeena Region. Smithers, B.C.

Baxter, J.S. 1997c. Kluatantan River steelhead: Summary of current data and status review, 1997. Skeena Fisheries Report SK-99.

BCCF. 1999. Morice detailed fish habitat/riparian/channel assessment for watershed restoration — Nanika and Lamprey sub-basins. British Columbia Conservation Foundation. Smithers, B.C.

B.C. Fisheries. 2002. Freshwater fishing regulations synopsis. Victoria, B.C.

B.C. Fisheries. 2005. 2005–2006 freshwater fishing regulations synopsis. Victoria, B.C.

BCFS. 1956. Babine Lake maps (5 sheets). 1:40 ch. scale. File P380-P530. British Columbia Forest Service. Victoria, B.C.

BCFWRP. 2001. Big Falls Creek watershed. In Bridge-Coastal Fish and Wildlife Restoration Program, vol. 2.

BC Hydro. 1983. Skeena-Rupert 500 kV transmission line. Environmental and socioeconomic assessment, vols. 1 and 2. Report no. ESS-20. Fisheries section prepared by D.B. Lister for BC Hydro.

B.C. Ministry of Environment, Lands and Parks, and Fisheries and Oceans Canada. 2001. Watershed-based Fish Sustainability Planning: Conserving B.C. Fish Populations and their Habitat. A guidebook for participants. Co-published by B.C. Ministry of Environment, Lands and Parks and Canada Dept. of Fisheries and Oceans.

B.C. MoE. Skeena River lake and stream management files. Various dates and authors. B.C. Ministry of Environment. Smithers, B.C.

B.C. MoF. 1981. Kispiox Timber Supply Area report. British Columbia Ministry of Forests, Planning Branch. Victoria, B.C.

B.C. MoF and B.C. MoE. 1999. Coastal watershed assessment procedure guidebook (CWAP). Interior watershed assessment procedure guidebook (IWAP). B.C. Ministry of Forests and B.C. Ministry of Environment. Victoria, B.C.

B.C. MED. 1977. The North West report '77. B.C. Ministry of Economic Development. Victoria, B.C.

B.C. Parks. 2000. Management direction statement for Babine River Corridor Provincial Park. Skeena District. Smithers, B.C.

BC Stats. 2003, 2005. www.bcstats.gov.bc.ca

Beacham, T.D., R.E. Withler, and A.P. Gould. 1985. Biochemical genetic stock identification of pink salmon (Oncorhynchus gorbuscha) in Southern British Columbia and Puget Sound. Can. J. Fish. Aquat. Sci. 42: 1474–1483.

Beacham, T.D., A.P. Gould, R.E. Withler, C.B. Murray, and L.W. Barner. 1987. Biochemical genetic survey and stock identification of chum salmon (Oncorhynchus keta) in British Columbia. Can. J. Fish. Aquat. Sci. 44: 1702–1712.

Beacham, T.D., R.E. Withler, and T.A. Stevens. 1996. Stock identification of chinook salmon (Oncorhynchus tshawytscha) using minisatellite DNA variation. Can. J. Fish. Aquat. Sci. 53: 380–394.

Beacham, T.D. and C.C. Wood. 1999. Application of microsatellite DNA variation to estimation of stock composition and escapement of Nass River sockeye salmon (Oncorhynchus nerka). Can. J. Fish. Aquat. Sci. 56(2): 297–310.

Beacham, T.D., S. Pollard, and K.D. Le. 2000. Microsatellite DNA population structure and stock identification of steelhead trout (Oncorhynchus mykiss) in the Nass and Skeena Rivers in Northern British Columbia. Marine Biotechnology, vol. 2, 6: 587–600.

Beacham, T.D., J.R. Candy, K.J. Supernault, T. Ming, B. Deagle, A. Schultze, D. Tuck, K.H. Kaukinen, I.R. Irvine, K.M. Miller, and R.E. Withler. 2001. Evaluation and application of microsatellite and major histocompatibility complex variation for stock identification of coho salmon in British Columbia. Trans. Amer. Fish. Soc. 130: 1116–1149.

Beacham, T.D., B. McIntosh, and C. MacConnachie. 2004. Population structure of lake-type and river-type sockeye salmon in transboundary rivers of northern British Columbia. Journal of Fish Biology. 65: 389–402.

Beacham, T.D., B. McIntosh, and C. MacConnachie. 2005. Population structure and stock identification of sockeye salmon (Oncorhynchus nerka) in coastal lakes in British Columbia, Canada. Can. J. Zool. 83: 834–844.

Beamish, R.J., G.E. McFarlane, and J.R. King. 1999. Fisheries climatology: Understanding decadal scale processes that naturally regulate British Columbia fish populations. In T. Parsons and P. Harrison, eds. Fisheries oceanography: A science for the new millennium.

Beaudry, P.G. and J.W. Schwab. 1990. Identification of fine sediment sources in the Telkwa watershed. Initial working plan. B.C. Forest Service, Forest Sciences Section. Smithers, B.C.

Beaudry, P.G., D.L. Hogan, and J.W. Schwab. 1990. Hydrologic and geomorphic considerations for silvicultural investments on the lower Skeena River floodplain. FRDA report 122. Victoria, B.C.

Beaudry, P.G., J.W. Schwab, and D. Septer. 1991. Suspended sediment, Telkwa River watershed. Data report. B.C. Forest Service, Forest Sciences Section. Smithers, B.C.

Beaudry, P.G. 1992. Sediment monitoring program, Nilkitkwa watershed working plan. Forest Sciences Section, Prince Rupert Forest Region, B.C. Ministry of Forests. Smithers, B.C.

Beaudry, P.G. and A. Gottesfeld. 2001. Effects of forest-harvest rates on stream-channel changes in the central interior of British Columbia. *In* D.A.A. Toews and S. Chatwin, eds. Watershed assessment in the southern interior of British Columbia. Res. Br., B.C. Ministry of Forests. Victoria, B.C. Working paper 57/2001.

Beere, M.C. 1991a. Radio telemetry investigations of Babine River steelhead, spring 1990. Skeena Fisheries Report SK-71. MELP, Skeena Region. Smithers, B.C.

Beere, M.C. 1991b. Radio telemetry investigations of steelhead tagged in the lower Bulkley River, 1989. Skeena Fisheries Report SK-70. Ministry of Environment. Smithers, B.C.

Beere, M.C. 1993. Juvenile steelhead surveys in the Kitwanga, Morice, and Zymoetz Rivers, 1993. Skeena Fisheries Report SK-90. MELP, Skeena Region. Smithers, B.C.

Beere, M.C. 1996. Movements of summer-run steelhead trout tagged with radio transmitters in the Babine River during spring, 1994. Skeena Fisheries Report SK-94. MELP, Skeena Region. Smithers, B.C.

Beere, M.C. 2002. Personal communication. WLAP, Skeena Region. Smithers, B.C.

Beere, M.C. 2003. Personal communication. WLAP, Skeena Region. Smithers, B.C.

Beere, M.C. 2005. Personal communication. Ministry of Environment, Skeena Region. Smithers, B.C.

Bennett, S. and K. Ohland. 2002. Expansion and recalibration of a benthic invertebrate index of biological integrity for the upper Bulkley watershed. Prepared for CFDC Nadina by BioLogic Consulting. Terrace, B.C.

Beschta, R.L., R.E. Bilby, G.W. Brown, L.B. Holtby, and T.D. Hofstra. 1987. Stream temperature and aquatic habitat: Fisheries and forestry interactions. *In* E.D. Salo and T.W. Cundy., eds. Streamside management: Forestry and fisheries interactions. University of Washington, Institute of Forest Resources, Contribution 57. 191–232.

Beschta, R.L., J.R. Boyle, C.C. Chambers, W.P. Gibson, S.V. Gregory, J. Grizzel, J.C. Hagar, J.L. Li, W.C. McComb, M.L. Reiter, G.H. Taylor, and J.E. Warila. 1995. Cumulative effects of forest practices in Oregon. Oregon State University. Corvallis, Oregon.

Bilton, H.T., E.A.R. Ball, and D.W. Jenkinson. 1967. Age, size, and sex composition of British Columbia sockeye salmon catches from 1912 to 1963. Fish. Res. Bd. Can. Circ. 25.

Bilton, T.H. and M.P. Shepard. 1955. The sports fishery for cutthroat trout at Lakelse Lake, British Columbia. Fish. Res. Bd. Can., Pacific Coast Stat. Rept. 104: 38–42.

Biolith Scientific Consultants Inc. 1999a. Juvenile salmonid population densities at index sites near Terrace, B.C.

Biolith Scientific Consultants Inc. 1999b. Level 1 detailed field assessment of aquatic and riparian habitat for the West Kitsuns Creek subbasin. Prepared for Gitsegukla Band Council.

Bocking, R. and M. Gaboury. 2002. Clear Creek (Milgeelde): Sockeye habitat condition and capability assessment. LGL Ltd. for Kitsumkalum Band Council. Sidney, B.C.

Boer, G.J., G. Flato, and D. Ramsden. 2000. A transient climate change simulation with greenhouse gas and aerosol forcing: Projected climate to the 21st century. Climate Dynamics. 16:6 427–450.

Bosch, J.M., and J.D. Hewlett. 1982. A review of catchment experiments to determine the effect of vegetation change on water yield and evapotranspiration. Journal of Hydrology, 55: 3–23.

Boyd, I.T., C.J.P. McKean, and R.N. Nordin. 1985. Kathlyn, Seymour, Round, and Tyhee Lakes water quality assessment and objectives. Ministry of Environment. Victoria, B.C.

Boyd, R. 1999. Indians, fire and the land in the Pacific Northwest. Oregon State University Press. Corvallis, Oregon.

Brett, J.R. 1952. Skeena River sockeye escapement and distribution. J. Fish. Bd. Can., 8 (7) 1952.

Brocklehurst, S. 1998. Historical data review on the upper Bulkley River watershed. Report prepared by J.O.A.T. Consulting for DFO. Smithers, B.C.

Brown, M. 1985. Commission evidence for *Delgamuukw et al v. the Queen*. Unpublished report on file at Gitxsan Treaty Office. Hazelton, B.C.

Brown, R.F. 1980. Groundwater reconnaissance for salmonid enhancement opportunities on Lakelse, Kitsumkalum, and Tseax River systems. Unpublished memo, DFO. Vancouver, B.C.

Brown, W. 1823. Summary report for New Caledonia 1822–23. Hudson's Bay Company Archives, B188/e/1.

Buckham, A.F. 1950. Indian Engineering. Canadian Geographical Journal. 174–181.

Buckham, A.F. and R. Latour. 1950. Groundhog Coalfields. Geological Survey of Canada. Bulletin No. 16.

Buirs, M. 2002. Personal communication. LIM, Morice Forest District.

Burgner, R.L. 1991. Life history of sockeye salmon *(Oncorhynchus nerka)*. In C. Groot and L. Margolis, eds. Pacific Salmon: Life Histories. 1991. UBC Press. Vancouver, B.C.

Burns, D.C. 1991. Cumulative effects of small modifications to habitat: AFS Position Statement. Fisheries 16 (1): 12–17.

Busack, C.A. and K.P. Currens. 1995. Genetic risks and hazards in hatchery operations: Fundamental concepts and issues. Am. Fish. Soc. Symp. 15: 71–80.

Bustard, D. 1975a. Proposed Smithers tree farm license, preliminary resource presentation: Fish and wildlife values. Ministry of Environment. Smithers, B.C.

Bustard, D. 1975b. Memo to file. Kluatantan flight notes, August 5, 1975. WLAP Skeena Region stream files. Smithers, B.C.

Bustard, D. 1983a. Juvenile salmonid winter ecology in a northern British Columbia river: A new perspective. Smithers, B.C.

Bustard, D. 1983b. 1982 investigations of adult coho salmon in the Telkwa River. Prepared for Crows Nest Resources.

Bustard, D. 1984a. Assessment of benthic invertebrate and juvenile fish populations in Goathorn and Tenas Creeks and the lower Telkwa River, September, 1983. Prepared for Crows Nest Resources Ltd. by David Bustard and Associates.

Bustard, D. 1984b. Assessment of benthic invertebrate and juvenile fish populations in Foxy and Buck Creeks, September, 1984. Prepared for Equity Silver Mines Ltd. by David Bustard and Associates.

Bustard, D. 1984c. Investigations of adult pink salmon in the Telkwa River. Prepared for Crows Nest Resources Ltd. by David Bustard and Associates.

Bustard, D. 1984d. Preliminary evaluation of fish utilization of Kasiks Channel and adjacent Skeena River side channels. Prepared for Ministry of Forests, Research Branch. Smithers, B.C.

Bustard, D. 1985. Aquatic resource assessment of the Telkwa Coal Project: 1984 studies. Prepared for Crows Nest Resources Ltd.

Bustard, D. 1986. Assessment of stream protection practices in the interior of the Prince Rupert Forest Region. Smithers, B.C.

Bustard, D. 1987. Babine Lake creel survey 1985–1986. Smithers, B.C.

Bustard, D. 1987–2002. Fish population monitoring in Foxy and Buck Creeks. A series of manuscript reports prepared for Equity Silver Mines Ltd.

Bustard, D. 1989. Assessment of rainbow trout recruitment from streams tributary to Babine Lake. Smithers, B.C.

Bustard, D. 1990. Sutherland River rainbow trout radio telemetry studies 1989. Prepared by David Bustard and Associates for B.C. Ministry of Environment.

Bustard, D. 1991. Lower Skeena River fisheries studies Exchamsiks River to Andesite Creek. Smithers, B.C.

Bustard, D. 1992a. Juvenile steelhead surveys in the Kitwanga, Morice, Sustut, and Zymoetz Rivers, 1991. Smithers, B.C.

Bustard, D. 1992b. Adult steelhead studies in the upper Sustut River 1992. Prepared for B.C. Environment (Habitat Conservation Fund). Smithers, B.C.

Bustard, D. 1993a. Lower Skeena River 1992 fisheries studies Exchamsiks to Andesite. Smithers, B.C.

Bustard D. 1993b. Juvenile steelhead surveys in the Kitwanga, Morice, Sustut, and Zymoetz Rivers, 1992. Smithers, B.C.

Bustard, D. 1993c. Fisheries assessment of bridge crossing sites in the upper Bear River.

Bustard, D. 1993d. Rainbow trout spawning studies in Foxy Creek 1993. Manuscript report prepared for Homestake Canada Ltd.

Bustard, D. 1993e. Summary of water quality and fish sampling in Esker gravel pit 1992–93. Prepared for Ministry of Transportation and Highways.

Bustard, D. 1994a. Steelhead spawning studies upper Sustut River 1993. Prepared for B.C. Environment (Habitat Conservation Fund). Smithers, B.C.

Bustard, D. 1994b. Evaluation of downstream fish movements in the upper Sustut River using a rotary screw trap May–June 1994. Prepared for Department of Fisheries and Oceans. Smithers, B.C.

Bustard, D. 1994c. Fisheries assessment of pipeline crossings in the Kasiks and Khyex Rivers. Prepared for Pacific Northern Gas Ltd. Terrace, B.C.

Bustard, D. 1996a. Winter measurements of dissolved oxygen at selected sites in the Bulkley watershed. Prepared for DFO, Northern Salmon Stock Assessment. Nanaimo, B.C.

Bustard, D. 1996b. Juvenile coho studies at the Telkwa River km 1011 ponds 1996. Prepared for Watershed Restoration Program. Province of B.C. and Pacific Inland Resources Ltd.

Bustard, D. 1996c. Fisheries assessment of the lower Clore River and tributaries. Preliminary report prepared for Skeena Cellulose Inc. Terrace, B.C.

Bustard, D. 1997a. Surveys for potential off-channel fish habitat development, Telkwa River. Prepared for the Watershed Restoration Program. Pacific Inland Resources.

Bustard, D. 1997b. Stream inventory, Thautil River watershed 1996. Prepared for Forest Renewal B.B. (Houston Forest Products Ltd.).

Bustard, D. 1997c. Assessment of juvenile coho populations in selected streams within the Skeena watershed 1997. Smithers, B.C.

Bustard, D. 1999. 1998 stream inventory report, Gosnell Creek watershed. Prepared for Forest Renewal B.C. (Northwood Inc.).

Bustard, D. 2000. Juvenile coho studies at the Telkwa River ponds Year 2000. Prepared for the Watershed Restoration Program, Pacific Inland Resources.

Bustard, D. 2002. Personal communication. Smithers, B.C.

Bustard, D. 2004. Gitnadoix River char studies 2003. Prepared for Habitat Conservation Trust Fund by David Bustard and Associates.

Bustard, D. and Associates and Limnotek Research and Development Inc. 1998. Aquatic resource baseline studies, Telkwa Coal Project 1997. Prepared for Manalta Coal Ltd.

Bustard, D. and C. Schell. 2002. Conserving Morice watershed fish populations and their habitat. Prepared for CFDC Nadina.

Campton, D.H. 1995. Genetic effects on hatchery fish on wild populations of Pacific salmon and steelhead: What do we really know? Am. Fish. Soc. Symp. 15: 353–377.

Carlander, K.D. 1969. Handbook of freshwater fishery biology. Vol. 1. Iowa State University Press. Ames, Iowa.

Carver, D. 2000. Using Indicators to Assess Hydrologic Risk. In Watershed Assessment in the Southern Interior of British Columbia: Workshop proceedings. Ministry of Forests Research Program. Victoria, B.C.

Chamberlin, T.W., R.D. Harr, and F.H. Everest. 1991. Timber harvesting, silviculture, and watershed processes. In W.R. Meehan, ed. Influences of forest and rangeland management on salmonid fishes and their habitats. Special Publication 19. Am. Fish. Soc. Bethseda, Maryland.

Charles, A.T., and M.A. Henderson. 1985. Chum salmon (Oncorhynchus keta) stock reconstructions for 1970–1982 Part 1: Queen Charlotte Islands, North Coast and Central Coast, British Columbia. Department of Fisheries and Oceans, Can. Manuscript Rpt. of Fish. Aquat Sci. No. 1814. 91 pp.

Chudyk, W.E. 1972a. Bear River report. On file at MELP. Smithers, B.C.

Chudyk, W.E. 1972b. Memo to file. Skeena Lake and stream management files. MELP. Smithers, B.C.

Chudyk, W.E. 1972c. Steelhead research: Sustut River watershed. Memo to file. Skeena Region stream files. Smithers, B.C.

Chudyk, W.E. 1974. Memo to file. WLAP Skeena Region stream files. Smithers, B.C.

Chudyk, W.E. 1978. Suskwa River steelhead trout: The 1977 inventory, creel survey, and life-history characteristics study leading to removal of a barrier on Harold Price Creek. SK-15. B.C. Fish and Wildlife Branch. Smithers, B.C.

Clague, J.J. 1978. Terrain hazards in the Skeena and Kitimat River basins, British Columbia. Geological Survey of Canada, Paper 78-1A. Vancouver, B.C.

Clague, J.J. and S.R. Hiscock. 1976. Sand and gravel resources of Kitimat, Terrace, and Prince Rupert, British Columbia. Geological Survey of Canada, Paper 76-1A. Vancouver, B.C.

Clague, J.J. 1983. Surficial geology Skeena River, Bulkley River area. Geological Survey of Canada, Map 1557A.

Clague, J.J. 1984. Quaternary geology and geomorphology, Smithers-Terrace-Prince Rupert area, British Columbia, Memoir 413, Geological Survey of Canada.

Cleugh, T.R. and B.C. Lawley. 1979. The limnology of Morice Lake relative to the proposed Kemano II power development. Volume 4. Fisheries and Marine Service, Department of the Environment. Vancouver, B.C.

Cleugh, T.R., C.C. Graham, and R.A. McIndoe. 1978. Chemical, biological and physical characteristics of Lakelse Lake, B.C. Fish. Mar. Ser. Man. Rep. 1472. Dept. of Fisheries and Environment. Vancouver, B.C.

Cleveland, M.C. 2000. Limnology of Kitwanga Lake: An attempt to identify limiting factors affecting sockeye salmon (Oncorhynchus nerka) production. Gitanyow Fisheries Authority. Kitwanga, B.C.

Cleveland, M.C. 2001. Kitwanga salmon enhancement program 2000–01. Unpublished report by Gitanyow Fisheries Authority for Habitat Restoration and Salmon Enhancement Program. Fisheries and Oceans Canada.

Cleveland, M.C. 2002a. Personnel communication. Gitanyow Fisheries Authority. Gitanyow, B.C.

Cleveland, M.C. 2002b. Kitwanga River fisheries treaty related measure #3: The 2001 adult steelhead/sockeye salmon enumeration and data gathering initiatives. Gitanyow Fisheries Authority. Kitwanga, B.C.

Collingwood, R. 1974. Steelhead Lodge data sheet. WLAP Skeena Region stream files.

Costella, A.C., B. Nidle, R. Bocking, and K.S. Shortreed. 1982. Limnological results from the 1980 lake enrichment program. Can. Man. Rept. Fish. And Aquat. Sci. No. 1635.

Cox-Rogers, S. 1985. Racial analysis of Skeena River steelhead trout by scale pattern analysis. M.Sc. thesis. University of British Columbia. 138 pp.

Cox-Rogers, S. 2000. Skeena sockeye and Nanika sockeye production trends. Memorandum. DFO. Prince Rupert, B.C.

Cox-Rogers, S. 2001. Sockeye presentation at 2001 Post-Season Review, North Coast areas 1–6. Prince Rupert, B.C.

Cox-Rogers, S., J.M.B. Hume, and K.S. Shortreed. 2004. Stock status and lake-based production relationships for wild Skeena River sockeye salmon. CSAS Research Document 2004/010.

Cox-Rogers, S. 2001. 2005 Skeena Sockeye Review. 2005 Post-season review. DFO. Prince Rupert, B.C.

Crouter, R.A. and R.N. Palmer. 1965. The status of Nanika-Morice sockeye population.

Croxall, D.A. and K.J. Wilson. 2001. 2001 Bulkley Landscape Unit integrated restoration plan for the Maxan, upper Foxy, Crow-Foxy, Day Lake and Broman subbasins within the Lakes Forest District. Prepared for FRBC by Ecofor Consulting Ltd. Fort St. James, B.C.

Culp, C. 2002. Personal communication. Deep Creek Hatchery Manager. Terrace, B.C.

Culp, J. 2000. 1999 adult coho enumeration program: Terrace area watersheds. Terrace Salmon Enhancement Society. Terrace, B.C.

Culp, J. 2002. Personal communication. Terrace Salmon Enhancement Society. Terrace, B.C.

Dams, R. and D. Bustard. 1996. Lower Skeena tributaries adult coho surveys and GSI sampling 1995.

David, J. 1985. Commission evidence for *Delgamuukw et al vs. the Queen*. Vols. 3 and 4. Moricetown, B.C.

David, M. 1981. Interview with L. George.

Davis, J.C. and I.G. Shand. 1976. Acute and sublethal copper sensitivity, growth and saltwater survival in young Babine Lake sockeye salmon. Fish Marine Ser. Tech. Rep. 847. 55 pp.

Dawson, G.M. 1881. Report of an exploration from Port Simpson on the Pacific Coast, to Edmonton on the Saskatchewan River, 1879. Dawson Brothers, Montreal.

Dayton and Knight Ltd. 2002. Village of Telkwa water quality study. Smithers, B.C.

DeGisi, J.S. 1997a. Gitnadoix River Provincial Recreation Area, Fisheries information summary. Prepared for B.C. Parks Skeena District. Smithers, B.C.

DeGisi, J.S. 1997b. Swan Lake Provincial Park, Fisheries information summary. Prepared for B.C. Parks Skeena District. Smithers, B.C.

DeGisi, J.S. 2000. Babine River Corridor Provincial Park, Fisheries information summary. Prepared for B.C. Parks Skeena District. Smithers, B.C.

DeGisi, J.S. and C. Schell. 1997. Reconnaissance inventory of Shea Lake watershed, code 460-6006508-005-283-01. Survey dates: August 9–10, 1996. Prepared for MELP, Fisheries Branch, Skeena Region.

DeLeeuw, A.D. 1991. Observations on cutthroat trout of the Lakelse River system, 1986, and implications for management. Skeena Fisheries Report SK-79. B.C. Environment, Fish and Wildlife Branch. Smithers, B.C.

DeLeeuw, A.D., Cadden, D.J., Abelson, D.H., Hatlevik, S. 1991. Lake trout management strategies for northern British Columbia, B.C. Environment, Fisheries Branch.

Department of Fisheries of Canada. 1964. Fisheries problems associated with the development of logging plans within the Morice River drainage system. Vancouver, B.C.

DFO. 1905–1949. BC 16 Salmon stream spawning reports. Prepared by various Fishery Officers for individual streams. BC 16 records on file at DFO, North Coast Division. Prince Rupert, B.C.

DFO. 1930–1960. Department of Marine and Fisheries, Annual Narrative Reports, Babine-Morice Area, District 2, B.C.

DFO. 1984. Towards a fish habitat decision on the Kemano Completion Project: A discussion paper. Fisheries and Oceans. Vancouver, B.C.

DFO and MoE 1984. Salmonid Enhancement Program. Annual Report 1984. Fisheries and Oceans Canada and Ministry of Environment. Province of B.C.

DFO. 1985. Pacific region salmon resource management plan, vol. 1. Technical report.

DFO. 1987. Timing of Bulkley/Morice coho stocks in the Skeena. Unpublished memorandum.

DFO. 1991a. Stream summary catalogue Subdistrict 4A Lower Skeena. Fish Habitat Inventory and Information Program. North Coast Division. Fisheries Branch.

DFO. 1991b. Stream summary catalogue. Fish habitat inventory and information program SISS Stream Summary Catalogue. Subdistrict 4B, Terrace. Department of Fisheries and Oceans. Vancouver, B.C.

DFO. 1991c. Fish habitat inventory and information program. Stream summary catalogue. Subdistrict 4C, Hazelton. North Coast Division, Department of Fisheries and Oceans.

DFO. 1991d. Fish habitat inventory and information program. Stream summary catalogue. Subdistrict 4D, upper Skeena-Babine. North Coast Division, Department of Fisheries and Oceans.

DFO. 1991e. Fish habitat inventory and information program SISS stream summary catalogue. Subdistrict 4D, Smithers (volume 2). Bulkley. North Coast Division, Department of Fisheries and Oceans. Prince Rupert, B.C.

DFO. 1999. Stock status of Skeena River coho salmon. DFO Science Stock Status Report D6-02 (1999).

DFO. 2000. 1999 Pacific region state of the ocean.

DFO. 2001. 2000 Pacific region state of the ocean. Ocean Status Report 2001-01.

DFO. 2002. Enhancement Support and Assessment database. Habitat and Enhancement Branch. Department of Fisheries and Oceans. Vancouver, B.C.

DFO. 2004. 2003 Pacific Region state of the ocean. Ocean Status Report 2004.

DFO. 2005. SEDS. (Salmon escapement data system) Department of Fisheries and Oceans. Stock Assessment Division. Prince Rupert, B.C.

Derrick, M. 1978. Adaawhl Gitanyaaw. Gitanyow history project. Gitanyow, B.C.

Diewert, R.E. 2000. Enumeration of adult steelhead in the upper Sustut River, 2000. Skeena Fisheries Report SK-128.

Dombroski, E. 1952. Sockeye smolts from Babine Lake in 1951. Pac. Prog. Rep. 91: 21–26.

Dombroski, E. 1954. The sizes of Babine Lake sockeye salmon smolt emigrants, 1950–53. Pac. Prog. Rep. 99: 30–34.

Donas, B.G. and R. Saimoto. 1999. Upper Bulkley River overwintering study 1998–2000, interim report. Prepared for Fisheries Renewal B.C.

Donas, B.G. 2000. Personal communication. DFO Community Advisor. Smithers, B.C.

Donas, B.G. 2001a. Upper Bulkley River coho assessment fence program report for 2001. Prepared for DFO, Strategic Stock Enhancement Program.

Donas, B.G. 2001b. Strategic Stock Enhancement Program report, upper Skeena: 1998/1999/2000. Prepared for DFO, Strategic Stock Enhancement Program.

Donas, B.G. and R. Saimoto. 2001. Upper Bulkley River and Toboggan Creek overwintering study 2000–2001. Prepared for Fisheries Renewal B.C.

Donas, B. 2002. Personal communication. DFO Community Advisor. Smithers, B.C.

Doublestar Resources Ltd. 2001. Sustut Copper Deposit information and scooping document. Vancouver, B.C.

Dubeau, P.R. and M.R.S. Johannes. 1996. Upper Sustut River rotary trap smolt assessment 1995. Report prepared for Department of Fisheries and Oceans, Pacific Biological Station. Nanaimo, B.C.

Drake, A. and L. Wilson. 1991. Eulachon: A Fish to Cure Humanity. Museum Note No. 32. UBC Museum of Anthropology. Vancouver, B.C.

Drewes, M. 2002. Personal communication. DFO Community Advisor. Terrace, B.C.

Dykens, T. and S. Rysavy. 1998. Operational inventory of water quality and quantity of river ecosystems in the Skeena Region. Prepared for MELP by B.C. Conservation Foundation. Smithers, B.C.

Dyson, J.B. 1949. Bulkley Falls investigation report.

Dyson, J.B. 1955. The Babine rock and earth slide. Thesis, UBC. Vancouver, B.C.

Eastwood, J. 1981. A survey of free-flowing rivers in northwestern British Columbia. Ministry of Forests. Prince Rupert, B.C.

Ellis, D.W. 1996. Net Loss: The salmon netcage industry in British Columbia. David Suzuki Foundation. Vancouver, B.C.

Envirocon Ltd. 1980. Kemano completion hydroelectric development environmental impact assessment. Volume 4: Fish Resource Studies, Morice System. MELP. Smithers, B.C.

Envirocon Ltd. 1984a. Physical and hydrological baseline information. Vol. 2. Environmental studies associated with the proposed Kemano Completion Hydroelectric Development. Aluminum Company of Canada. Vancouver, B.C.

Envirocon Ltd. 1984b. Fish resources of the Morice River system: Baseline information. In Environmental studies associated with the proposed Kemano Completion Hydroelectric Development. Vancouver, B.C.

Environment Canada. no date. Temperatures and precipitation 1941–1970, British Columbia. Atmospheric Environment Service. Downsview, Ont.

Environment Canada. 1979. Historical streamflow summary, British Columbia. Inland Waters Directorate, Water Resources Branch, Water Survey of Canada. Ottawa, Ont.

Environment Canada. 1991. Historical streamflow summary British Columbia to 1987. Inland Waters Directorate, Water Resources Branch, Water Survey of Canada. Ottawa, Ontario.

Environment Canada. 1993. Canadian Climate Normals 1961–1990, Vol. 1. British Columbia. Environment Canada. Ottawa, Ont.

Environment Canada. 2005. Climate Data Online. www.climate.weatheroffice.ec.gc.ca/climateData/canada_e.html

Evans, S.G. 1982. Landslides and surficial deposits in urban areas of British Columbia: A Review. Can. Geotech. J. 19: 269–87.

Evenchick, C.A., T.P. Poulton, H.W. Tipper, and I.G. Braidek. 2001. Fossils and facies of the northern two-thirds of the Bowser Basin, British Columbia. Geologic Survey of Canada. Open File 3956.

Evenchick, C.A., M.C. Hayes, K.A. Buddell, and K.G. Osadetz. 2002. Vitrinite and bitumen reflectance data and preliminary organic maturity model for the northern two-thirds of the Bowser and Sustut Basins, north-central British Columbia. Geological Survey of Canada, open file 4343. B.C. Ministry of Energy, Mines and Petroleum Resources Geology open file 2002-1.

Evenchick, C.A., Ferri, F., Mustard, P.S., McMechan, M., Osadetz, K.G., Stasiuk, L., Wilson, N.S.F, Enkin, R.J., Hadlari, T. and McNicoll, V.J. 2004. Recent results and activities of the integrated petroleum resource potential and geoscience studies of the Bowser and Sustut Basins Project, British Columbia. In Current Research, Geological Survey of Canada, 2003-A13, 11 pages.

Ewasiuk, J. 1998. Bulkley River fish fence report 1998. Prepared for DFO and Community Futures Development Corporation of Nadina.

Farina, J.B. 1982. A study of salmon migrating and spawning in the Nechako River system and Morice and Nanika Rivers. Alcan Smelters and Chemicals Limited. Kitimat, B.C.

Farstad, L. and D,G, Laird. 1954. Soil survey of the Quesnel, Nechako, Francois Lake and Bulkley-Terrace areas in the Central Interior of British Columbia. Report no. 4 of the British Columbia Soil Survey. Canada Department of Agriculture.

Fell, B. 2000a. Survey and design of tributary 1, Kitseguecla River South. Prepared for Ministry of Environment, Lands and Parks, Skeena Region.

Fell, B. 2000b. Survey and design of tributary 11, Kitseguecla River South. Prepared for West Fraser Mills Ltd, Pacific Inland Resources Division.

Fennelly, J.F. 1963. Steelhead Paradise. Mitchell Press. Vancouver, B.C.

Ferris, G. 2002. Personal communication.

Finnegan, B. 2002a. Personal communication. DFO, Stock Assessment Division. PBS. Nanaimo, B.C.

Finnegan, B. 2002b. Summary of water temperature data collected from the Skeena River drainage between June 1, 1994 and October 15, 1999. Department of Fisheries and Oceans. Nanaimo, B.C.

Finnegan, B. 2006. Morice Lake Sockeye Program. Unpublished data. DFO, Stock Assessment. Smithers, B.C.

Finnegan, B. 2007. Unpublished data. DFO, Stock Assessment. Smithers, B.C.

Fisheries Research Board. 1947. Annual reports of the Pacific Biological Station 1945–1948. Fisheries Research Board of Canada. Nanaimo, B.C.

Fisheries Research Board. 1948. Fisheries Research Board Pac. Prog. Rep. No. 74: 1948. Pacific Biological Station. Nanaimo, B.C.

FISS (Fisheries Information Summary System). 2002, 2003, 2005. British Columbia Ministry of Sustainable Resource Management, Fisheries Data Warehouse web site.

Foerster, R.E. 1968. The sockeye salmon, *Oncorhynchus nerka*. Bull. Fish. Res. Board Can. 162: 422 pp.

Foote, C.J., C.C. Wood, and R.E. Withler. 1989. Biochemical comparison of sockeye salmon and kokanee, the anadromous forms of *Oncorhynchus nerka*. Can. J. Fish. Aquat. Sci. 46: 149–158.

Foskett, D.R. 1948. Bear Lake. Appendix 5: Lakes of the Skeena River drainage. Skeena River salmon investigations interim report. Fisheries Research Board. Nanaimo, B.C.

Franklin, I.R 1980. Evolutionary change in small populations. *In* M.E. Soule and B.A. Wilcox, eds. Conservation biology: An evolutionary-ecological perspective. Sinauer. 135–150

Fremier, A.K., J.M. Mount, P.B. Moyle, and S.M. Yarnell, eds. 2004. Ecology and geomorphology of streams: The Skeena River study.

Freshwater Resources. 2001. Watershed assessment of the Kitseguekla community watershed. Prepared for Skeena Cellulose Inc.

Gabrielse, H., Monger, J.W.H., Tempelman-Kluit, D.J., and Woodsworth, G.J. 1991. Intermontane Belt. *In* H. Gabrielse and C.J. Yorath, eds. Geology of Canada no. 4, Geology of the Cordilleran orogen in Canada. Geological Survey of Canada. 591–625.

Gaherty, W., H. Temmenga, G. Higginson, and G. Johnstone. 1996. Review of cattle-community watershed conflicts in the Skeena Region. Report prepared for MELP. Smithers, B.C.

Galois, R. 1987. History of the upper Skeena 1850 to 1927. Summary of opinion evidence for *Delgamuukw et al v. the Queen*. Unpublished report on file at Gitxsan Treaty Office. Hazelton, B.C.

Garfield, V.E. 1939. Tsimshian clan and society. University of Washington Publications in Anthropology, vol. 7, no. 3. Seattle, WA.

Gartner Lee. 1993. Leachate assessment field manual and preliminary landfill leachate assessment reports. Report prepared for Skeena Region MELP, Smithers, B.C.

Gates, B.R. and D.S. Reid. 1985. British Columbia Railway Dease Lake Extension 1984 environmental inspection.

Industrial and Energy Project Evaluation, Planning and Assessment Branch. Ministry of Environment. Victoria, B.C.

Geertsema, M. and J.W. Schwab. 1995. The Mink Creek earth flow, Terrace, British Columbia. Proc. 48th Can. Geotech. Conf. 625–633.

Gelley, L.J. 1953. Report on the upper Bulkley River. Prepared for G.S. Reade. Supervisor of Fisheries. Prince Rupert, B.C.

Giesbrecht, S. and G. Grieve. 1997. Level 1 detailed assessment of fish and fish habitat in the south Gitsegukla River and its tributaries. Prepared for Gitsegukla Band Council. Biolith Scientific Consultants Inc. Terrace, B.C.

Giesbrecht, S., Grieve G. and M. Prins. 1998. Level 1 detailed assessment of fish and fish habitat in the south Kitwanga River and its tributaries. Biolith Scientific Consultants Inc. Terrace, B.C.

Gilchrist, A. 1998. Kitwanga River and Kitseguecla River Watershed Restoration Program: Hydrological and channel stability assessments of specific impact sites.

Gilchrist, A., G. Grieve, and D. Webb. 1999. Site survey and design for reach 1 of tributary 1, Kitseguecla River South subbasin. Prepared for the Gitsegukla Band Council.

Gilchrist, A. 2001. Clear Creek in the Kitsumkalum watershed: Stream channel stability assessment. Prepared for Kitsumkalum Band Council and DFO.

Ginetz, R.M.J. 1976. Chinook salmon in the North Coastal Division. Tech. Rept. Ser. No. PAC/T-76-12. Fish. Mar. Serv., Dept. Env. Vancouver, B.C.

Ginetz, R.M.J. 1977. A review of the Babine Lake Development Project 1961–1967. Can. Dept. Fish. Tech. Rept. Ser. T-77-6: 192 pp.

Giroux, P.A. 2002. Shelagyote River Bull Trout *(Salvelinus confluentus)* Life History. B.C. Ministry of Water, Land and Air Protection, Fish and Wildlife Science and Allocation Section. Smithers, B.C.

Gitanyow Fisheries Authority. 2005. Unpublished data. Gitanyow, B.C.

Gitsegukla Band. 1979. Adawkhl Gitsegukla. Gitsegukla, B.C.

Gitsegukla Band Council and Cedarvale Resources Ltd. 1998. Kitseguecla River Watershed Restoration Program Subbasins Andimaul and Shandilla — Modified overview assessment, modified sediment source survey, and access management plan. Prepared for the Ministry of Forests.

Gitksan and Wet'suwet'en Chiefs. 1987. Map 22 Fishing Sites. Gitksan and Wet'suwet'en Chiefs map atlas. Assembled by Canadian Cartographics. Coquitlam, B.C.

Gitxsan Watershed Authorities. In prep. Gitxsan traditional fisheries in the northern territories. Hazelton, B.C.

Glass, A. 1999. Bulkley River fish fence 1999. Prepared for DFO and Community Futures Development Corporation of Nadina.

Glass, A. 2000. Upper Bulkley River coho assessment fence 2000. Prepared for DFO and Community Futures Development Corporation of Nadina.

Godfrey, H., W.R. Hourston, J.W. Stokes, and F.C. Withler. 1954. Effects of a rock slide on Babine River salmon. Fisheries Research Board of Canada. Bulletin No. 101. Ottawa, Ont.

Godfrey, H. 1955. On the ecology of Skeena River whitefishes, Coregonus and Prosopium. J. Fish. Bd. Canada, 12 (4), 1955.

Godfrey, H. 1968. Ages and physical characteristics of maturing chinook salmon of the Nass, Skeena, and Fraser Rivers in 1964, 1965, and 1966. Fish Res. Bd. Man. Rept. Ser. No. 967.

Godin, B., M. Ross and M. Jones. 1985. Babine Lake data from chemical and biological surveys May 1983, May 1984, July 1984. Environmental Protection Service Canada, Pacific Yukon Region.

Godin, B., M. Hagen and G. Mitchell. 1992. Babine Lake monitoring June 19–22, 1990. Environmental Protection Service Canada, Pacific Yukon Region.

Golder and Associates. 1990. Report to Skeena Cellulose on geotechnical evaluation of landslide on Zymagotitz River logging road 15 km west of Terrace, B.C.

Gordon, D., A. Lorenz, and M. Friesen. 1996. Lakelse WRP Project, level 1 fisheries assessment. Terrace, B.C.

Gordon, D. 1999. Zymacord WRP overview fish and riparian assessment. Triton Environmental Consultants Ltd. Terrace, B.C.

Gottesfeld, A.S. 1985. Geology of the northwest mainland. Kitimat Centennial Museum Assoc. Kitimat, B.C. 114 pp.

Gottesfeld, A.S. 1995. Watershed hydrology and stream stability of the Suskwa River. Unpublished report. Prince George, B.C.

Gottesfeld, A.S. 1996. British Columbia flood scars: Maximum flood-stage indicators. Geomorphology, 14: 319–325.

Gottesfeld, A.S. 2001. 2000 Kitwanga River creel survey. Gitxsan Watershed Authorities, Hazelton, B.C.

Gottesfeld, A.S. and L.M.J. Gottesfeld. 1990. Floodplain dynamics of a wandering river, dendrochronology of the Morice River, British Columbia, Canada. Geomorphology, 3: 159–179.

Gottesfeld, A.S., C. Muldon, E. Plate, and R. Harris. 2000. Steelhead habitat utilization and juvenile density in streams of the Kispiox watershed 1998–99. Gitxsan and Wet'suwet'en Watershed Authorities. Unpublished report. Hazelton, B.C.

Gottesfeld, A.S., K. Rabnett and P. Hall. 2002. Conserving Skeena fish populations and their habitat. Skeena Fisheries Commission. Hazelton, B.C.

Gottesfeld A.S., Ryan, T, Rolston, D., and Procter, B. 2004. Sea lice and pink salmon smolts in the Skeena estuary, British Columbia. Skeena Fisheries Commission. Hazelton, B.C. 33 pp.

Gottesfeld A.S., Ryan, T, Rolston, D., and Procter, B. 2005. Sea lice and pink salmon smolts, North Coast of British Columbia. Proceedings of the 22nd Northeast Pacific Pink and Chum Salmon Workshop.

Gottesfeld A.S., Rolston, D., Procter, B. and Ryan, T. 2006. 2004–2005 North Coast Marine Baseline Survey and Sea Lice Research Program. Final Report. Skeena Fisheries Commission. Hazelton, B.C.

Gottesfeld, L.M.J. 1994a. Conservation, territory, and traditional beliefs: An analysis of Gitksan and Wet'suwet'en subsistence, Northwest British Columbia, Canada. Human Ecology, vol. 22, no. 4, 1994.

Gottesfeld, L.M.J. 1994b. Aboriginal burning for vegetation management in Northwest British Columbia. Human Ecology, vol. 22, no. 2.

Gottesfeld, L.M.J. and B. Anderson. 1988. Gitksan Traditional Medicine: Herbs and Healing. Journal of Ethnobiology 8(1): 13–33.

Goyette, D.E., D.E. Brothers, and D. Mill. 1970. Summary report of environmental surveys at Prince Rupert, 1961–1970. Dept. Fish. Memor. Rept.

Graham, C.C. and W. Masse. 1975. Presentation to Environment and Land Use Secretariat for Terrace-Hazelton forest resource study. Dept. of Environment, Fisheries and Marine Branch. Prince Rupert, B.C.

Grieve, G. 1996. Level 1 assessment of fish and stream habitat for the Kitsumkalum River east and the Cedar River and Clear Creek watersheds. Terrace, B.C.

Grieve, G.D. and D. Webb. 1997. Lakelse River steelhead: Summary of current data and status review, 1997. Skeena Fisheries Report SK-105. MELP, Skeena Region. Smithers, B.C.

Grieve, G.D. and D. Webb. 1999a. Kitsumkalum River steelhead: Summary of current data and status review, 1997. Skeena Fisheries Report SK-106. MELP, Skeena Region. Smithers, B.C.

Grieve, G.D. and D. Webb. 1999b. Kitwanga River steelhead: Summary of current data and status review. Skeena Fisheries Report SK-101. MELP, Skeena Region. Smithers, B.C.

Griffiths, J.S. 1968. Growth and feeding of the rainbow trout *Salmo gairdneri* and the lake trout *Salvelinus namaycush* from Babine Lake, British Columbia. University of Victoria, B.C.

Gulf Canada Inc. 1986. Annual report.

GWA. 2000. 1999 Kispiox watershed coho stock assessment. Gitxsan Watershed Authorities. Unpublished report. Hazelton, B.C.

GWA. 2001a. 2000 Kispiox watershed coho stock assessment. Gitxsan Watershed Authorities. Unpublished report. Hazelton, B.C.

GWA. 2001b. Unpublished data on file. Gitxsan Watershed Authorities. Hazelton, B.C.

GWA-GIS. 2002. Cultural resources of Luu Skaiyansit Lax Yip. Unpublished report on file at Gitxsan Watershed Authorities. Hazelton, B.C.

GWA. 2003a. Gitxsan cultural heritage: Suskwa and Middle Skeena watersheds. Unpublished database on file at Gitxsan Watershed Authorities. Hazelton, B.C.

GWA. 2003b. Skeena fish sites. Gitwangak to Burdick Creek. Unpublished map on file at Gitxsan Watershed Authorities. Hazelton, B.C.

GWA. 2003c. Skeena fish sites. Burdick Creek to Kispiox. Unpublished map on file at Gitxsan Watershed Authorities. Hazelton, B.C.

GWA. 2003d. Skeena fish sites. Kispiox to Utsun Creek. Unpublished map on file at Gitxsan Watershed Authorities. Hazelton, B.C.

GWWA. 1993. Preliminary Fiddler Creek stream assessment. Gitxsan-Wet'suwet'en Watershed Authorities. Unpublished report on file at Gitxsan Treaty Office. Hazelton, B.C.

Haas, G.R 1998. Indigenous fish species potentially at risk in B.C., with recommendations and prioritizations for conservation, forestry/resource use, inventory, and research. Ministry of fisheries management. Report no. 105.

Hackler, J.C. 1958. Factors leading to social disorganization among the Carrier Indians at Lake Babine. M.A. thesis, San Jose State College, California.

Haeussler, S. 1987. Ecology and berry chemistry of some food plant species used by Northwest British Columbia Indians. Opinion evidence for *Delgamuukw et al v. the Queen*. Unpublished report on file at Gitxsan Treaty Office. Hazelton, B.C.

Haggan N., Turner N., Carpenter J., Jones J.T., Mackie Q., Menzies, C. 2006. 12,000+ years of change: Linking traditional and modern ccosystem science in the Pacific Northwest. Fisheries Centre, University of British Columbia. Working paper series #2006-02, 28p.

Hall, P.E. 2004. Slamgeesh Lake Salmon Project 2003. Gitxsan Watershed Authorities. Hazelton, B.C. 87 pp.

Hall, P.E., and A.S. Gottesfeld. 2002. Kitwanga River-mouth creel survey. Gitxsan Watershed Authorities. Hazelton, B.C.

Hall, P.E., and A.S. Gottesfeld and J. Lucke 2002. Slamgeesh Lake Salmon Project 2000–2001. Gitxsan Watershed Authorities. Hazelton, B.C. 71 pp.

Hall, P.E., A.S. Gottesfeld. 2003a. Kitwanga and Bulkley River-mouth creel surveys. Gitxsan Watershed Authorities. Hazelton, B.C.

Hall, P.E., and A.S. Gottesfeld. 2003b. Slamgeesh Lake Salmon Project 2002. Gitxsan Watershed Authorities. Hazelton, B.C. 80 pp.

Hallam, R. 1975. A brief of EPS water quality monitoring and surveillance activities on Babine Lake, 1974–1975.

Halpin, M. and M. Seguin. 1990. Tsimshian Peoples. In Handbook of North American Indians: Northwest Coast. Volume 7. Smithsonian Institution. Washington, D.C.

Hancock, M.J., A.J. Leaney-East and D.E. Marshall. 1983. Catalogue of salmon streams and spawning escapements of Statistical Area 4 (lower Skeena River) including coastal streams. Can. Data. Rep. Fish. Aquat. Sci. 395: xxi + 422 pp.

Hancock, M.J., A.J. Leaney-East and D.E. Marshall. 1983. Catalogue of salmon streams and spawning escapements of Statistical Area 4 (upper Skeena River). Can. Data. Rep. Fish. Aquat. Sci. 394: xxiii + 324 pp.

Hancock, M.J. and D.E. Marshall. 1984. Catalogue of salmon streams and spawning escapements of Statistical Area 3 (Nass River) including adjacent streams. Can. Data. Rep. Fish. Aquat. Sci. 429: xxiii + 371 pp.

Hankin, D.G. and M.C. Healey. 1986. Dependence of exploitation rates for maximum yield and stock collapse on age and sex structure of chinook salmon (*Oncorhynchus tshawytscha*) stocks. Can. J. Fish. Aquat. Sci. 43: 1746-1759.

Hannigan, P., P.J. Lee, and G.G. Osadettz. 1995. Oil and gas resource potential of the Bowser-Whitehorse area of British Columbia. Geological Survey of Canada report for B.C. Energy, Mines and Petroleum Resources. Victoria, B.C.

Harding, D.R. 1969. The status of the Nanika-Morice sockeye salmon population and the Moricetown native food fishery in 1967 and 1968. Dept. of Fisheries. Vancouver, B.C.

Harper, G. 1977. Geology of the Sustut Copper deposit in B.C. Canadian Institute of Mining and Metallurgy Bulletin, vol. 70, no. 777, 97–104.

Hart, J.L. 1973. Pacific fishes of Canada. Bulletin 180, Fisheries Research Board of Canada. Ottawa, Ont.

Harr, R.D. 1976. Hydrology of small forest streams in western Oregon. USDA For. Serv. Pac. Northwest For. and Range Exp. Sta., Gen. Tech. Rept. PNW-55.

Harris, D.C. 2001. Fish law and colonialism: The legal capture of salmon in British Columbia. University of Toronto Press. 306 pp.

Harris, E.A. 1990. Spokeshute: Skeena River memory. Orca. Victoria, B.C.

Harrison, C. 2002. Personal communication. DFO, Fulton River Project.

Håstein, T and T. Lindstad. 1991. Diseases in wild and cultured salmon: Possible interaction. Aquaculture, 98: 277–288.

Hastings, N., A. Plouffe. L.C. Struik, R.J.W. Turner, R.G. Anderson, J.J. Clague, S.P. Williams, R. Kung, and G. Tacogna. 1999. Geoscape Fort Fraser, British Columbia; Geological survey of Canada. Miscellaneous report 66, 1 sheet.

Hatfield Consultants Ltd.. 1989. Babine Lake near Bell Mine: Water/sediment quality and benthic macroinvertebrate communities. Prepared for Noranda Minerals Inc.

Hatfield Consultants Ltd. 1998. Lakes District secondary lake inventory 1997 studies. Prepared for MELP, Skeena Region. Smithers, B.C.

Hatler, D. 1987. History and zoogeography of some selected mammals in Northern British Columbia. Opinion evidence for *Delgamuukw et al v. the Queen*. Unpublished report on file at Gitxsan Treaty Office. Hazelton, B.C.

Hatlevik, S.J., K. Diemert, and M.R. Whately. 1981. A creel survey of the Lakelse Lake cutthroat trout sport fishery, June–August, 1979. B.C Ministry of Environment, Fisheries Branch. Smithers, BC. SK report #28.

Hatlevik, S.J. 1999. Lake classification in the Fort St. James Forest District. MELP. Smithers, B.C.

Hay, D. and P.B. McCarter. 2000. Status of the eulachon *Thaleichthys pacificus* in Canada. Canadian Stock Assessment Secretariat, Research Document 2000/145. Fisheries and Oceans Canada. Nanaimo, B.C.

Hayes, M., F. Ferri, and S. Morii. 2004. Interior basins strategy. B. C. Ministry of Mines. Victoria, B.C.

Healey, M.C. 1980. The ecology of juvenile salmon in Georgia Strait, British Columbia. *In* W.J. McNeil and D.C. Himsworth, eds. Salmonid Ecosystems of the North Pacific. Oregon State University Press. Corvallis, Oregon. 203–229.

Healey, M.C. 1991. Life history of chinook salmon *(Oncorhynchus tshawytscha)*. In Pacific salmon life histories. C. Groot and L. Margolis, eds. UBC Press. Vancouver, Canada. 311–394.

Heard, W.R. 1991. Life history of pink salmon *(Oncorhynchus gorbuscha)*. In C. Groot and L. Margolis, eds. Pacific salmon life histories. UBC Press. Vancouver, B.C. 119–230.

Heath, D.D., S. Pollard and C. Herbinger. 2001. Genetic structure and relationships among steelhead trout *(Oncorhynchus mykiss)* populations in British Columbia. Heredity 96: 618–627.

Heath, D.D., C. Busch, J. Kelly, and D.Y. Atagi. 2002 Temporal change in genetic structure and effective population size in steelhead trout *(Oncorhynchus mykiss)*. Molecular Ecology 11(2), 197–214.

Helgerson, H. 1906. Thirty-Eighth Annual Report, 1906, Department of Marine and Fisheries. Ottawa, Ont.

Helmer, J. and B. Mitchell. 1972. The BCR Dease Lake Extension Survey. Report submitted to the Heritage Conservation Branch. Victoria, B.C.

Henson, S. 2004. Biogeochemical contributions to the water quality of the Skeena River.

Heuch P.A. and T.A. Mo. 2001. A model of salmon louse production in Norway: Effects of increasing salmon production and public management measures. Dis. Aquat. Org. 45: 145–152.

Hewes, G.W. 1973. Indian fisheries productivity in pre-contact times in the Pacific salmon area. Northwest Anthropological Research Notes 7(2): 133–155.

Higgins, M.J. and G. Munby. 2000. Pre-spawning mortality in sockeye salmon associated with gill parasites in the Babine Lake systems — 1999 observations. Unpublished report, Pacific Biological Station.

Higgins, M.J. 2001. Pre-spawning mortality in sockeye salmon associated with gill parasites in the Babine Lake systems during the 2000 spawning season. Unpublished report, Pacific Biological Station.

Higgins, M.J. 2002. Personal communication. Pacific Biological Station.

Higgins, R.J. and W.J. Schouwenburg. 1973. A biological assessment of fish utilization of the Skeena River estuary, with special reference to port development in Prince Rupert. Technical report 1973-1. Department of Environment, Fisheries and Marine Services. Vancouver, B.C.

Hilborn, R. 1992. Hatcheries and the future of salmon in the Northwest. Fisheries, Vol. 17 (1): 5–8.

Hiroshi A., B.A. Berejikian, M.J. Ford, and M.S. Blouin. 2008. Fitness of hatchery-reared salmonids in the wild. Evolutionary Applications 1(2): 342–355.

Hobenshield, E. 2002. Personal communication. Kitwanga, B.C.

Hogan, D.L. and J.W. Schwab. 1989. Floodplain and channel stability of the lower Skeena River. Unpublished report. Ministry of Forests. Smithers, B.C.

Holland, K. and W. Starr. 1999. Gitxsan fishing gear. Unpublished research report on file at Gitxsan Treaty Office. Hazelton, B.C.

Holland, S.S. 1976. Landforms of British Columbia. Bulletin 48. Queen's Printer. Victoria, B.C.

Hols, G. 1999. Marks of a Century. A history of Houston, B.C. 1900–2000. Published by District of Houston.

Holtby, L.B., R. Kadowacki, and L. Jantz. 1994. Update of stock status information for early run Skeena River coho salmon (through the 1993 return year). PSARC working paper S94-4: 44 pp.

Holtby, B., B. Finnegan, D. Chen, and D. Peacock. 1999. Biological assessment of Skeena River coho salmon. Canadian Stock Assessment Secretariat Research Document 99/140.

Hoos, L.M. 1976. The Skeena River estuary status of environmental knowledge to 1975. Report of the Estuary Working Group, Dept. of Environment. Regional Board Pacific Region. Special Estuary Series no. 3. Fisheries and Marine Service. West Vancouver, B.C.

Horn, H. 2001. Inventory assessment of the Morice LRMP area.

Horrall, R.M. 1981. Behavioral stock-isolating mechanisms in Great Lakes fishes with special reference to homing and site imprinting. Can. J. Fish. Aquat. Sci. 38: 1481–1496.

Houlden, G. and B. Donas. 2002. Chicago Creek Hatchery (operations) 2001–2002. Prepared for Chicago Creek Community Environmental Enhancement Society. South Hazelton, B.C.

Hudson, P. 2002. Watershed assessment of the Kispiox River watershed. (Kispiox Forest District). Draft for review by Round Table.

Hume, J. and K. Shortreed. 2006. Limnological and limnetic fish surveys of North Coast Area lakes in 2005. DFO. Cultus Lake, B.C.

Hutchison, W. W., H.C. Berg and A. V. Oukulitch (compilers) 1979. Geology – Skeena River, British Columbia – Alaska, Geological Survey of Canada Map 1385A. Ottawa, Ont.

IBM Business Consulting Services. 2006. Valuation of the wild salmon economy of the Skeena River watershed. Prepared for Northwest Institute for Bioregional Research. Smithers, B.C.

Imbleau, L.G.J. 1978. A creel survey of the Lakelse River cutthroat trout sport fishery. B.C. Ministry of Environment, Fisheries Branch, Smithers, B.C. SK Report #16.

Inglis, R., and G. MacDonald. 1979. Skeena River prehistory. Archaeological Survey of Canada. Mercury Series, no. 87. Ottawa.

Irvine, J.R., and N.T. Johnston. 1992. Coho salmon (Oncorhynchus kisutch) use of lakes and streams in the Keogh River drainage, British Columbia. Northwest Science 66(1): 15–25.

Jacobs, M. 1996. Suskwa Watershed Restoration Program 1996. Fisheries and riparian field assessments and prescriptions. Unpublished report. New Hazelton, B.C.

Jacobs, M. and T. Jones. 1999. Kitwanga River salmon recovery initiative 1998.

Jantz, L., B. Rosenburger and S. Hildebrandt. 1989. Salmon escapement and timing data for Statistical Area 4 of the North Coast of British Columbia. Unpublished MS, DFO, Prince Rupert, B.C.

Jantz, L., D. Wagner, D. Burnip, and S. Hildebrandt. 1989. Salmon escapement and timing data for Statistical Area 3 of the North Coast of British Columbia. Prince Rupert, B.C.

Jenness, D. 1937. The Sekani Indians of British Columbia. National Museum of Canada. Bulletin No. 84. Ottawa, Ont.

Jenness, D. 1943. The Carrier Indians of the Bulkley River: Their social and religious life. Anthropological Papers, No. 25. Smithsonian Institution, Washington, D.C.

Johnsen, B.O., and A.J. Jensen 1994. The spread of furunculosis in salmonids in Norwegian rivers. Journal of Fish Biology. 45: 47–55.

Johnson, L.M. 1997. Health, wholeness, and the land: Gitksan traditional plant use and healing. PhD. thesis. Edmonton, Alberta.

Johnson, W.E. 1956. On the distribution of young sockeye salmon (Oncorhynchus nerka) in Babine and Nilkitkwa Lakes, BC. J. Fish. Res. Board Can. 13: 695–708.

Johnson, W.E. 1958. Density and distribution of young sockeye salmon (Oncorhynchus nerka) throughout a multibasin lake system. J. Fish. Res. Board Ca. 15: 961–982.

Johnson, W.E. 1961. Aspects of the ecology of a pelagic, zooplankton eating fish. Verh. Internat. Verein. Limnol. 14: 727–731.

Johnson, W.E. 1964. Quantitative aspects of the pelagic, entomostracan zooplankton of a multibasin lake system over a 6-year period. Verh. Internat. Verein. Limnol. 15: 727–734.

Johnson, W.E. 1976. The morphometry of Alastair Lake. Manuscript Report Series, No. 819, Fisheries Research Board of Canada. PBS. Nanaimo, B.C.

Johnston, D. 2001. Gitsegukla planning unit restoration plan. Prepared for Ministry of Forests, Kispiox Forest District.

Joseph, G. 2001a. Interview with J. George. Smithers, B.C.

Joseph, W. 2001b. Personal communication. Wet'suwet'en Fisheries. Moricetown, B.C.

Jyrkkanen, J. 1996. Interfor North Coast chart watershed restoration project, level 1. Terrace, B.C.

Jyrkkanen, J. 1997. Interfor Watershed Restoration Program overview fisheries assessment.

Jyrkkanen, J., G. Wadley, D. Vegh, R. Collier, T. Lattie, G. Wilson, L. Petersen, and C. Hillis. 1995. Kispiox watershed restoration program project level 1 final report: The impact of logging on the Kispiox watershed and recommendations for level 2 restoration works.

Kerby, N. 1984. Greater Terrace Official Settlement Plan: Background studies and planning recommendations. Prepared for Regional District of Kitimat-Stikine. Terrace, B.C.

Kerby, N. 1997. Kispiox land use study, background report. Prepared for Kitimat-Stikine Regional District. Terrace, B.C.

Kindle, E.D. 1940. Mineral resources, Hazelton, and Smithers areas, Cassiar and Coast Districts, British Columbia. Geological Survey of Canada Memoir 223. Ottawa, Ont.

Kingston, D. 2002. Kitwanga River fisheries treaty related measure #1: Coho escapement and habitat relocation/ expansion; habitat studies and creel surveys. Gitanyow Fisheries Authority. Kitwanga, B.C.

Klohn Leonoff Consulting Engineers. 1985. Telkwa Coal Project. Stage II geotechnical, hydrogeological and hydrological design report. Prepared for Crows Nest Resources Ltd.

Kobrinsky, V. 1977. Ethnohistory and ceremonial representation of Carrier social structure. PhD thesis. UBC. Vancouver, B.C.

Kofoed, G. 2001. Personal communication. Retired Fishery Officer, Terrace, B.C.

Kofoed, G. 2002. Personal communication. Retired Fishery Officer. Terrace, B.C.

Kokelj, J. 2003. Lakelse Lake draft management plan. Prepared for Lakelse Watershed Society. MWLAP. Smithers, B.C.

Kokelj, J. 2004. Round Lake draft management plan. Prepared for Round Lake Watershed Enhancement Society. MWLAP. Skeena Region. Smithers, B.C.

Kondzela, C.M., C.M. Guthrie, S.L. Hawkins et al. 1994. Genetic relationships among chum salmon populations in southeast Alaska and northern British Columbia. Can. J. Fish. Aquat. Sci. 51(supl 1): 50–64.

Koski, W.R., Alexander, R.F., and K.K. English. 1995. Distribution, timing, fate and number of coho salmon and steelhead returning to the Skeena Watershed in 1994. Report by LGL Limited, Sidney, B.C. for Fisheries Branch, British Columbia Ministry of Environment, Lands, and Parks. Victoria, B.C.

Krkošek M., M.A. Lewis, and J.P. Volpe 2005. Transmission dynamics of parasitic sea lice from farm to wild salmon. Proceedings of the Royal Society of London Series B 272: 689–696.

Krkošek M., M.A. Lewis, Morton A, Frazer, L.N., and J.P. Volpe. 2006. Epizootics of wild fish induced by farmed fish. PNAS. 103 No. 42: 15506–15510.

Kussat, R. and K. Peterson. 1972. An assessment of the effects on the Morice and Bulkley River systems of a pulp mill at Houston, B.C. Fisheries Service, Dept. of Environment. Northern Operations Branch. Prince Rupert, B.C.

Kussat, R. 1973. Upper Skeena counting fences: Bear, Sustut and Johanson Rivers. DFO internal report. Prince Rupert, B.C.

Large, R.G. 1996. The Skeena: River of Destiny. Heritage House Publishing Co.

Larkin, P.A., and J.G. McDonald. 1968. Factors in the population biology of the sockeye of the sockeye salmon of the Skeena River. J. Anim. Ecol. 37, 229–258.

Levy, D.A. and K.J. Hall. 1985. A review of the limnology and sockeye salmon ecology of Babine Lake. Westwater Research Center, University if British Columbia. Westwater Tech. Rep. No. 27. Vancouver, B.C.

Lewis, A.F.J. 1998. Skeena eulachon study 1997. Prepared for Tsimshian Tribal Council. Triton Environmental Consultants Ltd. Terrace, B.C.

Lewis, A.F.J. 2000. Skeena steelhead and salmon: A report to stakeholders. Steelhead Society of British Columbia, Bulkley Valley Branch. Smithers, B.C.

Lewis, A.F.J. and S. Buchanan. 1998. Zymoetz River steelhead: summary of current data and status review, 1997. Skeena Fisheries Report SK-102. MELP, Skeena Region. Smithers, B.C.

Lewis, H.T. and T.A. Ferguson. 1988. Yards, corridors and mosaics: How to burn a boreal forest. Human Ecology 16(1): 57–77.

Lewynsky, V.A. and W.R. Olmstead. 1990. Angler use and catch surveys of the lower Skeena, Zymoetz (Copper), Kispiox, and Bulkley River steelhead fisheries, 1989. ESL Environmental Sciences Limited. Vancouver, B.C.

Liepens, S. 2002. Personal communication, MWALP. Prince Rupert, B.C.

Liknes, G.A. and P.J. Graham. 1988. Westslope cutthroat trout in Montana: Life history, status, and management. Amer. Fish Soc. Symp. 4: 53–60.

Lisle, T.E. and M. Church. 2002. Sediment transport-storage relations for degrading, gravel bed channels. Water Resources Research, vol. 38, no. 11.

Lindsey, C.C. and W.G. Franzin. 1972. New complexities in zoogeography and taxonomy of P. coulteri (Prosopium coulteri). J. Fish. Res. Board of Can. 29(12): 1772–775.

Loedel, M. and P. Beaudry. 1993. A study of forest interception at the Date Creek silvicultural systems project. Ministry of Forests, Forest Sciences. Smithers, B.C.

Lord, C.S. 1948. McConnell Creek map-area, Cassiar District, British Columbia. Geological Survey of Canada, Memoir 251. Ottawa, Ont.

Lorenz, A. 1998a. Reconnaissance (1:20,000) fish and fish stream inventory of Kitsuns Creek. Prepared for Skeena Cellulose Inc. and MELP.

Lorenz, A. 1998b. Reconnaissance (1:20,000) fish and fish stream inventory of West Kitsuns Creek. Prepared for Skeena Cellulose Inc. and MELP.

Lorenz, A. 1998c. Reconnaissance (1:20,000) fish and fish stream inventory of Skeena River tributaries south of Larkworthy Creek. Prepared for Skeena Cellulose Inc. and MELP.

Lorenz, A. 1998d. Reconnaissance (1:20,000) fish and fish stream inventory of Skeena River tributaries south of the Sicintine River. Prepared for Skeena Cellulose Inc. and MELP.

Lorenz, A. 1998e. Reconnaissance (1:20,000) fish and fish habitat stream inventory of Larkworthy Creek. Prepared for Skeena Cellulose Inc. and MELP.

Lorenz, A. 1998f. Reconnaissance (1:20,000) fish and fish habitat stream inventory of Deep Canoe Creek. Prepared for Skeena Cellulose Inc. and MELP.

Lough, M.J. 1979. Memo on file. (November 21, 1979). Skeena Region stream files. Ministry of Environment, Lands and Parks. Smithers, B.C.

Lough, M.J. 1980. Radio telemetry studies of summer run steelhead trout in the Skeena River drainage, 1979, with particular reference to Morice, Suskwa, Kispiox and Zymoetz River stocks. SK-29. B.C. Fish and Wildlife Branch. Smithers, B.C.

Lough, M.J. 1981. Commercial interceptions of steelhead trout in the Skeena River — radio telemetry studies of stock identification and rates of migration. B.C. Environment, Skeena Fisheries Report SK-32.

Lough, M.J. 1983. Radio telemetry studies of summer run steelhead trout in the Cranberry, Kispiox, Kitwanga, and Zymoetz Rivers and Toboggan Creek, 1980. Skeena Fisheries Report SK-33. MELP, Skeena Region. Smithers, B.C.

Lough, M.J. 1993. Memo on file. (February 22, 1993). Skeena Region stream files. Ministry of Environment, Lands and Parks. Smithers, BC.

Lough, M.J. and M.R. Whately. 1984. A preliminary investigation of Kitsumkalum River steelhead trout, 1980–81. Skeena Fisheries Report SK-81. MELP, Skeena Region. Smithers, B.C.

MacDonald, P.D.M. and H.D. Smith. 1980. Mark-recapture estimation of salmon smolt runs. Biometrics 36: 401–417.

MacKay, S., T. Johnston, and M. Jessop. 1998. Mid-Bulkley detailed fish habitat/riparian/channel assessment for watershed restoration. Report prepared by B.C. Conservation Foundation for CFDC Nadina. Houston, B.C.

MacLaren Plansearch Services Ltd. 1985. Water quality investigation: Telkwa Coal Project.

Maclean, D.B. 1983. Aldrich Lake: A data report on water quality and biological data from the receiving waters of the area around the abandoned Duthie Mine. MELP Waste Management Branch Skeena Region Rep. 83.05. Smithers, B.C.

Magdanz, H. 1975. Land use report for Inventory Project 331-E. TFL #40, E. 1975.

Malloch, G.S. 1912. Reconnaissance on the upper Skeena River, between Hazelton and the Groundhog Coalfields, British Columbia. Annual report of the British Columbia Minister of Mines. 2 George V., A. Sessional Paper No. 26.

Malloch, G.S. 1913. The Groundhog Coal Field, B.C. 2 George V., A. Sessional Paper No. 26.

Maloney, D. 1995. 1995 survey of TSS concentrations in headwater streams of the Nilkitkwa and Nichyeskwa watersheds. Forest Sciences Section, Prince Rupert Forest Region, B.C. Ministry of Forests. Smithers, BC.

Manalta Coal Ltd. 1997. Telkwa Coal Project. Application for a Project Approval Certificate. Volume 1. Submitted to the BC Environmental Assessment Office.

Martin, N.V. and C.H. Oliver. 1980. The lake char, Salvelinus namaycush. In E.K. Balon, ed. Chars: Salmonid fishes of the genus Salvelinus. Dr. W. Junk Publishing, The Hague, Netherlands.

Mark, C. 1923. Interview with M. Barbeau. (B-F-63.4) Field notes.

Matthews, D and I.R. Wilson. 2001. Archaeological inventory and impact assessment, Skeena Cellulose Inc. proposed forestry developments near New Hazelton, B.C. Heritage Conservation Branch permit 2000-186. Unpublished report on file at Heritage Resource Center, Ministry of Small Business, Tourism and Culture. Victoria, B.C.

Maxwell, I. 2003. Personal communication.

Maynard, D. 1994. Terrain classification, terrain stability, surface erosion potential, and sediment transfer capability of upper Zymoetz River watershed. Prepared for B.C. Ministry of Forests, Bulkley Forest District. Smithers, B.C.

McBean, G.A., O. Slaymaker, T. Northcote, P. Leblond, and T.S. Parsons. 1992. Review of models for climate change and impacts on hydrology, coastal currents, and fisheries in B.C. Climate Change Digest.

McCart, P.J. 1965. Growth and morphometry of four British Columbia populations of pygmy whitefish (Prosopium coulteri) J. Fish. Bd. Canada 22(5):1229–1255.

McCart, P.J. 1970. Evidence for the existence of sibling species of pygmy whitefish (Prosopium coulteri) in three Alaskan lakes. In C.C. Lindsey and C.S. Woods, eds. Biology of coregonid fishes. University of Manitoba Press. Winnipeg, Man.

McCarthy, M. 2000a. Comeau Creek fish and fish habitat assessment initiative, 1999. Prepared for Suskwa Restoration Society, FRBC, DFO, and 16–37 Community Futures.

McCarthy, M. 2000b. Suskwa coho synoptic survey 1999. Unpublished report. South Hazelton, B.C.

McCarthy, M., M. Cleveland and D. Kingston. 2002. The 2001 Kitwanga River chinook salmon enumeration initiative. Gitanyow Fisheries Authority. Kitwanga, B.C.

McDonald, J.G. 1973. Diel vertical movements and feeding habits of underyearling sockeye salmon (Oncorhynchus nerka) at Babine Lake, BC. Fish Res. Board Can. Tech. Rep. 378.

McDonald, J., and J.M. Hume. 1984. The Babine Lake sockeye salmon enhancement program: Testing some major assumptions. Can. J. Fish. Aquat. Sci. 40: 70–92.

McElhanney Consulting Services Ltd. 2001. Watershed restoration plan Kitwanga River watershed.

McElhanney Consulting Services Ltd., G.E. Bridges & Associates Inc., and Triton Environmental Consultants Ltd. 2001. Stewart Omineca Resource Road Impact Study. Prepared for the District of Stewart, B.C.

McKean, C.J.P. 1986. Lakelse Lake: Water quality assessment and objectives. Water Management Branch, MELP. Victoria, B.C.

McKinnell, S. and D. Rutherford. 1994. Some sockeye salmon are reported to spawn outside the Babine Lake watershed in the Skeena drainage. PSARC Working Paper S94-11. 52 pp.

McKinnell, S. and A.J. Thomson 1997. Recent events concerning Atlantic salmon escapees in the Pacific. ICES Journal of Marine Science. 54: 1221–1225.

McNeil, A.O. 1983. The Maxan Lake multi-land use study, a summary review. Project Number 271032. Canada–B.C. Subsidiary Agreement on Agriculture and Rural Development.

McNicol, R.E. 1999. An assessment of Kitsumkalum River summer chinook, a North Coast indicator stock. CSAS Res. Doc. 99/164. Fisheries and Oceans Canada. Ottawa, Ont.

McPhail, J.D. and C.C. Lindsey 1970. Freshwater fishes of northwestern Canada and Alaska. Fish. Res. Board Ca. Bull. 173: 381.

McPhail, J.D. and C.B. Murray. 1979. The early life history and ecology of Dolly Varden (Salvelinus malma) in the upper Arrow Lakes. University of British Columbia. Vancouver, B.C.

McPhail, J.D. and R. Carveth. 1993a. Field keys to the freshwater fishes of British Columbia. Aquatic Inventory task force of the Resources Inventory Committee. B.C. Ministry of Environment, Lands and Parks. Victoria, B.C.

McPhail, J.D. and R. Carveth. 1993b. A foundation for conservation: The nature and origin of the freshwater fauna of British Columbia. Queens Printer for B.C. Victoria, B.C.

Meidinger, D. and J. Pojar. 1991. Ecosystems of British Columbia. B.C. Ministry of Forests, Special Report, Series 6. Victoria, B.C.

MELP. 1997. Angling Use Plan, Babine River. Draft. Ministry of Environment, Lands and Parks. Smithers, B.C.

MELP. 2000. BC Freshwater fishing regulations synopsis. Ministry of Environment, Lands and Parks. Victoria, B.C.

Miles, M. 1990. Assessment of fluvial hazards; Zymoetz River section, Smithers to Terrace gas pipeline route. Prepared for Pacific Northern Gas. Vancouver, B.C.

Miles, M. 1991. Linear developments: Innovative techniques for sediment control and fisheries habitat recreation. In Proceedings of the B.C.-Yukon sediment issues workshop. T. Yuzyk and B. Tassone, eds. April 24–26, 1990. Vancouver, B.C. Water Resources Branch, Pacific and Yukon Region. Environment Canada. Vancouver, B.C.

Miles, M. and E. White. 1996. Habitat replacement study, Ayton Mainline Lower Skeena River.

Mills, A. and R. Overstall. 1996. The whole family lived there like a town. Anthropological summary and historical data forms. Prepared for Wet'suwet'en Cultural Heritage Resource Information Project.

Ministry of Energy, Mines and Petroleum Resources. 2002. 104A Minfile. www.em.gov.bc.ca/geology

Ministry of Environment. 1979. Assessment of the impact of the Kemano II proposal on the fish and wildlife resources of the Nanika-Morice systems. Victoria, B.C.

Ministry of Forests. 1988. Swan Lake/Brown Bear Lake wilderness designation process: Wilderness management options report. Kispiox Forest District and Regional Recreation Section, Prince Rupert Forest Region, Ministry of Forests. Smithers, B.C.

Ministry of Forests. 1996. Kispiox land and resource management plan. Amended 2001.

Ministry of Forests. 1999. Fort St. James land and resource management plan.

Ministry of Forests. 2000a. Access management direction for the Babine watershed. Kispiox Forest District. Hazelton, B.C.

Ministry of Forests. 2000b. Northern long term road corridors plan. Fort St. James Forest District coordinated access management plan.

Ministry of Forests. 2001a. Kispiox land and resource management plan. Amended.

Ministry of Forests. 2001b. Skeena-Bulkley Region resource management plan. Smithers, B.C.

Ministry of Forests. 2002. Timber supply review. Kispiox Timber Supply Area analysis report. Victoria, B.C.

Ministry of Lands, Parks and Housing. 1984. Kispiox valley crown land plan. Smithers, B.C.

Ministry of Sustainable Resource Management. 2002. Morice planning area background report. Victoria, B.C. www.luco.gov.bc.ca/lrmp/morice/techrpt/1.htm

Ministry of Water, Lands and Air Protection. 2001. Steelhead Harvest Analysis. Database maintained by the Fish and Wildlife Branch of the British Columbia Ministry of Water, Lands and Air Protection.

Mitchell, S. 1997. Riparian and in-stream assessment of the Bulkley River system: An examination and prioritization of impacts on the tributaries to the Bulkley River mainstem.

Mitchell, S. 1998. Station/Waterfall Creeks environmental assessment: An impact assessment of proposed water withdrawal options for the District of New Hazelton/ Hagwilget Band Council.

Mitchell, S. 2001a. A Petersen capture-recapture estimate of the steelhead population of the Bulkley/Morice River systems upstream of Moricetown Canyon during autumn 2000, including synthesis with 1998 and 1999 results. Report to Steelhead Society of British Columbia, Bulkley Valley Branch. 45 pp.

Mitchell, S. 2001b. Bulkley/Morice steelhead assessment, 2000. Submitted to Steelhead Society of British Columbia, Bulkley Valley Branch.

MoE. 1979. Aquatic biophysical maps (93M/5, 103P/9, 15). Resource Analysis Branch, Ministry of Environment. Victoria, B.C.

MoE. Various dates and authors. Skeena River lake and stream management files. B.C. Ministry of Environment. Smithers, B.C.

MoF. 2001. Kalum land and resource management plan. Ministry of Forests, April 2001. Terrace, B.C.

Monger, J.W.H. 1976. The Triassic Takla group in McConnell Creek Map-Area, North-Central British Columbia. Geological Survey of Canada Paper 76-29. 45 pp.

Moore, D. 1993. Babine River: Slide hazard assessment. BCHIL. Vancouver, B.C.

Moore, K.M.S. and S.V. Gregory. 1988. Summer habitat utilization and ecology of cutthroat trout fry (Salmo clarki) in Cascade Mountain streams. Can. J. Fish. Aquat. Sci., vol. 45: 1921–1930.

Morgan, J.D. 1985. Biophysical reconnaissance of the Kitsumkalum River system, 1975–1980. E.V.S. Consultants Ltd. For DFO, 1985.

Morice, A.G. 1978. The History of the Northern Interior of British Columbia. Interior Stationery. Smithers, B.C.

Morin, K.A. and N.M. Hutt. 2000a. Draft remediation plan and waste-reduction plan, former Topley concentrate shed site, Topley, B.C. Report prepared by Morwijk Enterprises Ltd. for Noranda Inc. Granisle.

Morin, K.A. and N.M. Hutt. 2000b. Draft remediation plan and waste-reduction plan, former Richfield Loop concentrate shed site, Topley, B.C. Report prepared by Morwijk Enterprises Ltd. for Noranda Inc. Granisle.

Morrell, M. 1985a. The Gitxsan and Wet'suwet'en fishery in the Skeena River system. Gitxsan-Wet'suwet'en Tribal Council. Hazelton, B.C.

Morrell, M. 1985b. The Gitxsan and Wet'suwet'en fishery in the Skeena River system. Gitxsan-Wet'suwet'en Tribal Council. Hazelton, B.C.

Morrell, M. 2000. Status of salmon spawning stocks of the Skeena River system. Northwest Institute for Bioregional Research. Smithers, B.C.

Morris, M. and B. Eccles. 1975. Kitseguecla River fisheries study. Ministry of Environment. Smithers, B.C.

Morten, K.L. 1999. A comparison of steelhead angler effort and catch estimates on the Bulkley River. Skeena Fisheries Report SK-123.

Morten, K.L. and C.K. Parken. 1998. A survey of Bulkley River steelhead anglers during the classified waters period of 1997. Skeena Fisheries Report #123. Cascadia Natural Resource Consulting. Smithers, B.C.

Morton, A. and R. Routledge. 2005. Mortality rates of juvenile pink (Oncorhynchus gorbuscha) and Chum (O. keta) salmon infested with sea lice (Lepeophtheirus salmonis) in the Broughton Archipelago. Alaska Fishery Research Bulletin 11: 143–149.

Mueter, F.J., R.M. Peterman, and B.J. Pyper 2002. Opposite effects of ocean temperature on survival rates of 120 stocks of Pacific salmon (Oncorhynchus spp.) in northern and southern areas. Can. J. Fish. Aquat. Sci. 59(3): 456–463, plus erratum printed in 60 757.

Mueter, F.J., R.M. Peterman, and B.J. Pyper. 2002. Opposite effects of ocean temperature on survival rates of 120 stocks of Pacific salmon (Oncorhynchus spp.) in northern and southern areas. Canadian Journal of Fisheries and Aquatic Sciences 59(3): 456–463 plus the erratum printed in Can. J. Fish. Aquat. Sci. 60: 757.

Muldon, C. 2003. Personal communication.

Naish, K.A., J.E. Taylor, P.S. Levin, T.P. Quinn, J.R. Winton, D. Huppert, and R. Hilborn. 2007. An evaluation of the effects of conservation and fishery enhancement hatcheries on wild populations of salmon. Advances in Marine Biology 53: 61–194.

Narver, D. 1970. Diel vertical movement and feeding of underyearling sockeye salmon and the limnetic zooplankton in Babine Lake, British Columbia. Fish Res. Board Can. 27: 281–316.

Nass, B.L., M.G. Foy, and A. Fearon-Wood. 1995. Assessment of juvenile coho salmon habitat in the Skeena River watershed by interpretation of topographic maps and aerial photographs. LGL Project No. EA#664. Report to DFO Pacific Region.

Naziel, W. 1997. Wet'suwet'en traditional use study. Office of the Wet'suwet'en. Moricetown, B.C.

Neilson, C. 2002. Personal communication. DFO. Prince Rupert, B.C.

Nelson, J.S. 1968. Distribution and nomenclature of North American kokanee, Oncorhynchus nerka. J. Fish. Res. Board Can. 25: 409–414.

Newell, D. 1993. Tangled webs of history: Indians and the law in Canada's Pacific Coast fisheries. University of Toronto Press. Toronto, Ont.

Nijman, R.A. 1986. Skeena-Nass Area, Bulkley River basin water quality assessment and objectives. Report and technical appendices. Water Management Branch. Victoria, B.C.

Nijman, R.A. 1996. Water Quality assessment and objectives for the Bulkley River headwaters. Draft. Water Quality Branch, Environmental Protection Dept. MELP. Victoria, B.C.

Norcan Consulting Ltd. 2000. Archaeological Impact Assessment report: Sustut bridge site. Heritage inspection permit: 2000-120.

Nortec Consulting. 1997. Kispiox watershed restoration project. Contract #CSK2087 CSK2072. Final report and appendices.

Nortec Consulting. 1998. Morice watershed restoration project level II: Report-assessment and survey and design. Smithers, B.C.

Northcote, T.G. and G.D. Taylor. 1973. Limnology and fish of the Gitnadoix River–Alastair Lake system in relation to ecological reserve considerations.

Northcote, T.G. and G.L. Ennis. 1994. Mountain whitefish biology and habitat use in relation to compensation and improvement possibilities. Reviews in Fisheries Science 2: 347–371.

Northwest Enhancement Society. 1985. Biophysical investigations of the Stikine River and its tributaries, 1985. Prepared for DFO and Great Glacier Salmon Ltd.

O'Dwyer, J.S. 1901. Report on the field operations performed during the season of 1899. Dept. of Railways and Canals, Canada, Ann. Rept. July 1, 1899 to June 30, 1900. 158–168.

Office of Wet'suwet'en. 2001. Field trip notes from Landscape Unit Planning sessions – Buck Creek, Deep Creek, and Bulkley Landscape Units.

O'Neill, M. 2003. Toboggan Creek Hatchery manager. Personal communication.

Orr, C. 2007. Estimated sea louse egg production from Marine Harvest Canada (Stolt) farmed salmon, Broughton Archipelago, British Columbia, 2003–2004. North American Journal of Fisheries Management 27(2): 187–197.

Osadetz, K.G., Evenchick, C.A., Ferri, F., Stasiuk, L., Obermajer, D.M. and Wilson, N.S.F. 2003. Molecular composition of crude oil stains from Bowser basins in geological fieldwork 2002. B.C. Ministry of Energy, Mines and Petroleum Resources. Paper 2003-1, 257–264.

Osadetz, K.G., Jiang, C., Evenchick, C.A., Ferri, F., Stasiuk, L.D., Wilson, N.S.F. and Hayes, M. 2004. Sterane compositional traits of Bowser and Sustut basin crude oils: Indications for three effective petroleum systems. Resource Development and Geoscience Branch, B.C. Ministry of Energy, Mines and Petroleum Resources. Summary of Activities 2004.

Ottens, R. 2002. Personal communication. Ministry of Forests, Kalum WRP.

Overstall, R. 2002. Gitxsan Treaty Office response to 2002 Timber Supply Review. Unpublished report on file at Gitxsan Treaty Office. Hazelton, B.C.

Overstall, R. and N.J. Sterritt. 1986. Gwalgwa maiy: Making berry cakes at Kispiox in the 1920s. (Based on interviews with Percy Sterritt, Neil B. Sterritt, and Gertie Morrison). Unpublished report on file at Gitxsan Treaty Office. Hazelton, B.C.

Paish, H. 1975. CN Meziaden project. Mile 0–75 environmental impact study. Vol. 1. Thurber Engineering Ltd. Smithers, B.C.

Paish, H. and Associates. 1983. A strategic overview of the Skeena and Nass Drainages. Prepared for the Department of Fisheries and Oceans.

Paish, H. and Associates. 1990. Gitnadoix River recreational fisheries assessment. Coquitlam, B.C.

Palmer, R.N. 1986b. Observations from studies of sockeye salmon stocks of the Nanika-Morice system, 1961–66. Unpublished note. DFO. Vancouver, B.C.

Parkin, C.K. and K.L. Morton. 1996. Enumeration of adult steelhead in the upper Sustut River 1995. Skeena Fisheries Report #94. Ministry of Environment, Lands and Parks. Smithers, B.C.

Patrick, T. 2001. Personal communication. Bear Lake, B.C.

Peacock, D. 2002. Personal communication. North Coast Stock Assessment, DFO, Prince Rupert.

Peacock, D., B. Spilsted, and B. Snyder, B. 1997. A review of stock assessment information for Skeena River chinook salmon. PSARC Working Paper S96-7.

Pendray, T. 1990. Habitat improvement and outplanting possibilities, upper Bulkley/Morice systems. Draft report for discussion purposes. Department of Fisheries and Oceans.

Pendray, T. 1993. Skeena River eulachon larvae sampling. Unpublished draft report. Department of Fisheries and Oceans. Prince Rupert, B.C.

People of K'san. 1980. Gathering what the great nature provided. Douglas and McIntyre, Vancouver, B.C., and University of Washington Press, Seattle, W.A.

PFRCC. 2001. Pacific Fisheries Resource Conservation Council. Annual report 2000–2001.

Pinsent, M.E. 1970. A report on the steelhead anglers of four Skeena watershed streams during the fall of 1969. Fish and Wildlife Branch. Smithers, B.C.

Pinsent, M.E. and W.E. Chudyk. 1973. An outline of the steelhead of the Skeena River system. B.C Fish and Wildlife Branch. Smithers, B.C.

Plate, E., C. Muldon, and R. Harris. 1999. Identification of salmonid habitat utilization and stock enumeration in streams of the Kispiox River watershed, 1998–99. Gitxsan and Wet'suwet'en Watershed Authorities. Hazelton, B.C.

Pojar, J. 2002. Personal communication. Smithers, B.C.

Pojar, J., F.C. Nuszdorfer, D. Demarchi, M. Fenger, T. Lea, and B. Fuhr. 1988. Biogeoclimatic and Ecoregion Units of the Prince Rupert Forest Region. Map.

Pollard, B.T. 1996. Level 1 fisheries assessment for the Zymoetz River. R.J.A. Forestry Ltd. Terrace, B.C.

Pollard, B.T. and S. Buchanan. 2000a. WRP overview fish and riparian assessment for the Exstew River watershed. Acer Resource Consulting Ltd. Terrace, B.C.

Pollard, B.T. and S. Buchanan. 2000b. WRP overview fish and riparian assessment for the Shames River watershed. Acer Resource Consulting Ltd. Terrace, B.C.

Portman, C. 2002. Water quality assessment and fish sampling on selected Bulkley River tributaries. Prepared for Environment Canada by Gaia Consulting. Smithers, B.C.

Portman, C. and T. Schley. 2001. Water quality sampling for the 2001 spring runoff in the Bulkley valley. Prepared for Environment Canada in conjunction with the Department of Fisheries and Oceans. Smithers, B.C.

Powell, L. 1995. Habitat and habitat management in the Skeena watershed. DFO. Prince Rupert, B.C.

Price, D. T., D.W. McKenney, D.W. Caya, and H. Côté. 2001. Transient climate change scenarios for high-resolution assessment of impacts on Canada's forest ecosystems. Report to Climate Change Action Fund and Canadian Institute for Climate Studies.

Pritchard, A.L. 1948. Skeena River salmon investigations interim report. Appendix 3. Fisheries Research Board, Nanaimo, B.C.

Pritchard, D.W. 1967. What Is an Estuary: Physical Viewpoint. In G.H. Lauff, ed. Estuaries. Amer. Assoc. Advancement Sci. Washington, D.C.

Psutka, J.F. 1996. Babine River slopes inspection. BCHIL. Vancouver, B.C.

Psutka, J.F., and P.A. Rapp. 1996. Babine River slide hazard assessment. BCHIL, Vancouver, B.C.

Quinn, T.P. 1993. A review of homing and straying of wild and hatchery-produced salmon. Fisheries Research 18: 29–44.

Rabnett, K. 1997a. Gitsegukla WRP: slope stabilization project at km 9.5 Kitsegukla FSR. Prepared for Gitsegukla Band Council.

Rabnett, K. 1997b. Jumbo Creek Status Report. Unpublished report. Suskwa Community Association.

Rabnett, K. 2000a. Cultural heritage survey on a portion of the McDonell Chart, SBFEP Bulkley-Cassiar Forest District. Unpublished report prepared for Ministry of Forests, Bulkley-Cassiar Forest District.

Rabnett, K. 2000b. Past into the present. Cultural heritage resources review of the Bulkley TSA. Hazelton, B.C.

Rabnett, K. 2001. Selected Wet'suwet'en cultural heritage within portions of Morice landscape unit planning areas. Office of the Wet'suwet'en. Smithers, B.C.

Rabnett, K. 2002. Cultural Heritage Resource Recce's and CMT Surveys. Unpublished report prepared for Ministry of Forests, Bulkley-Cassiar Forest District.

Rabnett, K. 2005. Morice-Nanika sockeye recovery plan backgrounder. Skeena Fisheries Commission. Hazelton, B.C.

Rabnett, K. 2006a. Morice-Nanika sockeye recovery. Evaluation of enhancement options. Technical report. Wet'suwet'en Fisheries. Moricetown, B.C.

Rabnett, K. 2006b. Lower Skeena fish passage assessment. Highway 16, 37S, and CN Rail. Skeena Fisheries Commission. Hazelton, B.C.

Rabnett, K. and R. Wright. 2001. Wet'suwet'en territorial stewardship plan: Cultural heritage inventory 2001. Prepared for Office of the Wet'suwet'en.

Rabnett, K., K. Holland and A. Gottesfeld. 2001. Dispersed traditional fisheries in the upper Skeena watershed. Gitksan Watershed Authorities. Hazelton, B.C.

Rabnett, K. and L. Williams. 2004. Highway 16 fish passage assessment in mid-Skeena watershed. Gitksan Watershed Authorities. Hazelton, B.C.

Rankin, D.P. 1977. Increased predation by juvenile sockeye salmon (Oncorhynchus nerka Walbaum) relative to changes in macrozooplankton abundance in Babine Lake, British Columbia. M.Sc. thesis, Dept. Zool. Univ. B.C.

Rankin, D.P., and H.J. Ashton. 1980. Crustacean zooplankton abundance and species composition in 13 sockeye salmon (Oncorhynchus nerka) nursery lakes in British Columbia. Can Tech. Rep. Fish. Aquatic Sci. 957.

Rankin, L. 1999. Phylogenetic and ecological relationship between giant pygmy whitefish (Prosopium spp.) and pygmy whitefish (Prosopium coulteri) in north-central British Columbia.

RDKS. 1991. Hazeltons Vicinity Official Community Plan. Regional District of Kitimat-Stikine. Terrace, B.C.

RDKS. 1996. Regional District of Kitimat-Stikine community and subregional populations.

RDKS. 2002. Regional District Kitimat-Stikine quick facts. Terrace, B.C.

Read Environmental and Planning Associates. 1982. Environmental overview of the Telkwa Project Area: Aquatic resource component. Draft. Prepared for Crows Nest Resources Ltd.

Reese-Hansen, L. 2001. Interim restoration plans for nine watersheds in the central Kalum Forest District. Kitsumkalum Band Council. Terrace, B.C.

Reese-Hansen, L. 2002. Personal communication. Terrace, B.C.

Regional District of Bulkley-Nechako. 1998. Houston/Topley/Granisle official community plan technical supplement. Burns Lake, B.C.

Remington, D. 1996. Review and assessment of water quality in the Skeena River watershed, British Columbia, 1995. Can. Data Rep. Fish. Aquat. Sci. 1003: 328 pp.

Remington, D. 1997. Survey of water quality and periphyton (attached algae) standing crop in the Bulkley River and tributaries, 1996. Prepared for DFO Habitat and Enhancement Branch, Skeena/Nass by Remington Environmental. Smithers, B.C.

Remington, D. 1998. Water quality and accumulation of periphyton (attached algae) in the Bulkley River and tributaries, 1997: Relationship with land use activities in rural watersheds. Prepared for DFO Habitat and Enhancement Branch, Skeena/Nass by Remington Environmental. Smithers, B.C.

Remington, D. 2002. Drinking water source monitoring: Skeena Region 2001. Prepared for B.C. WLAP by Remington Environmental. Smithers, B.C.

Remington, D., J. Wright and L.J. Imbleau. 1974. Steelhead angler-use survey on the Zymoetz, Kispiox, and Bulkley Rivers. Fish and Wildlife Branch. Smithers, B.C.

Remington, D. and B. Donas. 1999. Water quality in the Toboggan Creek watershed 1996–1998: Are land use activities affecting water quality and salmonid health? Prepared for DFO Habitat and Enhancement Branch, Skeena/Nass by Remington Environmental. Smithers, B.C.

Remington, D. and B. Donas. 2001. Nutrients and algae in the Upper Bulkley River Watershed 1997–2000. Prepared for the Community Futures Development Corporation of Nadina. Houston, BC. Amended 2001.

Remington, D. and J. Lough. 2003. Water quality assessment and objectives for Toboggan Creek and tributaries Report. Prepared for WLAP by Remington Environmental. Smithers, B.C.

Rescan Environmental Services. 1992. Bell 92 project closure plan support. Document H: Existing environmental conditions of Babine Lake in the vicinity of Bell Mine. Prepared for Noranda Minerals Inc.

Resource Analysis Branch. 1975–1979. Aquatic biophysical maps. 1:50,000. Min. of Environment. Victoria, B.C.

Richards, T.A. 1975. McConnell Creek Map Area (94D/E) Geology. Geological Survey of Canada Open File 342, 1:250,000.

Richards, T.A. 1990. Geology of Hazelton Map Area (93M). Geological Survey of Canada Open File 2322, 1:250,000.

Ricker, W.E. 1940. On the origin of kokanee, a freshwater type of sockeye salmon. Transactions Royal Society of Canada Series III, 34: 121–135.

Ricker, W.E. 1981. Changes in the average size and average age of Pacific salmon. Can. J. Fish. Aquat. Sci. 38: 1636–1656.

Ricker, W.E. and H.D. Smith. 1975. A revised interpretation of the history of Skeena River sockeye salmon (Oncorhynchus nerka). J. Fish. Res. Board Can. 32: 1369–1381.

Riddell, B. and B. Snyder. 1989. Stock assessment of Skeena River chinook salmon. PSARC working paper S89-18.

Riley, R.C. and P. Lemieux. 1998. The effects of beaver on juvenile coho salmon habitat in Kispiox River tributaries. Unpublished report for DFO. Smithers, B.C.

Runka, G.G. 1974. Soil resources of the Smithers-Hazelton area. Soil Survey Division, B.C. Department of Agriculture. Kelowna, B.C.

Rutherford D.T., C.C. Wood, M. Cranny, and B. Spilsted. 1999. Biological characteristics of Skeena River sockeye salmon (Oncorhynchus nerka) and their utility for stock compositional analysis of test fishery samples. Can. Tech. Rep. Fish. Aquat. Sci. 2295: 46 pp.

Ryan, B. 2000. Coal resources of Northwest B.C. and the Queen Charlotte Islands. B.C. Energy and Mines. Victoria, B.C.

Ryan, T. 2003. Kitselas: Skeena River eulachon survey 2002. Draft unpublished report. Tsimshian Tribal Council. Prince Rupert, B.C.

Rysavy, S. 2000a. Calibration of a multimetric benthic invertebrate index of biological integrity for the upper Bulkley River watershed. Prepared for CFDC Nadina by BioLogic Consulting. Terrace, B.C.

Rysavy, S. 2000b. Calibration of a multimetric benthic invertebrate index of biological integrity for the Kispiox River watershed. BioLogic Consulting. Terrace, B.C.

Rysavy, S. and I. Sharpe. 1995. Tyhee Lake management plan. B.C. Environment. Smithers, B.C.

Saimoto, R.K. 1994. Morice River watershed assessment 1994: Survey of logging related impacts on Cedric, Lamprey, Fenton, and Owen Creeks. SKR Consultants Ltd. Smithers, B.C.

Saimoto, R. 1993. Survey of Bulkley Lake and its inlet and outlet streams. Prepared for B.C. Fish and Wildlife Branch. Smithers, B.C.

Saimoto, R.S. 1996a. Telkwa watershed assessment: Fisheries, fish habitat, and riparian zone assessment. Prepared by SKR Consultants Ltd. for Pacific Inland Resources Ltd. Smithers, B.C.

Saimoto, R.S. 1996b. Toboggan Creek coho smolt enumeration 1996. Prepared for DFO, PBS. Nanaimo, B.C.

Saimoto, R. 2002. Personal communication. Fisheries biologist.

Saimoto, R.S. and M.O. Jessop. 1997. Assessment of overwintering habitat and distribution of coho salmon in the mid-Bulkley watershed (Houston to Bulkley Lake) January to March 1997. Prepared for DFO. Smithers, B.C.

Saimoto, R.S. and R.K. Saimoto. 2001. Coho salmon in the upper Bulkley River watershed. Unpublished report prepared for the Bulkley/Morice Salmonid Preservation Group, Nadina Community Futures, and Fisheries and Oceans Canada.

Saksida, S. 2003. Investigation of the 2001–2003 IHN epizootic in farmed Atlantic in British Columbia. Prepared for the B.C. Ministry of Food and Fisheries. 40 pp.

Sandercook, F.K. 1991. Life history of coho salmon (Oncorhynchus kisutch). In C. Groot and L. Margolis, eds. Pacific Salmon: Life Histories. UBC Press. Vancouver, B.C. 395–445.

Scarsbrook, J.R. and J. McDonald. 1975. Purse seine catches of sockeye salmon (Oncorhynchus nerka) and other species of fish at Babine Lake, British Columbia, 1973. Fish. Res. Board Can. Tech. Rep. 515.

Scarsbrook, J.R., P.L. Millar, J.M. Hume and J. McDonald. 1978. Purse seine catches of sockeye salmon (Oncorhynchus nerka) and other species of fish at Babine Lake, British Columbia, 1977. Fish. Res. Board Can. Data Rep. 69.

Schell, C. 2003. A brief overview of fish, fisheries, and aquatic habitat resources in the Morice TSA. Prepared for the Morice Land and Resource Management Plan.

Schug, S. 2002. Personal communication. Wet'suwet'en Fisheries. Smithers, B.C.

Schwab, J.W., M. Geertsema, and A. Stevens-Blais. 2004. The Khyex River flow slide, a recent landslide near Prince Rupert, British Columbia.

Schultze, G. 1983. Suskwa River steelhead: 1982 colonization of the upper Harold Price with steelhead fry. SK-40. B.C. Fish and Wildlife Branch. Smithers, B.C.

Scott, W.B. and E.J. Crossman. 1973. Freshwater fishes of Canada. Fisheries Research Board of Canada, Bull. 184.

Sedell, J.R. and F.J. Swanson. 1984. Ecological characteristics of streams in old-growth forests of the Pacific Northwest. In Fish and Wildlife Relationships in Old Growth Forests. Amer. Inst. Fish. Res. Bio. Morehead City, NC.

Sekerak, A.D., J.A. Taylor, and N. Stellard. 1984. Fisheries investigations in relation to the Mount Klappan coal project: Final report. LGL Ltd. Sydney B.C. for Gulf Canada Resources Inc. Calgary. 72 pp.

Septer, D. and J. W. Schwab. 1995. Rainstorm and flood damage: Northwest British Columbia 1891–1991. Ministry of Forests, Research Program. Victoria, B.C.

Shepard, B.B., K.L. Pratt, and P.J. Graham. 1984. Life histories of westslope cutthroat and bull trout in the upper Flathead River basin, Montana. EPA Contract No. R008224-01-5.

Shepherd, B.G. 1975. Upper Skeena chinook stocks. Evaluation of the Bear-Sustut, Morice, and lower

Babine stocks. Fisheries and Marine Service, Department of the Environment. Vancouver, B.C.

Shepherd, B.G. 1976. Upper Skeena chinook stocks in 1976. Fisheries and Marine Service, Department of the Environment. Vancouver, B.C.

Shepherd, B.G. 1978. Minnow traps as a tool for trapping and tagging juvenile chinook and coho salmon in the Skeena River system. *In* Proceedings of the 1977 Northeast Pacific Chinook and Coho Salmon Workshop. Fish. and Mar. Ser. Tech. Rep. No. 759.

Shepherd, B.G. 1979. Salmon studies associated with the potential Kemano II hydroelectric development, vol. 5. Salmon studies on Nanika and Morice River and Morice Lake. Dept. of Fish and Environ. Vancouver, B.C.

Shirvell, C., and B. Anderson. 1990a. Bear River trip report. PBS. Nanaimo, B.C.

Shirvell, C. and B. Anderson. 1990b. Sustut River and tributaries trip report. Salmon Habitat Section. Pacific Biological Station. Nanaimo, B.C.

Shirvell, C. and B. Anderson. 1991a. Chinook Salmon Freshwater Habitat Program. Sustut River and Johanson Creek, monthly report, August 1991. Salmon Habitat Section. Pacific Biological Station. Nanaimo, B.C.

Shirvell, C. and B. Anderson. 1991b. Chinook Salmon Freshwater Habitat Program. Sustut River and Johanson Creek. Monthly report, September 1991. Salmon Habitat Section. Pacific Biological Station. Nanaimo, B.C.

Shortreed, K.S. 2002. Personal communication, DFO limnologist.

Shortreed, K.S., J.M.B. Hume, K.F. Morton, and S.G. MacLellan. 1998. Trophic status and rearing capacity of smaller sockeye nursery lakes in the Skeena River system. Can. Tech. Rep. Fish. Aquat. Sci. 2240: 78 pp.

Shortreed, K.S. and K.F. Morton. 2000. An assessment of the limnological status and productive capacity of Babine Lake 25 years after the inception of the Babine Lake Development Project. Can. Tech. Rep. Fish. Sci. 2316.

Shortreed, K.S., K.F. Morton, K. Malange, and J.M.B. Hume. 2001. Factors limiting juvenile sockeye production and enhancement potential for selected B.C. nursery lakes. Canadian Science Advisory Secretariat. FOC. Cultus Lake, B.C.

Shortreed, K.S. and J.M.B. Hume. 2004. Report on limnological and limnetic fish surveys of North Coast area lakes in 2002 and 2003. Cultus Lake Salmon Research Laboratory. Cultus Lake, B.C.

Sigma Engineering Ltd. 1993. Brown Lake project environmental impact development. Vol. 1. Vancouver, B.C.

Silvicon Services Inc. 2001. Lower Telkwa Watershed Interim Restoration Plan.

Simpson, K., L. Hop Wo, and I. Miki. 1981. Fish surveys of 15 sockeye salmon nursery lakes in British Columbia. Can. Tech. Rep. Fish. Aquat. Sci. 1022: 87 pp.

Sinclair, W.F. 1974. The socioeconomic importance of maintaining the quality of recreational resources in northern British Columbia: The case of Lakelse Lake. DFO and Kitimat-Stikine Regional District. PAC/T-74-10 NOB/ECON 5-74.

Sinkewicz, K. 1999. 1998 adult coho enumeration program: Terrace area watersheds. Prepared for the Terrace Salmonid Enhancement Society. Terrace, B.C.

Sinkewicz, K. 2000. Juvenile salmonid density study of index sites in the mid-Skeena watershed 1999–2000. Gaia Environmental Consulting. Terrace, B.C.

Skeena Watershed Committee. 1996. Facing and forming the Future. Workshop proceedings. January 19 and 20, 1996. Prince Rupert, B.C.

SKR Consultants Ltd. 1998. Telkwa watershed assessment: Detailed habitat assessment of Howson Creek subunit road crossings. Prepared for Pacific Inland Resources. Smithers, B.C.

SKR Consultants Ltd. 1999. Toboggan Creek coho smolt enumeration 1999. Prepared for Department of Fisheries and Oceans. Nanaimo, B.C.

SKR Consultants Ltd. 2000. Buck Creek juvenile salmon trapping program. Project number nFi528-0-1702. Prepared for DFO. Smithers, B.C.

SKR Consultants Ltd. 2001. 2000 Steelhead tagging project at Moricetown Canyon. Data analysis and recommendations. Prepared for Wet'suwet'en Fisheries.

Small, M.P., R.E. Withler, and T.D. Beacham. 1998. Population structure and stock identification of British Columbia coho salmon, *Oncorhynchus kisutch*, based on microsatellite DNA variation. Fish. Bull. 96, 843–858.

Smith, B.D., B.R. Ward, and D.W. Welch. 2000. Trends in wild adult steelhead *(Oncorhynchus mykiss)* abundance in British Columbia as indexed by angler success. Can. J. Fish. Aquat. Sci. 57: 255–270.

Smith, G.R. and R.F. Stearly. 1989. The classification and scientific names of rainbow and cutthroat trouts. Fisheries 14: 4–10.

Smith, H.D., ed. 1973. Babine watershed change program. Annual report for 1974. Fisheries and Marine Service. PBS. Nanaimo, B.C.

Smith, H.D., ed. 1975. Babine watershed change program. Annual report for 1973. Fisheries and Marine Service. PBS. Nanaimo, B.C.

Smith, H.D., ed. 1976. Babine watershed change program. Annual report for 1972. Fisheries and Marine Service. PBS. Nanaimo, B.C.

Smith, H.D. and J. Lucop. 1966. Catalogue of salmon spawning grounds and tabulation of escapements in the Skeena River and Department of Fisheries Statistical Area 4. Fisheries Research Board of Canada, Manuscript Report Series No. 882. Biological Station. Nanaimo, B.C.

Smith, H.D. and J. Lucop. 1969. Catalogue of salmon spawning grounds and tabulation of escapements in the Skeena River and Department of Fisheries

Statistical Area 4. Fisheries Research Board of Canada, Manuscript Report Series No. 1046. Biological Station. Nanaimo, B.C.

Smith, H.D. and F.P. Jordan. 1973. Timing of Babine Lake sockeye salmon stocks in the North Coast commercial fishery as shown by several taggings at the Babine tagging fence and rates of travel through the Skeena and Babine Rivers. Fish. Res. Board Can. Tech. Rep. 418.

Smith, H.D., L. Margolis, and C.C. Wood. 1987. Sockeye salmon (Oncorhynchus nerka) population biology and future management. Can. Spec. Publ. Fish. Aquat. Sci. 96. 486 pp.

Smith, J.L. and G.F. Berezay. 1983. Biophysical reconnaissance of the Morice River system, 1979–1980. SEP Operations, Fisheries and Oceans Canada.

Smith, W. 2002. Personal communication. Retired Ranger. Morice Forest District.

SNDS. 1998. Skeena Native Development Society. 1998 Labour Market Census.

Southgate, D. 1978. Stream clearance report: Singlehurst Creek. DFO. Terrace, B.C.

Spence, B.C., G.A. Lomnicky, R.M. Hughes, and R.P. Novitzki. 1996. An ecosystem approach to salmonid conservation. Mantech Environmental Technology, Inc.

Spence, C.R. 1989. Rates of movement and timing of migrations of steelhead trout to and within the Skeena River, 1988. B.C. Ministry of Environment. Skeena Fisheries Report SK-62.

Spence, C.R., M.C. Beere, and M.J. Lough. 1990. Sustut River steelhead investigations 1986. Skeena Fisheries Report SK-64. B.C. Ministry of Environment. Smithers, B.C.

Spence, C.R., and R.S. Hooton. 1991. Run timing and target escapements for summer-run steelhead trout (Oncorhynchus mykiss) stocks in the Skeena River system. PSARC Working paper S91-07.

Spilsted, B. 2005. Personal communication.

Statistics Canada. 1996. Census information; Enumeration areas data.

Sterritt, G. 2001. 2001 Kispiox watershed coho stock assessment. Gitksan Watershed Authorities. Unpublished report. Hazelton, B.C.

Sterritt, G. 2003. Personal communication. Gitksan Watershed Authorities. Hazelton, B.C.

Sterritt, G. and A.S. Gottesfeld. 2002. 2001 Upper Kispiox sockeye stock assessment. Unpublished report. Gitksan Watershed Authorities.

Stewart, C. 2002. Personal communication. Ministry of Water, Lands and Air Protection. Smithers, B.C.

Stockner, J.G. and K.R.S. Shortreed. 1975. Phytoplankton succession and primary production in Babine Lake, British Columbia. J. Fish. Res. Board Can. 32: 2413–2427.

Stockner, J.G. and K.R.S. Shortreed. 1976. Babine Lake monitor program: Biological and physical data for 1974 and 1975. Fish. Res. Board Can. MS Rep. 1373: 34 pp.

Stockner, J.G. and K.S. Shortreed. 1979. Limnological Studies of 13 sockeye salmon (Oncorhynchus nerka) nursery lakes in British Columbia, Canada. Fish. Mar. Serv. Tech. Rept. No. 865. 125 pp.

Stokes, J.W. 1956. Upper Bulkley River survey, 1956.

Stokes, J.W. 1953. Pollution survey of the Watson Island area. Fish, Serv. MS. Rept.

Strimbold, F. 2002. Personal Communication. Long-time Topley resident.

Stuart, K.M. 1981. Juvenile steelhead carrying capacity of the Kispiox River system in 1980, with reference to enhancement opportunities. Min. of Environment, Fish and Wildlife Branch. Smithers, B.C.

Stumpf, A.J. 2001. Late quaternary ice flow, stratigraphy, and history of the Babine Lake–Bulkley River region, central British Columbia, Canada. PhD thesis, University of New Brunswick.

Suskwa Research. 2002. 'Ilk K'il Bin: Selected Wet'suwet'en cultural heritage recce 2002. Prepared for Office of the Wet'suwet'en and Ministry of Forests.

Sword, C.B. 1904. 1905 Annual Report, Department of Marine and Fisheries, Fisheries: 36 (1903) 254–255. Fisheries: 37 (1904) 238–239. Ottawa, Ont.

Swanson, F.J., S.V. Gregory, J.R. Sedell, and A.G. Campbell. 1982. Land-water interactions: The riparian zone. In R.L. Edmonds, ed. Analysis of coniferous forest ecosystems in the western United States. US/IBP Synthesis Series 14. Hutchinson Ross Publishing Co.

Takagi, K. and H.D. Smith. 1973. Timing and rate of migration of Babine sockeye stocks through the Skeena and Babine Rivers. Fish. Res. Board Can. Tech. Rep. 419.

Tallman, D. 1997. 1996 Kispiox River sport fishery survey summary report. J. O. Thomas and Associates, Vancouver, B.C.

Talon Development Services. 2002. Ned'u'ten Fisheries Strategic Plan 2002.

Tamblyn, G. and M. Jessop. 2000. Detailed fish habitat, riparian, and channel assessment for select central Bulkley River tributaries. Prepared for Bulkley-Morice Salmonid Preservation Group.

Tamblyn, G. and B. Donas. 2001. Healthy watersheds, healthy communities: Bulkley-Morice Salmonid Preservation Group draft strategic plan, phase 1.

Tautz, A.F., Ward, B.R., and R.A. Ptolemy. 1992. Steelhead trout productivity and stream carrying capacity for rivers of the Skeena drainage. PSARC working paper S92-6 and 8.

Taylor, G.D. and R.W. Seredick. 1968. Preliminary inventory of some streams tributary to Kispiox River. B.C. Fish and Wildlife Branch. Smithers, B.C.

Taylor, J.A. 1995. Synoptic surveys of habitat characteristics and fish populations conducted in lakes and streams within the Skeena River watershed, between 15 August and 12 September 1994. Unpublished report by J.A. Taylor and Associates for Fisheries and Oceans Canada.

Taylor, J.A. 1996. Assessment of juvenile coho population levels in selected lakes and streams within the Skeena river watershed, British Columbia, between 11 and 31 August 1995. Unpublished report by J.A. Taylor and Associates for Fisheries and Oceans Canada.

Taylor, J.A. 1997. Synoptic surveys of juvenile coho populations and associated habitat characteristics in selected lakes and streams within the Skeena River watershed, British Columbia, between 10 August and 2 September 1996. Unpublished report by J.A. Taylor and Associates for Fisheries and Oceans Canada.

Taylor, J.E. III. 1999. Making Salmon: An Environmental History of the Northwest Fisheries Crisis. University of Washington Press, Seattle. 421 pp.

Tetreau, R..E. 1982. Stream files, Ministry of Environment, Smithers, B.C.

Tetreau, R.E. 1999. Movement of radio tagged steelhead in the Morice River as determined by helicopter and fixed station tracking, 1994–95. Skeena Fisheries Report SK-125. MELP, Skeena Region. Smithers, B.C.

Thomas and Associates Ltd. 1995. 1995 Skeena River sport fish coho and steelhead catch and release study. J.O. Thomas and Associates Ltd. Vancouver, B.C.

Thomas and Associates Ltd. 1999. 1999 Lower Skeena River sport catch monitoring programs. J.O. Thomas and Associates Ltd. Vancouver, B.C.

Thomson, R. E. 1981. Oceanography of the British Columbia coast. Can. Spec. Publ. Fish. Aquatic. Sci. 56: 291 pp.

Thurber Engineering Ltd. 1992. Kemess South Project geotechnical evaluation of potential road corridors. Report to El Condor Resources Ltd. Vancouver, B.C.

Tipper, H.W. and T.A. Richards. 1976. Geology of Smithers map area, British Columbia. Geological Survey of Canada, Open File 351.

Toboggan Creek Hatchery. 1999. Annual report for Toboggan Creek Hatchery operations 1998–99. Unpublished report. Smithers, B.C.

Traxler, G.S., J. Richard, and T.E. MacDonald. 1998. *Ichthyophthirius multifilis* (ich) epizootics in spawning sockeye salmon in British Columbia, Canada. J. Aquat. Animal Health 10: 147–157.

Tredger, C.D. 1981 to 1986. Various Morice River stock assessment and fry monitoring reports. B.C. Environment. Smithers, B.C.

Tredger, C.D.1982. Upper Bulkley River reconnaissance with reference to juvenile steelhead carrying capacity. Habitat Improvement Section, B.C. Environment. Victoria, B.C.

Tredger, C.D. 1983a. Juvenile steelhead assessment in the Kispiox River (1980–1982). B.C. Ministry of Environment, Fish and Wildlife Branch. Smithers, B.C.

Tredger, C.D. 1983b. Kitsumkalum River reconnaissance report. Ministry of Environment stream files. Smithers, B.C.

Tredger, C.D. 1984. Skeena boat shocking program 1983. Fish Habitat Improvement Section, Ministry of Environment. Victoria, B.C.

Tredger, C.D. 1986. Upper Skeena steelhead fry population monitoring (Babine, Sustut, and Kluatantan Rivers). Prepared by the Fish Habitat Improvement Section of B.C. Ministry of Environment.

Tredger, D. 1981. Assessment of steelhead enhancement opportunities in the Morice River system. Progress in 1980. Fish Habitat Improvement Section, Ministry of Environment. Victoria, B.C.

Tredger, D. and P. Caw. 1974. Bulkley Lake survey data. On file in Skeena Region lake files. Water, Land and Air Protection. Smithers, B.C.

Trites, R.W. 1956. The oceanography of Chatham Sound, British Columbia. J. Fish. Res. Bd. Can. 13(3): 386–434.

Triton Environmental Consultants Ltd. 1996a. Kalum WPR Project, vol. 1: Level 1 fisheries assessment, level 1 riparian assessment. Terrace, B.C.

Triton Environmental Consultants Ltd. 1996b. Lakelse WRP project, final summary report. Unpublished report. Terrace, B.C.

Triton Environmental Consultants Ltd. 1997. Reconnaissance fish and fish habitat inventory for the Kispiox Forest District. Prepared for Repap Smithers Inc. and MELP. Terrace, B.C.

Triton Environmental Consultants Ltd. 1998a. 1:5,000 scale fish stream identification and RMA classification. Ayton Creek watershed. Cutting Permit #400, Blocks 4000, 4001, and 4005. Prepared for International Forest Products Ltd. Terrace, B.C.

Triton Environmental Consultants Ltd. 1998b. Reconnaissance-level fish and fish habitat inventory in the Bulkley T.S.A. [Working unit #8 – Kitseguecla]. Prepared for Pacific Inland Resources.

Triton Environmental Consultants Ltd. 1998c. Reconnaissance-level fish and fish habitat inventory in the Bulkley T.S.A. [Working unit #11 – Zymoetz]. Prepared for Pacific Inland Resources.

Triton Environmental Consultants Ltd. 1998d. Reconnaissance-level fish and fish habitat inventory in the Bulkley Forest T.S.A. [Working unit #13 – Telkwa]. Prepared for Pacific Inland Resources.

Triton Environmental Consultants Ltd. 1998e. Fisheries assessment, Gitnadoix River and Skeena Sidechannels. Prepared for Pacific Northern Gas. Triton Environmental Consultants Ltd. Terrace, B.C.

Triton Environmental Consultants Ltd. 1999. Upper Zymoetz (Copper River) WRP. Overview fish and riparian assessment. Terrace, B.C.

Triton Environmental Consultants Ltd. 2001. 2000–2005 Kispiox watershed restoration plan. Terrace, B.C.

Trosper, R.L. 2002. Northwest coast indigenous institutions that supported resilience and sustainability. Ecological Economics 41: 329–344.

Trusler, S. 2002. Footsteps among the berries: The ecology and fire history of traditional Gitxsan and Wet'suwet'en huckleberry sites. M.S. thesis, UNBC.

Trusler, S., A. George, S. Schug, K. Rabnett, and F. Depey. 2002. Wet'suwet'en Territorial Stewardship Plan. Final report. Prepared for the Office of the Wet'suwet'en.

Tully, O. and K.F. Whelan. 1993. Production of nauplii of Lepeophtheirus salmonis (Krøyer)(Copepoda: Caligidae) from farmed and wild salmon and its relation to the infestation of wild sea trout (Salmo trutta L.) off the west coast of Ireland in 1991. Fisheries Research 17: 187–200.

Turnbull, T. 1991. Personal communication. In DFO 1991.

Turner, N.J. 1998. Plant technology of First Peoples in British Columbia. UBC Press Vancouver, B.C.

Varnavskaya, N.V., C.C. Wood, and R.E. Everett. 1994. Genetic variation in sockeye salmon (Oncorhynchus nerka) populations of Asia and North America. Can. J. Fish. Aquat. Sci. 51. 132–146.

Vernon, E.H. 1951. The utilization of spawning grounds on the Morice River system by sockeye salmon. B.A. thesis, UBC. Vancouver, B.C.

Volpe, J.P., E.B. Taylor, D.W. Rimmer, and B.W. Glickman. 2000. Natural reproduction of aquaculture escaped Atlantic salmon (Salmo salar) in a coastal British Columbia river. Conservation Biology 14: 899–903.

Volpe, J.P., B.R. Anholt, and B.W. Glickman. 2001. Competition among juvenile Atlantic salmon (Salmo salar) and steelhead trout (Oncorhynchus mykiss): Relevance to invasion potential in British Columbia. Can. Jour. Fish. and Aquat Sci. 58: 197–207.

Wadley, G. and L. Gibson. 1998. Kispiox River channel assessment. Unpublished report prepared for Ans'payaxw Development Corporation.

Wagner, D. 2002. Personal communication. DFO, Area 4. Resource Management, Prince Rupert.

Waldichuk, M. 1962. Observations in the marine waters of the Prince Rupert area, particularly with reference to pollution from the sulphite pulp mill on Watson Island, Sept. 1961. Fish. Res. Bd. Can. MS. Rept. (Biol.) (733).

Waldie, W.F. and W.H. van Heek. 1969. Working plan number five, 1970–1974. The Port Edward Tree Farm License, Tree Farm License #1. Prepared for Skeena Kraft Limited.

Walters, C.J. 1988. Mixed-stock fisheries and the sustainability of enhancement production for chinook and coho salmon. In W.J. McNeil, ed. Salmon production, management, and allocation, Oregon State University Press. Corvallis, Oregon.

Waples, R.S. 1990. Conservation genetics of Pacific salmon: Effective population size and the loss of genetic variability. J. Heredity 81: 267–276.

Waples, R.S. 1991. Genetic interactions between hatchery and wild salmonids; Lessons from the Pacific Northwest. Can. J. Aquat. Sci. 48 (Supl. 1): 124–133.

Waples, R.S. 1995. Evolutionary significant units and the conservation of biological diversity under the Endangered Species Act. American Fisheries Society Symposium 17: 8–27.

Ward, B.R., A.F. Tautz, S. Cox-Rodgers, and R.S. Hooton. 1993. Migration timing and harvest rates of the steelhead populations of the Skeena River system. PSARC working paper S93-6.

Water Survey of Canada. Environment Canada, hydrometric station 08EB004.

Weiland, I. 1993. Sediment source mapping in the Nilkitkwa River and Nichyeskwa Creek area. Smithers, B.C.

Weiland, I. 1995. In Suskwa Watershed Restoration Program, L1. Fish, fish habitat, stream channel, and riparian assessments. Oikos Ecological Services, Irene Weiland and Wildfor Consultants.

Weiland, I. 2000a. Road construction upslope of unstable terrain: Effects on downslope hydrology and terrain stability, McCully and Date Creek watersheds, Kispiox Forest District. Weiland Terrain Sciences. Smithers, B.C.

Weiland, I. 2000b. Reconnaissance sediment source mapping, Kispiox River watershed, Kispiox Forest District. Weiland Terrain Services. Smithers, B.C.

Weiland, I. and J.W. Schwab. 1991. Nilkitkwa area, Bulkley TSA. Slope stability and surface erosion assessment. Forest Sciences Section, Prince Rupert Forest Region, B.C. Ministry of Forests. Smithers, B.C.

Weiland, I. and J.W. Schwab. 1992. Floodplain stability, Morice River, between Owen Creek and Thautil River.

Weiland, I., and J. W. Schwab. 1996. Floodplain erosion hazard assessment, lower Zymoetz River, British Columbia. Unpublished report for B.C. Ministry of Forests. Smithers, B.C.

Weiland, I. and D. Maloney. 1997. Review of surface erosion potential ratings Nilkitkwa area, Bulkley TSA. Smithers, B.C.

Weiland, I. and S. Bird. 2007. Sediment source mapping, detailed channel assessment, and reconnaissance sediment budget for Williams Creek for the period 1949 to 2001.

West, C.J., and J.C. Mason. 1987. Evaluation of sockeye salmon (Oncorhynchus nerka) production from the Babine Lake Development Project. In H.D. Smith, L. Margolis, and C.C. Wood, eds. Sockeye salmon (Oncorhynchus nerka) population biology and future management. Can. Spec. Publ. Fish. Aquat. Sci. 96: 176–190

Wet'suwet'en Chiefs. 2001. Wet'suwet'en landscape unit planning process. Office of the Wet'suwet'en. Smithers, B.C.

Wet'suwet'en Fisheries. 2003. Unpublished data on file at Wet'suwet'en Fisheries. Moricetown, B.C.

Wet'suwet'en Treaty Office. 1996. Morice watershed restoration project. Smithers, B.C.

Whately, M.R. 1975. Memo to file (January 29, 1975). WLAP Skeena Region stream files. Smithers, B.C.

Whately, M.R. 1977. Kispiox River steelhead trout. B.C. Technical Fisheries Circular No. 36.

Whately, M.R., W.E. Chudyk, and M.C. Morris. 1978. Morice River steelhead trout: The 1976 and 1977 sportfishery and life-history characteristics from anglers' catches. Fish. Tech. Circ. No. 36. Smithers, B.C.

Whately, M.R. and W.E. Chudyk. 1979. An estimate of the number of steelhead trout spawning in Babine River near Babine Lake, spring, 1978. Skeena Fisheries Report SK-23. MELP, Skeena Region. Smithers, B.C.

Whelpley, M.C. 1983. Lakelse River project, 1982–83. Stream files, Ministry of Environment. Smithers, B.C.

Whelpley, M.C. 1984. Lakelse River project, 1983–84. Stream files, Ministry of Environment. Smithers, B.C.

Whelpley, M.C. 2002. Personal communication. Terrace, B.C.

White, E.R. 1988. A wetlands mitigation proposal to compensate for Highway 16 reconstruction near Prince Rupert, B.C. Prepared for B.C. Ministry of Transportation and Highways. Victoria, B.C.

White, E.R. 1990. Khyex-Tyee: Foreshore habitat reconstruction works. Prepared for B.C. Ministry of Transportation and Highways. Victoria, B.C.

White, E.R. 1997. 1997 Foreshore habitat assessment for the Ayton Mainline, lower Skeena River.

White, W.H. 1953. Supplementary geological report on the Babine slide.

White, W.H. 1964. Re-examination of the Babine slide.

Whitwell, T. 1906. Skeena River hatchery. Annual Report. Department of Marine and Fisheries. Fisheries: 38 (1905): 257–259.

Wild, J. 1991. Babine River slide investigation. Letter to file 8030-B8.

Wild Stone Resources. 1995. Level 1 assessment for the Kitseguecla watershed. Prepared for Skeena Cellulose Inc.

Wildstone Resources Ltd. 1995. Level 1 assessment for the Kitwanga watershed. Vol. 1.

Wilford, D.J. 1985. A forest hydrology overview of the Kispiox watershed. Ministry of Forests. Smithers, B.C.

Wilford, D.J. 1987. Watershed workbook: Forest hydrology sensitivity analysis for coastal British Columbia watersheds. B.C. Ministry of Forests. Smithers, B.C.

Wilford, D., M. Sakals, H. DeBeck, and G. Marleau. 2000. Tsezakwa Creek fan. Ministry of Forests, Smithers, B.C.

Wilkes, B. and R. Lloyd. 1990. Water quality summaries for eight rivers in the Skeena River drainage, 1983–1987: The Bulkley, upper Bulkley, Morice, Telkwa, Kispiox, Skeena, Lakelse and Kitimat Rivers. Skeena Region MELP, Environmental Section Report 90-04.

Wilton, D.H.C. and Sinclair, A.J. 1988. Ore petrology and genesis of a strata-bound disseminated copper deposit at Sustut, British Columbia. Economic Geology 83: 30–45.

Williams, B. and D. Gordon. 2001. Assessment of select CN stream crossings near Terrace. Recommendations and strategies for restoring fish access. Triton Environmental Consultants Ltd. Terrace, B.C.

Williams, G.L. 1991. Prince Rupert area coastal fish habitat bibliography.

Williams, H., D. McLennan, and K. Klinka. 2000. Classification and interpretation of hardwood dominated ecosystems in the dry cool Sub-boreal Spruce (SBSdk) subzone and moist cold Interior Cedar Hemlock (ICHmc2) variant of the Prince Rupert Forest Region. Unpublished report prepared for the Prince Rupert Forest Region. Smithers, B.C.

Williams, I.V., T.J. Brown, and G. Langford. 1994. Geographic distribution of salmon streams of British Columbia with an index of spawner abundance. Can. Tech. Rep. Fish. Aquat. Sci. 1967: 200 pp.

Williams, R.A., G.O. Stewart, and P.R. Murray. 1985. Juvenile salmonid studies in the Sustut and Bear Rivers, B.C. Envirocon Ltd. Report for D.F.O.

Williams, S. 1987. Research notes of N.J. Sterritt. Vol. 4, Notebook 7. Unpublished report on file at Gitxsan Treaty Office. Hazelton, B.C.

Williamson, C.J. 1998. The enumeration of adult and juvenile salmon in the upper Sustut River 1998. Can. Man. Rep. Fish. Aquat. Sci. XXXX.

Wilson, H. 1982. Interview by V. Smith. Ex. 974-26. *Delgamuukw et al. v. the Queen.* 1989 Supreme Court of B.C. Pre-trial evidence including affidavits, interrogatories, commission evidence and notes from witness preparation.

Wilson, M.F. 1997. Variation in salmonid life histories: Patterns and perspectives. U.S. Department of Agriculture Forest Service, Pacific Northwest Research Station, Res. Pap PNW-RP-498. 50 pp.

Wilson, R.D. and R.D. Marsh. 1975. Climatic constraints to logging and climatic suitability for recreation in the Terrace-Hazelton forest resources study area. ELUC Secretariat. Victoria, B.C.

Wilson, T., and A.S. Gottesfeld. 2001. Juvenile coho population assessment in selected streams within the Gitxsan Territories 2001. Gitksan and Wet'suwet'en Watershed Authorities, Hazelton, B.C. 32 pp.

Wilson, T. 2004. Personal communication.

Wilson, T. 2004. Juvenile coho population assessment in selected streams within the Gitxsan Territories 2003. Gitksan Watershed Authorities, Hazelton, B.C. 26 pp.

Wilton, D.H. and A.J. Sinclair. 1988. Ore petrology and genesis of a strata-bound disseminated copper deposit at Sustut, British Columbia. Economic Geology 83: 30–45.

Winther, I. 2002. Personal communication. DFO, North Coast Stock Assessment.

Winther, I. 2006. North Coast Chinook Salmon 2005: Stock status. DFO, North Coast Stock Assessment.

Wisley, W.A. 1919. Report of the Commissioner of Fisheries for the year ending December 31, 1919. Victoria, B.C.

Withler, R.E., K.D. Le, J. Nelson, K.M. Miller, and T.D. Beacham. 2000. Intact genetic structure and high levels of genetic diversity in bottlenecked sockeye salmon (Oncorhynchus nerka) populations of the Fraser River, British Columbia. Can. J. Fish. Aquat. Sci. 57: 1985–1998.

Woloshyn, P. 2002. Personal communication. DFO Fisheries Officer. New Hazelton, B.C.

Woloshyn, P. 2003. Personal communication. DFO Fisheries Officer. New Hazelton, B.C.

Wood, C.C. 1995. Life-history variation and population structure in sockeye salmon. American Fisheries Society Symposium 17: 195–216.

Wood, C.C. 2001. Managing biodiversity in Pacific salmon: The evolution of the Skeena River sockeye salmon fishery in British Columbia.

Wood, C.C., B.E. Riddell, D.T. Rutherford, and R.E. Withler. 1994. Biochemical genetic survey of sockeye salmon (Oncorhynchus nerka) in Canada. Can. J. Fish. Aquat. Sci. 51: 114–131.

Wood, C.C., and C.J. Foote. 1996. Evidence for sympatric genetic divergence of anadromous and nonanadromous morphs of sockeye salmon (Oncorhynchus nerka). Evolution 50: 1265–1279.

Wood, C., D. Rutherford, D. Bailey and M. Jakubowski. 1997. Babine Lake sockeye salmon: Stock status and forecasts for 1998. CSAS Research Document 97/45. Fisheries and Oceans Canada, PBS. Nanaimo, B.C.

Wood, C.C., and L.B. Holtby. 1999. Defining conservation units for Pacific salmon using genetic survey data. In B. Harvey, C. Ross, D. Greer, and J. Carolsfeld, eds. Action before extinction: An international conference on conservation of fish genetic diversity. 233–250. World Fisheries Trust, Victoria, Canada.

Credits

Photos

42 A. Gottesfeld • 50 A. Gottesfeld • 65 A. Gottesfeld • 66 Superfly • 67 Royal Canadian Mounted Police • 69 Superfly • 79 K. Rabnett • 87 E. Sutherland • 84 Superfly • 92 Superfly • 92 Superfly• 99 D. Gordon • 100 B.C. Government • 102 D. Gordon • 104 M. Whelply • 105 S. Devcic • 106 Dept. Fisheries & Oceans - Fisheries Research Board (DFO-FRB) • 107 D. Gordon • 110 D. Gordon • 113 D. Gordon • 115 D. Gordon • 120 D. Gordon • 121 L. Miller • 122 L. Miller • 123 Superfly • 124 D. Gordon • 130 A. Gottesfeld • 136 K. Rabnett • 138 Unknown • 144 Unknown • 147 A. Gottesfeld • 148 Public Archives Canada • 149 L. Shotridge • 150 M. Cleveland • 151 M. Cleveland • 154 A. Gottesfeld • 155 A. Gottesfeld • 157 K. Rabnett • 161 A. Gottesfeld • 164 G. Webb • 170 Superfly • 171 A. Gottesfeld • 176 A. Gottesteld • 177 A. Gottesfeld • 179 A. Gottesfeld • 180 A. Gottesfeld • 187 A. Gottesfeld • 188 K. Rabnett • 191 DFO-FRB • 193 Public Archives Canada • 194 K. Rabnett • 195 DFO • 202 A. Gottesfeld • 203 A. Gottesfeld • 204 A. Gottesfeld • 209 P. Hall • 209 P. Hall • 210 A. Gottesfeld • 211 A. Gottesfeld • 212 A. Gottesfeld • 214 A. Marshall (Ecofor) • 218 T. Wilson • 223 M. Beere • 224 A. Gottesfeld • 226 M. Beere • 227 O'Dwyer 1899 • 229 T. Wilson • 230 T. Wilson • 234 A. Gottesfeld • 235 Unknown • 236 DFO FRB • 237 DFO FRB • 242 A. Gottesfeld • 243 J. Allen • 244 A. Gottesfeld • 245 C. Evenchick, Geological Survey of Canada • 247 T. Wilson • 248 T. Wilson • 251 A. Gottesfeld • 254 K. Rabnett • 255 Unknown • 256 K. Rabnett • 257 K. Rabnett • 258 J. Schwab • 260 W. Joseph • 264 Public Archives Canada • 265 Public Archives Canada • 266 DFO FRB • 274 A. Gottesfeld • 276 A. Gottesfeld • 278 K. Rabnett • 283 T. Wilson • 293 M. Bahr • 295 D. Bustard • 298 M. Bahr • 300 M. Bahr • 304 A. Gottesfeld • 337 (left) J. Gottesfeld • 337 (right) A. Gottesfeld

Maps

All maps by Lance Williams, GIS Technician, Gitxsan Watershed Authorities, Hazelton, B.C.

Illustrations

12, 18–20, 24 T. Knepp • 14, 27 NOAA Historic Fisheries Collection

Figures

Figure 1: A. Gottesfeld, data source Environment Canada HYDAT • Figure 2: A. Gottesfeld, data source Environment Canada HYDAT • Figure 3: A. Gottesfeld, data source Environment Canada HYDAT • Figure 4: A. Gottesfeld, data source Environment Canada HYDAT • Figure 5: After Ivan.Winther (DFO) 2006 • Figure 6: After Ivan Winther (DFO) 2006 • Figure 7: A. Gottesfeld, data source DFO SEDS Database • Figure 8: A. Gottesfeld, data source DFO SEDS Database • Figure 9: Data source Pacific Salmon Commission Northern Boundary Committee, after Steve Cox-Rogers (DFO) 2000 • Figure 10: A. Gottesfeld, data source DFO SEDS Database • Figure 11: Data source Joel Sawada 2007 • Figure 12: Data source Joel Sawada 2007 • Figure 13: Data source Mark Beere (B.C. Ministry of Environment) • Figure 14: Data source B.C. Ministry of Environment • Figure 15: L. Williams, A. Gottesfeld • Figure 16: A. Gottesfeld, data source Environment Canda HYDAT • Figure 17: After Museum of Northern British Columbia, Prince Rupert • Figure 18: L. Williams, A. Gottesfeld • Figure 19: A. Gottesfeld, data source DFO SEDS Database • Figure 20: A. Gottesfeld, data source DFO SEDS Database

Tables

Table 1: Data from B.C. Ministry of Environment FISS 2002, and Godfrey 1955 • Table 2: Data source Rabnett et al. 2001 • Table 3: Data source Gitksan Watershed Authorities unpublished 2007 • Table 4: Data source B.C.Ministry of Environment 2006 • Table 5: Data sources Gitksan and Wet'suwet'en Chiefs 1987, Mills 1998, Morten 1999

Note: Toponym usage and spelling was guided by the B.C. Geographical Names Information System (BCGNIS) and the Geographical Names Board of Canada (GNBC).

Index

About the Authors

Allen S. Gottesfeld is head scientist for the Skeena Fisheries Commission, a consortium of Skeena First Nations fisheries management groups in northern British Columbia. He has been a professor of watershed management at the University of Northern British Columbia and taught at the University of California, Berkeley, and the University of Alberta. Allen has lived in the Skeena watershed for 30 years.

Ken A. Rabnett is a fisheries researcher widely familiar with the grey literature on the resources of the Skeena watershed. He leads fisheries research projects for the Skeena Fisheries Commission and has been a resident of the Skeena watershed for 25 years.